METHODS IN MOLECULAR BIOLOGY™

Series Editor
John M. Walker
School of Life Sciences
University of Hertfordshire
Hatfield, Hertfordshire, AL10 9AB, UK

For further volumes:
http://www.springer.com/series/7651

Protein-Ligand Interactions

Methods and Applications

Second Edition

Edited by

Mark A. Williams and Tina Daviter

ISMB Biophysics Centre
Institute of Structural and Molecular Biology
Birkbeck, University of London
London, United Kingdom

Humana Press

Editors
Mark A. Williams
ISMB Biophysics Centre
Institute of Structural and Molecular Biology
Birkbeck, University of London
London, United Kingdom

Tina Daviter
ISMB Biophysics Centre
Institute of Structural and Molecular Biology
Birkbeck, University of London
London, United Kingdom

ISSN 1064-3745 ISSN 1940-6029 (electronic)
ISBN 978-1-4939-5873-3 ISBN 978-1-62703-398-5 (eBook)
DOI 10.1007/978-1-62703-398-5
Springer New York Heidelberg Dordrecht London

Printed on acid-free paper

Humana Press is a brand of Springer
Springer is part of Springer Science+Business Media (www.springer.com)

Preface

Proteins are the cell's workers, their messengers and overseers. In these roles proteins specifically bind small molecules, nucleic acid and other protein partners. Cellular systems are closely regulated, and changes in the populations of particular protein complexes or products of protein-mediated reactions by as little as a factor of 2 can switch cells from one state to another, from growth to stasis, from replication to apoptosis. Such changes in populations correspond to very small effects on thermodynamics or kinetics of reactions. Consequently, detailed characterization of protein interactions is of paramount importance in a quantitative and integrative biology, which aims to understand biological systems in terms of their molecular components.

Further, interfering with the interactions of proteins is the dominant strategy in the development of new pharmaceuticals. The discovery of novel small-molecule ligands, the characterization of their interactions with protein targets and the use of that information in guiding development of an inhibitor into a drug is a key component of the early stages of creating new medicines.

This volume aims to provide a complete introduction to common and emerging procedures for characterizing the interactions of individual proteins. All stages of the research process are covered—from the initial discovery of natural substrates or potential drug leads to the detailed quantitative understanding of the mechanism of interaction. We focus on those techniques that are, or are anticipated to become, widely accessible; that are performable with mainstream commercial instrumentation.

Much of this volume is aimed particularly at researchers new to the field of biophysical characterization of protein interactions—whether beginning graduate students or experts in allied areas of molecular cell biology, microbiology, pharmacology, medicinal chemistry or structural biology—who need to characterise their protein's interactions in greater detail. There is a particular emphasis on obtaining good quality data and helping the researcher understand whether or not they have succeeded in doing so.

We hope that the breadth of coverage and detailed consideration of technical issues will also serve as a reference for the professional molecular biophysicist "straying" outside their area of specific expertise.

London, United Kingdom ***Mark A. Williams***
Tina Daviter

Contents

Part IV Ligand Discovery

Part V Molecules in Native Environments

Part VI Structural and Computational Methods

Contributors

ALEX BREEZE • *AstraZeneca R&D, Macclesfield, UK*
MARK D. CALMIANO • *Department of Medicinal Chemistry, UCB Pharma, Slough, UK*
A.W. EDITH CHAN • *Medicinal Chemistry, Wolfson Institute for Biomedical Research, University College London, London, UK*
NIKOLA CHMEL • *Department of Chemistry, University of Warwick, Coventry, UK; Warwick Centre for Analytical Science, University of Warwick, Coventry, UK*
ALESSIO CIULLI • *College of Life Sciences, University of Dundee, Division of Biological Chemistry and Drug Discovery, The Wellcome Trust Biocentre, Scotland, UK; Department of Chemistry, University of Cambridge, Cambridge, UK*
NICOLA COLCLOUGH • *AstraZeneca R&D, Macclesfield, UK*
LUMINITA DAMIAN • *Microcal Products, GE Healthcare, Little Chalfont, UK*
GARETH DAVIES • *AstraZeneca R&D, Macclesfield, UK*
BEN DAVIS • *Vernalis Ltd (R&D), Great Abington, Cambridge, UK*
TINA DAVITER • *ISMB Biophysics Centre, Institute of Structural and Molecular Biology, Birkbeck, University of London, London, UK*
PAUL EMSLEY • *Structural Studies, MRC Laboratory of Molecular Biology, Cambridge, UK*
RÉMI FRONZES • *Unité G5 Biologie structurale de la sécrétion bactérienne, Institut Pasteur, Paris, France*
ÅSA FROSTELL • *GE Healthcare Bio-Sciences AB, Uppsala, Sweden*
PAUL J. GANE • *Medicinal Chemistry, Wolfson Institute for Biomedical Research, University College London, London, UK*
KLAUS GERWERT • *Lehrstuhl für Biophysik, Ruhr-Universität Bochum, Bochum, Germany*
STEFAN GESCHWINDNER • *AstraZeneca R&D, Mölndal, Sweden*
GEOFFREY HOLDGATE • *AstraZeneca R&D, Macclesfield, UK*
HEIKO KELLER • *BIOTEC, Dresden University of Technology, Dresden, Germany*
MICHAEL A. KING • *Department of Medicinal Chemistry, UCB Pharma, Slough, UK*
CARSTEN KÖTTING • *Lehrstuhl für Biophysik, Ruhr-Universität Bochum, Bochum, Germany*
BORIS KROEPLIEN • *Department of Medicinal Chemistry, UCB Pharma, Slough, UK*
PETER N. LOWE • *Biomolecular Interactions Consultancy, Hertford, UK*
KATHERINE LUBY-PHELPS • *Department of Cell Biology, UT Southwestern Medical School, Dallas, TX, USA*
STEPHEN R. MARTIN • *Division of Physical Biochemistry, MRC National Institute for Medical Research, London, UK*
ERICK MENESES • *Department of Chemistry, McGill University, Montreal, QC, Canada*
PETER MICHAELY • *Department of Cell Biology, UT Southwestern Medical School, Dallas, TX, USA*
ANTHONY MITTERMAIER • *Department of Chemistry, McGill University, Montreal, QC, Canada*

GABOR MOCZ • *Pacific Biosciences Research Center, University of Hawaii, Honolulu, HI, USA*
WILLIAM R. PITT • *Department of Medicinal Chemistry, UCB Pharma, Slough, UK*
SHANICA N. POMPEY • *Department of Cell Biology, UT Southwestern Medical School, Dallas, TX, USA*
LILIANA QUINTANAR • *Departamento de Química, Centro de Investigación y de Estudios Avanzados, Mexico City, Mexico*
LINA RIVILLAS-ACEVEDO • *Departamento de Química, Centro de Investigación y de Estudios Avanzados, Mexico City, Mexico*
ALISON RODGER • *Department of Chemistry, University of Warwick, Coventry, UK; Warwick Centre for Analytical Science, University of Warwick, Coventry, UK*
JUSTIN A. ROSS • *Queensland Institute of Medical Research, Herston, QLD, Australia*
MARIA J. SCHILSTRA • *Biological and Neural Computation Group, School of Computer Science, University of Hertfordshire, Hatfield, UK*
PETRA SCHWILLE • *Max Planck Institute of Biochemistry, Am Klopferspitz 18, Martinsried, Germany; BIOTEC, Dresden University of Technology, Dresden, Germany*
HANS SJÖBOM • *GE Healthcare Bio-Sciences AB, Uppsala, Sweden*
RICHARD D. TAYLOR • *Department of Medicinal Chemistry, UCB Pharma, Slough, UK*
DAVID TEMESI • *AstraZeneca R&D, Macclesfield, UK*
ANDREW P. TURNBULL • *CRT Discovery Laboratories, Department of Biological Sciences, Birkbeck, University of London, London, UK*
JAMES P. TURNER • *Department of Medicinal Chemistry, UCB Pharma, Slough, UK*
CARA K. VAUGHAN • *Department of Biological Sciences, Institute of Structural and Molecular Biology, Birkbeck, University of London, London, UK*
LENA VINTERBÄCK • *GE Healthcare Bio-Sciences AB, Uppsala, Sweden*
LARA WARD • *AstraZeneca R&D, Macclesfield, UK*
MARK A. WILLIAMS • *ISMB Biophysics Centre, Institute of Structural and Molecular Biology, Birkbeck, University of London, London, UK*
REMIGIUSZ WORCH • *Institute of Physics, Polish Academy of Sciences, Warsaw, Poland; BIOTEC, Dresden University of Technology, Dresden, Germany*

Part I

Introduction and Overview

Chapter 1

Protein–Ligand Interactions: Fundamentals

Mark A. Williams

Abstract

Here are described the basic mechanisms governing the interactions between proteins and their natural or manmade ligands, together with the principles underlying their analysis. The consequences of these principles are detailed for the simplest case of one-to-one binding. The general features of experimental measurements of biomolecular interactions arise from properties of the molecules involved and, thus, are common to many methods of detection. Consequently, an understanding of these principles greatly simplifies adoption and comparison of experimental methods and provides the rationale underlying many common protocols. In seeking to understand and interpret the results of experiments or identify possible sources of error these fundamental ideas are a constant guide.

Key words Protein complexes, Thermodynamics, Kinetics, Nonlinear regression, Specificity, Drug design

1 Introduction

1.1 Protein Interactions Are Key to Biological Function

Proteins have many functions: as enzymes that accelerate vital chemical reactions, as regulatory molecules whose binding to other proteins inhibits or activates their function, as components of the microscopic structures of the cytoskeleton, as engines whose conformational changes act as molecular motors, as signals via their own chemical modification, as detectors of signals through binding of small molecules, peptides or proteins, as selective channels that enable transport through cellular membranes. In all of these roles, proteins bind small molecules, nucleic acids, and other protein partners forming a transient or long-lived complex through noncovalent interactions between the components of the complex. These interactions are often very specific in that a particular protein binds only to one or a few other molecules in the cell. Many cellular processes are performed by a group of proteins which act in a coordinated manner, for example, in metabolic pathways in which an initial molecule is modified by several enzymes in a series of steps, or in a signaling pathway in which binding of an extracellular

Mark A. Williams and Tina Daviter (eds.), *Protein-Ligand Interactions: Methods and Applications*, Methods in Molecular Biology, vol. 1008, DOI 10.1007/978-1-62703-398-5_1, © Springer Science+Business Media New York 2013

signal causes a conformational change which enables a kinase to become active and phosphorylate a target (or targets), which then leads to their binding other proteins and so on. Such networks of interactions between proteins are closely regulated. Changes in the populations of particular protein complexes, or changes in populations of the products of protein mediated reactions, of as little as a factor of two can switch cells from one state to another; from growth to stasis, from replication to apoptosis.

Such small variations in molecular populations correspond to very small differences in the thermodynamics or kinetics of reactions. Consequently, detailed characterization of protein interactions with their binding partners is of paramount importance in a quantitative and integrative biology which aims to understand biological systems in terms of their molecular components. Perhaps the ultimate goal is to characterize the thermodynamic and kinetic behavior of components of living systems in sufficient detail that the response of those systems to stimuli (be they natural substrates or drugs) can be modeled with sufficient accuracy that the behavior of the system may be predicted or modified as desired.

1.2 Important Characteristics of Protein–Ligand Interactions

Key to the function of individual proteins and the network of interactions to which they belong are three factors (1) how much protein and ligand are present, (2) how much complex is formed, and (3) how quickly do the complexes form and break apart. Knowing these facts for each protein and complex enables, in principle, the prediction of behaviors of whole networks of interactions. It also enables the prediction of the outcome of any modification in the behavior of one reaction (e.g., as a result of mutation or addition of a competitive ligand which changes the amount of a complex formed) on the behavior of the system as a whole (1, 2). Biophysical approaches to protein–ligand interactions are concerned with identifying the molecular components of complexes (*see* Fig. 1) and quantifying their equilibrium populations and kinetics (rates) of association and dissociation. The need for information on these characteristics is common to all types of interaction (i.e., between any pair of—protein, small molecule, and nucleic acid). Enzymes are a special case, which, in addition to being characterized by each of their binding reactions with substrates, products, cofactors, activators, and inhibitors, are characterized by the degree of acceleration of each reaction that they enable. Only their binding reactions will be considered here.

Let us consider the simple binding reaction

$$P + L \rightleftarrows PL, \tag{1}$$

in which each protein, P, binds one ligand, L, to form a complex, PL, in a reversible manner.

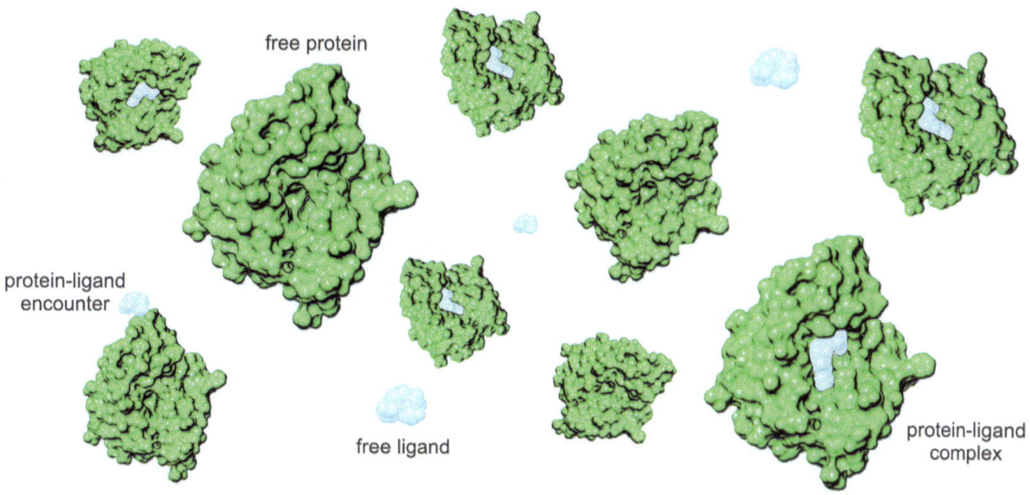

Fig. 1 Molecular-scale view of a binding equilibrium. In all circumstances there is a mixture of each of the molecular species, free (unbound) protein, free ligand, and the protein–ligand complex that becomes populated as the ratio of ligand to protein increases. The protein–ligand complex will comprise ligand bound at a particular site on the protein's surface. Other "encounter" complexes of distinct conformation will form transiently, but no one of these conformations will have a significant population

At equilibrium, by definition, there are no changes in the concentration of any components of the solution, which means that the rate of association of free protein and free ligand molecules to form complexes is equal to the rate of dissociation of complexes into their constituent molecules.

The overall rate of association will depend on the concentration of both the free ligand, [L], and free protein, [P], present in solution, because the more ligand and protein molecules there are, the more likely they are to come close to (encounter) each other and associate. A process involving collision of two molecules is described by "second order" kinetics, the second order association rate constant, $k_{on,}$ in units of M^{-1} s^{-1}, describes the rate of association for a (hypothetical) standard state of 1M protein and 1M ligand. In an idealized, purely diffusion controlled reaction, the on-rate constant depends only upon the size of the molecules involved (their translational and rotational diffusion coefficients) and the size of the binding site; $k_{on,}$ is ~10^8 M^{-1} s^{-1} for a medium size protein and a small-molecule ligand (3). The initial encounter rate will be greater for larger proteins, because of their greater surface area, perhaps by a factor of 10–100 (4, 5). For a particular real protein–ligand interaction, $k_{on,}$ is an intrinsic property of the individual molecules involved, which depends not only upon their size but also the microscopic details of their interaction. Not all close encounters are productive in terms of forming the final complex, but it seems that many proteins have evolved to enhance the chance of their cognate ligand binding (i.e., to increase k_{on}) once protein and ligand are near to each other. If a charged protein and ligand approach through random diffusion to within a few nanometres, long-range electrostatic interactions can then pull them toward each other and steer the charged ligand toward the binding site (3, 6, 7). Alternatively, the arrangement of surface chemical groups on the protein may guide a ligand that collides anywhere on the protein surface toward the binding site (8). In contrast, k_{on} may be reduced in a situation where only a subset of the protein or ligand molecules are in a binding competent conformation at any one time or which have unfavorable electrostatics. Such enhancement or reduction of k_{on} may serve a functional purpose in native interactions or may be manipulated by the experimenter through chemical modification or mutation to achieve some desired outcome (9, 10).

The dissociation rate constant, k_{off}, is simply a rate per second, and does not depend on concentration because it represents the probability with which each individual complex that is present in solution dissociates in the next second (a process dependent on only a single species is called first order). Dissociation is usually stochastic and depends upon the ligand having excess local energy (gained through random "collisions" with surrounding protein atoms or solvent) sufficient to escape from the protein's binding site. Although the actual lifetime of any particular complex will vary randomly, the mean lifetime $\tau = 1/k_{off}$ averaged over many complexes will be well-defined and can be used to describe behavior of populations of molecules. k_{off} is strongly dependent on the nature of interactions in the bound state. It is presumed that where there are a few individual strong interactions between chemical groups of the ligand and protein, they create a single substantial activation barrier to release, which is rarely crossed because it is rare that a sufficiently large amount of energy will be available, and consequently k_{off} is small. Conversely, in a case where there are many weak interactions contributing to binding, these can each be broken successively with relatively small amounts of energy and k_{off} is relatively large.

The rates of the association and dissociation reactions observed in a particular experiment are equal to the rate constants multiplied by the appropriate concentrations of each species. At equilibrium there is no change in the system and the rate of association equals the rate of dissociation, i.e.,

$$k_{on}[L][P] = k_{off}[PL]. \quad (2)$$

At equilibrium, the ratio of concentrations of the free protein and ligand molecules to the bound complex defines the association constant (K_a or binding affinity). Rearranging Eq. 2 gives

$$K_a = \frac{k_{on}}{k_{off}} = \frac{[PL]}{[P][L]}, \quad (3)$$

where the units of K_a are M^{-1}. Equivalently, the dissociation constant K_d is also used

$$K_d = \frac{k_{off}}{k_{on}} = \frac{[P][L]}{[PL]}. \quad (4)$$

The units of K_d are Molar, and are frequently in the range of μM to nM, with the latter representing a higher affinity, less dissociable interaction. For a protein–ligand interaction with a diffusion-limited association rate constant of 10^8 M^{-1} s^{-1} and a typical dissociation constant of 1 μM this implies a k_{off} of 100 s^{-1}. On-rate constants for protein–ligand interactions range from 10^4 M^{-1} s^{-1} to 10^{10} M^{-1} s^{-1} and k_{off} from 10^{-4} s^{-1} to 10^5 s^{-1} (*11, 12*).

1.3 Drugs Act by Modifying a Protein's Biological Interactions

Almost all pharmaceuticals act by binding to particular proteins, their "targets", interfering with the protein's biological function. Where the target is an enzyme, drugs are usually inhibitors (antagonists) targeted to bind in such a way that they, at least partially, block the binding-site for the natural substrate, thus inhibiting the enzyme from performing its function. Occasionally, a drug may act in an allosteric manner; binding elsewhere on the protein, but in such a manner as to disturb the (average) conformation of the substrate site so modifying its affinity for the substrate. With other types of protein target, many drugs are agonists; mimics of the natural biological ligand that binds and activates the protein.

1.4 Thermodynamics of Protein–Ligand Interactions

The experimentally determined equilibrium and rate constants for protein–ligand interactions can be used to infer some features of the energetics of the interaction between protein and ligand on the microscopic scale. The equilibrium dissociation constant determined under normal experimental conditions of constant temperature and pressure is related to the difference between the free energy, *G*, of protein and ligand molecules alone in solution and when bound together i.e., the binding free energy change ΔG_{bind} given by

$$\Delta G_{bind} = G^{PL}_{solution} - \left(G^{P}_{solution} + G^{L}_{solution}\right). \quad (5)$$

The relationship between ΔG_{bind} and K_d is usually given (*see* **Note 1**) as

$$\Delta G_{bind} = RT \cdot \ln(K_d), \quad (6)$$

where *R* is the universal gas constant and *T* the temperature (*see* **Note 2**). Negative free energy changes favor the reaction; a $K_d = 1$ μM corresponds to $\Delta G_{bind} = -34.2$ kJ mol^{-1}. In principle, this free energy change can be related to the redistribution of energy among the atoms of the protein and ligand molecules, and the solvent which surrounds them, upon formation of the complex.

$$\Delta G_{\text{bind}} = RT \cdot \ln\left(\frac{\sum\limits_{\text{freestates}} e^{-\varepsilon_i/RT}}{\sum\limits_{\text{boundstates}} e^{-\varepsilon_i/RT}}\right), \tag{7}$$

where ε_i are the possible microscopic energy states of the protein, ligand, and the surrounding solvent. Protein, ligand, and water (and other solvent) molecules are in constant motion on the Ångstrom length scale and although the "bound state" may present as single experimental observable (e.g., in respect of properties such as its UV absorption or fluorescence spectrum or NMR frequencies), it is in fact an average over the many bound molecules in a typical sample (i.e., 1 ml of 10 μM protein contains ~ 6×10^{15} protein molecules). Both the bound and the free states of the molecules are composed of many distinct possible conformations of protein and ligand and surrounding solvent molecules, each of which will have a particular energy that will contribute to the observed binding free energy. This microscopic heterogeneity of bound and free states is one of the factors that make prediction of thermodynamic and kinetic properties of protein–ligand interactions from the structures of molecules a still unsolved problem (*see* Subheading 1.8 for further discussion).

The free energy change itself can be decomposed into two terms, the enthalpy change, ΔH, and the entropy change, ΔS, of the reaction

$$\Delta G_{\text{bind}} = \Delta H_{\text{bind}} - T\Delta S_{\text{bind}}. \tag{8}$$

Rearrangements of the molecules upon binding lead to changes in the energy states of the system of molecules, and heat is either taken up or released into the surroundings. In the microscopic view (*see* **Note 3**) the molar enthalpy change is

$$\Delta H_{\text{bind}} \approx \Delta E_{\text{bind}} = \frac{\sum\limits_{\text{boundstates}} \varepsilon_i \cdot e^{-\varepsilon_i/RT}}{\sum\limits_{\text{boundstates}} e^{-\varepsilon_i/RT}} - \frac{\sum\limits_{\text{freestates}} \varepsilon_i \cdot e^{-\varepsilon_i/RT}}{\sum\limits_{\text{freestates}} e^{-\varepsilon_i/RT}}. \tag{9}$$

If the surroundings of the reaction volume consist of a microcalorimeter, then the heat transfer can be measured and used to monitor the reaction and the total heat lost or gained per mole of complex formed is ΔH_{bind} (*see* Chapter 4 for further details). It is important to remember that the enthalpic and entropic changes separately do not determine the outcome of the reaction, only the free energy change corresponds to the changes in population of bound and free species. For any given reaction, either or both of ΔH_{bind} or $-T\Delta S_{\text{bind}}$ may be a favorable (i.e., have negative value) or unfavorable contribution. However, because, by definition, reported interactions have a negative ΔG_{bind} it is most common that both ΔH_{bind} and $-T\Delta S_{\text{bind}}$ are favorable.

The ΔH_{bind} and $-T\Delta S_{bind}$ values are additional characteristics of a reaction that may in principle provide some mechanistic insight. For example, because the measured heat transfer associated with binding arises from all the events that occur in the reaction vessel and because changes in protonation state cause heat changes, it is straight forward to detect protonation state changes that accompany binding using calorimetry. The reaction is simply carried out several times at different pH values or in buffers with different heats of ionization and an observed change in the heat of reaction implies that a protonation change occurs. Less straightforwardly, there are good theoretical reasons to suppose that precisely complimentary interactions between protein and ligand will generate large favorable enthalpy changes upon binding. Thus, modifications of a ligand that make ΔH_{bind} more negative may indicate improved complimentary, which would likely also mean greater specificity of interaction (*see* Subheading 1.10), which is a desirable property in drugs (13). Practical application of this scheme in drug development is, however, complicated by the role of water molecules, which could have large positive or negative enthalpy contributions in any particular protein–ligand reaction, and by the phenomenon of enthalpy-entropy compensation.

Experimental evidence shows that changes in ΔH_{bind} and $-T\Delta S_{bind}$ due to modification of an interaction (e.g., through changes to ligand chemistry or amino-acid mutation) tend to be compensated (i.e., change in opposite directions) for most protein–ligand interactions, leaving the ΔG_{bind} only slightly changed (14). Enthalpy-entropy compensation and the poorly understood role of water are significant (and possibly related) barriers to straight-forwardly interpreting or utilizing the components of the free energy in an explanatory manner at the present time (this does not prevent many publications purporting to do this!). The origin of enthalpy-entropy compensation is not precisely clear, but must lie in the limited number of ways in which energy can be distributed in a particular protein binding site (15). Consequently, the major present use of calorimetry is simply as a method to obtain affinities and the ΔH_{bind} and $-T\Delta S_{bind}$ data are largely being accumulated as a challenge to computational modeling and for their potential to provide insight in the future.

1.5 Typical Properties of Natural Ligands and Drugs

Most biological interactions are required to be functionally reversible in vivo. For example, where a protein binds to an activating phosphopeptide and a signal is propagated (e.g., by conformational change and binding to another protein), phosphatase activity may then increased via some cellular feedback circuit, and the activating phosphopeptide must then escape from its binding site in order to be dephosphorylated so that signal can be switched off. In the case of enzymes, substrate must bind rapidly and product must be released fairly soon after its formation. Further, many enzyme cofactors (ATP, NADH, FADH) require regeneration by another enzyme and must be released promptly to solution to enable regeneration. Consequently, most proteins are evolved so that their natural substrates are able to bind and be released, and rebind reasonably frequently at cellular concentrations. This means, for example, that k_{off} should not be much slower than 1 s^{-1} if the system is required to respond to changes on the timescale of seconds or not slower than 1,000 s^{-1} if it operates on a millisecond timescale. It is also necessary that regulated variation in protein concentration (e.g., through control of gene expression or of protein degradation) substantially and proportionately modulates the population of complexes present. As will be seen below in Subheading 2.1, this means that the K_d of a protein–ligand interaction needs to be similar to the cellular concentration of that protein. Experimentally it is found that almost all protein interactions with their natural substrates have a K_d between 50 nM and 1 mM (16), commensurate with cellular protein concentrations which are in the range 5 nM to 100 μM in the vast majority of cases (17).

Any drug whose mode of action is to prevent binding of the natural substrate, must replace more than 90 % of substrate. If substrate and drug are present at the same concentration, then the drug must have a tenfold greater affinity than substrate. If the substrate concentration is tenfold greater than that of the drug, the drug must have a 100-fold greater affinity. Of course, if substrate concentrations are relatively low or the drug is designed to supplement natural substrate activity, then its affinity may in principle be more substrate-like. However, often there are also advantages to lowering drug concentration as this may reduce side-effects. Creating high-affinity ligands is thus an explicit goal in the pharmaceutical industry. Consequently, drugs typically have greater affinity for their protein targets than natural substrates, most frequently K_d is in the 100 nM to 10 pM range. Drugs may also be developed to have a slower off rate constant, thus prolonging activation. Furthermore, whereas most interactions with natural ligands have ΔH_{bind} as the largest favorable contribution, drugs on average have a larger favorable $-T\Delta S_{bind}$ term and binding is typically driven by both favorable enthalpy and entropy of binding—this is likely a consequence of drugs being on average more apolar than natural ligands, and bias in the drug design process toward apolar compounds in order to generate affinity or to improve absorption through the cell membrane (16).

1.6 Experimental Monitoring of Protein–Ligand Interactions

In principle, any signal that changes in proportion to the amount of complex formed can be used to monitor binding and determine thermodynamic and kinetic characteristics of the reaction. Consequently, there are many techniques in use to characterize protein binding and new methods emerge every few years. The methodology chosen to investigate a particular protein system will depend upon factors such as the quantities of protein and ligands available, the specific properties of the protein and any known ligands (e.g., are they fluorescent?), and whether precise quantitation is required or simply a yes/no answer to the question "Do they form a complex?" Key to any useful experimental method is an ability to detect signal variation upon complex formation using only small quantities of protein and ligand. Many experimental strategies rely on changes in spectral properties of protein or ligand e.g., UV or visible light absorption, circular dichroism, or fluorescence spectra. Such approaches are naturally limited to those protein–ligand systems in which there are substantial changes in such spectral properties and that is by no means all systems. In contrast, nuclear magnetic resonance (NMR) spectra are extremely sensitive to interactions between chemical groups and can be relied upon to change when a complex is formed. Consequently, NMR is very widely used in routine screening applications as well as investigation of detailed structural features of complexes. The main drawback to NMR is the relatively large quantities of protein required for experiments (usually several milligrams). Because approximately 95 % of protein–ligand reactions will generate sufficient heat change as to be reliably detected by modern calorimeters, isothermal titration calorimetry is also nearly universally applicable, and requires somewhat less protein than NMR. A wide range of methods have also been developed that do not rely upon a change in the properties of the components of the complexes, but instead immobilize one of the putative binding partners and detect the change in location of the other as it binds. Examples of these methods are surface plasmon resonance and optical wave guide technologies. These location-based methods also have near universality and typically require very small quantities of protein, albeit they rely on being able to tether one component to a surface without interfering with the binding event of interest. The applicability of particular experimental methods and common experimental scenarios are considered in more detail in Chapter 3 and specific details of individual methods of detection are given many other chapters.

Although experiments may vary greatly in their mechanism of detection, the main factors in the design of experimental protocols and evaluation of data have many characteristics in common because they depend upon general features of binding reactions and not the instrumentation. Experimental strategies can be developed for detection of formation of a complex via monitoring a change in a ligand or protein property on binding, or through detection and quantitation of the populations of free ligand or free protein. Obviously, analyses of these three types of signal are all simply related by Eq. 3 and can in principle be converted from one to the other provided total ligand and total protein concentration are known. However, in a typical experimental scenario in which ligand is titrated into protein, for a high-affinity interaction almost all titrated ligand will bind to protein during the initial points of a titration and free ligand concentration will only become substantial as the protein binding sites become saturated late in the titration. In such a scenario, errors are proportionately smaller for methods which directly detect the amount of complex formed or the reduction in free protein. Consequently, only analysis of these types of measurement will be considered in detail in Subheading 2.

1.7 Structural Basis of Protein–Ligand Interactions

Proteins usually bind their biological ligands at a single site (rarely more) on a structured domain. Where a biological function requires a protein to bind several distinct ligands, the protein will likely consist of several distinct domains each of which binds a specific "cognate" ligand. There are some notable exceptions to this "rule", such as those enzymes that have cofactor and substrate binding at neighboring sites on a single domain (although it could be argued that these neighbors form a single extended binding site). Ligand-binding sites are often only a very small part of a protein's surface (*see* Fig. 2a) and modification by mutation may change the binding site and affect the affinity or specificity for ligands with little impact on the structure of the domain as a whole. The possibility of combining distinct binding domains in a new protein through gene fusion and the mutability of binding sites means that evolutionarily related proteins can bind distinct ligands generating beneficial new functions (18, 19). A fundamental problem of both molecular biology and drug design is to understand the physical origin of the affinity and specificity of the interactions made by protein domains in terms of interaction of chemical groups of the protein, of the ligand and with surrounding water molecules, and thus to fully understand both the ability of proteins to discriminate between ligands and the physical factors limiting evolution.

In structures of protein complexes, complementarity of the shape (steric complementarity) and of the chemical nature of the groups on the ligand and protein surfaces is generally observed. Thus, complementarity appears to be a key ingredient in the formation of stable complexes and underlies many computational efforts to predict which ligands will bind with high affinity to a given protein. Apolar groups tend to be brought together with apolar groups, hydrogen bond donors match acceptors and charge groups on a ligand are frequently neutralized by a nearby, oppositely charged protein sidechain (*see* Fig. 2b). However, it should always be borne in mind that the precise thermodynamics and kinetics of association are determined not only by the changes in protein–ligand interactions as the protein and ligand come together but also by the changes in protein–water and

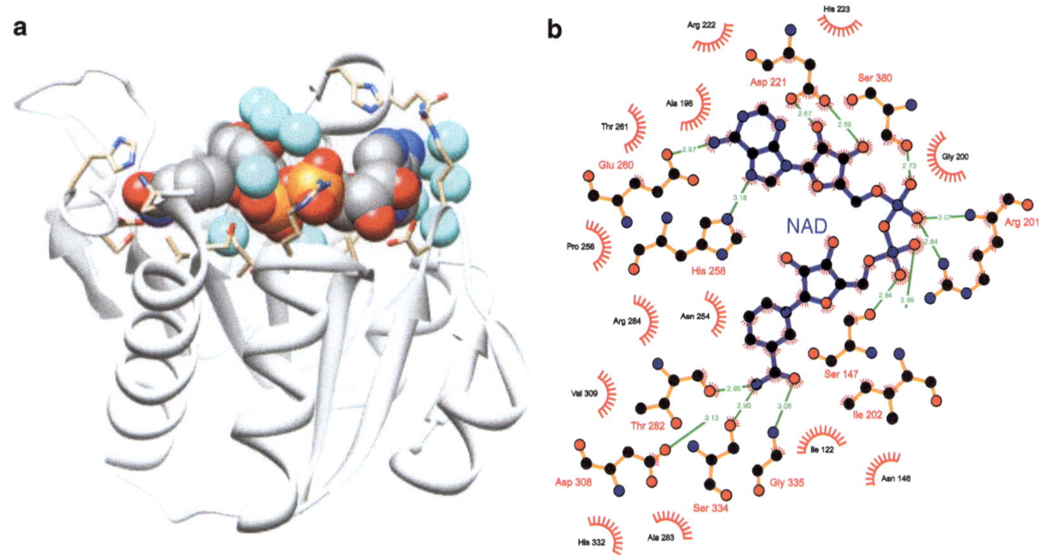

Fig. 2 The structural basis of ligand recognition by proteins is illustrated here by NAD binding to formate dehydrogenase (PDB structure: 2NAD). (**a**) The path followed by the protein backbone is represented by the ribbon, highlighting the Rossmann-fold architecture common to many nucleotide binding proteins. At the periphery of the protein, conserved amino acids (stick representation) far apart in sequence are found close in space and envelope the dinucleotide (*spheres*), making more than 40 interactions directly or indirectly via crystallographically observed water molecules (*cyan spheres*). (**b**) The direct interactions are annotated in a schematic of the binding site (adapted from PDBsum). Amino acids making hydrogen bonds (*green*) to the ligand are drawn as atoms + bonds, and those making only van der Waals interactions are represented by "eye-lash" symbols (*see* Table 1 for references)

ligand–water interactions. Consequently, it is the net effect of changes in interactions and not complementarity alone which determines ligand binding (20).

It has been inferred from analysis of structures of the free and bound states of protein and ligand, together with experimental thermodynamic data on formation of the complex, that in most cases the largest contribution to affinity is provided by bringing apolar groups (usually CH_2, CH_3, aromatic, and other carbon-ring structures) of protein and ligand into close proximity (16). The interactions between protein and ligand apolar groups (largely van der Waals interactions) may themselves be energetically more favorable than those they make with water in the free state, but the association is thought to be largely driven by the favorable free energy contribution from water molecules being displaced from the apolar surface into the bulk water where they can form additional hydrogen bonds with other water molecules. It has often been assumed that the driving force of this contribution to the affinity is the gain of entropy due to water release. However, the experimental support for this idea is not strong and it is probably an oversimplification (14). In practice, there is typically a fine balance between the enthalpic and entropic contributions of the changes in interactions of many individual chemical groups and solvent molecules to overall affinity (16).

The formation of hydrogen bonds between protein and ligand appears to be crucial to the specificity of ligand recognition by proteins. Proteins have many medium-strength hydrogen bond donor and acceptor groups (e.g., N–H, C=O, O–H, COO−, NH_3+) and also groups that are able to form weaker bonds (e.g., S–H, C–H, delocalized electrons of aromatic rings) (21). In the free state, almost all of the protein's medium strength hydrogen bond donor and acceptor groups will either make intra-protein hydrogen bonds or hydrogen bond to water molecules (22, 23). Mutational analysis of protein–ligand complexes shows that displacement of water from an acceptor or donor group as a result of binding, without the formation of a new hydrogen bond in the complex, produces a substantial (~5 kJ mol^{-1}) unfavorable contribution (24). Hydrogen bonds are also directional (unlike apolar interactions which merely require that atoms are brought close together) and chemical groups from ligand and protein must be oriented appropriately in order to make the bond. Because donor and acceptor groups are generally a minor component of the ligand, there will be only a few relative orientations of protein and ligand which can maximize hydrogen bonding. This directionality and the energetic penalty of not forming hydrogen bonds means that they are key discriminators between ligands and a considerable source of specificity in interactions (25). However, the requirement for complementarity of hydrogen bonding between protein and ligand is not absolute, as interactions between protein and ligand can also be bridged by water molecules, e.g., a ligand NH may donate to water which may then either donate or accept from a polar group on the protein.

1.8 Computational Approaches to Prediction of Protein–Ligand Interactions

Although, much has been learnt about the structural features of protein–ligand interactions over the past 30 years, this knowledge has yet to be used with full effect in the interpretation and control of protein function. Statistical analyses of the structures of proteins, protein–ligand complexes, and organic chemicals more widely, have led to a good understanding of the intra-molecular geometries of molecules and also of the intermolecular spatial distribution of interacting chemical groups. Such analyses are an important input to software used to predict the structures of protein complexes and aid in designing chemical modifications of ligands. Molecular modeling tools for "docking" prospective ligands into known active sites and screening for ligands that may be able to form complimentary interactions are now widely and successfully used in ligand discovery and drug design (*see* Chapters 18 and 19).

Presently, it is the case that quantitative analysis of many systems has led to the ability to frequently predict realistic poses (conformations) for small molecules bound to proteins (26) using one of several distinct empirical scoring schemes that partition interacting groups and molecular properties into broad categories (e.g., polar/apolar). These empirical schemes are parameterised to estimate thermodynamic quantities as a weighted sum of changes in types of interaction between the bound and the free state, and are somewhat successful at predicting binding affinities of ligands (27). However, they remain subject to a substantial margin of error in ΔG_{bind} (~15 kJ mol^{-1}), do not, or poorly, quantitate the enthalpic or entropic components and take no or little account of environmental effects. Consequently, a major remaining challenge is to accurately relate the three-dimensional structures of proteins and ligands to the kinetics and thermodynamics of their interaction.

Future insight will probably arise from developments in more computationally intensive models involving physics-based simulation of atomic interactions and motions and direct averaging over the microscopic states of the protein–ligand system. Such simulations also have the advantage of providing information on the flexibility of the protein target and alternative conformations of the binding site that may not be apparent from a crystal structure. These physics-based methods have been a very active and complex area of research (28) for three decades, but they have not achieved a clear advantage over the cruder empirical methods in predicting affinity or affinity changes. A predictive accuracy of better than ± 1 kJ mol^{-1} would be needed to replace experimental study and this still seems some way off.

1.9 The Extended Interface: Long-Ranged and Environmental Effects on Ligand Binding

Protein complex formation is usually monitored in vitro in particular buffer and salt solutions that have been found to keep the protein and/or ligand soluble, and in vivo proteins are found in the distinct dense soups of the cellular cytoplasm or blood plasma. Many of the interactions between protein, ligand, and water are sensitive to differences between these environments. The measured affinity and kinetics of protein–ligand interactions depend upon the ionic environment, on pH and on crowding effects due to surrounding nonparticipant proteins or other solvent components. Changes in K_d by a factor of 10–100 in response to a change of 1 pH unit or of ± 100 mM NaCl (cellular concentrations are often ~200 mM, but very dependent on the organism) are commonplace (29). Substantial differences in binding affinity between mutant proteins or between different ligands may sometimes be due to their effect on increasing or diminishing the sensitivity to some environmental factor and thus will not correspond to a readily identifiable structural feature of the protein–ligand interaction. Environmental effects are, generally, rather poorly understood and the field as a whole would benefit from more data.

One practical consequence of this sensitivity to environmental factors is that assays of protein–ligand binding must be performed under precisely controlled conditions so that variations in behavior can be reliably attributed to changes in the protein or ligands themselves. A significant feature in the binding of many charged ligands (including common cofactors such as ATP) is change in the protonation state of the ligand or protein on binding. For example, if there is a possibility to form a hydrogen bond in the complex that is stronger than that made with water, then the proton involved in that hydrogen bond will be stabilized in the complex. In a hypothetical case where the hydrogen bonding group has pKa = 6 in the free state and in the complex pKa = 8, then at any pH between these values, protons will migrate from the solvent to form the hydrogen bonds in the complexes. In general, protonation effects will have an effect both on the free energy and enthalpy of binding, whose pH dependence will reveal the existence and details of any such process for a protein–ligand system of interest. Such mechanistic understanding of environmental influences may lead to more reliable data, identification of better assay conditions (e.g., further from the relevant pKas or with higher buffer concentrations) or insight that may guide further modification of the ligand in a drug development program (e.g., to change its charge or pKas).

1.10 Specificity

It is common to encounter discussions of specificity of interactions (sometimes called selectivity), which encompass several distinct but related ideas concerning the relative likelihood of observing particular complexes. Firstly, the question "Is there a "specific" binding site on the protein for a particular ligand?" means is there a particular location at which almost all the molecules of particular ligand bind during a titration before a population of protein complex is found with the ligand at another location. Kinases specifically bind ATP as a cofactor at an evolutionarily well-conserved site. Secondly, the question may be "Is the protein specific for a particular ligand?" i.e., are there no other small molecules (in the cell) which will bind at the same site. In the case of kinases other adenine-based nucleotides usually bind to the same site with similar affinity so kinases are not specific for ATP, but they may well not bind GTP or any other cellular components, so they may often be specific for adenine-based nucleotides. Thirdly, there are perhaps one thousand different proteins in humans that bind to ATP, and consequently ATP is not a specific ligand for a particular protein. On the other hand, a key feature of drug development is to obtain ligands that not only have high affinity but also bind to the target protein with much greater affinity than any other protein i.e., it is usually required that drugs are specific in this third sense.

Attaching a number to the specificity of a protein or ligand is difficult, or at least time consuming, as it requires comparative analysis. It is necessary to screening many ligands against a protein or many proteins against a ligand. Often a ligand will bind to several evolutionarily or functionally related proteins and the practical investigation of specificity becomes a one of establishing how the ligand partitions between the desired target and each one of those related proteins. A level of specificity of 100-fold greater binding to the target than any other protein, where the proteins are at equal total concentration, would be considered "highly specific."

It is thought that some features of ligands and proteins confer specificity. A distinct arrangement of localized and directional interactions such as hydrogen bond donors/acceptors (discussed above) or charged groups are less likely to be complemented by chance than an array of apolar groups. Shape complementarity can also be a key factor in achieving specificity, with steric clashes created by single chemical groups of otherwise complementary molecules strongly discriminating against binding (30).

1.11 Nonspecific Binding

"Nonspecific" binding is used more narrowly to describe any binding of a ligand to sites other than that (or those) of interest; it might be more accurately called off-target binding. Typically, nonspecific/off-target binding is a consequence of the ligand having low affinity for a large number of binding-sites that exist in the experimental system. This results in binding curves having two components; a signal due to the ligand binding to the specific site and a signal due to binding to the nonspecific/low affinity sites. The first component has normal hyperbolic binding (*see* Subheading 2.1) reaching saturation at several-fold excess of ligand over protein and the second component shows a very slowly changing signal that saturates at much higher excess (100-fold) or may not appear to saturate at all within the range of concentrations studied. Nonspecific binding often gives false positive results in assays that are not subject to sophisticated analysis (many weak interactions can still bind a substantial amount of ligand) and may create errors or difficulties with analysis of titration data. However, nonlinear regression fits of binding curves (*see* Subheading 3) to a two component model can usually separate the components (31) if the off-target binding is substantial. Methods that measure an effect that is unique to a particular binding site (e.g., NMR chemical shifts of signals arising from an amino acid in the binding site) or competition experiments (i.e., in which a reference ligand known to bind exclusively at the site of interest is displaced by a test ligand) are relatively immune to nonspecific effects.

1.12 Chemical Modification and Fragment Approaches

From a basic research and biophysical perspective, the most valuable data is often obtained by studies of series of related reactions (e.g., changes that arise in the thermodynamics or kinetics of binding a particular ligand due to mutation of residues in the protein or through binding many related ligands to a particular protein). Focusing on differences arising from small changes gives a greater chance of finding a plausible structural explanation for binding affinity or specificity for a particular ligand. At a superficial level it has become popular to perform alanine scanning mutagenesis on a protein (changing all its surface amino-acids one by one to alanine) and observing the effect on affinity of each mutation. It is found that residues in the binding site have most effect (and this is way of identifying the binding site in the absence of a structure) and that among those a small number of amino-acids will have the greatest effect—these are designated "hot-spot" residues (32). A similar approach can be taken from the ligand perspective. Either making modifications to series of ligands or fragmenting a ligand into each of its chemical substructures and binding each to the protein enables some assessment of the contribution of each part of the ligand to binding (33). However, it should be remembered that any change to the binding partners may be accompanied by changes in conformation and solvation of both the free and complexed molecules. Consequently, it cannot simply be assumed that large changes in affinity as a result of chemical changes definitively identify those groups that are the most important contributors in the native state; further structural characterisation is required as corroboration.

Fragments (molecules of <200 Da molecular weight representing common substructures of natural substrates and drugs) have also become a key tool in characterization of a protein's binding capabilities and in the development of novel ligands and drugs (34, 35). Distinct fragments that bind at neighboring sites on the surface of a protein may be chemically linked to create a high-affinity ligand or a fragment may be incrementally grown to include chemical groups providing additional specificity or affinity.

1.13 Data Resources for Protein–Ligand Interactions

There has been a strong trend to make the results of government-funded research available as widely as possible both to scientists and the public. Also web-resources have been created recently which replace some of the functions of commercial products, which used to be the main source of chemical information, with the aim of lowering the total cost to public sector scientists of obtaining information to further their research. Given an interest in a particular protein or ligand there is a good chance that something is already known about its interactions and much laboratory time can potentially be save by some desk-bound research. Table 1 describes some of the principal sources of information that are often used in studies of protein–ligand interactions.

Table 1
Major sources of information on protein–ligand interactions

Resource	Contents	Utility	Reference
Blast	Sequences of all know proteins/genes	Search for proteins of similar sequence. Similarity to a region of a protein with known function may help infer ligand binding possibilities	blast.ncbi.nlm.nih.gov (36)
Pfam	Sequence motifs corresponding to functional and/or structural domains of proteins	Tools are available to identify domains in a user supplied protein sequence. Such domains typically have a conserved or related binding function	pfam.sanger.ac.uk (37)
CSD	3D structures of >500,000 chemical compounds + tools and analyses based on these structures	Obtaining parameters to describe small molecules; Statistical analyses of likelihood and geometry of interactions between chemical groups (Isostar); Database of protein–ligand complexes derived from the PDB organized from a small molecule perspective to enable comparison of ligands and their interactions (Relibase)	www.ccdc.cam.ac.uk (38) isostar.ccdc.cam.ac.uk (39) relibase.ccdc.cam.ac.uk (40)
PDB	More than 80,000 protein structures, including >10,000 structures of protein-small molecule complexes	Searchable by protein sequence, ligand name and chemical substructures. Essential for modeling of ligand interactions, ligand docking, pharmaphocore development and attempts to understand results of mutation of protein or variation of ligand	www.pdb.org (41)
PDBsum	Graphical summaries of PDB entries	Maps of ligand interactions, structure descriptions and assessment of structure quality	www.ebi.ac.uk/pdbsum/ (42)
BindingDB	Binding data for >800,000 protein–ligand binding experiments (published and unpublished)	Comparative analysis for identical or homologous proteins or ligands, links to PDB structures where available	www.bindingdb.org (12)

(continued)

Table 1 (continued)

Resource	Contents	Utility	Reference
ChEMBL	Binding properties and bioactivities of >1,000,000 drug-like small molecules	Properties of compound or target of experimental interest; Identification of related ligands by structure similarity searching, comparative analysis of binding or pharmacokinetic behavior	www.ebi.ac.uk/chembl (43)
Brenda	Curated database of properties of enzymes organized by reaction type	Substrate specificity, affinity, inhibition constants, temperature and pH dependence, literature sources for previous work on the enzyme	www.brenda-enzymes.info/ (44)
Pubchem	Nonredundant database of available or previously synthesized compounds	Search for structurally related molecules, provides bioactivity and chemical data	pubchem.ncbi.nlm.nih.gov (45)
Zinc	Database of commercially available compounds including computer generated 3D structures	Search for structurally related molecules, create datasets for virtual screening	zinc.docking.org (46)

2 Analysis of Interactions

This Subheading considers the definitions and concepts which are common to most of the experimental procedures that are discussed in later chapters. The discussion is specific to the simplest case of one-to-one binding, but all of the issues raised are applicable to more complicated cases (multiple binding sites, allostery, etc.).

2.1 Equilibrium Properties

In most experimental methods, we measure a signal, S, arising from the complex, which changes with the fraction of total protein $[\mathrm{P}]_\mathrm{T}$ bound to ligand (or the closely related quantity of total amount of complex).

$$S = S_0 + f_\mathrm{b} S_\mathrm{sat}, \tag{10}$$

where S_0 is the signal with no ligand, S_sat the signal when protein is saturated with ligand and the fraction bound is

$$f_\mathrm{b} = \frac{[\mathrm{PL}]}{[\mathrm{P}]_\mathrm{T}}. \tag{11}$$

The total amount of amount of protein and ligand is simply the amount of free species plus bound species

$$[\mathrm{P}]_\mathrm{T} = [\mathrm{P}] + [\mathrm{PL}] \tag{12}$$

$$[\mathrm{L}]_\mathrm{T} = [\mathrm{L}] + [\mathrm{PL}] \tag{13}$$

Consequently, the fraction bound can also be written

$$f_\mathrm{b} = \frac{[\mathrm{P}]_\mathrm{T} - [\mathrm{P}]}{[\mathrm{P}]_\mathrm{T}} = \frac{[\mathrm{L}]_\mathrm{T} - [\mathrm{L}]}{[\mathrm{P}]_\mathrm{T}}, \tag{14}$$

which means that the concentrations of free species can be written in terms of fraction bound and total protein and ligand concentrations

$$[\mathrm{P}] = [\mathrm{P}]_\mathrm{T}(1 - f_\mathrm{b}) \tag{15}$$

$$[\mathrm{L}] = [\mathrm{L}]_\mathrm{T} - f_\mathrm{b}[\mathrm{P}]_\mathrm{T}. \tag{16}$$

Substituting these quantities into the definition for K_d (Eq. 4) gives

$$K_\mathrm{d} = \frac{[\mathrm{P}]_\mathrm{T}(1 - f_\mathrm{b})([\mathrm{L}]_\mathrm{T} - f_\mathrm{b}[\mathrm{P}]_\mathrm{T})}{[\mathrm{P}]_\mathrm{T} f_\mathrm{b}} = \frac{(1 - f_\mathrm{b})([\mathrm{L}]_\mathrm{T} - f_\mathrm{b}[\mathrm{P}]_\mathrm{T})}{f_\mathrm{b}} \tag{17}$$

And rearranging gives a quadratic formula for fraction bound at equilibrium in terms of K_d and the total protein and ligand concentrations

$$f_\mathrm{b}{}^2[P]_\mathrm{T} - f_\mathrm{b}(K_\mathrm{d} + [\mathrm{L}]_\mathrm{T} + [\mathrm{P}]_\mathrm{T}) + [\mathrm{L}]_\mathrm{T} = 0. \tag{18}$$

This quadratic equation has two solutions, but only one gives physically reasonable results (i.e., that $f_\mathrm{b} = 0$ when $L_\mathrm{T} = 0$):

$$f_\mathrm{b} = \frac{(K_\mathrm{d} + [\mathrm{L}]_\mathrm{T} + [\mathrm{P}]_\mathrm{T}) - \sqrt{(K_\mathrm{d} + [\mathrm{L}]_\mathrm{T} + [\mathrm{P}]_\mathrm{T})^2 - 4[\mathrm{L}]_\mathrm{T}[\mathrm{P}]_\mathrm{T}}}{2[\mathrm{P}]_\mathrm{T}}. \tag{19}$$

Equation 19 (often called the tight-binding equation) describes the changes in fraction of protein bound to ligand for an interaction of given K_d as total ligand and/or protein concentration are changed. Equations 10 and 19 form the basis of all analysis of single-site protein–ligand titrations where they give a hyperbolic curve which tends to S_sat as the concentration of one component becomes relatively large (*see* Fig. 3).

2.2 EC_{50}

The total concentration of ligand at which half of the protein is bound is called the EC_{50} (50 % effect concentration) of the reaction. The EC_{50} is widely used to report results of assays since estimating the midpoint of a titration by inspection is straightforward. At the midpoint of a titration when half of the protein is bound to ligand [P] = [PL] and $f_\mathrm{b} = ½$, substituting into Eq. 4 gives

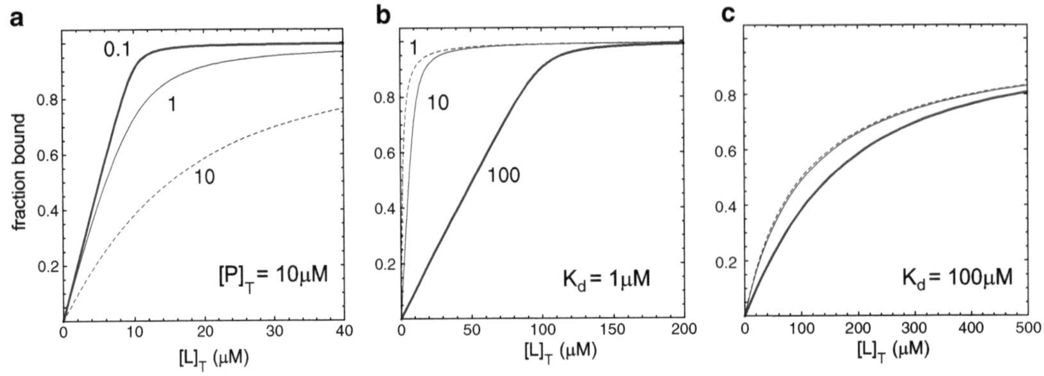

Fig. 3 Properties of binding curves. The fraction of total protein bound to ligand (determined from Eq. 19) is shown as a function of the total ligand concentration, $[L]_T$, in a variety of circumstances. (**a**) For a fixed protein concentration ($[P]_T = 10\ \mu M$) and $K_d = 0.1$, 1 and 10 μM. At the highest affinity, almost all added ligand binds to protein and the point of inflection (saturation) of the curve at ~10 μM ligand indicates the stoichiometry. Only when K_d is similar to the total protein concentration is it likely to be accurately measurable using Eq. 21. (**b**) For a fixed and moderately high affinity, the fraction bound changes almost linearly with ligand/protein ratio for total ligand concentrations (($[P]_T = 1, 10, 100\ \mu M$) above the K_d. (**c**) For a relatively low affinity interaction, fraction bound is almost independent of protein concentration for total protein concentration below the K_d and does not saturate even at very high ligand to protein ratio ($[P]_T = 1, 10, 100\ \mu M$ as in **b**)

$$K_d = \frac{[P][L]}{[PL]} = [L]_{mid}. \tag{20}$$

The concentration of free ligand at which half the protein is bound is equal to the K_d. Consequently the EC_{50} of a titration is often thought of as a reasonable estimate of K_d, however this is only the case if the free ligand and total ligand concentrations are similar. An alternative equation is obtained by substitution into Eq. 17. which gives

$$K_d = \frac{(1 - f_b)([L]_T - f_b[P]_T)}{f_b} = [L]_{T,mid} - \frac{[P]_T}{2} = EC_{50} - \frac{[P]_T}{2}. \tag{21}$$

The marginally simpler Eq. 20 is usually given in introductory textbooks as it is historically important and its theoretical treatment is more easily generalized to multiple binding sites. However, in many, if not most, experimental setups [L], the free ligand concentration, is not measured nor can it be well estimated from total ligand concentration, and Eq. 21, written in terms of total ligand concentration is much more useful. It is easy to see from Eq. 21 that using the EC_{50} as an estimate of K_d would not be correct unless the protein concentration was very much less than the K_d (*see* Fig. 3c). At high protein concentration, the value of the EC_{50} of the binding reaction has no special significance and simply tends to one half of the protein concentration, with the dissociation constant having little influence unless the interaction is very weak. Of course, it is straightforward to obtain K_d from the EC_{50} by subtraction of $[P]_T/2$, but this can be subject to considerable error, e.g., if $[P]_T$ is 200 μM and K_d is 1 μM the midpoint should be 101 μM ligand—if due to experimental error the measured ligand concentration is 103 μM at the midpoint (only a 2 % error) then the inferred K_d of 3 μM is a factor of three (200 % error) from the true value. Consequently, effective assay setups require a protein concentration near the K_d value to maximize the ability to detect weak binding while giving a reliably measurable K_d. Furthermore, fitting K_d for the full titration curve is always more accurate than estimates from the midpoint as more data points are used to generate the result.

2.3 Competitive Binding and the IC_{50}

In many circumstances it is valuable to assess the competition between two ligands for binding to a protein. If one of these is a "reference" ligand with a known binding site and if titrating in another "test" ligand diminishes binding of the "reference," then it means that the test ligand either binds at the same site as the "reference" or at an allosteric site which influences the reference's binding site. Use of such a competition experiment largely rules out interference from nonspecific effects. Furthermore, it is sometimes very useful to construct assays around modulation of binding of a known ligand, since that will have consistent properties (e.g., UV or fluorescence spectrum) in each test reaction, rather than having to understand the properties of each of the many test ligands. Consequently, competitive inhibition (a.k.a. displacement) assays are very common in a medium to high-throughput contexts. If the reference compound is the protein's biological substrate or an analog thereof, then inhibition of substrate binding by the test ligand may form the basis of a drug discovery program.

Consider a situation of competitive binding where an inhibitor, I, directly competes with the reference ligand for a single binding site, i.e., the reaction

$$PL \rightleftarrows P + I + L \rightleftarrows PI. \tag{22}$$

At equilibrium, both reactions obey Eq. 4 and if the dissociation constant for the inhibitor is K_I then

$$[P] = \frac{K_d[PL]}{[L]} = \frac{K_I[PI]}{[I]}, \tag{23}$$

and, thus

$$\frac{[PL]}{[PI]} = \frac{K_I[L]}{K_d[I]}, \tag{24}$$

Equation 24 shows that the effect of the inhibitor is simply proportional, the greater the affinity or concentration of free inhibitor [I], the lower the ratio of ligand is bound.

However, since [I] will itself depend upon the K_I and K_d and the relative concentrations of I and L, the equilibrium situation in terms of the total concentrations of the reactants is more complicated than the single ligand case in Subheading 2.1. Only relatively recently has a full analysis been made of the cubic equations, the work of Wang (47) means that it is now possible to carry out nonlinear fitting using computer software for competitive binding. However, such analysis is as yet rather infrequently carried out in practice.

Probably as a result of tradition and its superficial simplicity, it is most common to characterize competitive inhibition in terms of an IC_{50} value. Reflecting the definition of EC_{50}, the IC_{50} of an inhibitor is the concentration of inhibitor at which the maximal effect produced by reference ligand binding is reduced by 50 %. This makes for a simple assay format in which a signal from the ligand complex is monitored and inhibitor titrated until the signal is reduced by half and this practical simplicity explains the wide use of IC_{50} in studies of inhibitor series. However, although the relative values of IC_{50} for a particular assay setup do provide a useful ranking of the relative affinity of a series of ligands, it is generally difficult or impossible to convert IC_{50} values to K_I or ΔG_{bind} values. Commonly used expressions for making such conversions (48) are subject to very restrictive conditions under which they are valid, and although alternatives may be applicable in wider circumstances (49) such conversion is generally not recommended for quantitative study of protein–ligand binding. Having identified "interesting" inhibitors in a high-throughput assay, experimenters should separately measure their binding reactions in a noncompetitive experiment or investigate and analyze the concentration dependence in a competition experiment more fully (47, 50) to extract the inhibitor's intrinsic binding parameters. Further details on the theoretical analysis and practical aspects of competition experiments are given in Chapter 3.

2.4 Approach to Equilibrium

Consider a situation in which an amount of ligand is added to protein. Initially, every protein molecule will be unbound ($[P] = [P]_T$) and encounters between ligand and protein frequently lead to binding. As complexes are formed then there are fewer free ligands, and those ligands will less frequently encounter free protein, consequently the rate of complex formation will slow. Simultaneously, the number of complexes is increasing, and since each complex dissociates with a constant rate k_{off}, the total dissociation rate rises. The association rate continues falling and dissociation rate rising until equilibrium is reached when association and dissociation rates are equal. At any point in time, t, the rate of formation of the complex is the difference between association and dissociation rates

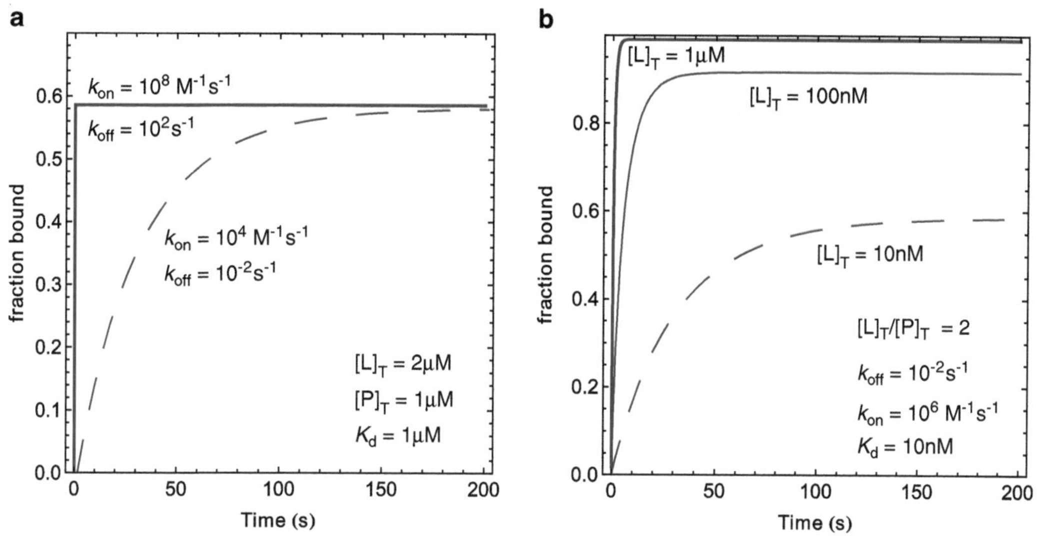

Fig. 4 The approach to equilibrium. (**a**) For a typical $K_d = 1\ \mu M$ the time to achieve equilibrium changes almost in proportion to the on rate constant from ~0.01 s to ~100 s as it is varied from a typical $10^8\ M^{-1}\ s^{-1}$ to a slow $10^4\ M^{-1}\ s^{-1}$, such as might be encountered if binding required a significant conformational change of the protein. (**b**) For a typical high-affinity ligand, time to equilibrium slows dramatically at concentrations near (or are below) K_d

$$\frac{d[PL]}{dt} = k_{on}[P][L] - k_{off}[PL] = k_{on}([P]_T - [PL])([L]_T - [PL]) - k_{off}[PL]$$
$$= k_{on}[P]_T[L]_T - (k_{on}([P]_T + [L]_T) + k_{off})[PL] + k_{on}[PL]^2. \tag{25}$$

Shortly after ligand is injected little complex will have formed, thus the term in $[PL]^2$ can be ignored, and the rate of formation of complex approximately falls in direct proportion to amount of complex present. Consequently, at least initially or in any circumstance where [PL] is small, the approach to equilibrium is described by a simple exponential, i.e.,

$$f_b(t) \propto 1 - e^{-(k_{on}([P]_T + [L]_T) + k_{off})t}. \tag{26}$$

The time constant for the approach to equilibrium, $\tau_{relax} = 1/(k_{on}([P]_T + [L]_T) + k_{off})$, turns out to be a good approximation to the true value in a wide variety of circumstances (*see* **Note 4**). The fraction of protein bound will usually achieve more than 98 % of its equilibrium value in 4 τ_{relax} and in practice this time often depends largely upon k_{on} and the total concentration of protein and ligand since k_{off} is relatively small.

Experiment design for equilibrium measurements should take into account the time taken to reach equilibrium. For a typical protein–ligand binding assay with 1 μM protein, 2 μM ligand and a K_d of 1 μM the time to equilibrium can vary dramatically with the kinetic properties of the protein (*see* Fig. 4). Whereas for a typical k_{on}

of 10^8 M^{-1} s^{-1} and a k_{off} of 100 s^{-1}, equilibrium is achieved in approximately 0.01 s and is unlikely to be of practical importance, for a system with the same K_d but with slow kinetics equilibrium may not be achieved until more than 100 s have elapsed (*see* Fig. 4). Working at very low protein concentration also extends the time to equilibrium. In the case of a competition experiment where, at the start of the experiment, the reference ligand is already bound to the protein, then the initial on-rate of the test ligand is diminished in proportion to the degree of saturation and also may substantially depend on the off-rate of the reference compound. Because off-rates are usually slow, competition experiments are particularly prone to inaccurate results due to failure to achieve equilibrium. Consequently, when establishing an experiment design it is important to try several delays before measurements are made in order to detect any possible issues (*see* Chapter 3 for further discussion).

3 Dealing with Binding Data

3.1 Nonlinear Regression Analysis

All of the equations that describe binding reactions in terms of total protein and total ligand concentrations are inherently "nonlinear" in respect of their important parameters (i.e., doubling K_d does not halve fraction bound) (**Note 5**). The standard computational methods for fitting nonlinear equations to data (called nonlinear regression) use iterative approaches which proceed from an initial guess of the parameter values and systematically modify them to minimize the deviation between experimental datapoints and the values given by the parameterized equation. Although not the only statistically justifiable approach, in an overwhelming majority of cases a least-squares criterion is used to evaluate the minimum deviation between the experimental values ($d_{i,expt}$) and those calculated with the current set of parameters ($d_{i,calc}$), i.e.,

$$\mathrm{Min}\left[\sum_i \frac{\left(d_{i,expt} - d_{i,calc}\right)^2}{\sigma_i^2}\right], \tag{27}$$

is sought, where σ_i is the experimental standard deviation of the *i*th datapoint (**Note 6**).

Computer programs for iterative refinement of the parameter values most commonly implement the Levenberg-Marquardt (LM) algorithm (51). This algorithm relies on being able to calculate how the least-squares deviation values change with small variations in the current parameters (i.e., the partial derivatives of the equation with respect to the parameters) and making changes in the parameters in the direction determined to lead to most rapid reduction in the deviation. The LM method is an adaptive combination of a straightforward gradient minimization and the Gauss-Newton (GN) approach (which is also sometimes encountered) and is designed to be more efficient than either (51, 52). It is important to realize that no nonlinear regression method is guaranteed to find the parameters giving the minimum deviation in all circumstances. Both LM and GN methods generally work well, but if one fails to find a good fit then the other may succeed since they will follow different routes in searching for the minimum.

All nonlinear regression methods rely on having a reasonably good initial guess for the parameters and on the equation being fitted actually being relevant to the real behavior of the system being studied. Any program may fail to find a genuinely good fit if either of these conditions are not met. Usually, it is the user of the program who needs to spot any problems. It is vitally important not to blindly believe the best-fit parameters. A program will (almost) always give an answer for "best-fit" parameters, but "best" in this sense means "best that the program could do in the circumstances"! It is always necessary to inspect the curve described by the best fit parameters, to see if it passes through the data points. The residuals (i.e., the values of $d_{i,expt} - d_{i,calc}$) should be inspected to see if they vary in any systematic way with concentration during a titration. If they do, then sources of systematic error in the experiment (Subheading 3.2) and alternative binding models should be considered. One caveat however, is that more complex binding models almost always give better fits to data with smaller residuals; consequently improved residuals alone do not make the more complicated model better. Statistical testing is required to establish which is the most likely model in those cases where the simplest model is unsatisfactory. Such decisions are considered in detail in the excellent introductions by Tellinghuisen (53) and Johnson (54).

3.2 Sources of Error in Estimated Binding Parameters

Experiments and their analysis can be afflicted by several sources of systematic and/or sampling errors. Systematic errors can include miscalibration of instruments, failure to observed protocols or instrument instructions precisely, loss of protein activity (e.g., due to progressive aggregation), inaccurate affinities for the reference compound in competition experiments, variation in solution conditions between batches of protein and failure to achieve equilibrium. Sampling errors arise from variations in instrument performance from moment to moment due to electrical noise or mechanical variation (e.g., in pipetting).

Systematic errors can often be identified by the availability of one or more control experiments. In fact, one should not consider carrying out biophysical analysis without first establishing a positive control experiment for the instrument (e.g., using some inexpensive commercially available protein and a known ligand). Protocols should be practised with the control until you can achieve consistent results. Any instrument variation over time can also be checked with this control experiment. Negative controls should also be performed for the "target" protein, i.e., carry out the binding protocol without ligand, this should show up any effect from protein aggregation or interaction of protein with buffer components. If a positive control is available for the target protein, i.e., a known ligand is available in sufficient quantity, then that is also valuable as it can help identify any batch-to-batch variation in the protein preparation.

In terms of absolute measurements of binding constants, the most common sources of error are concentration measurements of the protein and/or ligand, e.g., *see* ref. 55. Protein concentration measurement is discussed in Chapter 2. The critical point is that most concentration measurement methods have some level (perhaps 5–10 %) of error in the measurement of absolute protein concentration due to their measured signal being dependent on the protein sequence in a not entirely predictable manner. In critical cases, a sample of the stock protein solution should be sent for commercial quantitative amino acid analysis and any routine laboratory method benchmarked against the results of that analysis. Ligands are most commonly received from synthesis as dry powders. Because it is difficult to accurately and consistently weigh small quantities of ligand for individual experiments, stock solutions sufficient for a whole series of experiments should be made wherever possible. For commercially purchased ligands, it may be worthwhile having compounds supplied as solutions of certified composition.

Even if concentrations of protein and ligands are known accurately, it is not necessarily the case that all protein or all ligand in the sample are competent to bind; there could be impurities in the ligand or a portion of the protein could be misfolded. This can be difficult to detect as small (<20 %) losses of activity may not be easily distinguished. Quantitation of the active proportion of the reagent is best supported by some orthogonal technique, e.g., ligand purity can be checked by HPLC or mass spectrometry, and NMR spectra of a protein with and without a control ligand can eliminate the possibility of misfolding. In general, the best guard against errors from loss of activity in binding is a thorough physical characterization of the protein. If a positive control ligand-binding reaction is available for the target protein, then the stoichiometry of that reaction (e.g., as determined in ITC) may enable percent activity to be measured and other results can then be suitably corrected. Furthermore, it is often the case that the absolute values of K_d, etc., are less important than the relative values in a series of ligands being examined. For example, the relative affinity or the ranking of affinities is important for decision making in drug development. In such cases, when working with a protein stock that has a consistent active concentration, the relative systematic errors in parameters will often be smaller than the absolute systematic error.

Data analysis becomes more susceptible to measurement error at extremes of high- and low-affinity binding (as the binding hyperbola becomes very rectangular at high affinity, small measurement errors can lead to large changes in best-fit parameters; for low affinity, the binding curve becomes very shallow and it is difficult to saturate the protein or distinguish any nonspecific binding). In the high-affinity case, working at lower concentration of reactants may improve the quality of fit, but at the expense of signal to noise. Equally, low affinity interactions are more readily quantitated at higher concentration. There is an optimal compromise for each protein–ligand system and experimental method and this may be worth identifying for large scale assays and difficult cases.

3.3 Analysis of Random Errors

Random sampling errors in individual measurements naturally give rise to deviations of the best-fit parameters from their true values. It is important to establish the likely size of those deviations if the best-fit parameters are to be used in decision making or in validating computational models of individual protein–ligand interactions or in constructing models of systems of interactions. Ideally, titration experiments should be repeated several times and the values of the best-fit parameters determined for each titration averaged and the variation of each value used to estimate the error in mean value of each parameter following standard statistical methods (53). Certainly, publication quality studies will often require duplicates or triplicates.

Restrains on time in a high-throughput context, or availability of reagents more generally, may mean that no or only a few experiments are repeated for a given set of experimental conditions. In this circumstance some estimate of likely error in the derived parameters is still both possible and necessary. Such estimates can be obtained even from a single titration dataset by virtue of the fact that each measurement likely contains some random error, and thus the data points are unlikely to lie precisely on the true binding curve, but will be scattered about it. Thus, the scatter of the data points around the curve described by the best-fit parameters, which is our best-estimate of the true curve, will provide a measure of error in the data points and, via further analysis, the errors in the parameters.

For nonlinear equations, there is no exact theory that relates errors in the data to errors in the parameters. Most, if not all, commercial data analysis packages give a linear approximation called the "asymptotic standard error" as a measure of parameter error. As a rule of thumb, only where the asymptotic error is small (say <10 % of the parameter value) is it likely to be a reasonable estimate of the overall error in the parameters (53). Preferably, and certainly where sampling errors are large, a more robust analysis via Monte Carlo (MC) resampling or bootstrapping should be performed (**Note 7**). The principle of Monte Carlo error analysis is simple; artificial data sets are created by adding random errors to each experimental data point and then the best fit equation to each artificial data set is found. The errors in individual data points may be estimated from their residuals, which will naturally underestimate the error in some data points and overestimate it in others, but if there are sufficient datapoints this will "average out" (a related alternative approach where errors on datapoints are expected to be similar is to repeatedly randomly swap some of the residuals—an approach known as bootstrapping (52)). Each regression analysis of the artificial datasets will give slightly different parameters to the original. Repeat this process many times and you will find many values for parameters that best fit these artificial data. Their mean values will be the same as the fit to the original experimental data, and their variance/standard deviation will be a good estimate of the errors in the parameters that arise from errors in the data points. However, it should still be borne in mind that the errors estimated in this way will only reflect the measurement stage of the protocol, and replicates of the entire procedure may highlight variations that occur elsewhere.

3.4 Conclusion

The three most valuable in attributes for the experimental analysis of protein-ligand complexes are caution, curiosity and experience. The methods themselves are usually straight forward. Instruments come with manuals, and other chapters in this book provide further details of protocols and experimental strategies that are important to achieving a successful outcome. Data analysis routines are implemented in a variety of software and are generally straightforward to use. Although, the output of software must be treated with caution and some understanding of its meaning in order to enable realistic reporting of results. In practice, most difficulties arise from features of the protein's behavior, such as instability or multiple binding sites, sensitivity to pH, salt or other buffer components. Experience of "normal" experiments is essential to enable the experimenter to identify the "unexpected" results and formulate hypotheses as to their origin. In this regard, replicates and control experiments are invaluable for accurate quantitative work and generating results that future users of biophysical data can rely on.

4 Notes

1. Whereas Eq. 6 is almost universally cited, because taking the logarithm of K_d (which has units of Molar) is not strictly allowed, some authors modify Eq. 6 to include an additional term within the logarithm so that ΔG_{bind} is defined at a hypothetical 1 M standard state. In practice, provided Molar units of concentration are used throughout, this difference has no impact on the numerical values.

2. Most protein–ligand experiments are conducted under standard conditions for chemistry, i.e., 25 °C (298.15K) and 1 atmosphere pressure. Under these conditions $RT \approx 2.4789569$ kJ mol^{-1}. For conversion of joules to calories, which are still widely used in the USA, divide by 4.184. It is advisable not round any numbers until the very final step of any calculation—rounding errors account for inaccuracies in more than 10 % of publications (16).
3. Formally, the enthalpy change, ΔH, includes both the change in internal energy or heat (ΔE) and the work done against external pressure by volume changes resulting from the reaction. In biomolecular reactions, the external work is usually very small compared to ΔE and, as here, is usually ignored.
4. The general solution for the approach to equilibrium described by Eq. 25 is

$$f_b(t) = \frac{1}{2[\mathrm{P}]_\mathrm{T}}\left(K_\mathrm{d} + [\mathrm{L}]_\mathrm{T} + [\mathrm{P}]_\mathrm{T} - \mathrm{Re}\left(\gamma \tan\left(\arccos\left(\frac{1}{2\sqrt{[\mathrm{L}]_\mathrm{T}[\mathrm{P}]_\mathrm{T}}}\gamma\right) - \frac{k_\mathrm{on}t}{2}\gamma\right)\right)\right), \tag{28}$$

where $\gamma = \sqrt{4[\mathrm{L}]_\mathrm{T}[\mathrm{P}]_\mathrm{T} - (K_\mathrm{d} + [\mathrm{L}]_\mathrm{T} + [\mathrm{P}]_\mathrm{T})^2}$ and Re (..) signifies the real part of the expression. Equation 28 converges to the equilibrium expression for f_b (Eq. 19) at large values of the time, t, however, it is rather complicated to compute. If we are only interested in the initial rate or situations when concentration of complex formed is low compared to the total reactants, then an approximate solution is

$$f_b(t) = \frac{k_\mathrm{on}[\mathrm{L}]_\mathrm{T}}{k_\mathrm{on}([\mathrm{P}]_\mathrm{T} + [\mathrm{L}]_\mathrm{T}) + k_\mathrm{off}}\left[1 - \mathrm{e}^{-(k_\mathrm{on}([\mathrm{P}]_\mathrm{T}+[\mathrm{L}]_\mathrm{T})+k_\mathrm{off})t}\right]. \tag{29}$$

This does not converge to the correct equilibrium value for f_b except in the weak binding limit, however, the rate of approach is rather similar to the true value across a wide range of conditions. Another approximation is frequently encountered in the enzyme literature (56), i.e., that the protein concentration is negligibly small and thus

$$f_b(t) = \frac{k_\mathrm{on}[\mathrm{L}]_\mathrm{T}}{k_\mathrm{on}[\mathrm{L}]_\mathrm{T} + k_\mathrm{off}}\left(1 - \mathrm{e}^{-(k_\mathrm{on}[\mathrm{L}]_\mathrm{T}+k_\mathrm{off})t}\right). \tag{30}$$

However, in general, Eq. 30 is a much less accurate reflection of the true rate than Eq. 29, and, since it is also not much simpler, its use is discouraged.
5. Whereas it is essential to utilize the general (nonlinear) binding or kinetic equations in most quantitative studies of protein–ligand interactions, in some cases, notably when studying enzymes,

experimental circumstances prevail such that some simplifications can be made to analysis. Many enzyme reactions are carried out in the presence of a large excess of ligand and/or where the concentration of protein is very low (<0.01 × K_d), and consequently the equations describing reactions can be simplified to assume that free ligand concentration is very little change by formation of the complex. This is the scenario in which much of the classical analysis of enzymes was developed in the period 1910–1940, and these methods of analysis still prevail in certain application areas. These linearised approaches, which provide for simpler graphical or linear regression analysis, are described in detail in Copeland's excellent textbooks (56, 57). Be careful to ensure that the conditions for their validity are met before using them (52, 58).

6. In the χ^2 measure (Eq. 27) used to identify the parameters giving the best fit to the data, each term in the sum is weighed by the experimental standard deviation in each data point (σ_i). In principle, this means that values which are experimentally more precisely known are more important in the fitting process, which is a good idea. In practice, however, the σ_i may not be accurately known (e.g., the error in the estimate of the standard deviation is 50 % for three repeats). Consequently, in cases where the σ_i values are unknown or vary by less than a factor of 2–3, it is more reasonable to use a uniform weight of $\sigma = 1$ in identifying the best-fit. Conversely, if there is some *a priori* knowledge that there will be a systematic variation in the deviation (e.g., that σ_i depends on [PL]) then appropriate relative weights can be used without determining the actual standard deviations.
7. Monte Carlo analysis is accessible via the scripting capabilities of several commercial data analysis packages (GraphPad Prism, GraphPad Software; Kaliedagraph, Synergy Software; Origin 8, OriginLab), and can be readily implemented in general mathematical packages (Mathematica, Wolfram Research; Matlab, TheMathWorks) or even in a spreadsheet program (59).

References

1. Andrews SS, Bray D (2004) Stochastic simulation of chemical reactions with spatial resolution and single molecule detail. Phys Biol 1:137–151
2. Resat H, Petzold L, Pettigrew MF (2009) Kinetic modeling of biological systems. Methods Mol Biol 541:311–335
3. Davis ME, Madura JD, Sines J, Luty BA, Allison SA, McCammon JA (1991) Diffusion-controlled enzymatic reactions. Methods Enzymol 202:473–497
4. Northrup SH, Erickson HP (1992) Kinetics of protein–protein association explained by Brownian dynamics computer simulation. Proc Natl Acad Sci U S A 89:3338–3342
5. Camacho CJ, Weng Z, Vajda S, DeLisi C (1999) Free energy landscapes of encounter complexes in protein–protein association. Biophys J 76:1166–1178
6. Schreiber G, Fersht AR (1996) Rapid, electrostatically assisted association of proteins. Nat Struct Biol 3:427–431

7. Held M, Noé F (2011) Calculating kinetics and pathways of protein–ligand association. Eur J Cell Biol 91:357–364
8. Schreiber G, Haran G, Zhou HX (2009) Fundamental aspects of protein–protein association kinetics. Chem Rev 109:839–860
9. Kiel C, Selzer T, Shaul Y, Schreiber G, Herrmann C (2004) Electrostatically optimized Ras-binding Ral guanine dissociation stimulator mutants increase the rate of association by stabilizing the encounter complex. Proc Natl Acad Sci U S A 101:9223–9228
10. Gabdoulline RR, Wade RC (2002) Biomolecular diffusional association. Curr Opin Struct Biol 12:204–213
11. Gutfreund H (1995) Kinetics for the life sciences: receptors, transmitters and catalysts. Cambridge University Press, Cambridge, UK
12. Liu T, Lin Y, Wen X, Jorrisen RN, Gilson MK (2007) BindingDB: a web-accessible database of experimentally determined protein–ligand binding affinities. Nucleic Acids Res 35: D198–D201
13. Ladbury JE, Klebe G, Freire E (2010) Adding calorimetric data to decision making in lead discovery. Nat Rev Drug Discov 9:23–27
14. Olsson TSG, Ladbury JE, Pitt WR, Williams MA (2011) The extent of enthalpy-entropy compensation in protein–ligand interactions. Protein Sci 20:1607–1618
15. Dunitz JD (1995) Win some, lose some: enthalpy-entropy compensation. Chem Biol 2:709–712
16. Olsson TSG, Williams MA, Pitt WR, Ladbury JE (2008) The thermodynamics of protein–ligand interaction and solvation: Insights for ligand design. J Mol Biol 384:1002–1017
17. De Godoy LM et al (2008) Comprehensive mass-spectrometry-based proteome quantification of haploid versus diploid yeast. Nature 455:1251–1254, data in Supp. Fig. 5 with haploid cell volume of 42 μm^3
18. Pawson T, Nash P (2003) Assembly of cell regulatory systems through protein interaction domains. Science 300:445–452
19. Kavanagh KL, Jörnvall H, Persson B, Oppermann U (2008) Medium- and short-chain dehydrogenase/reductase gene and protein families: the SDR superfamily: functional and structural diversity within a family of metabolic and regulatory enzymes. Cell Mol Life Sci 65:3895–3906
20. Kahraman A, Richard J, Morris RJ, Laskowski RA, Favia AD, Thornton JM (2010) On the diversity of physicochemical environments experienced by identical ligands in binding pockets of unrelated proteins. Proteins 78:1120–1136
21. Williams MA, Ladbury JE (2003) Hydrogen bonds in protein–ligand complexes. In: Boehm H-J, Schneider G (eds) Molecular recognition in protein–ligand interactions. Wiley, New York
22. McDonald IK, Thornton JM (1994) Satisfying hydrogen bond potentials in proteins. J Mol Biol 238:777–793
23. Williams MA, Goodfellow JM, Thornton JM (1994) Buried waters and internal cavities in monomeric proteins. Protein Sci 3:1224–1235
24. Fersht AR et al (1985) Hydrogen bonding and biological specificity analysed by protein engineering. Nature 314:235–238
25. Hann MM, Leach AR, Harper G (2001) Molecular complexity and its impact on the probability of finding leads for drug discovery. J Chem Inf Comput Sci 41:856–864
26. Ferrara P, Gohlke H, Price DJ, Klebe G, Brooks CL 3rd (2004) Assessing scoring functions for protein ligand interactions. J Med Chem 47:3032–3047
27. Sotriffer CA, Sanschagrin P, Matter H, Klebe G (2008) SFCscore: scoring functions for affinity prediction of protein–ligand complexes. Proteins 73:395–419
28. Jorgensen WL (2009) Efficient lead discovery and optimization. Acc Chem Res 42:724–733
29. Ladbury JE, Williams MA (2004) The extended interface: measuring non-local effects in biomolecular interactions. Curr Opin Struct Biol 14:562–569
30. Koh JT (2002) Engineering selectivity and discrimination into ligand-receptor interfaces. Chem Biol 9:17–23
31. Mendel CM, Mendel DB (1985) "Non specific" binding a problem and a solution. Biochem J 228:269–272
32. Zerbe BS, Hall DR, Vajda S, Wjitty A, Kozakov D (2012) Relationship between hot spot residues and ligand binding hot spots in protein–-protein interfaces. J Chem Inf Model. doi:10.1021/ci300175u
33. Ciulli A et al (2006) Probing hot spots at protein–ligand binding sites: a fragment-based approach. J Med Chem 49:4992–5000
34. Ciulli A, Abell C (2007) Fragment-based approaches to enzyme inhibition. Curr Opin Biotechnol 18:489–496
35. Hadjuk PJ, Greer J (2007) A decade of fragment-based drug design: strategic advances and lessons learned. Nat Rev Drug Discov 6:211–219
36. Altschul SF, Gish W, Miller W, Myers EW, Lipman DJ (1990) Basic local alignment search tool. J Mol Biol 215:403–410

37. Punta M et al (2012) The Pfam protein families database. Nucleic Acids Res 40:D290–D301
38. Allen FH et al (2002) The Cambridge structural database: a quarter of a million crystal structures and rising. Acta Cryst B58:380–388
39. Bruno IJ et al (1997) Isostar: a library of information about non-bonded interactions. J Comput Aided Mol Des 11:525–537
40. Günther J, Bergner A, Hendlich M, Klebe G (2003) Utilising structural knowledge in drug design strategies: applications using Relibase. J Mol Biol 326:621–636
41. Berman HM et al (2000) The protein data bank. Nucleic Acids Res 28:235–242
42. Laskowski RA (2009) PDBsum new things. Nucleic Acids Res 37:D355–D359
43. Gaulton A et al (2011) ChEMBL: a large-scale bioactivity database for chemical biology and drug discovery. Nucleic Acids Res 40: D1100–D1107
44. Scheer M et al (2011) BRENDA, the enzyme information system in 2011. Nucleic Acids Res 39:670–676
45. Bolton E, Wang Y, Thiessen PA, Bryant SH (2008) PubChem: integrated platform of small molecules and biological activities. In: *Annual Reports in Computational Chemistry*, Vol 4, American Chemical Society, Washington
46. Irwin JJ, Shoichet BK (2005) ZINC—a free database of commercially available compounds for virtual screening. J Chem Inf Model 45:177–182
47. Wang Z-X (1995) An exact mathematical expression for describing competative binding of two different ligands to a protein molecule. FEBS Lett 360:111–114
48. Cheng Y-C, Prushoff WH (1973) Relationship between the inhibition constant (K_I) and the concentration of inhibitor which causes 50 % inhibition (IC_{50}) of an enzymatic reaction. Biochem Pharm 22:3099–3108
49. Lazareno S, Birdsall NJM (1993) Estimation of antagonist Kb from inhibition curves in functional experiments: alternatives to the Cheng-Prushoff equation. Trends Pharm Sci 14:237–239
50. Sigurskjold BW (2000) Exact analysis of competition ligand binding by displacement isothermal titration calorimetry. Anal Biochem 277:260–266
51. Press WH, Teukolsky SA, Vetterling WT, Flannery BP (2007) Numerical recipes: the art of scientific computing, 3rd edn. Cambridge University Press, New York
52. Johnson ML (2008) Non-linear least-squares fitting methods. Meth Cell Biol 84:781–805
53. Tellinghuisen J (2008) Stupid statistics! Meth Cell Biol 84:739–780
54. Straume M, Johnson ML (1992) Analysis of residuals: criteria for determining goodness of fit. Meth Enzymol 210:87–105
55. Myszka DG et al (2003) The ABRF-MIRG '02 Study: assembly state, thermodynamic, and kinetic analysis of an enzyme/inhibiter interaction. J Biomol Tech 14:247–269
56. Copeland RA (2005) Evaluation of enzyme inhibitors in drug discovery: a guide for medicinal chemists and pharmacologists. Wiley, New York
57. Copeland RA (2000) Enzymes: a practical introduction to structure, mechanism and data analysis. Wiley, New York
58. Johnson ML (1992) Why, when, and how biochemists should use least squares. Anal Biochem 206:215–225
59. Delahunty MD, Mack JPG (1993) A general method of curve-fitting and error analysis using a spreadsheet: determination of the binding constants of tight binding ligands in variable volume assays. Comput Appl Biosci 9:127–131

Chapter 2

Protein Sample Characterization

Tina Daviter and Rémi Fronzes

Abstract

Most biophysical experiments require protein samples of high quality and accurately determined concentration. This chapter attempts to compile basic information on the most common methods to assess the purity, dispersity, and stability of protein samples. It also reminds of methods to measure protein concentration and of their limits. The idea is to make aware and remind of the range of methods available and of commonly overlooked pitfalls. The aim is to enable experimenters to fully characterize their preparations of soluble or membrane proteins and gain reliable and reproducible results from their experimental work.

Key words Purity, Concentration, Solubility, Homogeneity, Electrophoresis, Absorbance

1 Introduction

The main characteristics anyone must know about their sample are its purity and concentration, stability, and solubility. Most biophysical techniques require certain buffers, a certain sample concentration and further require very pure protein. Hence, a lot of problems can be prevented and time saved by checking basic sample properties in the relevant buffers in advance. The methods and assays covered in this chapter are well known, maybe too well: knowledge about them is taken for granted, but rarely comprehensively communicated. Protocols of them are ubiquitous so we will concentrate on reminding of their principles, capabilities, limitations, and pitfalls.

The methods covered in this chapter are various types of gel electrophoresis, spectroscopic methods of concentration determination using dyes or intrinsic absorbance, we refer to dynamic light scattering, mass spectrometry, and analytical ultracentrifugation to determine the heterogeneity of protein samples and we refer to fluorescence spectroscopy, circular dichroism, and differential scanning fluorimetry to determine the foldedness of proteins.

Membrane proteins have moved more and more out of a niche and have become mainstream objects of science and research.

Mark A. Williams and Tina Daviter (eds.), *Protein-Ligand Interactions: Methods and Applications*, Methods in Molecular Biology, vol. 1008, DOI 10.1007/978-1-62703-398-5_2, © Springer Science+Business Media New York 2013

However, their handling remains more difficult than that of soluble proteins. We try to point out the differences and address some of the issues related to using the techniques described here, particularly in the Subheading 2 of this chapter.

1.1 Assessing Protein Purity, Dispersity, and Oligomeric State

To assess the purity of a protein sample means to detect the impurities in it. The main problem in all methods described here is the detection limit. To detect a contaminant present at a low concentration, a much higher concentration of your sample will have to be analyzed than to detect the protein itself. Many biophysical techniques described in this book require at least 95 % purity.

1.1.1 Gel Electrophoresis

Proteins are commonly run through gels of polymerized and cross-linked polyacrylamide, a method called gel electrophoresis. Proteins will migrate when an electric field is applied and depending on the method used, migration depends on parameters like charge, size, and shape. Subsequent staining of the proteins within the gel will reveal their location.

Most of the protocols can be adapted for membrane proteins; some have been specifically designed for them. We will point this out in Subheading 2.

The most commonly used method is denaturing *SDS-polyacrylamide gel electrophoresis* (SDS-PAGE) first described by Laemmli (1). Its advantages are simplicity, speed, and sample economy. It is used to monitor a protein sample throughout purification and experiments and to detect contaminating proteins.

In principle, proteins are incubated in a reducing and denaturing sample buffer and are heated for a few minutes at >95 °C, so they are unfolded monomers and bind negatively charged SDS (sodium dodecyl sulfate) homogenously along the amino-acid chain. The intrinsic charge of the proteins is abolished and all proteins carry a negative charge proportional to their length, their migration in a gel therefore mostly depends on their size.

Staining of the gel afterwards makes the protein species visible, which have migrated as separate bands. Their size (molecular weight, not dimensions) can be compared to standard proteins applied on the same gel. There are different ways to stain a gel. Silver staining is more sensitive than Coomassie blue staining, and fluorescent stains even more so (*see* Subheading 2).

In *native PAGE* proteins are not denatured and protein–protein interactions are not disrupted if they are of high enough affinity. This allows to monitor purity and homogeneity and to estimate the size of protein complexes. As no SDS is present all buffers must relate to the isoelectric point (p*I*) of the protein/protein complex to ensure that it is charged and will migrate. Usually, the pH and acrylamide concentration for a native PAGE protocol have to be optimized specifically for the sample under investigation.

Blue-native gel electrophoresis is more robust than native PAGE in that it uses a negatively charged blue dye that binds to the protein samples but does not denature them. This means that migration in the gel is independent of their p*I* but dependent only on their molecular weight (size) and shape. For these gels, standards are available.

Furthermore, *isoelectric focussing* is a way to run gels that separates proteins based on their isoelectric point. This can resolve proteins of very similar size, for example a phosphorylated and a nonphosphorylated form of the same protein. It is often used as the first step of 2D gel electrophoresis.

In *2D gel electrophoresis* two gel separation methods are combined. For example, a strip of an isoelectric focusing gel or a lane from a native gel is cut, placed horizontally on a denaturing gel (SDS-PAGE) and run.

Many more variations of gel electrophoresis exist and as they are inexpensive, quick, and do not require much specialist equipment, they are well worth exploiting. Good descriptions can be found in manuals that come with commercial gel apparatuses or with precast gels and on suppliers' Websites.

1.1.2 Gel Filtration

Gel filtration (GF), also called size exclusion chromatography (SEC) is often the first characterization step as well as being a purification step. The column matrix (gel) is a porous resin. It separates the proteins purely on the basis of their molecular mass and shape. Small proteins are able to travel into the porous matrix of the gel. As bigger proteins will be excluded, they will pass through the gel much more quickly, so their elution volume will be smaller. It is important to keep in mind that the elution volume of any protein or protein complex also depends on its shape. Elongated molecules will run faster than expected suggesting a larger than real molecular weight.

One symmetrical peak in a gel filtration chromatogram indicates either the presence of one pure protein or protein complex (dimer, trimer, etc) or it may also be a mix of several species in equilibrium, e.g. monomer and dimer, if association and dissociation are faster than separation of the species in the matrix.

Gel filtration columns can be calibrated with molecular weight standards, and comparison of their elution volumes with that of a given protein sample will give an idea of the size of the protein sample and therefore also indicate if it forms complexes, as well as indicate significant impurities.

The separation efficiency of a column is not buffer-dependent, so the stability of protein complexes in different buffers can be tested with this method.

If detergent is present in the gel filtration buffer, this method can be applied to membrane proteins. However, it is important to keep in mind that the nature of the detergent could modify the oligomeric state of a protein or protein complex. From one detergent to another, the results obtained may differ. As membrane proteins are embedded in the detergent micelle it will travel through the gel filtration column at a smaller elution volume, i.e. seem of higher MW.

A few rules have to be obeyed for successful gel filtration runs, which are summarized in Subheading 2 as well as basic instructions on molecular weight estimation, column calibration, and the assessment of association equilibria resulting in polydispersity.

1.1.3 Dynamic Light Scattering

Light passing through a protein solution will be scattered. The scattering intensity is proportional to the molecular weight and the weight concentration of the scattering particle; so light scattering (LS) methods are good at detecting large contaminating species like aggregates (2). In static LS, the signal is averaged over time, whereas in dynamic LS (DLS, also called quasi-elastic LS or photon correlation spectroscopy) fluctuations of light-scattering intensity are measured over ns to ms time scales. These fluctuations are due to Brownian motion, so these are measurements of the diffusion coefficient D of the scattering species. The Stokes-Einstein equation (Eq. 1) offers a way to derive the hydrodynamic radius R_h from measurement of the diffusion coefficients:

$$R_h = \frac{k_B T}{6\pi\eta D}, \qquad (1)$$

where k_B is the Boltzmann constant, T the absolute temperature in Kelvin and η the solvent viscosity. The hydrodynamic radius is very useful to interpret the peak positions in gel filtration, particularly of large protein complexes and associating systems (3), so the two methods are perfectly complementary and light scattering detectors are often put in line with gel filtration chromatography (size exclusion chromatography-multi angle light scattering or SEC-MALS).

Detergent micelles scatter light in the same way as any macromolecule. Therefore, LS cannot easily be applied to membrane proteins, as there will always be a mix of scattering species in the solution.

1.1.4 Mass Spectrometry

The strengths of mass spectrometry are supreme sensitivity as well as accuracy of molecular weight determination. In all mass spectrometry techniques, the sample is vaporized, ionized, and accelerated through an electric field. The trajectory that molecules fly on depends on their molecular weight and their charge. The accuracy of this is close to the Dalton, so that even impurities arising from digestion of a few amino acids or posttranslational modifications become apparent (4, 5).

Specialized methods of mass spectrometry exist to analyze protein complexes (6). Performed at milder conditions, the proteins in a complex stay together and the sample flies as intact assembly. This is a great method to determine the purity and molecular weight of hetero- or homo-oligomers as it will detect the heterogeneity arising from having one or more of the components in excess, and whether the complex is dissociating. However, some complexes will not survive the conditions of the experiment and fall apart. This may be due to a low affinity, but also depends on the type of interactions holding them together. When there is heterogeneity detected in the mass spectrometry experiment controls are necessary to see if a protein complex was forced apart during the experiment or whether this is a true representation of the complex in solution. In some cases, mass spectrometry will have to be complemented with other methods for clarification.

Because detergent has to be removed, it is commonly admitted that mass spectrometry is incompatible with membrane proteins. However, some techniques have been developed to overcome this problem (7, 8). Denaturing as well as native mass spectrometry can now be applied to membrane proteins.

1.1.5 Analytical Ultracentrifugation

In both modes of analytical ultracentrifugation, sedimentation velocity (SV) and sedimentation equilibrium (SE), samples will stay at equilibrium in their native solution conditions. Unlike in mass spectrometry, there are no dissociating forces acting on the sample. However, it is more time-consuming and will even more likely require specialist help.

An analytical ultracentrifuge allows monitoring the sedimentation process of a sample while moving in the centrifugal field. Absorbance or interference measurements are most common, and fluorescence detection is becoming more available. Sedimentation velocity experiments are done at high speed to analyze the movement of the sedimenting boundaries until all sample collects at the bottom of the cell. In sedimentation equilibrium experiments only the final distribution is of interest. To obtain a nice distribution throughout the centrifugal cell, sedimentation equilibrium experiments are done at a set of lower speeds (9).

SE experiments are more accurate for MW determination; however, to analyze the heterogeneity of a sample, velocity experiments are more suitable. In SV, contaminants and aggregates become visible as separate species. To exclude concentration-dependent association or dissociation, at least three different dilutions of sample are run. Similar to gel filtration a fast equilibrium between species will result in one peak, the sedimentation coefficient of which will change with concentration, i.e. as the equilibrium shifts. This one peak represents the weight-average sedimentation coefficient of the interchanging species (10).

Both SV and SE are in principle able to analyze detergent-bound membrane proteins. Several protocols exist to measure the molecular weight of the protein (rather than protein–detergent micelle) more accurately (11, 12).

1.2 Concentration Measurements

Knowing the concentration of your protein is absolutely critical. Most biophysical techniques described in this book rely on an accurate concentration value when it comes to the stage of data analysis. As the simplest example, any titration in dependence on the protein concentration will give you results that have at least the same error as that of the concentration measurement itself. Lack of reproducibility, e.g. when measuring a dissociation constant, are often problems in determining the protein concentration reproducibly. It is therefore very important to be aware of and estimate the error associated with each method to determine the sample concentration.

If determining the concentration of a homo-oligomer, it will be important to keep track of whether you determine the concentration of monomer or oligomer. Of course, deducing an oligomer concentration makes sense only if it is known that the complex is fully formed and stable at the particular concentration and if the stoichiometry is known.

On a very practical note, many protein preparations have nucleic acid contamination (*see* **Note 1**). In this case, simple absorbance measurements at 280 nm will lead to overestimation of the protein concentration, whereas dye-based methods will not. It may be useful to compare results from different techniques, also as most of them vary in their accuracy from protein to protein.

Dye-based assays are available as kits with extensive documentation on conditions and accuracy, and reading their handbooks is highly recommended.

1.2.1 Absorbance at 280 nm

To measure the absorbance of a protein solution is quick and easy, and it is the only method in which the solution is not consumed, but recoverable for later use. Absorbance (or absorption) is most correctly measured in a peak or trough of a spectrum, i.e. where the slope is (close to) zero. The aromatic amino acids Phe, Tyr, and Trp as well as oxidized Cys have absorption peaks around 280 nm and this wavelength is normally used to measure protein concentrations. The extinction coefficients of these amino acids do not simply add up in the protein environment, so algorithms have been developed to estimate the extinction coefficient of proteins. The ProtParam Webtool on the Expasy servers (13) estimates the extinction coefficient, the pI, and other basic parameters from the sequence of a protein.

To estimate the extinction coefficient of hetero-oligomers by summation, the exact stoichiometry of the different components of the complex has to be known. Detergent micelles often interfere with absorbance measurements through light scattering, so this method is usually incompatible with membrane proteins.

Calculation of the concentration from absorbance (A) uses the Beer-Lambert law (Eq. 2):

$$A = \varepsilon\ c\ d. \qquad (2)$$

Hence

$$c = \frac{A}{\varepsilon\, d},$$

where ε is the extinction coefficient (in $M^{-1}\ cm^{-1}$), c is the concentration (in M) and d the pathlength (in cm). A is dimensionless, also referred to as optical density (OD), and is actually the decadic logarithm of the fraction of light passing through a sample:

$$A = \log_{10}\frac{I_0}{I} = \log_{10}\frac{1}{T},$$

where I and I_0 are the intensities after and before the sample, respectively, and T stands for transmission.

1.2.2 Dye-Based Absorbance Assays

There are absorbance assays that use dyes that change their spectral properties upon binding to protein, like the Bradford (14), Lowry (15), or BCA assays (16). These are particularly useful when the buffer absorbs or scatters light (e.g. in the presence of detergent), when the protein does not contain aromatic residues or when the extinction coefficient of a protein is unknown. For example, the Bradford assay uses Coomassie Brilliant Blue G-250 under acidic conditions. The absorption maximum of the dye will shift from 465 to 595 nm upon binding to protein.

Particularly the Bradford and Lowry assays are vastly dependent on the nature of the protein with some proteins being stained better than others. Sample preparation has to follow a strict protocol and there is only a limited time and concentration window in which the measurement will be accurate. It may be advisable to use one of the commercially available kits to increase reproducibility.

An advantage of these assays is that contaminating nucleic acids or nucleotides do not influence the result as they do in normal absorbance measurements at 280 nm.

1.2.3 Dye-Based Fluorescence Assays

Several fluorescent dyes have been found to have a shifted emission peak or changed intensity after binding to protein. Therefore, fluorescence intensity can be used to quantitate proteins. Fluorescamine, NanoOrange, naphthalene-2,3-dicarboxaldehyde (NDA), and *o*-Phthalaldehyde (OPA) are some of them. They detect much smaller quantities of protein than any of the above methods, some down to a concentration of 10 ng/ml. As with dye-based absorbance assays, care must be taken in the preparation of protein samples and standard curves must be obtained in the same experiment for best results, but again, available kits usually make these assays relatively quick and easy. If the instructions are followed closely, they can be very reproducible.

1.2.4 Quantitative Amino Acid Analysis

An in-house or commercial service usually carries out quantitative amino acid analysis (QAA), which involves the complete hydrolysis of a protein into amino acids followed by their separation, detection, and quantitation. QAA is the only truly accurate way to measure the concentration of a sample, so in critical cases this should be done. QAA can also be used to determine the correction factor of any other method for concentration measurement, if they have a systematic error but are otherwise reproducible.

1.3 Protein Stability: Solubility, Degradation, and Structural Integrity

There are at least three aspects of protein stability, the solubility, the stability of its fold (three-dimensional structure) and the stability against degradation. They can be related to each other, i.e. partial unfolding can expose surfaces for aggregation or precipitation.

Complexes that are composed of more than one protein can simply fall apart into their components. If this is irreversible, the relevant sample is essentially lost.

1.3.1 Protein Solubility

Protein solubility often dramatically depends on solution conditions, particularly on the pH and on the salt concentration. This relates to the protein surface, the pK_a and charge of exposed side chains or the presence of hydrophobic patches. Most notably, membrane proteins, which have large hydrophobic areas to interact with membrane in the cellular context, require detergents to keep them in solution when they are purified, and often only certain detergents will work.

Small amounts of precipitate or some aggregates might not be visible by eye, their formation might be reversible or not. In all cases precipitation or aggregation reduce the actual concentration of protein in solution. It is a different case if protein specifically or functionally "aggregates" or assembles to form dimers, oligomers, or polymers.

Many experiments require specific buffers and salt concentrations and a certain concentration of protein, so therefore it is important to check whether a protein is soluble at the required conditions. As a basic test, protein solubility should be tested at the required temperature in dependence of protein concentration, pH, and salt. Varying two parameters at a time and plotting the occurrence of precipitate in dependence of these parameters gives a basic, two-dimensional phase diagram that will suggest suitable buffer conditions for the protein to survive the measurements.

Often, partial unfolding will expose hydrophobic side chains that act as aggregation surfaces or target sites for proteases. Assessing the structural stability more specifically might give crucial information on how to keep a protein in solution (see below).

1.3.2 Protein Degradation

Protein degradation is normally due to proteases. They are either copurified, and even the smallest invisible quantities are sufficient to degrade a sample, or arise from microbial growth within a sample. To slow down any degradation, samples should be kept on ice all the time as it slows both enzymes and growth. Many proteins can even be frozen for storage and thawed before use without losing their activity. As long as this is done properly, this may be the simplest way to avoid degradation.

Protease inhibitors can prevent digestion and are often added to the very early stages of protein purification, i.e. cell lysate. Sodium azide is typically used to stop growth in the purified sample, and some protein preparations might require protease inhibitors until the end of the procedure. This is particularly true with membrane proteins that are highly sensitive to proteases.

Both protease inhibitors and sodium azide have disadvantages and may be incompatible with the methods described in this book. It is preferable to improve the purification protocol with the aim of reducing the presence of proteases. Sometimes purifying more stable constructs of the protein is the way forward.

1.3.3 Assessing Structural Stability

Various methods exist to establish protein folding and the stability of its fold or structural integrity. Almost certainly, the structural integrity—like the solubility—depends on solution conditions (buffer, pH, salt, reducing agents, detergents) and also on the temperature. Various methods exist to determine structural changes, e.g. the CD spectrum, and the NMR spectrum undergo characteristic changes upon unfolding and the fluorescence spectrum is likely to change, too (*see* **Note 2**). Protein denaturation (unfolding, melting) experiments are usually performed as temperature scans (thermal denaturation), or as titrations with chaotropic agents like urea or guanidine hydrochloride (*see* Subheading 2.3.1).

Circular dichroism is the property to absorb left- and right-handed circularly polarized light to a different extent. All chiral molecules have this property and proteins are chiral. CD spectroscopy measures this property. Certain secondary structural elements have certain CD signals, so the method can be used to assess the fold of protein samples. CD instruments are capable of temperature scanning and can be used to measure protein melting more accurately, but one condition at a time (*see* Chapter 8 for details on the method).

A very simple and convenient way to measure thermal denaturation is *differential scanning fluorimetry.* It is often used for screening of many conditions simultaneously and very quickly. The assay uses a fluorescent dye, the emission peak of which shifts and increases massively upon binding to hydrophobic patches on proteins when they get exposed through protein unfolding (17). Detergents interfere somewhat with the assay, but membrane proteins have nonetheless been tested successfully by this method (18).

Differential scanning calorimetry (DSC) provides a thorough thermodynamic characterization of protein fold stability (19), but requires a lot of sample, is quite slow and might be more sophisticated than required if the aim is just to obtain soluble protein sample.

1.3.4 Stability of Multimeric Protein Complexes

Finally, multimeric protein complexes can dissociate over time. This can happen to a purified complex, if dissociation is very slow, but thermodynamically favorable. It might be worth trying different solution conditions, as association could be favored in other conditions.

Dissociation can also be caused by degradation or unfolding in any of the single components. However, the dissociation might be the only or first observable change.

The methods to detect dissociation are the same as discussed for assessment of purity and monodispersity, e.g. various types of gel electrophoresis, gel filtration, DLS, AUC, and mass spectrometry.

1.3.5 Membrane Proteins and the Role of Detergents

Membrane proteins display large patches of hydrophobic residues required for their insertion into or interaction with membranes. In solution, these patches promote protein aggregation through strong hydrophobic interactions. Detergents are required to prevent aggregation and to keep membrane proteins soluble in aqueous solutions. Detergents are amphiphilic molecules that usually contain both hydrophobic and hydrophilic parts. Their hydrophobic portion binds the protein and their hydrophilic portion interacts with the aqueous solution. Therefore, they make hydrophobic patches at the surface of proteins inaccessible to the solution, preventing protein aggregation through these patches.

Detergents are classified with regards to the nature of their hydrophilic or hydrophobic moieties. The hydrophilic moieties can be nonionic, anionic (negative charge), cationic (positive charge), or zwitterionic (both charges). Nonionic or zwitterionic detergents are the most commonly used detergents for membrane protein purification. The hydrophobic moiety of detergents can be classified by size (i.e. length of the alkyl chain) and flexibility (i.e. linear alkyl chain, branched aromatic, or aliphatic ring moieties).

The physicochemical properties of detergents in solution are described by several characteristics such as the critical micellar concentration (CMC), the aggregation number, or the size of the micelles:

– The *CMC* is the concentration below which monomers of detergent are soluble. Above this concentration, micelles form spontaneously. In these aggregates, the hydrophilic part of the detergent monomers are in contact with surrounding solution while the hydrophobic part is sequestered in the micelle center. The CMC depends on the physicochemical properties of each detergent, but also on the temperature, ionic strength, or pH of the solution. Some methods exist to determine precisely the CMC of a given detergent in a particular buffer condition (20). Usually suppliers indicate the CMC in water of commercially available detergents. It is in most cases safe to work with at least twofold this value.

- The *aggregation number* gives the average number of detergent monomers within the micelle. The micelle size is the average size of the detergent micelle.
- Some suppliers (Novagen) provide tables where the micelle sizes are expressed in kDa. This corresponds to the molecular weight of a protein with the hydrodynamic radius equal to that of the detergent micelle. This value is particularly useful to implement the techniques described below, many of which are size dependent.

It is very often necessary to adjust buffer conditions or concentrate the sample during protein purification or for a given assay. When detergents are present in the sample buffer, it complicates the basic procedures that are usually used for this. How this can be resolved, or how the detergent can be exchanged, is summarized in Subheading 2.3.5.

2 Materials and Methods

2.1 Methods to Determine Purity, Dispersity, and Oligomeric State

2.1.1 Gel Electrophoresis

Many manufacturers produce gel-running equipment. Protein gels are usually run vertically between glass plates with the top immersed in anode buffer and the bottom dipping into the cathode buffer. Some manufacturers offer precast gels, and they will very often fit only into their own running equipment.

We have chosen the most common protein gel methods to present here. There are many more and if the standard protocols do not give good separation or clear bands it is well worth investigating the literature and optimizing protocols for your own application.

Modes of Gel Electrophoresis

The most common gels run are *denaturing polyacrylamide gels (SDS-PAGE)*, where SDS is kept in sample and running buffer and is included in the gel to keep the protein entirely denatured.

The classic protocol (1) involves a stacking (focusing) and a resolving gel as two layers in the gel, which differ in acrylamide concentration as well as pH (discontinuous gel electrophoresis). The acrylamide concentration, ratio of mono- and bis-acrylamide and concentration of the crosslinker TEMED determine the pore size of polymerized acrylamide and is chosen depending on the sizes of proteins that need to be resolved. Gradient gels with the acrylamide concentration rising towards the bottom make separation and visualization of small (10 kDa) and large (200 kDa) proteins on the same gel possible. They retain fast moving small proteins in the bottom of the gel while large proteins can still enter at the top. Commercially available precast gels are expensive, but extremely reproducible and give nice bands.

Importantly, some general rules need to be followed when preparing and running gels:

- Some of the components need to be prepared fresh (*see* **Note 3**).
- Gels can be poured in bulk and stored, if only for a limited time (*see* **Note 4**).
- The lower detection limit is around 100 ng (Coomassie) or 5–10 ng (silver stains/fluorescent stains) of protein. To detect small amounts of contaminants, the gel should be overloaded and/or sensitive staining methods used.
- Rinse out the wells after removing the comb, so samples can sink in more easily. Typical minigels of 1 mm thickness and 15 wells take about 15 μl sample.

It is worth trying the following simple variations, certainly if results are unexpected:

- Try some samples twice so you can compare them boiled and not boiled (*see* **Note 5**).
- Try leaving out reducing agent from the sample buffer and also adding more/fresh reducing agent.

Boiled and reduced proteins should run at monomer molecular weight. Bands at multiples of the molecular weight indicate Cys-bridge formation, which might be specific and present in the original sample (*see* **Note 6**) or be formed unspecifically during sample denaturation or electrophoresis (*see* Fig. 1).

Running buffer can often be reused; however, if the current or voltage drops when running the gel this indicates the running buffer has gone off and needs replacing. With old buffer run times are much longer and the bands are blurry.

SDS-PAGE can be used for most membrane proteins. However, they are more likely to be resistant to SDS-denaturation or more heat sensitive than soluble proteins. Therefore, the results of SDS-PAGE have to be analyzed with caution with this type of protein. For example, the β-barrel fold of outer membrane proteins is SDS-resistant and if samples are not boiled the protein appears to have smaller molecular weight on SDS-PAGE. This phenomenon is well described for OmpA (21, 22).

Even slightly modifying a gel protocol may make a big difference to the outcome with a particular sample. This can be a problem if inaccurate work leads to lack of reproducibility, but it can also be a blessing and consciously be used to characterize the sample (*see* Fig. 1).

In *native PAGE* the denaturant SDS is omitted, so proteins retain their fold and migrate according to shape as well as size. Also, gels are continuous and run at just one constant pH with identical gel and running buffer. Standards exist, but will only be approximate due to inevitable differences in shape.

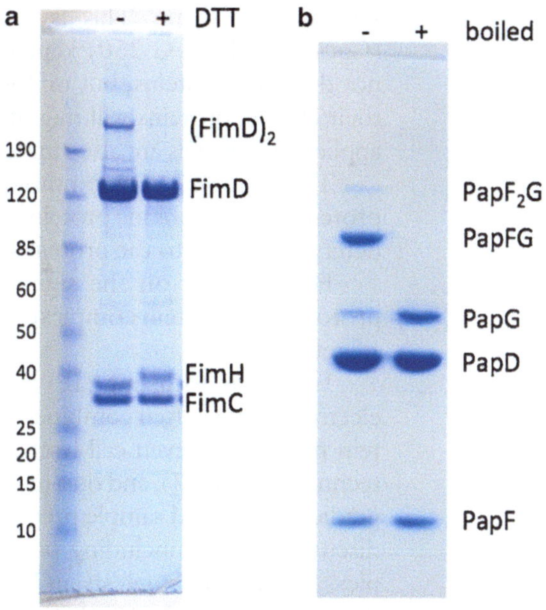

Fig. 1 Differences due to treatment of samples for SDS-PAGE. (**a**) FimD (92 kDa) membrane protein complex with FimC-FimH (23 and 29 kDa, respectively) run with and without 10 mM DTT on SDS-PAGE (NuPAGE run in supplied MES buffer, Invitrogen). There are four Cys-residues per monomer capable of forming two intramolecular disulfide bridges in the oxidizing environment of the outer membrane. Possibly because of the over-expression condition, there is a small fraction of reduced Cys-residues available to form dimers that are most likely nonfunctional (40). The standard is known to be inaccurate. NB FimH also shows changed behavior when DTT is present. (**b**) PapD-PapF ("PapDF") and PapD-PapG ("PapDG") preincubated and run on a standard, 12 % SDS-PAGE with and without prior boiling of samples. In boiled sample all proteins, PapD, PapF, and PapG run as separate bands. However, in sample that has not been boiled, a heavier band appears while the PapG band is nearly depleted, indicating formation of a PapDFG complex with an SDS-resistant PapF-PapG interaction. PapDF was used in excess, and some PapDF$_2$G seems to have formed as indicated by another, faint heavier band that likely consists of two PapF and one PapG. (*Panel (a) courtesy of Gilles Phan, panels (a, b) with the kind agreement of Gabriel Waksman, Institute of Structural and Molecular Biology, London, UK*)

The critical issue is to find a buffer that suits a particular protein, but buffers of a wide range of pH values suitable for proteins of a wide range of p*I* values are known (23). Native gels are run for a long time, at high voltage, and on ice/in the coldroom to keep them reasonably cool.

This method can be applied to detergent-solubilized membrane proteins if the specific detergent required by the protein is present in the gel and running buffer.

The blue-native PAGE method (24) is designed to keep proteins in their native state, but to ensure that migration is predominantly by size, not by shape. This is achieved with a charged, blue dye (Coomassie Blue G-250) kept in the running buffer, which does not denature proteins, but uniformly binds to their surface. Due to their charge, proteins will migrate in the gel once an electric field is applied. These gels are also run for a long time and on ice.

This method is very commonly used for the analysis of membrane proteins and membrane protein complexes. The blue dye replaces the detergent bound to the protein and keeps the protein soluble.

For a review on the significance and methods of native gel protocols for protein complexes and membrane protein complexes, *see* ref 25.

Isoelectric focussing (IF) is mainly used as one step in 2D gel electrophoresis, often combined with SDS-PAGE for analysis of protein mixtures or even cell lysates. There are many reviews on these techniques (26, 27), and on a newer development of immobilized pH gradients (28) and sample preparation for isoelectric focusing gels is discussed in (29) including preparation of membrane protein samples. As the name suggests, in IF gels a protein will stop migrating in the pH region of the gel that is identical to their own p*I* as it is then uncharged and unresponsive to the electric field. This is what constitutes the separation power between proteins with a charge difference, e.g. phosphorylated and nonphosphorylated isoforms.

Gel Staining Methods

Visualization of proteins in gels can be achieved using organic dyes, silver stains, or fluorescent dyes and the choice will depend on the required sensitivity, reproducibility, and downstream applications like Western blotting (30, 31).

The standard gel staining procedure uses solutions containing Coomassie Brilliant Blue G-250 or R-250, but other organic dyes are suitable as well. A standard protocol would use 0.1 % Coomassie brilliant blue R-250 in 40 % ethanol and 10 % acetic acid as a staining solution, destain with a solution of 40 % ethanol and 10 % acetic acid, and use a solution of 20 % ethanol and 10 % glycerol for storage. Variations of this try to dispense of the destaining step or increase sensitivity. Typically, 15 μg of protein can be quantitatively stained in a band, and down to 1 μg detected, however as the dye interacts with certain amino acids and not others, staining is sequence and hence protein dependent.

Silver staining is of roughly 100-fold higher sensitivity than Coomassie staining. However, it is more protein dependent and less reproducible due to varying amounts of background staining. Many different protocols exist.

Finally, fluorescence staining can reach the high sensitivities of silver staining at higher reproducibility, but requires a fluorescence imaging system for visualization. Different specialist fluorescent dyes exist to specifically stain posttranslational modifications.

The efficiency of all these stains is protein-dependent (*see* **Note 7**) (30, 31), so quantitation by densitometry requires comparison with samples of the same protein at known concentrations, as well as on even staining throughout the gel.

2.1.2 Gel Filtration Chromatography

Gel filtration (size exclusion) chromatography separates macromolecules purely based on their size and shape (*see* **Note 8**). The molecular weight of the species to be separated will have to be considered when choosing the right column matrix. A variety of matrices is commercially available. Depending on their pore size, the separation range will vary. This range is usually expressed in terms of molecular weight (32).

Molecules too large to diffuse into the pores will elute with the void volume V_0, the volume equivalent to the mobile phase, these molecules do not interact at all with the matrix. The smallest molecules will elute at a volume close to the total column (bed) volume V_t, i.e. mobile (V_0) plus stationary phase volumes (V_s) together (with $V_s = V_t - V_0$) as they can occupy the pores in addition to the mobile phase. Intermediate size molecules will have part of the stationary phase available to diffuse into and therefore elute at volumes V_e that are larger than V_0 and smaller than V_t. An elution volume smaller than the void volume indicates channeling through the matrix and means the column needs to be repacked. An elution volume larger than the total column volume indicates unspecific interaction with the matrix and a different resin should be chosen.

Not only the matrix but also the column geometry and the sample size will determine how well it separates different species. Generally, longer columns better separate species of similar molecular weight and are also better with larger sample volumes. Columns for analytical use will be a smaller diameter and shorter length than those for preparative use. They run faster, but have a much smaller sample capacity. The sample volume may have to be as small as 0.5 % of the column bed volume in any gel filtration application to achieve good separation.

The sample concentration can be very high, but above protein concentrations of 70 mg/ml the sample viscosity will impede separation (32). There is an upper limit of the speed (usually given in ml/min) specific for each column matrix. Above this speed, separation will suffer as sample diffusion between the stationary phase (buffer inside the pores) and mobile phase (moving buffer outside the pores) is not given enough time. Furthermore, excess speed leads to column pressure that will collapse the pores, irreversibly damaging the separation power of the matrix (32).

Using Gel Filtration to Determine the Molecular Weight

A standard curve is required to determine the molecular weight of a protein. This is done using a set of standard proteins of known molecular weight (available as calibration mixes). The distribution coefficients K_{av} of these standard proteins are calculated from Eq. 3

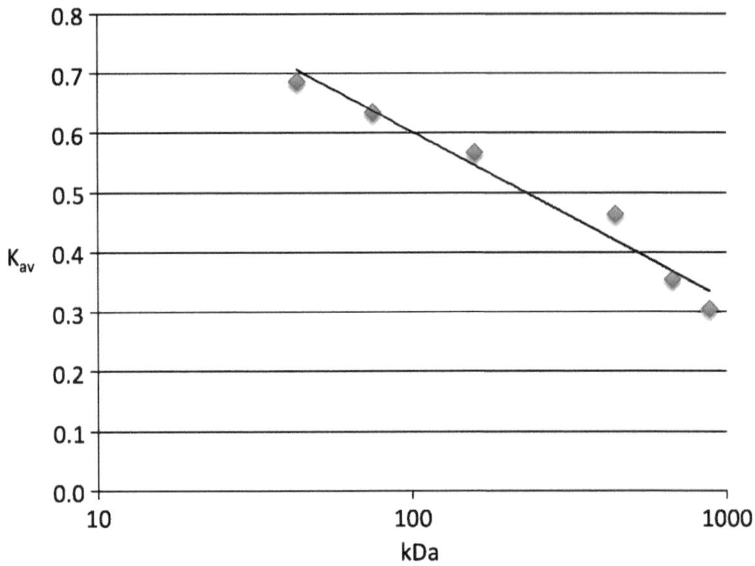

Fig. 2 Gel filtration calibration (Superose 6). Ovalbumin (43 kDa), conalbumin (75 kDa), aldolase (158 kDa), ferritin (440 kDa) and thyroglobulin (669 kDa) are injected on the gel filtration column. Their elution volume is recorded (V_e). The void volume (V_0) of the column is measured by injecting Dextran 2000 onto the column. To obtain the standard curve, K_{av} is calculated (*see* Eq. 3) and plotted against the molecular mass of the standard proteins

$$K_{av} = \frac{V_e - V_0}{V_t - V_0}, \tag{3}$$

where V_e is the elution volume for each of them, V_0 the void volume and V_t to total column volume. These distribution coefficients are plotted against the logarithm of their molecular weights and should lie on a straight line (*see* Fig. 2). V_0 can be determined by running 1 mg/ml of blue Dextran 2000 over the column. V_e of the sample in question will then give a molecular weight estimate by reading off the calibration plot or by calculating from the equation for the fitted regression line.

Any deviation from the true molecular weight is likely due to either a difference in shape of the protein species in the sample and those in the calibration mix or it might be due to an association reaction resulting in a peak representing one or even more different species.

Using Gel Filtration to Assess Protein Dispersity

Polydispersity is immediately obvious if there is more than one peak in a GF chromatogram of a sample that looked clean on SDS-PAGE. Any peaks in the void volume suggest vary large species. Depending on the separation range this is most likely unspecific aggregate. Any peaks at multiples of the monomer MW are likely homo-oligomers. If two or more peaks exist, this indicates that the species are not in a fast equilibrium with each other, i.e. they are kinetically stable. At such a slow equilibrium of associating species there will be a peak per species. To investigate a shift in the equilibrium, samples have to be pre-incubated at their different concentrations to equilibrate entirely before they are run. A lower concentration sample should give higher peaks for the species of lesser MW (higher elution volume), and vice versa, than for samples preincubated at a higher concentration. If peak heights do not change at all, the protein species could either be extremely stable or they might have lost their ability to interchange.

But also a single peak can suggest polydispersity. If the MW is unexpectedly high for a monomeric species this could not only be due to an elongated shape but alternatively (or additionally!) to a fast associative equilibrium with dimers, trimers, or higher order oligomers. In this case, lowering the concentration of a sample would shift the equilibrium towards the smallest stable species, and therefore the peak in a gel filtration chromatogram towards higher elution volumes, i.e. lower MW, and vice versa. The peak represents all species present in the sample. As they are in fast interchange the different species will never separate.

In any of these cases it will be informative to run peak fractions on SDS-PAGE and native PAGE concurrently to confirm the identities of these complexes.

2.1.3 Notes on Dynamic Light Scattering Experiments

Samples for DLS can be prepared in nearly any buffer that does not itself contain large scattering particles like detergent micelles, colloids, or crowding agents. A relatively large concentration of about 1–10 mg/ml is required to obtain a good signal for most proteins of 50–100 kDa. The volume required varies massively from instrument to instrument, from only a few μl to a couple of ml. It is hugely important that the sample does not contain dust, so only filtered buffers should be used, or the sample itself be filtered before the experiment.

In theory, the concentration and diffusion coefficient can be obtained for each of a mix of species. But in practice distinct species are visible only if they are at least factor eight apart in molecular weight (*see* **Note 9**), which would be the case for aggregates, but not for small oligomers.

If investigating an associating system, samples of different concentration should be tested to check whether the suggested molecular size moves. In that case the size given by any one of these experiments is an average size of the mix of species present. If the species are sufficiently different in size, like a monomer and a hexamer, then even a fast equilibrium between these species will give a peak for each species as the measurement is done on such small time scales (different from gel filtration where these would be averaged).

2.1.4 Sample Preparation for Mass Spectrometry

It is absolutely critical that MS samples do not contain any salt or detergent. These would bind to the macromolecules to be analyzed in different ratios and unspecifically, so that peaks are enormously broadened and resolution is lost. For some techniques, the sample is transferred into a volatile, organic solvent, for others, ammonium acetate can be used as a buffer as it evaporates in the experiment.

Standard protocols for buffer exchange can be used like several volume exchanges in a spin concentrator, dialysis, or gel filtration. Most convenient might be the use of a small desalting column. Most importantly, this has to be done thoroughly. In particular, also detergents have to be removed entirely e.g. when running membrane proteins. Newer, special mass spectrometry methods allow keeping detergents in the protein sample. The removal of the detergent from the protein is performed within the mass spectrometer itself (7, 8).

2.1.5 Notes on Analytical Ultracentrifugation Experiments

In principle, AUC experiments can be performed in any buffer compatible with absorption or interference spectroscopy. Total absorbance of buffer plus protein should not exceed about 1.2 OD, and if wavelengths shorter than 280 nm are to be used reducing agents have to be carefully chosen or avoided not to absorb at the required wavelength. Interference optics is even more versatile, however even the sedimentation of salts will be detected if sample and reference buffer do not match perfectly. Extensive dialysis is recommended. Buffers containing glycerol or a high concentration of salt may form concentration gradients, so that the sedimenting protein will encounter different solution properties at different positions in the cell. This will require special treatment during data analysis (33), so should be avoided if possible.

To discover large aggregates or precipitate the first scans are of tremendous importance and scans may have to be taken particularly quickly as heavy aggregates will sediment very quickly.

After fitting the traces for several concentrations, the sedimentation coefficient should be plotted over the concentration: if it goes up with concentration components in the sample associate (*see* **Note 10**). If the sedimentation coefficient goes down with rising concentration this indicates nonideal behavior and more dilute samples should be run (34, 35).

2.2 Concentration Determination

Any of the following methods can only be as accurate as the pipetting, e.g. when making dilutions (*see* **Note 11**).

All of the following methods have systematic errors that are protein dependent and entirely reproducible. If accurate concentrations are needed, QAA should be used to determine a correction factor for the day-to-day method of concentration measurements for each protein.

2.2.1 Absorbance Measurements

There are two common mistakes made during concentration measurements using absorbance at 280 nm. The first is to ignore the limits of the UV/Vis spectrometer (*see* **Note 12**) in relation to the

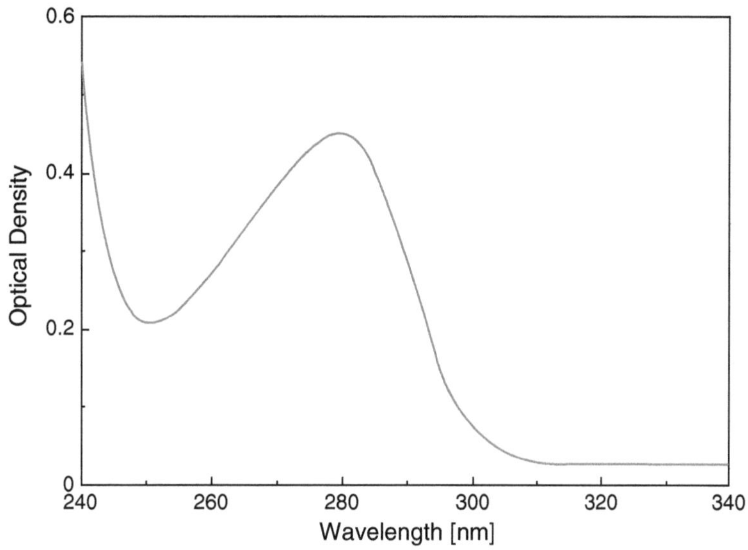

Fig. 3 UV/Vis-absorbance (OD) spectrum of a membrane protein. Recorded in 50 mM Tris pH 8, 200 mM NaCl and 5 mM LDAO taken on Jasco V-650 spectrometer in a 0.05 mm capillary. A largely flat baseline between 340 and 320 nm indicates little contribution from scattering to the peak at 280 nm. However, the baseline is offset and also, there is a tiny bump at 260 nm suggesting a small contribution of nucleotides/nucleic acids to the main peak (*Courtesy of Natalia De Val, Institut Pasteur, Paris, France*)

total absorbance of sample, buffer, and cuvette taken together. The sensitivity of the instrument should be checked in the manual, and the absorbance of empty cuvette and buffer should all be measured separately as well as that of the sample.

The second common mistake is to overlook the presence of contaminants. While the contribution from contaminating proteins will have to be estimated from other methods, other contaminants would often be apparent if a spectrum was run and inspected (*see* Fig. 3). Usual protein spectra have a fairly symmetrical peak at 280 nm resulting in a trough at 250 nm before absorbance becomes very high below 240 nm. Contaminating nucleic acids (for their removal *see* **Note 1**) would show extra absorbance around 260 nm and reducing agents would also show extra absorbance within a spectrum and quite possibly at and below 280 nm. Certainly many detergents absorb at relevant wavelengths, so in fact the method is often inappropriate for membrane proteins. If the baseline rises from 400 nm towards the peak at 280 nm this indicates light scattering by large particles like aggregate or micelles. The contribution of scattering can be subtracted from the absorbance due to soluble protein (*see* **Note 13**).

This is a thorough, recommended procedure to measure a protein spectrum for concentration determination:

1. Switch on the instrument and allow it to warm up for 20 min.
2. Run a baseline with nothing in it from 340 nm down to 190 nm (at least 240 nm).
3. Insert cuvette and take a spectrum of the empty cuvette to identify contamination and to make sure it is transparent at the required wavelengths (*see* **Note 14**).
4. Add buffer and run a spectrum to measure the absorbance of cuvette plus buffer. The absorbance will go up at low wavelengths as carbonyl and amide bonds as well as salts absorb here, DTT and β-mercaptoethanol absorb at even higher wavelengths.
5. Run a baseline with buffer.
6. Remeasure buffer to make sure it is consistently at 0 Abs, this also checks whether the lamp is warm and excludes lamp intensity fluctuations.
7. Run a spectrum with the protein solution and inspect the spectrum shape: peak at 280 nm, trough at 250 nm, then absorbance going up steeply towards 200 nm. Any absorbance decrease below 250 nm is likely an artifact due to very low and therefore noisy light levels (poor signal to noise ratio).
8. Calculate from the two spectra whether $Abs_{280,\ prot}$ + $Abs_{280,\ buffer\ +\ cuvette}$ is below the specified dynamic range of the instrument (*see* **Note 12**). If not, dilute protein to repeat or preferably use a less absorbent buffer.

Use the A_{280} and the extinction coefficient to calculate the concentration of the protein dilution in the cuvette. Bear in mind that both the measurement and the extinction coefficient contain errors, e.g. preparing dilutions inevitably introduces significant pipetting errors. Also, the extinction coefficient varies with solution conditions, so use a buffer close to that suggested on the ProtParam site on the ExPasy server (13).

If the extinction coefficient and the protein sequence are unknown, the protein concentration might be estimated using Eq. 4 (36):

$$[P] = 1.55A_{280} - 0.76A_{260}, \quad (4)$$

where [P] is the protein concentration in mg/ml, and A_{280} and A_{260} the absorbance at 280 and 260 nm, respectively. This relies on the sample being free of nucleic acids or nucleotides which absorb at 260 nm. Similarly inaccurate is to assume what is the "average" protein extinction coefficient of 1 OD per 1 mg/ml sample. Both these methods do not take into account that different proteins contain a different number of aromatic side chains and can therefore be grossly wrong.

2.2.2 Using the Bradford Assay

The Bradford assay, and all kits based on it, requires that your protein is soluble at acidic conditions, as only soluble protein binds dye and is measureable. The blank measurement (dye plus buffer, no protein) usually has an absorbance of 0.4–0.5 already, so also with these methods care has to be taken not to overstretch the dynamic range of the photo spectrometer (*see* **Note 12**). The assay itself also has a limited linear range.

Preparation of a standard curve is usually done with BSA, but this may not be a good standard for your protein as they might not stain to the same extent. The best standard is a preparation of your purified protein of known concentration. Quantitative amino acid analysis (see below) could be used to establish this standard.

2.2.3 Using Fluorescent Dyes for Protein Concentration Determination

Dyes are available from companies such as Molecular Probes (Invitrogen) and Promega. It is essential to follow the instructions closely as some have maximum tolerance of salts or other buffer components and as some are dependent on temperature and other conditions. Only small volumes of protein solution are usually required and the suitable concentration range varies from kit to kit.

As with dye-based absorbance assays, the accuracy of these fluorescence assays for the case of your protein is worth checking once by comparison with quantitative amino acid analysis. The reproducibility of results is usually high.

Of course measurement requires a fluorescence spectrometer. If none is available, some manufacturers of these assays sell a relatively cheap mini-fluorometer designed for assay use (Quant-iT assays, Molecular Probes). These are set up to generate standard curves and calculate results from sample measurements.

2.3 Protein Stability

2.3.1 Assessing Protein Solubility

A systematic check of protein solubility would at least involve varying the pH from about 4 to about 9, trying 0–1,000 mM NaCl and varying the temperature from 4 to 37 °C or higher if the protein is from a thermophilic organism. Not only the concentration but also the nature of the salt may be important for protein solubility. While many ions might be able to shield surface charges, a specific ion might be required for correct protein folding and therefore its stability. A few different monovalent and divalent salts should be tried. Bear in mind also, that some buffers require pH adjustment with bases, which will contain a metal ion that might itself influence solubility.

The solubility of some proteins depends on glycerol, but its presence generally is a disadvantage as it makes concentrating difficult, is hard to remove and interferes with many experiments.

Often, solution conditions are changed systematically during protein purification, i.e. a salt gradient is applied for elution from an ion exchange column. Careful observation at this stage gives an initial idea of protein solubility. A more systematic variation of the pH, salt concentrations, and protein concentration with small aliquots of the purified protein in a dedicated experiment will give a wider picture of solubility. Plotting the protein state in dependence of two (or more) conditions gives a two (or more)-dimensional phase diagram which is of prime interest to protein crystallographers, but very informative to establish safe conditions to keep a protein soluble (37, 38).

Precipitation is often visible by eye as white fluff or streaks. But some aggregates are soluble or small amounts of precipitate might not be visible by eye. Spinning samples in a tabletop centrifuge (25,000–30,000 RCF) might visualize a small pellet, and supernatant can be removed carefully (*see* **Note 15**). Importantly, the concentration of the cleared protein solution will have to be measured again as it will be lower and to determine protein loss.

Other methods are intrinsically more quantitative: Gel filtration will show larger amounts of aggregate which would not enter the column matrix and go straight through with the void volume and appear as a peak in the beginning of the chromatogram. Dynamic light scattering is far more sensitive, detecting the smallest quantities of precipitates or unspecific aggregates. The early scans in a sedimentation velocity AUC experiment would also indicate heavy species in protein samples. All these methods are discussed in the paragraphs on "monodispersity".

2.3.2 Detecting Degradation

Most methods to detect protein degradation have already been introduced previously in this chapter. Gels or gel filtration will often give the first indication, particularly when multiple bands or peaks appear or the molecular weight appears to be smaller than expected. But these methods will not detect small amounts of breakdown products or those only different in a few amino acids' length.

Mass spectrometry will give unequivocal information on the molecular weight and is therefore the method of choice in case of suspicion. *N-terminal sequencing* will identify the N-terminal amino acids of each breakdown product, i.e. identify the cleavage site. N-terminal sequencing is done by Edman degradation and often provided as in-house service or available as commercial service.

2.3.3 Preventing Degradation

Protease inhibitors and mixes of inhibitors are available from many of the standard companies. Many, but not all contain EDTA, which might not be wanted (*see* **Note 16**).

Azide will prevent microbial growth in protein samples which eliminates a source of proteases.

Snap freezing is done in small protein aliquots (20–50 μl) in thin-walled PCR tubes in liquid nitrogen to prevent ice crystal formation that can damage proteins. For slow freezing, glycerol can be added as cryoprotectant. However, glycerol is not easy to remove and might not be suitable in future experiments. Whether freezing is appropriate needs to be tested for each protein by checking for precipitation and aggregation and investigating whether the protein has retained its fold after thawing. Ideally, a functional test should be used to assess damage through freeze–thawing or at least one of above methods to establish structural integrity. Often, a protein aliquot should not be frozen a second time.

If the cleavage site of a protein is known this information can sometimes be used to express just that fragment of the protein to ensure that the sample is homogeneous. Of course this only makes sense if this fragment retains the overall fold and function of the full-length protein. Sometimes, adding a protein tag (or leaving it on after purification) can hinder degradation from the N- or C-termini.

2.3.4 Assessing Fold Stability

To assess the stability of a protein's fold with respect to chaotropic agents, a titration should be performed, e.g. from 0 to 6 M urea in 0.5 M steps. CD or fluorescence readings could be taken if the chaotropic agent is pure and does not interfere with these methods. The protein concentration should be identical in all these samples and generally needs to suit the measurement method chosen, e.g. to get enough, but not an oversaturated signal in a fluorimeter.

Differential scanning fluorimetry to asses thermal fold stability is often set up in a real-time PCR instrument programmed to ramp up the temperature slowly while taking frequent fluorescence measurements. 25–50 μl protein solution at 1–10 mg/ml in different conditions are tested usually in 96 (or 384)-well plate format. For screening, conditions to be tested may be pH and salt concentration initially as described for solubility studies. A dilution of SyproOrange (*see* **Note 17**) is added which becomes fluorescent upon binding to hydrophobic patches exposed in denaturation. In a melting event the fluorescence rises sharply, and the melting temperature can be determined. If this varies for different solution conditions it suggests that they affect protein stability.

2.3.5 Handling of Detergent-Solubilized Membrane Proteins

For exchange or removal of detergent, many techniques can be used:

- Affinity chromatography is an easy way to exchange or remove detergent. If it is possible to bind the protein on an affinity column (ion exchange, hydroxyapatite, Ni-NTA, streptactin, etc...) removal or exchange of detergent can be achieved by extensive washing by a buffer with no or new detergent. However, it is important to note that this method could lead to the removal of bound lipids from the purified membrane protein. This removal could affect its properties.
- Depending on the size of the protein, gel filtration could be used to exchange or remove detergent. However, if the

detergent micelles have a similar size than the sample, this method will not work.

- Dialysis is very often used to remove and exchange detergents. However, several criteria have to be taken into account: The size of the protein, the size of the detergent micelles and the CMC of the detergent(s). If it is possible to use a dialysis membrane with a cut-off large enough to let the micelles pass through it but not the protein, then dialysis will be very efficient. If only detergent monomers can pass through the dialysis membrane, then this technique will be effective only for detergents with high CMC (>1 mM).
- Finally, detergent can be removed by adsorption to hydrophobic beads. This technique is particularly useful to remove low concentrations of detergents. However, it has to be noted that membrane proteins can also be adsorbed to the same surface.

To concentrate protein–detergent mixes, two techniques can be used:

- Ultrafiltration is a widely used method to concentrate protein samples. Water and small molecules will pass through the filter. The pore size of the filter membrane has to be adjusted in function of the protein size. If the detergent micelles are much smaller (at least 2–3 times smaller) than the pore size, then detergent will also be filtrated and only the protein will be concentrated. However, because the size difference between the protein and the detergent micelles is often not big enough, it is very common that ultrafiltration also leads to concentration of the detergent.
- Affinity chromatography can be used to concentrate a protein sample. Large amounts of protein can be bound on the column and eluted in a small volume. This method will not concentrate the detergent but the protein will be in the elution buffer. A dialysis step may be required to remove some reagents that could interfere with following assays.

3 Notes

1. Nucleic acids can be removed from protein samples in many ways, the safest of which is an anion exchange step (DEAE sepharose) in the purification. Alternatively, use of DNase during cell lysis is recommended, and the small nucleic acid fragments and nucleotides should then be removed in a gel filtration or dialysis step. RNA contamination is less common.
2. Proteins that contain tryptophan or tyrosine, and to a lesser extent phenylalanine, are fluorescent. Fluorescence is usually strongly dependent on the close environment of the

fluorophore (the fluorescent group or molecule). So when a protein unfolds its intrinsic fluorescence may change (*see* Chapter 7 for details on the method).

3. The APS stock needs replacing every few days and if gels fail to set APS is the prime suspect. TEMED should be kept at 4 °C or on ice at all times for safety reasons.
4. Gels can be poured in bulk and stored wrapped in damp paper towel and cling film for a limited amount of time. The acrylamide network will deteriorate, proteins move faster and protein bands loose their sharpness. In discontinuous gels the pH difference between stacking and resolving gel will blur. Many kinds of gels can be purchased as precast gels, these usually have a longer shelf life, but are expensive.
5. For better denaturation, samples should be incubated at 90 °C for a few minutes. However, some proteins may precipitate, particularly membrane proteins or highly concentrated proteins. It needs to be tested which procedure is better. Also, some protein–protein interactions are SDS-resistant and will only be dissociated if the sample is denatured entirely by boiling.
6. Cys-bridges will be native to the protein if it is active in a nonreducing cellular compartment, e.g. the periplasm in bacteria or the Golgi apparatus in eukaryotes. It would have to be assessed whether experiments are better done at oxidizing or reducing conditions.
7. Not all proteins stain equally well. This is true for any known method. In addition, a larger protein will likely take up more dye than a small one. So the darkness of bands cannot be compared in terms of the molar concentrations of different proteins, but only in terms of the weight concentrations (mg/ml).
8. Gel filtration really is a hydrodynamic method, i.e. the hydrodynamic radius of any protein species determines how fast it moves through the resin. The hydrodynamic radius is larger for molecules of a larger molecular weight as well as for more extended molecules of the same molecular weight. In rare cases, proteins may be unspecifically adsorbed by the column material with or without permeating into the pores. This will interfere with the method.
9. For spherical proteins, an eightfold difference in molecular weight only translates into a twofold difference in diameter, which is what limits the separate detection of different species.
10. Sedimentation velocity analysis is an equilibrium method. However, analysis of a complex at a concentration near its K_D will not resolve the components into its various species, unless the rates of association and dissociation are slow compared to

the time scale of the experiment (hours). This is a similar phenomenon as in gel filtration. So here also, the experiment may give a single peak with a particular sedimentation coefficient, but this will be a mix of all species present. Importantly, a control of different concentrations is required to see whether this peak shifts in a concentration dependent manner.

11. Never pipette less than 2 μl, preferably 5 μl and never make dilutions higher than 1:100 in one go, i.e. to reach 1:1,000 do two sequential dilutions. Check the accuracy of pipettes frequently to avoid errors due to lack of calibration. Doing all measurements in triplicate will identify random errors.
12. In any photo spectrometer, a certain amount of light has to get to the detector for the measurement to be accurate. In many instruments the measurement should not be below 0.1 or above 1.0. Importantly, the upper limit of 1.0 is for the total sample, i.e. including buffer absorbance. The dynamic range of an instrument depends on the light intensity, the quality of the optics (stray light etc) and the detector sensitivity. They range from 0.7 absorbance units (Nanodrop-type instruments) to 6 absorbance units (on CD instruments), but it is usually safe to work below 1 absorbance unit of total absorbance, i.e. cuvette, buffer plus sample taken together. Check the instrument manual for the high absorbance limit.
13. It is recommended to take protein scans from at least 340 nm down, so that light scattering can be detected between 320 and 340 nm, in which region the baseline should be flat. The contribution of scattering to the overall absorbance at 280 nm can be calculated from scans starting at 400 nm or higher using the near-linear relationship between log $A_{scatter}$ and log λ for extrapolation (39).
14. The range of cuvettes available is huge. Most plastic cuvettes are not transparent in the UV range. Quartz cuvettes are certainly best. When choosing a small-volume cuvette, bear in mind the beam size and that the center height of the cuvette needs to match the location of the beam. If the window is smaller than the beam, then blacked out cuvettes only give a correct measurement; however, these cuvettes have absorbance on their own as they block part of the beam, so this needs to be added to buffer and protein absorbance to calculate the maximum sample concentration acceptable before the Beer-Lambert law breaks down. Cuvettes need frequent and thorough cleaning. Hellmanex is a specialist product to clean quartz cuvettes. Use a dilution per instructions and do under no circumstance incubate for longer than the advised time, a few minutes are nearly always sufficient, additional sonication will help to remove contaminants.

15. Protein pellet might be light and fluffy and be stirred up easily, so remove samples from the centrifuge carefully. Use a small tip and draw up slowly to remove supernatant to avoid picking up the pellet at the same time. Bear in mind in centrifuges that are not cooled samples heat up during spinning, so there is also a temperature effect.
16. Protease inhibitors are mixes of inhibiting agents and often contain EDTA, which will chelate all bivalent metal ions. This is to inhibit metallo-proteases, but will also inhibit most proteins that bind nucleotides and nucleic acids and others that depend on bivalent ions. EDTA-free protease inhibitors exist and are preferable. Both protease inhibitors and sodium azide are toxic and have spectral properties that prohibit their use for certain methods. Think carefully before adding them to your final purified sample.
17. SyproOrange is usually used at a final dilution of 1:500 to 1:5,000. Make a solution of dye stock of which to add 1 μl to every 25 μl sample. Final protein concentrations of 0.5 to 5 mg/ml usually give a good signal depending on the size of the hydrophobic core. Discard any leftover of the dilution as it will lose activity within hours.

References

1. Laemmli UK (1970) Cleavage of structural proteins during the assembly of the head of bacteriophage T4. Nature 227:680–685
2. Philo JS (2006) Is any measurement method optimal for all aggregate sizes and types? AAPS J 8:E564–E571
3. Harding SE, Jumel K (2001) Light scattering. Curr Protoc Protein Sci Chapter 7:Unit 7.8
4. McLafferty FW, Breuker K, Jin M, Han X, Infusini G, Jiang H, Kong X, Begley TP (2007) Top-down MS, a powerful complement to the high capabilities of proteolysis proteomics. FEBS J 274:6256–6268
5. Domon B, Aebersold R (2006) Mass spectrometry and protein analysis. Science 312:212–217
6. Aebersold R, Mann M (2003) Mass spectrometry-based proteomics. Nature 422:198–207
7. Cadene M, Chait BT (2000) A robust, detergent-friendly method for mass spectrometric analysis of integral membrane proteins. Anal Chem 72:5655–5658
8. Barrera NP, Di Bartolo N, Booth PJ, Robinson CV (2008) Micelles protect membrane complexes from solution to vacuum. Science 321:243–246
9. Cole JL, Lary JW, Moody TP, Laue TM (2008) Analytical ultracentrifugation: sedimentation velocity and sedimentation equilibrium. Methods Cell Biol 84:143–179
10. Howlett GJ, Minton AP, Rivas G (2006) Analytical ultracentrifugation for the study of protein association and assembly. Curr Opin Chem Biol 10:430–436
11. Ebel C (2011) Sedimentation velocity to characterize surfactants and solubilized membrane proteins. Methods 54:56–66
12. Fleming KG (2008) Determination of membrane protein molecular weight using sedimentation equilibrium analytical ultracentrifugation. Curr Protoc Protein Sci Chapter 7:Unit 7.12.1–7.12.13
13. ExPASy server, ProtParam tool: http://web.expasy.org/protparam/
14. Bradford MM (1976) A rapid and sensitive method for the quantitation of microgram quantities of protein utilizing the principle of protein-dye binding. Anal Biochem 72:248
15. Lowry OH (1951) Protein measurement with the Folin phenol reagent. J Biol Chem 193:265
16. Smith PK, Krohn RI, Hermanson GT, Mallia AK, Gartner FH, Provenzano MD, Fujimoto

EK, Goeke NM, Olson BJ, Klenk DC (1985) Measurement of protein using bicinchoninic acid. Anal Biochem 150(1):76–85, Erratum in: Anal Biochem 163:279
17. Cummings MD, Farnum MA, Nelen MI (2006) Universal screening methods and applications of ThermoFluor. J Biomol Screen 11:854–863
18. Kean J, Cleverley RM, O'Ryan L, Ford RC, Prince SM, Derrick JP (2008) Characterization of a CorA Mg2+ transport channel from Methanococcus jannaschii using a Thermofluor-based stability assay. Mol Membr Biol 25:653–663
19. Cooper A, Nutley MA, Wadood A (2000) Differential scanning microcalorimetry. In: Harding SE, Chowdhry BZ (eds) Protein–ligand interactions: hydrodynamics and calorimetry. Oxford University Press, Oxford
20. Chattopadhyay A, London E (1984) Fluorimetric determination of critical micelle concentration avoiding interference from detergent charge. Anal Biochem 139:408–412
21. Dornmair K, Kiefer H, Jähnig F (1990) Refolding of an integral membrane protein. OmpA of Escherichia coli. J Biol Chem 265:18907–18911
22. Schweizer M, Hindennach I, Garten W, Henning U (1978) Major proteins of the Escherichia coli outer cell envelope membrane. Interaction of protein II with lipopolysaccharide. Eur J Biochem 82:211–217
23. McLellan T (1982) Electrophoresis buffers for polyacrylamide gels at various pH. Anal Biochem 126:94–99
24. Schägger H, von Jagow G (1991) Blue native electrophoresis for isolation of membrane protein complexes in enzymatically active form. Anal Biochem 199:223–231
25. Krause F (2006) Detection and analysis of protein–protein interactions in organellar and prokaryotic proteomes by native gel electrophoresis: (Membrane) protein complexes and supercomplexes. Electrophoresis 27:2759–2781, Erratum in: Electrophoresis (2008) 29:5067
26. Dunn MJ, Corbett JM (1996) Two-dimensional polyacrylamide gel electrophoresis. Methods Enzymol 271:177–203
27. Righetti PG (1988) Isoelectric focusing as the crow flies. J Biochem Biophys Methods 16:99–108
28. Dunn MJ, Corbett JM (1994) Two-dimensional polyacrylamide gel electrophoresis using immobilized pH gradients in the first dimension. Methods Mol Biol 32:87–96
29. Shaw MM, Riederer BM (2003) Sample preparation for two-dimensional gel electrophoresis. Proteomics 3:1408–1417
30. Syrový I, Hodný Z (1991) Staining and quantification of proteins by polyacrylamide gel electrophoresis. J Chromatogr 596:175–196
31. Steinberg TH (2009) Protein gel staining methods: an introduction and overview. Methods Enzymol 463:541–563
32. Gel filtration, principles and methods. GE Healthcare Handbook 18-1022-1018. https://www.gelifesciences.com/gehcls_images/GELS/Related Content/Files/1314807262343/litdoc18102218AK_20110831220049.pdf
33. Schuck P (2003) A model for sedimentation in inhomogeneous media. I. Dynamic density gradients from sedimenting co-solutes. Biophys Chem 108:187–200
34. Ralston G Introduction to analytical ultracentrifugation. Beckman booklet, https://www.beckmancoulter.com/wsrportal/bibliography?docname=361847.pdf
35. McRorie DK, Voelker PJ Self-associating systems in the analytical ultracentrifuge. Beckman booklet, https://www.beckmancoulter.com/wsrportal/bibliography?docname=362784.pdf
36. Layne E (1957) Spectrophotometric and turbidimetric methods for measuring proteins. Methods Enzymol 3:447–455
37. Kueltzo LA, Ersoy B, Ralston JP, Middaugh CR (2003) Derivative absorbance spectroscopy and protein phase diagrams as tools for comprehensive protein characterization: a bGCSF case study. J Pharm Sci 92:1805–1820
38. Rozhkov SP, Goryunov AS (2010) Thermodynamic study of protein phases formation and clustering in model water-protein-salt solutions. Biophys Chem 151:22–28
39. Winder AF, Gent WLG (1971) Correction of light-scattering errors in spectrophotometric protein determinations. Biopolymers 10:1243–1251
40. Phan G, Remaut H, Wang T, Allen WJ, Pirker KF, Lebedev A, Henderson NS, Geibel S, Volkan E, Yan J, Kunze MB, Pinkner JS, Ford B, Kay CW, Li H, Hultgren SJ, Thanassi DG, Waksman G (2011) Crystal structure of the FimD usher bound to its cognate FimC-FimH substrate. Nature 474:49–53

Chapter 3

Measurement of Protein–Ligand Complex Formation

Peter N. Lowe, Cara K. Vaughan, and Tina Daviter

Abstract

Experimental approaches to detect, measure, and quantify protein–ligand binding, along with their theoretical bases, are described. A range of methods for detection of protein–ligand interactions is summarized. Specific protocols are provided for a nonequilibrium procedure pull-down assay, for an equilibrium direct binding method and its modification into a competition-based measurement and for steady-state measurements based on the effects of ligands on enzyme catalysis.

Key words Complex formation, Equilibrium binding, Kinetics, Affinity, Stoichiometry, Interaction, Pull-down assay, Competition, Steady state

1 Introduction

This subheading gives an overview of the basis for measuring interactions between a ligand, which can be anything from a low-molecular-weight species to a macromolecule, and its binding partner, termed here the target (*see* **Note 1**). To detect the formation of a complex between the ligand and the target protein, a measureable response, or signal, and an appropriate detection technology are required. To provide absolute measurement of affinity, additionally a simple mechanistic model is required. Complex formation can be measured either directly ("direct binding") or indirectly via competition experiments. Measurements can be performed at a single time point, typically when equilibrium has been reached, or by following the time courses of complex formation and dissociation.

1.1 Types of Signal

The measured signal can come from the ligand, the target, the ligand–target complex, or a combination of them all. The signal can be either intrinsic to the ligand or target or to the interaction itself or a label (probe) that is attached to one of the species.

Experiments using intrinsic signals are often referred to as "label-free." In some procedures, a good example being ITC, the

Mark A. Williams and Tina Daviter (eds.), *Protein-Ligand Interactions: Methods and Applications*, Methods in Molecular Biology, vol. 1008, DOI 10.1007/978-1-62703-398-5_3, © Springer Science+Business Media New York 2013

interacting species are unperturbed in free solution. In others, the interacting species may be perturbed, for example with biosensors in which one species must be immobilized. Examples of intrinsic signals would include:

- Intrinsic fluorescence of a ligand (*see* Chapter 7)
- Mass, which can be detected by a range of technologies such as mass spectrometry, biosensors (*see* Chapter 6), and gel staining (*see* Subheading 2.1).
- Enthalpy of interaction, detected by ITC (*see* Chapter 4)
- Free energy of interaction, detected by thermal shift technologies (*see* **Note 2**)
- Chirality, detected by CD (*see* Chapter 8)

Labelling is required in the absence of intrinsic signals, but can also be advantageous in giving more sensitivity or specificity. The introduced label is designed to provide a signal whilst not interfering with the interaction partners or the interaction itself (*see* **Note 3**). Labelling is most frequently accomplished by covalent attachment of a low-molecular-weight probe to the target or the ligand, or via a tight-binding interaction mediated via an antibody or biotin/streptavidin interactions. Covalent attachment can be performed chemically in vitro, or via recombinant modified proteins such as to incorporate an appropriate tag (biotin, epitope for an antibody such as FLAG, GST, or enzyme such as HALO or SNAP). As compared to intrinsic signals, labelling not only requires the additional labelling procedure but also has a potential disadvantage in altering the ligand or the target. However, increased sensitivity (higher signal) and specificity (i.e., low background) and a wider range of detection technologies can make labelling a method of choice. Examples of signals from labels include the following:

- Fluorescence (intensity, polarization, fluorescence resonance energy transfer (FRET), fluorescence lifetime) (*see* Chapter 7) (*see* **Note 4**).
- Radioactivity, detected by autoradiography or more easily quantified with a scintillation counter. Scintillation proximity methods are particularly suited to measuring complex formation (*see* **Note 5**).
- Electrochemiluminescence (ECL) (*see* **Note 6**).

1.2 Detection and Quantification of Signals

Different signals and detection technologies, as well as experimental procedures, can result in differences in what is actually measured, and/or in how data needs to be interpreted and analyzed. In the conceptually simplest procedure, the system is homogeneous in solution, a signal is only detectable upon formation of the complex, the magnitude of the signal is directly proportional to the amount of complex formed, and the system is in equilibrium.

In practice this is often not the case. For example, using a scintillation proximity assay one component must be immobilized (*see* **Note 5**); using fluorescence polarization the signal is not unique to the complex (*see* Subheading 2.2) and is not directly proportional to its concentration; using a pull-down assay (*see* Subheading 2.1) the system will not be at equilibrium.

For quantitative analysis of the system that is to be employed, one needs to know the following:

- Whether the signal change derives from the (*see* **Note 7**):
  - Complex alone, or its formation (e.g., ITC)
  - Sum of signals from all species containing the ligand
  - Sum of signals from all species containing the target
  - Target alone
  - Ligand alone
- Whether the signal derives from an extensive property dependent on concentration (e.g., fluorescence intensity) or is from an intensive property that is independent of concentration (e.g., fluorescence polarization) (*see* **Note 8**)
- Whether the system will have reached equilibrium (*see* **Note 9**) at the time of measurement
- Whether the detection procedure can perturb the equilibrium
- Whether the system is heterogeneous, for example with an immobilized component

Where the system either has not reached equilibrium or the detection procedure perturbs the equilibrium, absolute affinities cannot generally be measured (*see* **Note 10**). However, if the distortions are not too great an apparent affinity can be obtained that might approximate the true value. Particularly when comparing a series of related interactions, relative apparent affinities can closely approximate the true relative affinities.

With immobilized systems, it is very often the case that the measured apparent affinity will approximate the affinity as measured in homogeneous solution. However, this certainly cannot be assumed without appropriate measurements with controls or standards, or without comparison with a true solution procedure (1–3).

1.3 Direct Binding Assays: Affinity, Stoichiometry, and Mechanism

The standard measurement of affinity is K_d (*see* Chapter 1). In general, this can be determined by:

- Establishing (or assuming) the binding model/mechanism and stoichiometry. Most commonly (simply) this will be of the form $P + L \leftrightharpoons PL$, in which the stoichiometry is 1:1.
- Varying the concentration of one component, at a fixed concentration of the other, such as to vary the concentration of complex from zero to that of the fixed-concentration component.

- Waiting until equilibrium, or better follow the signal with time until equilibrium is reached, and measure the signal at each concentration.
- Curve-fitting the data using nonlinear regression to an appropriate equation for the binding model. Raw response data, as opposed to molar concentrations, can usually be directly used with an algorithm appropriate for the type of response and detection technology.

1.4 Complex Formation Determined in Competition Experiments

Complex formation can also be inferred indirectly using competition procedures. A common situation would be that a standard ligand has been defined for a target, and by labelling that ligand direct binding to the target can be measured. Test ligands can then be added to a mixture of target and labelled standard ligand, and any decrease in signal detected. Competition procedures, in contrast to direct binding procedures, only detect ligands that perturb the affinity of the labelled standard ligand. Test ligands can act competitively by binding at the same site as the standard ligand, or by binding elsewhere (allosteric binding), thereby modulating the affinity of the standard ligand. If they bind mutually exclusively the labelled standard ligand can be totally displaced. Ligands can also partially displace the labelled ligand, or can increase the affinity of the labelled ligand such as to increase its binding.

1.5 Complex Formation Determined from Binding Kinetics

In most of the above, i.e., measurements at, or close to equilibrium, the measured signal is time independent. Alternatively, the time course (kinetics) of complex formation and disassociation can be recorded. This can give detailed mechanistic information, and also increase the range of affinities that are amenable. One downside is that often more sophisticated technology is required (*see* Chapter 5) or measurements must be made on immobilized systems (3); also *see* Chapter 6. It is important to realize that to go from raw measurements of signal with time, via rate constants, to K_d values correct kinetic models are required (*see* **Note 11**). With anything other than the simplest kinetic mechanism, obtaining the overall K_d for a binding reaction can be surprisingly involved.

Despite these limitations, determination of dissociation and association rate constants is of increasing interest because ligands can differ in their residence time on their targets (4). For drugs, this can affect their efficacy, longevity of action, and rate of metabolism.

1.6 Complex Formation Determined from Enzyme Activity

With enzymes, their catalytic properties give an alternative way of measuring complex formation through the effect of ligand binding on catalytic function (*see* Subheading 3). The formation of product (or depletion of substrate) is measured over time to determine the enzymatic rate. Then, the formation of a complex between a ligand and the enzyme can be determined from the effect of a ligand on

Table 1
Detection methods for binding experiments

	Signal and evaluation	Requirements/labels for detection	Throughput[a]	Specialist equipment[b]	References
Nonequilibrium: Requires low dissociation rate (K_d typically << 10 μM); excellent for "yes/no" binding detection					
Pull-down	Co-precipitation with protein	Protein and proteinaceous ligands detected by SDS-PAGE. Small MW ligands require detection by R, F or Abs in the same fractions as protein	Low	n	Subheading 2.1
Gel filtration	Co-elution with protein		Low	n	(2, 5)
Gradient centrifugation	Co-localization with protein		Medium	y	(2, 6)
Co-pelleting	Co-pelleting with protein		Low-medium	y	
Filter binding	Ligand retained only via protein	R of ligand	Medium	y	
Gel shift	Shift of protein band in native PAGE	Ligand changes mass/charge of band	Medium	y	
Equilibrium: Homogeneous solution conditions, no interference with equilibrium; excellent for K_d measurements					
Dialysis	Co-localization with protein	R, F, Abs, or MS to detect ligand	Low	n	(2, 7)
Fluorescence intensity[c]	Intensity change (environment)	F: Intrinsic or label	Low-high	y	Chapter 7, (7)
Fluorescence anisotropy[c]	Anisotropy change (size)	F: Label	Low-high	y	Chapter 7, (7)
FRET[c, d]	Intensity change (distance)	Donor and acceptor F: Labels	Low-high	y	Chapter 7, (7, 8)
ITC[e, f]	Binding heat (enthalpy)	>0.5 mg of protein	Low	y	Chapter 4, (7, 9)
NMR[g]	Magnetic resonance (environment)	^{15}N label or intrinsic signals	Low-medium	y	Chapters 9 and 14 (7, 9)
Quasi-equilibrium: Measurement may influence equilibrium[h]; for relative or apparent K_d					
Scintillation proximity	Radioactivity (distance)	Protein immobilized, R on ligand	Low-high	y	(1, 7)
AUC[e]	Colocalization of ligand and protein	Abs to detect ligand	Low	y	(2, 6)
Thermal shift, optical	Change in T_m (stabilization)	F: Intrinsic or dye, CD	Low-high	y	Chapters 2 and 13, (9, 10)

(continued)

Table 1 (continued)

	Signal and evaluation	Requirements/labels for detection	Throughput[a]	Specialist equipment[b]	References
Thermal shift, DSC	T_m and heat capacity changes	Unfolding ideally reversible	Low	y	
Nano-electrospray MS[e]	Mass of complex	No label	Low-high	y	(7, 11)
SPR, resonant waveguide[c, e]	Refractive index (mass on surface)	One component immobilized	Low-high	y	Chapter 6, (7, 12)
Alpha screen[d]	Luminescence (distance)	Attached to beads; tags/labels	Low-high	y	(8, 13)

ITC isothermal titration calorimetry, *DSC* differential scanning calorimetry, *SPR* surface plasmon resonance, ***n*** stoichiometry, T_m melting temperature, *R* radioactivity, *F* fluorescence, *Abs* absorbance, *MS* mass spectrometry, *FRET* fluorescence resonance energy transfer, *AUC* analytical ultracentrifugation, *NMR* nuclear magnetic resonance, *CD* circular dichroism

[a]This depends upon the equipment utilized, but a guideline is low <50 samples/day; medium 50–200 samples/day; high >200 samples/day

[b]Specialist equipment is anything beyond the normal setup of a molecular biology laboratory which includes gel running and Western blotting equipment, chromatography equipment, and tabletop centrifuges. Please refer to the references given for instrumentation required for these methods

[c]Provides additional information on association and dissociation kinetics

[d]Can be difficult to vary or utilize concentrations such that the K_d can be measured. However, if K_d is known, K_i for competitive ligands can be measured

[e]These methods can directly provide information on the stoichiometry of an interaction

[f]Provides additional information on ΔH, ΔS, and ΔG

[g]There are many NMR-based techniques (e.g., *see* Chapter 9), the technique of choice being dependent upon whether labelled protein is available and what information is required: yes/no binding, K_d, or identification of the binding site

[h]The apparent affinities measured by these systems can be perturbed from those measured in homogeneous solution at ambient temperature: Immobilization on a surface can affect functional activity, association and dissociation kinetics, and hence the equilibrium position. In a thermal shift experiment, the label can perturb the system, and apparent affinities are typically measured at elevated temperatures. For nano-electrospray-MS, the measurements are made in the gas phase in which the relative strengths of electrostatic and hydrophobic interactions are perturbed

Table 2
Detection methods for enzyme assays

Method	Readout	Requirements	Mode[a]	Equipment	Throughput[b]
Direct		Does not require detection reagents or chemically altered substrate			
MS	MW/charge	Unique *m/z*; MS-compatible buffer or desalting	q	Mass spectrometer	Low-high
HPLC	Abs/F/MS/R	Column/buffer system to separate substrate and product	q	HPLC system	Low-medium
CE	Abs/F/MS	Mobility shift; intrinsic signal or label (for Abs, F)	q	CE system	Low-high
Radiometric	R (scintillation)	Radiolabelled substrate, SPA, or substrate/product separation	q/c	Scintillation counter	Low-high
Calorimetric	Heat released	Suitable buffer; sufficient heat change	c	ITC	Low
Optical	Abs/F	Distinguishable optical signal from substrate and product	c	Abs/F spectrometer	Low-high
Indirect		Requires secondary detection reagent or chemical label			
Coupled	Abs/F	Coupled enzyme reaction that uses product formed to create optical signal	c	Abs/F spectrometer	Low-high
Affinity capture	Abs/F/R/L	Capture via antibody or other molecule that recognizes product[c]	q/c		
FRET	F	Donor and acceptor incorporated into a substrate that is cleaved[d]	c		Low-high

MS mass spectrometry, *HPLC* high-pressure liquid chromatography, *Abs* absorbance, *F* fluorescence, *R* radioactivity, *CE* capillary electrophoresis, *ITC* isothermal titration calorimetry, *FRET* fluorescence resonance energy transfer, *SPA* scintillation proximity assay

[a]The suitability of the method to be able to continuously monitor the reaction: q indicates that the reaction is quenched and then analyzed, c indicates that the reaction can be followed continuously in real time, and q/c indicates that it depends upon the system and the specific detection procedure

[b]This depends upon the equipment utilized, but as a guideline low <50 samples/day; medium 50–200 samples/day; high >200 samples/day

[c]Subsequent detection could be by ELISA, fluorescence methods (polarization, FRET), radiometry, luminescence

[d]Suited to proteases in which suitable donor and acceptor labels are incorporated either side of the cleavable peptide bond. The donor/acceptor pair needs to be selected based on the distance between them

Table 3
Some commonly used enzyme assays

Products	Methods and readouts	Mode[a]	Reagent (supplier)	References
Phosphate[b]	Binding to F-labelled phosphate-binding protein	c	PhosphoWorks™	(14)
Phosphate[b]	Coupling to purine nucleoside phosphorylase (Abs or F)	c	EnzChek® (Invitrogen)	
Phosphate[b]	Coupling to maltose phosphorylase and Amplex Red (Abs or F)	c	PiPer™ (Invitrogen)	
Phosphate[b]	Reaction with malachite green (Abs)	q		(15)
ATP, ADP	L (luciferase)			(13, 16)
ADP	F intensity, F polarization or FRET (competition with labelled ADP for ADP-binding antibody)	c		(16)
ADP	Coupling with pyruvate kinase and lactate dehydrogenase, measuring loss of NADH (Abs or F)[c]	c		(16)
Phosphoprotein	F polarization or FRET (nanoparticle binding (IMAP®) to fluorophore-tagged phosphopeptide)	q/c		(16)
Phosphoprotein	Anti-phosphoprotein (ELISA, HTRF, AlphaScreen), R (SPA, filter-binding), CE (F-labelled peptide)	q/c		(16)

Abs absorbance, *F* fluorescence, *R* radioactivity, *CE* capillary electrophoresis, *L* luminescence, *FRET* fluorescence resonance energy transfer, *ELISA* enzyme-linked immunosorbent assay, *HTRF* homogeneous time-resolved fluorescence, *SPA* scintillation proximity assay

[a]The suitability of the method to be able to continuously monitor the reaction: q indicates that the reaction is quenched and then analyzed, c indicates that the reaction can be followed continuously in real time, and q/c indicates that it depends upon the system and the specific detection procedure

[b]Pyrophosphate, a product of DNA- and RNA-polymerases, and adenylate- and guanylate-cyclases, can be detected by coupling with pyrophosphatase

[c]*see* **Note 57**

the enzymatic rate at steady state (*see* **Note 9**). In this case, the data is very similar to that in Subheading 1.4 with the "enzyme rate" being the measured signal (*see* **Note 12**). The study of inhibiting or activating ligands via effects on catalytic properties can give detailed information on the mechanism of a ligand, differentiating competitive from other mechanisms.

1.7 Technologies to Detect Complex Formation

A wide variety of detection technologies is available; the choice of which to use is dependent not only on the biological system (molecular weights, affinity, kinetics of dissociation, stability) but also on practical and economic aspects (e.g., amounts of material required, impact and ease of labelling, cost of consumables, cost and maintenance requirements of instrumentation, specialist knowledge, and required throughput and precision). Table 1 summarizes many of the most widely used technologies to detect and quantify complex formation. It focuses on direct binding of a ligand to a protein. Most procedures can also be used to measure the effects of a competitive ligand. For quantifying the affinity of a competitive ligand, thermal and gel shift, analytical ultracentrifugation, gel filtration, and fluorescence intensity are less useful, or will only work well with labels or a specific detection principle. Tables 2 and 3 focus on measuring concentrations of substrates and products of enzyme catalysis.

2 Methods Based on Formation of Protein–Ligand Complexes

2.1 Example of a Nonequilibrium Method

2.1.1 Pull-Down Assay for Identification of Protein–Ligand Interactions

A pull-down assay using affinity tags can be a straightforward qualitative method to determine whether two proteins interact with one another. As a nonequilibrium technique, it is mainly for identification of interactions and not suitable to measure affinities (*see* **Note 13**). The protocol detailed in this chapter relies on the availability of purified protein components whose interaction is to be assessed. One of these components, sometimes called "bait," must have a suitable affinity tag (e.g., His_6, GST, or Strep) (17, 18) for which affinity resins are available. Interaction partners of this tagged protein, called "prey," are "pulled down" with it. In principle, any tagged bait can be used and often peptides are used instead of full-length proteins to pull down potential interaction partners.

With little variation, the method can also be applied to pull down interacting proteins from protein mixtures or even cell lysate. There are many related techniques, e.g., protocols are available elsewhere for immunoprecipitation experiments using antibodies with cell lysates (19, 20).

This approach has been used extensively, typically in combination with mass spectrometry techniques (21, 22), to map protein–protein interaction networks in a variety of organisms (23–25).

In general, this method works well for high-affinity interactions. With care, it can also be used for interactions of an affinity in the lower micromolar range. The consequences will be pointed out in Subheading 2.1.4 below.

2.1.2 Pull-Down Assay: Materials

1. Purified protein components of known concentration, one of which must have an affinity tag. The other protein must not have the same affinity tag.
2. An unrelated protein with the same affinity tag to be used as a control (*see* **Note 14**). This is not essential but will provide a better control for the specificity of your experiment.

3. Pull-down buffer: PDB

   This can be your standard buffer (*see* **Note 15**) provided it is compatible with the affinity resin to be used (e.g., without reducing agent for nickel resins). PBS is a good starting point in the absence of a preferred buffer for your proteins. A small amount of detergent is included to reduce nonspecific binding. We use:

   50 mM Tris–Cl pH 7.5

   150 mM NaCl

   0.1 % (v/v) Igepal CA-630 (NP40)

4. 10× Pull-down buffer: 10×PDB

   500 mM Tris–Cl pH 7.5

   1.5 M NaCl

   1 % (v/v) Igepal CA-630 (NP40)

5. Blocking buffer: BB

   This is used to block the beads to reduce nonspecific binding. We use:

   PDB + 1 mg/ml BSA (from Sigma)

6. Affinity resin: The resin chosen will depend on the nature of the prey's affinity tag. In the example below, nickel–NTA resin could be used for a His-tagged bait. Note that different extents of nonspecific binding may be observed with different resins, and this can be investigated if nonspecific binding is a problem. There are a wide variety of suppliers for most affinity resins.

7. Standard SDS-PAGE equipment

2.1.3 Pull-Down Assay: Methods

Each reaction is carried out in an Eppendorf tube. All centrifugation steps are carried out at 1,500 × *g*.

Preparation

1. Calculate the reaction volumes required for the experiment as shown in Table 4. This example tests the putative interaction of two proteins, His_6-A (bait) and B (prey). The unrelated tagged protein His_6-X is used as a control. We use 50 μg of each protein per reaction (final concentration in the range 1–10 μM) (*see* **Note 16**).

2. Provided the proteins are at a suitably high concentration (less than 10 % total reaction volume) it is sufficient to use 10×PDB to ensure appropriate buffer components for the experiment. If this is not possible all proteins should be buffer exchanged into PDB prior to the experiment.

Table 4
Pipetting scheme for the pull-down assay

Components	Stock concentration µg/µl	Volume (µl) for 50 µg	Control 1	Control 2	Expt
His_6-X	25	2		+	
His_6-A	25	2			+
B	25	2	+	+	+
10× PDB	10×	20	+	+	+
H_2O	–	–	178	176	176
Total volume (µl)			200		

Equilibrate and block resin

3. Aliquot 100 µl of resuspended affinity resin slurry into each reaction tube for a final packed resin volume of 50 µl.
4. Pellet the resin by gentle centrifugation for 1.5 min. Aspirate and discard the storage buffer.
5. Equilibrate the resin with 3 × 500 µl of PDB buffer: Pipette 500 µl of PDB buffer into the tube. Ensure that it is fully mixed; if necessary gently and briefly vortex. Pellet the resin. Aspirate 500 µl and discard. Repeat three times.
6. Add 200 µl of BB; ensure that it is mixed and allow tumbling at 4 °C while preparing the next step (approximately 30–60 min).

Incubate samples with resin

7. Prepare the protein mixtures as in Table 4 (*see* **Note 17**), making up to a total volume of 200 µl. Incubate for 60 min with tumbling at 4 °C.
8. Prepare a tenfold diluted "Load" sample for the final gel (*see* **Note 18**) by removing 5 µl from the reaction mix to a separate tube containing 45 µl PDB.
9. Pellet the resin from **step 6**. Aspirate and discard the blocking buffer (*see* **Note 19**).
10. Add the remaining 195 µl of sample to the resin; mix gently. Incubate with tumbling at 4 °C for 2 h (*see* **Note 20**).

Wash resin and elute bound proteins

11. Pellet the mix by centrifugation. Aspirate and discard the unbound fraction.
12. Wash the resin with 3× 1 ml PDB ensuring full mixing each time using the same technique as in **step 3** (*see* **Note 21**). Discard the wash fractions (*see* **Note 22**).

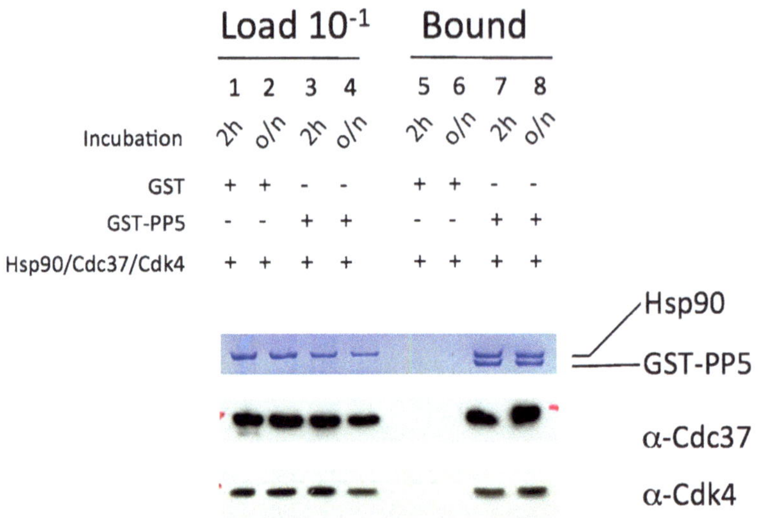

Fig. 1 An example pull-down assay examining the consequences of addition of an interacting partner for the integrity of a molecular chaperone complex. The interaction of GST-tagged bait with untagged prey is assessed using glutathione sepharose as the affinity resin. The pull-down assay shows that the GST-tagged phosphatase PP5 can simultaneously bind a preformed molecular chaperone complex of Hsp90, the cochaperone Cdc37, and the kinase Cdk4, without displacement of any of the constituent proteins. GST alone is used as a control (*Lanes 1*, *2*, *5*, *6*). There is no nonspecific binding of the Hsp90/Cdc37/Cdk4 complex to the resin in the absence of phosphatase PP5 (*Lanes 5* and *6*). The direct interaction between Hsp90 and GST-PP5 is detected using Coomassie staining, whereas the cochaperone Cdc37 and kinase Cdk4 binding was detected using Western blot and detection with mouse monoclonal anti-Cdc37 (C1, Affinity Bioreagents), rabbit anti-Cdk4 (H303, Santa Cruz Biotechnology), and horseradish peroxidase-linked anti-mouse and anti-rabbit (GE Healthcare)

13. Add 100 μl of concentrated SDS-PAGE gel loading buffer (GLB) to the resin (i.e., do not dilute to 1×, use your preferred stock solution; we use 3× GLB). Heat at 100 °C for 5 min to elute the "Bound" proteins (*see* **Note 23**).

Analysis

14. For each reaction run the "Bound" and tenfold diluted "Load" samples on an SDS-polyacrylamide gel (*see* **Note 24**).

For a typical result *see* Fig. 1.

2.1.4 Pull-Down Assay: Further Applications

Comparison with Mutants

A small-scale pull-down assay gives a qualitative insight into whether two proteins interact with each other. Of course this means that it is an excellent means to assess whether certain mutants will abolish binding or to test for the effect of buffer components or additives, e.g., salt or nucleotides, on binding. Given the assay is purely qualitative, all that can usually be obtained is a yes/no answer. However, should the experiment be set up to use protein concentrations in the K_d range, then gel bands of different strength might be tentatively interpreted with respect to the relative amount of protein bound and therefore obtain relative binding affinities (*see* **Note 13**).

Protein Complexes

This protocol can also be used for three- and more component systems; in our experience Western blot will normally be required for detection of the third component. As the above protocol is for low-affinity interactions, binding of the first prey may not be stoichiometric, and binding of the second will be even lower.

Catalytic Complexes

The unbound fraction can be analyzed to study other aspects of the interaction. For example, if the tagged component is a kinase, whose activity on the untagged protein can be assessed by an antibody, a Western blot of the unbound fraction can be used to probe whether the kinase is actively phosphorylating the pool of added untagged protein. Again this is purely qualitative.

Scaling Up

The possibility of scaling this up to large scale for the preparation of a purified complex is dependent on the equilibrium and dissociation rate constants of the complex, and may not be straightforward. Again, particularly low-affinity interactions might be identified in this assay, but the wash regime of a column run is an entirely different situation with larger dilutions and often longer separation times. Again, for high-affinity interactions, scaling up to preparative format should be easier.

2.2 Equilibrium Binding Methods

This subheading will describe in general the measurement of a binding constant of a ligand to a target (*see* Subheading 2.2.3) and then describe how this knowledge can be used to measure binding of competing ligands to the same target (*see* Subheading 2.2.4). Competition assay formats give information on whether one ligand competes for binding with another ligand (*see* **Notes 25** and **26**), and can also provide measurement of absolute or relative (*see* **Note 27**) affinities of competing ligands.

2.2.1 Introduction to Direct and Competition Binding Assays

Competition binding assays typically utilize a standard, fixed ligand and one or many test ligands, in conjunction with an assay technology that can measure the amount of free standard ligand or the amount of complex of the standard ligand with the target (*see* **Note 28**). Commonly, the standard ligand is labelled with a radiolabel or fluorophore to facilitate detection, though many label-free methods can be utilized (*see* Table 1).

To measure the absolute affinity of a test ligand, the affinity of the standard ligand for the target must be known or determined. Also the active molar concentrations of the standard ligand and the test ligand are required.

Competition binding assays are illustrated here using a fluorescence polarization assay and a standard ligand labelled with a fluorophore (26, 27). The principles of assay design will be the same with many other detection technologies. Part 1 (*see* Subheading 2.2.3) of the protocol demonstrates how the assay is initially used to determine the K_d for the standard ligand binding to the target, and this same basic procedure would be used to determine the affinity of any ligand binding to a target. The same assay format is then used to determine the K_d for competitive test ligands as described in Part 2 (*see* Subheading 2.2.4).

In this assay, the fluorescence polarization of the free ligand increases upon binding to the target. The measured fluorescence polarization is the average of the polarization levels of free and bound ligand, weighted according to their individual concentrations (*see* **Note 29**).

2.2.2 Competition Binding Assay: Materials

- Instrument to measure fluorescence polarization at or near the peak excitation and emission wavelengths of the fluorophore (*see* **Notes 30** and **31**).
- Assay buffer chosen to give appropriate functional activity and stability of the target, and to be compatible with the ligands and minimize assay interference (*see* **Note 32**).
- Fluoroligand, i.e., standard ligand (*see* **Note 33**) labelled with fluorophore (*see* **Note 34**), dissolved in assay buffer (*see* **Note 35**). The minimum assay concentration is chosen to be such that the fluorescence intensity is sufficient to allow measurement of the fluorescence polarization with less than 10 % error (*see* **Note 36**).
- Test ligand(s) dissolved in assay buffer (*see* **Note 35**).
- Target at a known molar concentration in assay buffer.
- Cuvettes or microplates in which to perform the assay.
- Optional: Standard ligand, nonfluorescent. Ideally the unlabelled form of the same standard ligand. This is used for assay validation or for measuring of affinities relative to its affinity.

2.2.3 Competition Binding 1: Measurement of the K_d of Standard Ligand

This is a direct binding assay, but always required as a first step in a quantitative competition binding experiment (*see* **Note 37**). The fluorescence polarization of the fluoroligand alone, and in the presence of various concentrations of target (*see* **Note 38**), is measured (*see* **Note 39**). The concentrations are chosen so as to define the shape of the dose–response curve, in particular covering the range of concentrations between $0.2K_d$ and $5K_d$. Even with the knowledge of an estimated K_d an initial experiment will unlikely be in exactly this range, so it may have to be repeated to obtain a curve suitable for evaluation. The time course of changes in polarization should also be recorded so that the approach to final equilibrium

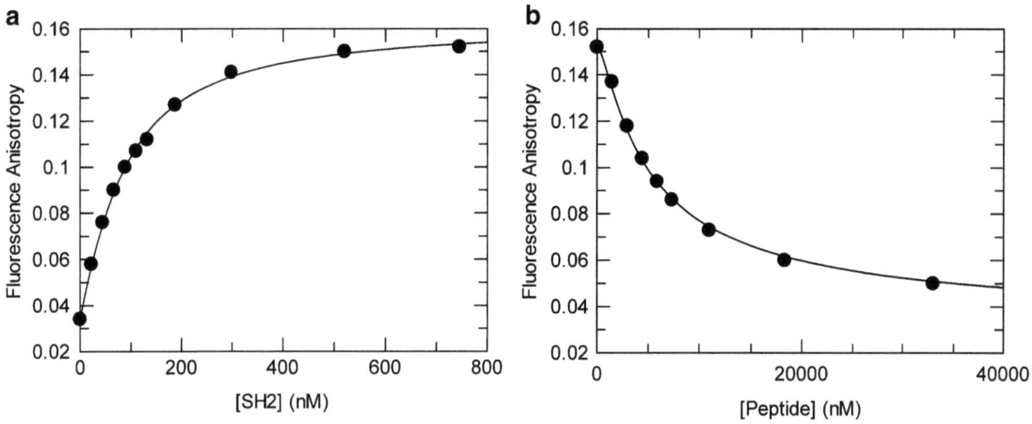

Fig. 2 Example of an equilibrium binding assay followed by validation through a competition assay, in which binding of a fluorescent peptide to a protein is monitored by fluorescence anisotropy. (**a**) Direct binding of fluoropeptide to the Src-SH2 domain. The fluorescence anisotropy of 54 nM fluoropeptide (carboxyfluorescein-QY(P)EEIPIYL-NH2) is measured in the presence of varying concentration of the Src-SH2 domain. The data is fitted to Eq. 1 substituted with Eq. 2 to yield the K_d for the interaction. (**b**) Competition between unlabelled peptide and fluoropeptide for binding to the Src-SH2 domain. The fluorescence anisotropy is measured at fixed concentrations of fluoropeptide (54 nM) and of SH2 domain (745 nM) in the presence of varying concentrations of the unlabelled peptide, QY(P)EEIPIYL-NH_2. The data is fitted to Eq. 1, substituted with Eq. 3 (*see* Subheading 2.2.4), to yield the K_i for inhibition by that peptide

can be followed. This will minimize the risk of measurements being made prior to equilibrium having been obtained, and hence underestimating the affinity. (N.B. The time to reach equilibrium increases with decreasing reactant concentration and in the presence of competitor ligand; *see* Chapter 1 for further details.)

Figure 2a shows such an experiment for the binding of the SH2-domain of the Src protein to the fluoroligand, carboxyfluorescein-QY(P)EEIPIYL-NH2. This is a cognate phospho-tyrosine-containing synthetic peptide, which was labelled at the N-terminus with carboxyfluorescein (11). This initial experiment was performed in a cuvette of a Perkin-Elmer LS50B spectrofluorimeter using a titration approach in which small volumes (0.5–2.5 μl) of 8,800 nM Src-SH2 were added sequentially to 200 μl of 54 nM probe such that the final concentration of Src-SH2 was varied between 0 and 745 nM, without changing the overall volume by more than 10 %. After each addition, the total fluorescence intensity, $(I_{\|} + 2I_{\perp})$, and the fluorescence anisotropy, $(I_{\|} - I_{\perp})/(I_{\|} + 2I_{\perp})$, were calculated, where $I_{\|}$ and $I_{\perp}$ are the parallel and perpendicular intensities. The total intensity did not change.

The signal *S*, i.e., in this case equilibrium fluorescence anisotropy data, is plotted at each concentration of target (including zero). This plot is called a binding isotherm and will have the usual hyperbolic shape, and the data can be fitted to Eq. 1 to give the K_d for the interaction:

$$S = S_0 + (S_{sat} - S_0)\frac{[PL]}{[P]_T}, \qquad (1)$$

where S_0 is the signal in the beginning of the titration, S_{sat} is the maximum (saturated) signal on the binding plateau, $[L]_T$ is the known, added concentration of ligand, and $[P]_T$ is the known total concentration of protein and where [PL] is substituted by Eq. 2:

$$[PL] = \frac{(K_d + [L]_T + [P]_T) - \sqrt{(K_d + [L]_T + [P]_T)^2 - 4[L]_T[P]_T}}{2}, \quad (2)$$

which is the solution to the quadratic one-to-one binding equation (*see* Chapter 1 for a derivation). Equations 1 and 2 are not specific to a signal due to fluorescence anisotropy, but will apply to any normal titration for a reversible one-to-one binding reaction, P + L ⇆ PL, where the signal changes proportionally to [PL] (including non-anisotropy uses where the signal may reduce upon complex formation). Modification of the equations is, however, required if the stoichiometry is not 1:1.

In the experiment presented here, $[P]_T$ is the varied total concentration of the Src-SH2 domain (target) which is used to titrate a fixed concentration of fluoroligand $[L]_T$. S_0 is the anisotropy of fluoroligand alone and S_{sat} is the anisotropy of the ligand in complex with the protein.

Figure 2a shows the curve fit which yielded anisotropy values of 0.034 and 0.163 for the fluoroligand and the complex of fluoroligand with Src-SH2, respectively, and a K_d of 58 nM for the interaction.

2.2.4 Competition Binding 2: Validation and Optimization of the Assay

The assay should be validated by establishing a dose–response relationship using the unlabelled standard ligand, or another non-fluorescent standard ligand for which the K_d is known, to compete with the fluoroligand (*see* **Note 37**). At the end of the titration shown in Fig. 2a, small volumes (0.5–2.5 μl) of 500 μM unlabelled peptide, QY(P)EEIPIYL-NH_2, were added sequentially such that the final concentration of unlabelled peptide was varied between 0 and 33 μM, again without changing the overall volume by more than 10 %. The measured anisotropy is plotted against the concentration of unlabelled, competing ligand (inhibitor) as shown in Fig. 2b, resulting in a dose–response curve.

The change in anisotropy is proportional to the change in [PL], which in turn depends on [I], the concentration of inhibitor. A known concentration of I is added to known, fixed concentrations of the protein, $[P]_T$, and the ligand, $[L]_T$. The affinity of the protein for the ligand is defined by K_d. The affinity of the inhibitor for the protein is defined by K_i. Binding of inhibitor and ligand are assumed to be mutually exclusive (purely competitive). Data analysis and the equations used rely on these principles.

In general, the equation to fit the competition data and obtain the K_i, i.e., the affinity of the inhibitor for protein, is the solution of a cubic equation (9, 26–28). The measured anisotropy signal is again given by Eq. 1; in this case the following substitutions are made: $[PL] = [P][L]_T/(K_d + [P])$ and [P] is further substituted, reflecting its dependence on the total concentration of inhibitor (*see* Fig. 2b), ligand and protein:

$$[P] = \frac{2}{3}\sqrt{(a^2 - 3b)}\cos\left[\frac{1}{3}\arccos\left[\frac{-2a^3 + 9ab - 27c}{2\sqrt{(a^2 - 3b)^3}}\right]\right] - \frac{a}{3}, \quad (3)$$

with $a = [I]_T + [L]_T - [P]_T + K_d + K_i$, $b = K_d\ ([I]_T - [P]_T) + K_i\ ([L]_T - [P]_T) + K_dK_i$, and $c = -[P]_TK_dK_i$. In the quite frequent case where at every concentration the inhibitor is in large excess over the protein, we can assume that the free and total concentrations of inhibitor are approximately equal and the data obtained can then be fitted to a somewhat simpler equation (*see* **Note 40**). In either case, the total concentrations of fluoroligand $[L]_T$ and of target $[P]_T$ and the K_d for their interaction are known (from data in Subheading 2.2.3), and the equations can be readily fitted by modern software to obtain K_i (*see* **Note 41**).

Ideally, the K_d for the unlabelled and fluoro-labelled standard ligand should be identical; however, the label may perturb the affinity (*see* **Note 42**). In the system illustrated in Fig. 2, the K_d for the fluoroligand was 54 nM (*see* Fig. 2a), whereas the competition experiment with the unlabelled ligand in Fig. 2b yielded a K_i of 390 nM (this K_i is of course the K_d for the unlabelled standard ligand), i.e., the addition of the fluorescent probe enhanced the affinity of the peptide. This might be because labelling removes the free amino N-terminus, or because the fluorophore provides extra binding surface, possibly through its hydrophobic bulk. It should always be considered whether or not such a perturbation caused by the label will require any changes to the assay design.

The initial experiments were performed at a relatively high concentration of fluoroligand as dictated by the sensitivity of the plate reader used. Further, the competition experiment shown in Fig. 2b was performed with a large excess of the Src-SH2 domain (745 nM) over the fluoropeptide (54 nM) in order to ensure a high level of saturation, which resulted in the measured IC_{50} (5,000 nM) for displacing unlabelled peptide being more than tenfold greater than its K_d (*see* **Note 43**).

These starting conditions would usually be further optimized for specific purposes. In this case, a higher throughput and greater sensitivity to inhibitors were required. For higher throughput, the volume was reduced to 50 μl to run in 96-well microplates, and the concentration of fluoroligand was dropped to 5 nM in line with the increased detection sensitivity of an Analyst™ plate reader (*see* **Note 44**). Simulations (26, 27) such as in Fig. 3 were then used to choose a concentration of protein that increased sensitivity to variation in concentration of competitive ligand, yet still gave a substantial signal. A concentration of 90 nM Src-SH2 was

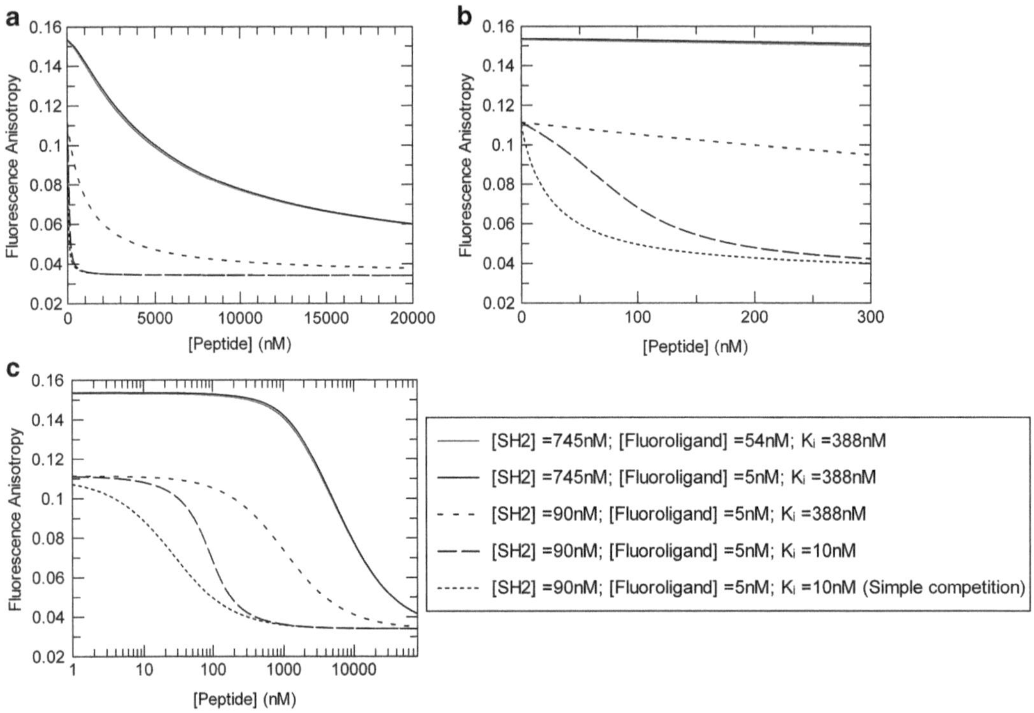

Fig. 3 Simulations to visualize effects of changing the concentrations of SH2 domain (745 and 90 nM) and fluoroligand (54 and 5 nM) on the behavior of a competition assay in which the concentrations of either unlabelled standard peptide (K_i =388 nM) or a higher affinity inhibitory ligand (K_i = 10 nM) were varied. These were created from the solution to the exact cubic equation describing competitive inhibition (Eq. 1 substituted with Eq. 3), except for the curve labelled "simple competition," which used Eq. 1 substituted with the simpler equation in **Note 40** in which depletion of inhibitor by binding to SH2 domain has been ignored. The curves were simulated using K_d for the fluoroligand binding to SH2 domain = 58 nM. Panels (**a**), (**b**), and (**c**) are different visualizations of the same data; (**a**) and (**b**) both use a linear concentration scale, with Panel (**b**) expanding the lower concentration range, whereas Panel (**c**) is on a logarithmic scale

ultimately selected. Figure 3 shows that reducing the fluoroligand concentration from 54 to 5 nM has little effect. Reducing the Src-SH2 concentration from 745 to 90 nM has two effects. Firstly, the anisotropy in absence of competitive ligand decreased from 0.153 to 0.110 due to a lower proportion of the fluoroligand being bound to the Src-SH2. Secondly, the IC_{50} for a competitive ligand is also greatly decreased, e.g., here the IC_{50} for the unlabelled standard peptide (K_d = 388 nM) changed from 4,500 to 1,100 nM, i.e., less ligand is required to obtain an IC_{50} value.

For a balance of high sensitivity to competitive inhibitors, low consumption of target, and acceptable fluorescence polarization signals, the concentration of the target for the competition assay is often chosen so that 50 % of the fluoroligand is bound, i.e., at the target's EC_{50} concentration (*see* **Note 43**).

The simulations in Fig. 3 also illustrate the situation in competition experiments in which the free concentrations of ligand/

inhibitor cannot be approximated by their total concentration. In a simulation with the simplified competition equation (*see* **Note 40**) and assuming a K_i of 10 nM and that $[I] = [I]_T$, the result is a classical hyperbolic dose–response curve with an apparent IC_{50} of about 20 nM. In contrast, the actual results would follow the clearly sigmoidal curve in Fig. 3b and the IC_{50} is found to be over four times greater. Even with inhibitors of weaker affinity, significant deviation from the classic hyperbolic dose–response curve is seen. These deviations are less easy to see in the semilogarithmic visualizations (Fig. 3c) commonly used when fitting for EC_{50}. However, high slope factors will be found and in extreme cases the data will not fit well to the standard four-parameter logistic (*see* **Note 43**).

2.2.5 Competition Binding 3: Determination of Affinity of Test Ligands for the Target

Assay solution containing target and ligand at the concentrations determined in Subheading 2.2.4 is mixed with a fixed volume of test ligand. The mixture is then left long enough to reach equilibrium (*see* Subheading 2.2.3).

To obtain K_i values, full dose–response data are obtained and then fitted using Eqs. (1) and (3). A complete set of data at varying concentrations of the test ligand that evenly span the range from zero to that necessary to reach a plateau should be obtained. It is convenient, and will increase the precision of the fitted K_i values, to fix any well-defined parameters as constants. In the Src-SH2 example, these would be the concentrations of fluoroligand and of Src-SH2, the K_d for that interaction, and the fluorescence anisotropies of the fluoroligand alone and in complex with the SH2 domain (*see* **Note 45**).

Experiments can also be performed with a single concentration of test ligand, to obtain relative inhibition data; however this will be less accurate and leave uncertainty as to the inhibition mechanism. In particular single-point data cannot distinguish a fully competitive from a partially competitive ligand (*see* **Note 46**).

IC_{50} values can be obtained by fitting to a four-parameter logistic equation (*see* **Note 43**). Relative affinities can then be calculated as the ratio of IC_{50} values relative to that of the unlabelled standard ligand (*see* **Note 47**).

3 Methods Involving Enzyme Catalysis

3.1 Introduction to Enzyme Catalysis

This subheading deals with experimental approaches based on measuring the rate of enzyme activity (catalytic rate, turnover) for a simple enzymatic reaction:

$$E + S \leftrightarrows ES \rightarrow EP$$

Enzyme reactions are preferentially carried out under steady-state conditions in which the substrate is in great excess over the enzyme and rates of reaction are consequently constant. These conditions greatly simplify data analysis since the equations describing steady state are much simpler than the general case. Two main types of experiment are performed:

(a) Those in which information is gained from a system, which only contains enzyme, required cofactors, substrates, products, and detection reagents. Such experiments can be used to characterize the catalytic process and enzyme substrate/product interactions, e.g., determining K_m, k_{cat}, and enzyme mechanism.

(b) Those which additionally contain other ligands that modulate enzyme activity, but are not altered by the enzyme. Such experiments can be used to gain further detail on the enzyme mechanism, or to characterize inhibitors or activators of the enzyme. Most commonly the experiments would be run under steady-state conditions with respect to the substrate/enzyme, and under equilibrium conditions with respect to the modulatory ligand. In this scenario, the catalytic rate can serve as a method for measuring the binding of a modulatory ligand, just like the methods discussed in Subheading 2.2.

In either case the key requirement is to measure the catalytic rate or turnover as described in Subheadings 3.2 and 3.3.

3.2 Materials to Measure Enzyme Turnover

- Enzyme. Some enzymes, such as kinases, may require prior activation.
- Substrates.
- Cofactors or regulatory factors.
- Detection technology, and any required instrumentation, to measure the loss of substrate, appearance of product, or associated heat change (*see* Tables 2 and 3).
- Assay buffer chosen to give appropriate functional activity and stability of the enzyme.
- Cuvettes or microplates in which to perform the assay, preferably thermostatted to a fixed temperature, as catalytic rate can be highly temperature sensitive. Temperature can be selected for sufficient stability and to give a convenient length of time course.

3.3 Quantitation of Enzyme Turnover

Turnover is quantified by monitoring the decrease in the concentration of substrate, the increase in the concentration of product, the decrease in substrate, and the increase in product (or ratio thereof) or the catalytic process itself. Most commonly the appearance of product is measured (*see* **Note 48**). This measurement can be direct or made indirectly via a coupled reaction (*see* **Note 49**). Measurement by disappearance of substrate may be possible, but is generally problematic as it requires detection of a small change in a large initial concentration (*see* **Note 50**).

In all cases, blanks and controls need to be included to demonstrate that the measurement indeed measures turnover catalyzed by the enzyme under study. Simple omission of the enzyme preparation is insufficient, as this does not control for contaminating activities. Confidence may be gained through the use of specific inhibitors, control enzyme preparations (e.g., without recombinant enzyme), or combining detection methods (e.g., for a protease—combining a fluorescence procedure with an LC-MS method).

Commonly used methods for measuring substrate or product are summarized in Tables 2 and 3, with the latter focusing on methods for some common species: phosphate, phosphorylated protein, ATP and ADP. Separation methods such as HPLC and capillary electrophoresis, and detection methods such as electrospray mass spectrometry, are capable of quantifying both the substrate and the product. Calorimetry is unique in allowing the whole catalytic process to be quantified without regard to substrate or product chemistry by measuring the accompanying heat change, though it has the corresponding disadvantage of lacking any specificity other than that imposed by the substrate provided.

3.3.1 Measurement of Initial Catalytic Rate

The simplest approach to measure the catalytic rate is to ensure that the system is running at a steady state and to use "classical" steady-state kinetics. Sufficient time must be allowed to reach the steady state, although for most enzyme reactions this is short (ms to second timescale). It also requires measurement of the so-called initial rate prior to any effects of substrate depletion, product inhibition, approach to equilibrium, enzyme inactivation, etc.

In the first instance, the amount of turnover (e.g., the amount of product formed) is measured and plotted against time to obtain a "progress curve" or "time course" (*see* **Note 51**). Ideally, the initial part of the progress curve should be linear. The initial rate is the tangent to the progress curve at "zero time." In practice, it is more convenient to measure the rate as the amount of turnover between two time points near the beginning of the curve rather than as a tangent. In the simulated progress curves shown in Fig. 4, measurement of the gradient in the period between 0 and t1 approximates to the true initial rate. However, if the gradient is measured between t1 and t2, the true initial rates are measured for some but not for all of the progress curves (*see* **Note 52**).

3.3.2 Determination of K_m and k_{cat}

Under steady-state conditions, the measured enzymatic rate (or turnover) of a single-substrate enzyme, v, is related to the concentration of substrate, S, by the Michaelis–Menten equation (Eq. 4):

$$v = \frac{V_{\mathrm{max}}[\mathrm{S}]}{[\mathrm{S}] + K_{\mathrm{m}}}, \quad (4)$$

where K_{m}, the Michaelis constant, is the substrate concentration that gives half the maximal rate, and V_{max} is the maximum rate of

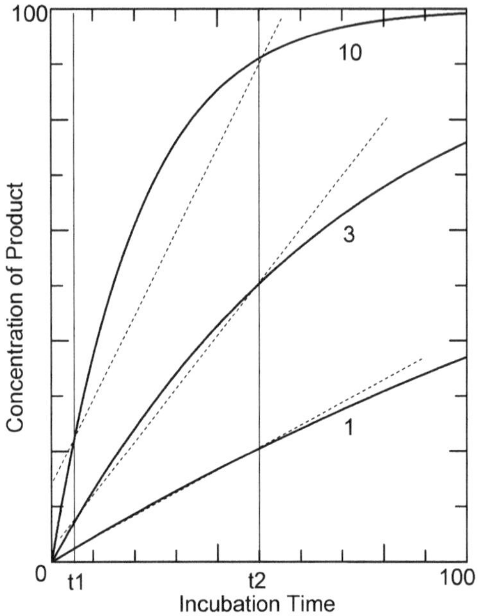

Fig. 4 Obtaining initial rates from enzyme progress curves. Progress curves were simulated at high (10), medium (3), and low (1) amounts of enzyme. These curves could also represent equal amounts of enzyme in the presence of low, medium, and high concentrations of an inhibitor, respectively. Between 0 and t1, the dead time, it is not possible to measure the amount of product formed, and the system has not yet reached steady state. Measurements of product accumulated, p1 and p2, are made at t1 and t2, respectively. The *dotted lines* indicate the slopes determined by (p2 − p1)/(t2 − t1) to show how as a result of substrate depletion they can deviate from the true initial rate, which is the rate at t1, most accurately represented by a tangent through t1 indicating the slope in that point. As substrate depletion sets in, the initial rate is underestimated, particularly at higher enzyme concentration (curves 3 and 10), at lower substrate concentration, and/or if t1 and t2 are further apart

the enzyme, i.e., the rate at a saturating concentration of substrate. The K_m of a substrate is always greater than the K_d of substrate binding to the enzyme (*see* **Note 53**). It is assumed that substrate is in large excess over enzyme and that consequently [S] can be taken to be approximately equal to the total substrate concentration. For details on the other assumptions, derivation, and limitations of this equation and those that describe more complex enzyme reactions (multiple substrate, product inhibition, in the presence of inhibitors, etc.) *see* refs. 2, 29–32.

The catalytic parameters of an enzyme are determined by varying the concentration of substrate, at a fixed concentration of enzyme (*see* **Note 54**). Similarly to the case of binding reactions, $[S]$ should ideally cover a range of at least $0.2K_m$ to $5K_m$ while ensuring that initial rates can be measured at each concentration (*see* **Note 55**). At low substrate concentrations, curvature of progress

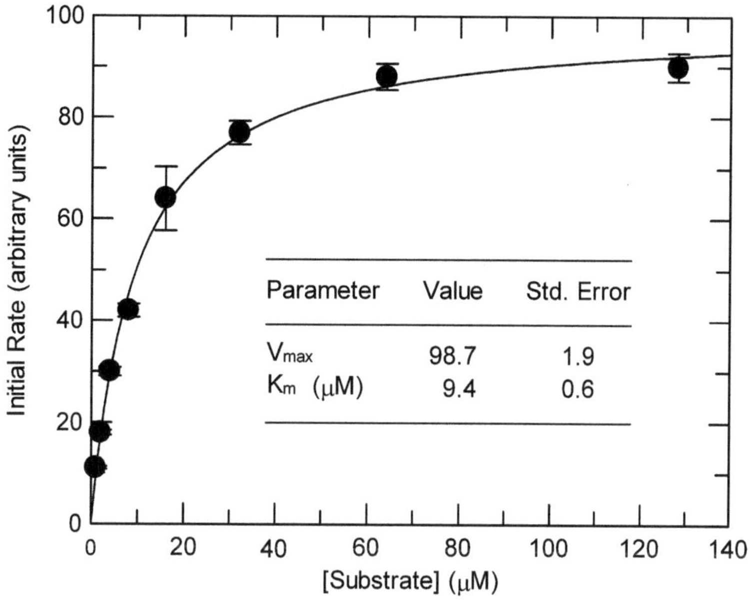

Fig. 5 Determination of K_m and V_{max} by curve fitting to the Michaelis–Menten equation (Eq. 4). A series of enzyme assays are performed at a range of substrate concentrations spanning at least from $0.2K_m$ to $5K_m$. The initial rate of product formation measured at each substrate concentration is plotted in arbitrary units of product formed in unit time. Curve fitting yields an estimate of the two parameters and their associated standard errors. The *error bars* represent the standard deviation of replicates

curves is inevitable. This can only be avoided by using a more sensitive procedure to measure turnover such that a good signal is obtained before too much substrate is used up (*see* **Note 56**). Measuring initial rates in the shortest reasonable time also reduces the potential impact of product inhibition (*see* **Notes 56** and **57**). Additionally, optimization of the assay medium (*see* **Note 58**), its pH (*see* **Note 59**), and its temperature (*see* **Note 60**) can help to achieve sufficient linearity.

The initial rates of turnover that have been obtained at several substrate concentrations are then plotted against the respective concentrations to yield a plot such as in Fig. 5. A nonlinear curve fit to Eq. 4 yields the K_m and the V_{max}. If the enzyme concentration is known, the turnover number (k_{cat}) can be calculated:

$$V_{max} = k_{cat}[E]_T \tag{5}$$

(*see* **Note 61**).

3.4 Determination of K_i and Inhibitor Mechanism

The IC_{50} of an inhibitor is a simple measure of its potency (*see* Subheading 2.2). It can be obtained by measuring the initial rates at any suitable single fixed concentration of substrate whilst varying the inhibitor concentration over a sufficient range so as to define a full dose–response curve. The data is then fitted by nonlinear regression to an IC_{50} logistic (*see* **Note 43**), using S_0 = minimum (with saturating ligand) and S_{sat} = maximum (without ligand) initial rates. The relationship between the measured IC_{50} and the chosen substrate concentration is dependent upon the enzyme mechanism and the mode of inhibitor interaction (28) (*see* **Note 43**).

However to obtain the affinity of the inhibitor, information on the kinetic mechanism by which the inhibitor affects catalytic rate is also required. For a single-substrate enzyme (*see* **Note 62**), this is usually obtained from a family of dose–response curves at varying [I] and [S].

In the absence of any prior knowledge, this would consist of a matrix of varying [S] and [I], with at least 25 data points in total. Ideally [S] would be varied between at least $0.2K_m$ and $5K_m$ at every [I] such as to obtain at least five well-distributed points at every [I]. In practice, the impact of [I] on K_m may not be known, and hence trial experiments are usually required to get an initial idea of how [I] is affecting the V_{max} and K_m values. Simulations are invaluable in selecting optimal concentrations of substrate and inhibitor to give good definition of the curves.

The family of curves is then fitted by nonlinear regression to equations representing alternative models of inhibition. Commonly these models would be the following:

- Competitive inhibition in which substrate and inhibitor binding is mutually exclusive, i.e., ES and EI but not ESI can form. The affinity of I is determined by K_{ic}.
- Uncompetitive inhibition in which inhibitor can ONLY bind to the ES complex to form ESI, and EI is not formed. The affinity of I is determined by K_{iu}.
- Mixed inhibition in which inhibitor can bind to either E or ES, to form BOTH EI and ESI. The affinity of I is determined by two constants, K_{ic} and K_{iu}.

A rare occurrence in simple enzymes is noncompetitive inhibition, which is a specific case of mixed inhibition in which K_{ic} is equal to K_{iu}. Given the rarity of this situation, one should check that results are not due to the inhibitor being slow or tight binding (*see* Subheading 3.5).

Figure 6 exemplifies how experimental data is analyzed by fitting it to each of the standard models and assessing the quality of fit. The quality is checked statistically, by the standard error of each parameter and the magnitude of overall χ^2 values (*see* Chapter 1), and visually, via linear transformations into Lineweaver–Burk plots of 1/rate versus 1/[S] which can show systematic deviations more clearly. Where there are sufficient replicate measurements, the mean and SD can be used (as in Fig. 6) and the curve fits weighted according to the errors associated with each data point. The replicates can also be treated as independent experiments so as to result in independent estimates of the parameters.

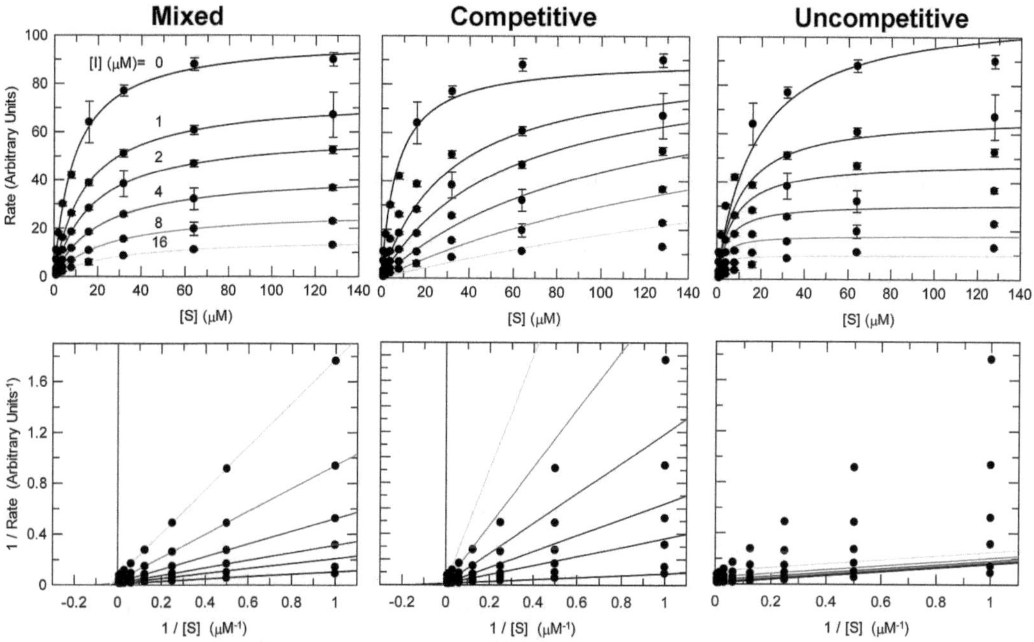

Parameter	Value±SE
V_{max}	99±1
K_m (μM)	9.4±0.3
K_{ic} (μM)	0.94±0.04
K_{iu} (μM)	3.06±0.09

Parameter	Value±SE
V_{max}	90±4
K_m (μM)	7.0±1.2
K_{ic} (μM)	0.28±0.04
K_{iu} (μM)	-

Parameter	Value±SE
V_{max}	111±5
K_m (μM)	16.8±2.1
K_{ic} (μM)	-
K_{iu} (μM)	1.54±0.15

Fig. 6 Determination of the mechanism of inhibition. The enzyme initial rates in arbitrary units of product formed in unit time, obtained at each substrate concentration ([S] = 1, 2, 4, 8, 16, 32, and 64 μM) and at each concentration of inhibitor ([I] = 0, 1, 2, 4, 8, 16 μM), were fitted by nonlinear regression to equations for three single-substrate models: mixed, competitive, and uncompetitive inhibition. The *error bars* represent the standard deviation of replicates. The *lines* represent the best fits to each model. Each set of data is plotted both as rate versus concentration of substrate, and as a double reciprocal Lineweaver–Burk plot. The curve-fitting parameters, with their associated standard errors (SE), are given for each model. The reduced χ^2 were 0.7, 27, and 26, respectively. These show that the fit is effectively of equal quality for competitive and for uncompetitive models. As the measured data points are generally reproducible, visually the best-fit lines do not follow the experimental data, and the standard errors on parameters are large. It is highly likely that neither of these models are appropriate. In contrast, both visually and statistically, the fits to a mixed model are good, yielding low standard errors for each parameter. As a further test, one can compare the value of K_m (9.4 ± 0.6 μM) and V_{max} (99 ± 2 arbitrary units) obtained in the absence of inhibitor (*see* Fig. 5) with that obtained from the global fit to the inhibition model. They should match well, and do so for the mixed model but deviate significantly for the other models

3.5 Determination of an Enzyme's Kinetic Mechanism

Many enzymes utilize more than one substrate, and the system will be defined by a catalytic parameter akin to K_m for each substrate. Experimentally, the methods are as described in Subheading 3.3.1, except that experiments need to be performed in which the concentrations of both substrates are varied and additionally product inhibition studies may be required (2, 30–32). To obtain the K_m

parameters the kinetic mechanism needs to be established, as data needs to be fitted to the appropriate model. Knowledge of this mechanism is required to understand how, both mechanistically and in terms of affinity, other ligands interact with an enzyme (33).

3.6 Deviation from Assumptions: Slow and Tight-Binding and Pre-steady-State Kinetics

The previous subheadings described the experimental and analytical procedures when the catalytic process is at steady state, and in which additional ligands are bound at equilibrium with their concentrations negligibly depleted by binding to the enzyme. Experiments performed under pre-steady-state conditions (*see* **Note 54**) are done in an entirely different way and are described in detail in Chapter 5. Where the enzyme is in steady state with regard to substrate and product, but the ligand is not in equilibrium, experiments "as if" in equilibrium can still give some information. The typical cases, described below, are slow-binding ligands and ligand depletion.

A slow-binding ligand is one that has slow association with an enzyme relative to substrate association and product dissociation. Where slow binding of a ligand has resulted in measurements being made before equilibrium has been obtained, values of IC_{50} will not reflect potency and attempts to determine the mechanism of an inhibitor or its K_d will fail. Theoretically, the remedy is to allow a longer equilibration time before measurements are made, as was done for simple binding reactions in Subheading 2. However, in practice this may not be possible for a catalytic measurement because substrate depletion, product inhibition, or functional instability may occur at longer times. It is important to note that pre-equilibration of a slow-binding ligand with enzyme (in absence of substrate) will not relieve the problem. Once substrate is added, the equilibrium position will change and that timescale of adjustment will be predominantly determined by the ligand's dissociation rate constant, which for a slowly associating ligand will necessarily be extremely slow. With slow binding, if true equilibrium in the presence of the substrate cannot be achieved, then either a catalytic method is inappropriate or an approach in which the rate of change of turnover, i.e., the kinetics of inhibitor binding, is monitored needs to be utilized (29, 34–36).

Tight binding refers to situations in which depletion of free ligand occurs such that [L] is significantly less than $[L]_T$ (2, 29, 31, 32, 36). This typically occurs when K_i is similar to or lower than $[E]_T$. Lowering $[E]_T$ will turn a tight-binding ligand into a standard ligand. However, as lowering $[E]_T$ results in lower turnover, more sensitive detection technologies may be required.

The sensitivity created in a multi-turnover catalytic assay typically allows $[E]_T$ to be in the fM to nM range. Consequently, any ligand that has tight-binding behavior will necessarily have a sub-nM K_d and a low dissociation rate constant such that tight-binding ligands are commonly also slow-binding. In this situation, the catalytic activity becomes essentially a measure of free enzyme concentration and this may limit the possibility of obtaining mechanistic information.

In particular, an unrecognized tight-binding or slow-binding inhibitor will often appear to be purely uncompetitive when analyzed by the procedures in Subheading 3.3.2. In such circumstances, an alternative kinetic approach is to measure the rate of decrease of the catalytic rate at a fixed concentration of a tight-binding inhibitor while varying the concentration of substrate. This will often reveal any competition of ligand with substrate (2, 29).

4 Notes

1. In this chapter, we refer to "target" in the sense that target is the binding partner for a specified ligand, and not in the sense of the target for a drug. We generally assume that the target is a macromolecular protein, but the experimental approaches are either directly appropriate or can be modified, when the target is another type of macromolecule, or a peptide or oligonucleotide.
2. In thermal shift experiments, either the rate or the extent of unfolding (denaturation) is measured. Target–ligand complex formation stabilizes the target and so it is more resistant to heat-induced unfolding. Unfolded protein can be detected by dyes that fluoresce when bound to it. Also, differential scanning calorimetry (DSC) or circular dichroism (CD, *see* Chapter 8) give thermodynamic or structural measurement, respectively, of an unfolding process.
3. In practice, labels that alter the interaction can be appropriate, particularly in competition systems.
4. The ability to choose the excitation and emission wavelengths and the intensity or other fluorescence properties of the signal are major advantages of fluorescent probes, as compared with intrinsic fluorescence. This often outweighs the disadvantage that labelling can induce. FRET occurs when a suitable donor and acceptor pair is sufficiently close, and thereby has a further advantage of detecting changes in distance. The signal can be detected by spectral and intensity changes. Time-resolved FRET gives a higher signal-to-background ratio, but requires more specialist donors and detection equipment.
5. Most common scintillation proximity methods use flashplates or scintillation proximity assay (SPA) beads (1, 7). An unlabelled species is captured on a solid surface into which a scintillant has been incorporated during manufacture. This scintillant emits light when stimulated by ionizing radiation. A radiolabelled species (short path length emitters such as ^{3}H, ^{33}P, ^{125}I) is added in solution. When the radiolabel is sufficiently close to the surface, such as when a complex forms, scintillation will occur. However the radiolabel in bulk solution is too far from

the embedded scintillant to give a signal, as the path length of emission is short. As complex, but not ligand or target, is detected, no separation is required.

6. Electrochemiluminescence (ECL), e.g., using kits by Meso Scale Discovery, is a proximity procedure. An unlabelled species is captured on a solid surface attached to an electrode. A species with a suitable label, e.g., ruthenium, is added in solution. When the label is close to the surface, as when a complex forms, light is emitted; this is sustained and amplified by inclusion of a regenerating agent in solution. ECL is characterized by both high sensitivity and a large dynamic range.
7. Signals from ligand alone or target alone can be obtained with MS and NMR, antibody detection, pull-down, and enzyme catalysis procedures. Signals from complex alone can be seen with scintillation-proximity and pull-down assays. The sum of signals from all species containing the ligand or the target is commonly seen when a ligand or a target is labelled.
8. When the signal derives from an intensive property, the free species with that property will have a different magnitude of signal as compared to that of the complex. Where a mixture of free and complex is present, the measured signal will be a weight-average of the free and complex signals.
9. For this purpose, the steady-state situation seen in flow-based SPR (*see* Chapter 6) or in enzyme kinetics (*see* Subheading 3) can be treated as equivalent to unperturbed equilibrium.
10. For systems that do not reach equilibrium, it will only be possible to measure absolute affinities where BOTH the mechanism of interaction is known AND the association and dissociation rate constants can be measured (*see* Chapter 5).
11. Very often, it is convenient to make unsupported assumptions, e.g., commonly the simplest possible reaction $P + L \leftrightharpoons PL$ is assumed, rather than a reaction with distinct bound forms such as $P + L \leftrightharpoons PL \leftrightharpoons PL^*$. However, if these assumptions do not in fact conform to the reaction in hand, very significant errors in both meaning and numerical values of derived parameters can arise (29).
12. If the time course to reach steady state is to be analyzed (for example with slow-binding ligands), the data is similar to that in Subheading 1.5, but requires the appropriate equations to account for the raw measurement being product versus time, whereas the desired measurement is enzymatic rate versus time.
13. Relative affinities for a series of ligands can be obtained when the K_d for each ligand is directly proportional to the respective dissociation rate constant (i.e., the association rate constant is the same) and other factors such as concentrations and molecular weight are the same.

14. If an unrelated tagged control protein is not available simply use a control of the untagged protein alone to check for non-specific binding, aggregation, or precipitation on the resin (Control 1 in Table 4).
15. The buffer components can be varied to enhance interaction, e.g., salt concentration can be lowered, and glycerol can be added.
16. The concentrations used in this specific protocol work to identify interactions of a K_d of around 10 μM. During the wash steps samples become diluted, which means that Western blotting is required to detect the remaining bound protein.
17. The order of addition can be varied, e.g., the tagged protein can be incubated with resin prior to addition of partner protein, instead of incubating the two proteins first. Your choice here may depend on the size and position of the tag relative to the anticipated interaction interface. The protein ratio can be altered to change the amount of untagged protein relative to tagged protein. For a first experiment, we typically use an equal weight (50 μg) of each protein, but this may not be appropriate depending on their molecular weight and what is already known about affinity and stoichiometry of interaction.
18. The samples to be loaded are typically labelled "Load," the proteins that remain bound to the resin after washing are "Bound," and the "Unbound" fraction is the sample that is aspirated after incubation of tagged and untagged proteins with the resin.
19. Take care during aspiration of the pelleted resin. Try not to disturb resin surface while removing as much buffer as possible. This takes a little bit of practice. This helps to ensure that equal amounts of protein are retained in each reaction of the experiment.
20. The incubation time of the proteins with the resin can be increased, e.g., we commonly also incubate overnight. We recommend a minimum incubation time of 1 h.
21. These washes are performed without extra incubation time. When testing low-affinity interactions this is critical as the wash volume dilutes the sample to below K_d. Incubation time during washing steps can be lengthened to allow proteins more time to diffuse from the resin, e.g., if testing cell lysate for high-affinity interactions extensive washes with 30-min incubation are used.
22. Unless the interaction being probed is of very high affinity (approximately nanomolar K_d) there will be a significant amount of untagged protein in these fractions, so running them on a gel will not be informative. The efficiency of washing of the resin can be assessed by checking the control lanes on the gel (*see* Fig.1).

23. Direct denaturation/elution into GLB ensures minimal dilution to detect even small amounts of bound protein. This can be important for low-affinity interactions.
24. It may be appropriate to detect bound protein using Western blot. This has advantages over Coomassie staining in that it allows for more sensitive detection and can specifically detect one protein, which can be particularly helpful in cases where the prey is in cell lysate. Typically, Western blotting is at least ten times more sensitive than Coomassie staining, although this will vary depending on the quality of the antibody and the detection method being used. The more sensitive detection methods include chemiluminescence (recorded using X-ray film or a phosphorimager) and chemifluorescence (recorded using a fluorescence imager). Like with Coomassie staining, quantitation of Western blots requires calibration of the signal observed for known amounts of protein under a standard set of reaction conditions. However, performing a Western blot does take considerably longer than Coomassie staining a gel.
25. Most competition assay formats cannot distinguish a pair of ligands binding at the same site from a pair that bind at different sites such that each reduces the affinity of the other (allosteric binding). Hence, the terminology "mutually exclusive" binding is often used to describe competitive binding at a single site. Where two ligands bind at allosteric sites, or indeed at partially overlapping sites, the reduction of affinity can be complete, causing full competitive behavior or incomplete, causing partial inhibition in some assay formats.
26. To establish that two ligands bind at the same site technologies that give structural information such as X-ray crystallography or 2D-NMR are required.
27. Relative affinities can be obtained by comparing IC_{50} values under conditions in which (a) the concentrations of the test ligand are not significantly depleted by binding to the protein, and (b) the interaction is fully competitive.
28. With the appropriate technology (*see* Table 1), assays can be set up with any species, protein/peptide, nucleic acid, carbohydrate, etc.
29. Any accompanying fluorescence intensity change must be also taken into account (37).
30. The wavelength can be altered to minimize interference from assay components or to suit the instrumentation, e.g., when fixed wavelength filters must be employed.
31. The preferred unit of measurement is anisotropy, though polarization values (mP) can be converted. With instruments for which the g-factor is not close to unity a correction is required.

For details on the use of the *g*-factor and conversion of polarization to anisotropy values *see* Chapter 7.

32. Addition of low concentrations of detergent, e.g., 0.001–0.01 % Tween 20, NP40, or Triton X100, can reduce nonspecific binding and improve pipetting quality, particularly with low-volume microplates.
33. The standard ligand should be chosen on the basis of its affinity and physicochemical properties; high aqueous solubility and low hydrophobicity tend to reduce nonspecific binding and higher molecular weight maximizes the change in anisotropy signal when binding to target.
34. The fluorophore's fluorescence properties, e.g., lifetime, wavelengths, and intensity, must suit the target and assay technology (*see* Chapter 7). Alternatively, the standard ligand might have intrinsic fluorescence of appropriate intensity and wavelength such that it can be directly used as the fluorescent ligand.
35. If poorly water soluble, a higher concentration of ligand is made in 100 % DMSO, and then diluted to final assay concentration into the assay buffer.
36. Higher concentrations of standard ligand can be employed, but as the ligand concentration approaches its K_d for target binding the sensitivity to competitive test ligands will diminish.
37. If only simple detection of binding is required, and not affinities, Parts 1 and 2 could be omitted. However, inclusion of these parts of the protocol is important to validate the assay, in determining the concentrations of target to be used and the sensitivity of the assay to competitive ligands.
38. This variation in concentration can be obtained via a dilution series (1 in 2 for best precision in estimating K_d, although 1 in 3 may suffice) either in assay buffer or assay buffer supplemented with the assay concentration of standard ligand. Alternatively, a sequential titration is made within the assay vessel by adding increasing increments of target. In all cases the assay should be performed under conditions where there is no significant change in the total concentration of ligand.
39. As fluorescence polarization is an extensive property, measured by summation of the fluorescence intensities of free probe and of complexed probe, the simplest procedure is to vary the concentration of target at a fixed concentration of fluoroligand. With other technologies, it is more appropriate to keep the target concentration fixed and vary the ligand concentration. For some procedures, one can choose which is best based on other factors such as availability or nonspecific signals.
40. In a competition binding assay for which one can make the approximation that the total concentration of inhibitor, $[I]_T$, does not differ significantly from the free concentration of

inhibitor at equilibrium, [I], i.e., the inhibitor is in large excess, then the cubic binding equation simplifies to a quadratic form and the following, relatively simple, equation can be used for the concentration of complex during the titration:

$$[\mathrm{PL}] = \frac{K_d[\mathrm{I}]_T + K_i(K_d + [\mathrm{L}]_T + [\mathrm{P}]_T) - \sqrt{(K_d[\mathrm{I}]_T + K_i(K_d + [\mathrm{L}]_T + [\mathrm{P}]_T))^2 - 4K_i^2[\mathrm{L}]_T[\mathrm{P}]_T}}{2K_i}.$$

The measured signal is given by substituting this expression for [PL] into Eq. 1.

41. GraphPad Prism (www.graphpad.com), Origin (www.originlab.com), Mathematica (www.wolfram.com), and GraFit (www.erithacus.com) are examples of curve-fitting software that allow multiline equations with substitutions (often termed temporary or intermediate variables). By using temporary variables equation definitions are easier to read and check, and can speed up the curve fitting or simulation as these values need to be only calculated once. GraFit 7.0 has been used here for simulations and curve fitting.

42. Labelling can decrease or increase affinity. Fluorophores are noted for increasing affinity by adding extra hydrophobic binding potential. If this is problematic a different fluorophore should be selected. This is a key advantage of using an intrinsically fluorescent ligand.

43. EC_{50} is the concentration that gives 50 % of maximal effect under some particular experimental conditions. The relationship between EC_{50} and K_d is straightforward for a direct binding assay: the concentration of ligand at which half the protein is bound is $EC_{50} = K_d + 0.5[\mathrm{P}]_T$ (alternatively, the concentration of protein at which half the ligand is bound is $K_d + 0.5[\mathrm{L}]_T$). If $K_d >> [\mathrm{P}]_T$ (or $[\mathrm{L}]_T$) then $EC_{50} \approx K_d$. EC_{50} values are commonly obtained by plotting dose–response data on a semilogarithmic plot (response against log[L]) which will give an S-shaped curve that can be fit to the four-parameter equation $S = S_0 + (S_{sat} - S_0)/(1 + [\mathrm{L}]/EC_{50})^n$. This four-parameter logistic is defined by the response in the absence of ligand (S_0), the response in the presence of infinite ligand (S_{sat}), the EC_{50}, and the slope factor (n). For standard single-site binding, for weak interactions, the value of n should be 1. For tight binding, or cooperative multisite binding, interactions, typically $n > 1$. When some parameters are well defined, typically S_0 and/or S_{sat}, their values can be fixed to simplify fitting and increase precision of the remaining variables. IC_{50} is commonly used to indicate the EC_{50} of an inhibitor. For a competition assay, the relationship between the IC_{50} of the test ligand and its K_i cannot be simply stated except under specific circumstances (28). For a competitive inhibitor, whose free

concentration is insignificantly depleted by binding to the protein (i.e., "weak" binding), $IC_{50} = 0.5[P]_T + K_i (1 + ([L]/K_d))$.

44. Lower concentrations would have given adequate signals, but a high concentration was chosen in order to give high fluorescence intensity, and thereby reduce interference when testing compounds that had high absorbance or fluorescence.
45. This protocol is also suitable for measuring relative affinities, and for single concentration testing. For these measurements any quantitative comparison requires the knowledge that the ligands are fully competitive, and that neither standard nor test ligands are "tight" binding.
46. Apparent partial inhibition can also be seen where the test ligand is not fully soluble.
47. The EC_{50} logistic will not give a good fit to the dose–response curve if the EC_{50} is close to the concentration of the target, and relative EC_{50} values will have little meaning (*see* Fig. 3).
48. The choice of what to detect and the detection technology can have a significant effect on likely success: When two products are formed, one may choose which product to measure. This choice is dependent not only upon the ease of detection of each product but also on any other reactions that may be occurring, such as the likelihood of contaminant enzymes (e.g., a phosphatase in a kinase assay) or instability of any product. Furthermore, if product inhibition occurs, then it may be advantageous to use a detection procedure that removes product as it is formed. Product instability can be overcome by a rapid detection procedure or by capture procedures.
49. Detection via a coupled reaction needs to be set up so as to ensure that the rate of the coupled reaction is sufficiently fast not to be rate limiting (38). An advantage of coupled reactions is that the product is continuously removed, and in some cases can even be regenerated into substrate (*see* Subheading 3.3.1).
50. If the substrate can be destroyed via other processes than the desired catalytic reaction, then measurements through loss of substrate may not be appropriate.
51. This is usually easiest using "real-time" detection methods that allow continuous monitoring of catalysis over the period of interest. If this is not possible, the reaction needs to be quenched at defined time points. Quenching is often most easily achieved by removing samples at intervals from a bulk reaction and adding to a quench solution. The method of quenching is dependent upon the detection procedure and the reagents, and includes pH shift, temperature shift, removal of cofactors such as metal ions, addition of excess unlabelled

reagent, and denaturation. Controls would need to be performed to ensure that the reaction is completely stopped.

52. There are many causes for curvature. Substrate depletion will always result in curvature. The higher the concentration of substrate, above K_m, the longer the progress curve can be considered effectively linear. For substrate concentrations below K_m, the progress curve is never linear. Measurements with <10–20 % conversion of substrate often give acceptable approximation to the true initial rates. Other common reasons for curvature are approaching equilibrium, product inhibition, enzyme instability or inactivation, time-dependent inhibition, and assay artifacts such as nonlinear detection.

53. The Michaelis constant K_m is determined by the rate constants of substrate binding (k_{on}) and dissociation (k_{off}) as well as substrate turnover (k_{cat}) (*see* ref. 2 for derivation of the Michaelis–Menten equation): $K_m = (k_{off} + k_{cat})/k_{on}$. Since $K_d = k_{off}/k_{on}$, K_m is always greater than the K_d for the binding of substrate to the enzyme.

54. The assumptions in the standard steady-state analysis described here require that $[E]_T << [S]_0$ (the initial substrate concentration) and that $[E]_T << K_m$. If $[E]_T >> [S]_0$, then a single turnover occurs. Data interpretation can be easier but often requires rapid measurement procedures (*see* ref. 39 and Chapter 5).

55. The concentration of enzyme should be chosen such as to give linearity over a sufficient monitoring time at all substrate concentrations of the titration.

56. Another approach is to regenerate the substrate from the product, as in the use of phosphoenolpyruvate and pyruvate kinase to regenerate ATP from ADP (*see* Table 3).

57. Product inhibition can be reduced by procedures that capture the product, or that can regenerate substrate (*see* **Note 50**).

58. The assay medium should be chosen to ensure optimal stability, and hence linearity of the progress curves. Alterations in the buffer (e.g., phosphate versus Tris versus HEPES), and its ionic strength (e.g., ±150 mM NaCl), and inclusion of reducing agents (e.g., DTT or the non-thiol, tris(2-carboxyethyl) phosphine (TCEP)), glycerol, detergents, or proteins (e.g., BSA, casein), can enhance stability and linearity. Additionally, some proteins might require specific cofactors or metal ions.

59. Alteration of the assay pH will affect the enzyme's catalytic parameters. Sufficient buffering power is essential to ensure that the pH does not change during the progress of the

reaction. The effect of pH can be used to advantage by altering the pH to maximize k_{cat}, and/or decrease K_m.

60. Assay temperature should be kept constant, as the catalytic rate can be highly temperature sensitive. Lowering the assay temperature can slow the reaction and increase linearity of progress curves, and thus allow more convenient assay times.
61. $[E]_T$ is the total molar concentration of *functional* catalytic sites, rather than the total protein concentration. This can be obtained by measuring the total concentration of protein, establishing what percentage is functional, and knowing the molecular weight of the enzyme and the number of catalytic sites within each molecule. If an inhibitor is available with sufficient affinity that its EC_{50} is similar to that of $[E]_T$, then an active-site titration can be performed which directly gives the molar concentration of inhibitor-binding sites (usually assumed to be the same as the number of functional catalytic sites) (29). With appropriate turnover detection methods, the range of possible inhibitors can be increased by raising the $[E]_T$ to be close to the K_d of the inhibitory ligand, yet still keeping $[S]_0 >> [E]_T$. A non-catalytic alternative is to use ITC to determine the concentration of binding sites (*see* Chapter 4).
62. For multi-substrate enzymes, information on the kinetic mechanism is required (*see* Subheading 3.5), and the data analysis and interpretation need to take into account the K_m of each substrate. However, apparent affinities and some mechanistic information on inhibitors can be obtained by keeping one substrate at a fixed concentration, and just varying the concentration of the other.

Acknowledgements

We gratefully acknowledge the expert advice and information provided by Pirthipal Singh (Singh Consultancy) in the preparation of this chapter.

References

1. Thompson G, Owen D, Chalk PA, Lowe PN (1998) Delineation of the Cdc42/Rac-binding domain of p21-activated kinase. Biochemistry 37:7885–7891
2. Bisswanger H (2008) Enzyme kinetics: principles and methods, 2nd edn. Wiley, Chichester
3. Sigmundsson K et al (2002) Determination of active concentrations and association and dissociation rate constants of interacting biomolecules: an analytical solution to the theory for kinetic and mass transport limitations in biosensor technology and its experimental verification. Biochemistry 41:8263–8276
4. Copeland RA, Pompliano DL, Meek TD (2006) Drug-target residence time and its implications for drug optimization. Nat Rev Drug Discov 5:730–739
5. Winzor DJ (2001) Quantitative characterization of ligand binding by chromatography. In: Harding SE, Chowdhry BZ (eds) Protein-ligand interactions: hydrodynamics and calorimetry. Oxford University Press, Oxford

6. Harding SE, Winzor DJ (2001) Sedimentation velocity and sedimentation equilibrium ultracentrifugation. In: Harding SE, Chowdry BZ (eds) Protein-ligand interactions: hydrodynamics and calorimetry. Oxford University Press, Oxford
7. Chung CW, Lowe PN (2007) Biophysical methods: mechanism of action studies. In: Jhoti H, Leach AR (eds) Structure-based drug discovery. Springer, Dordrecht, Netherlands
8. Xu Z et al (2009) Development of high-throughput TR-FRET and AlphaScreen assays for identification of potent inhibitors of PDK1. J Biomol Screen 10:1257–1262
9. Chung CW (2007) The use of biophysical methods increases success in obtaining liganded crystal structures. Acta Crystallogr D Biol Crystallogr 63:62–71
10. Crowther GJ et al (2009) Buffer optimization of thermal melt assays of Plasmodium proteins for detection of small-molecule ligands. J Biomol Screen 6:700–707
11. Bligh SWA, Haley T, Lowe PN (2003) Measurement of dissociation constants of inhibitors binding to Src SH2 domain protein by non-covalent electrospray ionization mass spectrometry. J Mol Recognit 16:139–147
12. Cooper MA (ed) (2009) Label-free biosensors: techniques and applications. Cambridge University Press, Cambridge
13. Tanega C et al (2009) Comparison of bioluminescent kinase assays using substrate depletion and product formation. Assay Drug Dev Technol 7:606–614
14. Okoh MP, Hunter JL, Corrie JE, Webb MR (2006) A biosensor for inorganic phosphate using a rhodamine-labeled phosphate binding protein. Biochemistry 45:14764–14771
15. Harder KW et al (1994) Characterization and kinetic analysis of the intracellular domain of human protein tyrosine phosphatase beta (HPTP beta) using synthetic phosphopeptides. Biochem J 298:395–401
16. Singh P, Ward WHJ (2008) Alternative assay formats to identify diverse inhibitors of protein kinases. Expert Opin Drug Discov 3:819–831
17. Sambrook J, Russell DW (2001) Molecular cloning: a laboratory manual. Chapter 18: Protein interaction technologies, Protocol #3: Detection of protein-protein interactions using the GST fusion protein pull-down technique, 3rd edn. Cold Spring Harbor Laboratory Press, Plainview, New York
18. Zhou Z et al (2011) Structural basis for recognition of centromere histone variant CenH3 by chaperone Scm3. Nature 471:234–237
19. Bonifacino JS, Dell'Angelica EC, Springer TA (2006) Immunoprecipitation. In: Crawley JN (ed) Current Protocols in Neuroscience. Wiley, New York
20. Kaboord B, Perr M (2008) Isolation of proteins and protein complexes by immunoprecipitation. Methods Mol Biol 424:349–364
21. Brymora A, Valova VA, Robinson PJ (2004) Protein-protein interactions identified by pull-down experiments and mass spectrometry. In: Bonifacino JS (ed) Current Protocols in Cell Biology. Wiley, New York
22. Gingras AC, Gstaiger M, Raught B, Aebersold R (2007) Analysis of protein complexes using mass spectrometry. Nat Rev Mol Cell Biol 8:645–654
23. Butland G et al (2005) Interaction network containing conserved and essential protein complexes in Escherichia coli. Nature 433:531–537
24. Gavin AC et al (2006) Proteome survey reveals modularity of the yeast cell machinery. Nature 440:631–636
25. Krogan NJ et al (2006) Global landscape of protein complexes in the yeast Saccharomyces cerevisiae. Nature 440:637–643
26. Roehrl MH, Wang JY, Wagner G (2004) A general framework for development and data analysis of competitive high-throughput screens for small-molecule inhibitors of protein-protein interactions by fluorescence polarization. Biochemistry 43:16056–16066
27. Roehrl MH, Wang JY, Wagner G (2004) Discovery of small-molecule inhibitors of the NFAT–calcineurin interaction by competitive high-throughput fluorescence polarization screening. Biochemistry 43:16067–16075
28. Copeland RA (2003) Mechanistic considerations in high throughput screening. Anal Biochem 320:1–12
29. Fersht A (1999) Structure and mechanism in protein science. Freeman, New York
30. Segel IH (1975) Enzyme kinetics. Wiley, New York
31. Copeland RA (2000) Enzymes: a practical introduction to structure, mechanism and data analysis. Wiley, New York
32. Copeland RA (2005) Evaluation of enzyme inhibitors in drug discovery: a guide for medicinal chemists and pharmacologists. Wiley, New York
33. Yang J, Copeland RA, Lai Z (2009) Defining balanced conditions for inhibitor screening assays that target bisubstrate enzymes. J Biomol Screen 2:111–120
34. Venäläinen JI et al (2004) Slow-binding inhibitors of prolyl oligopeptidase with different

functional groups at the P1 site. Biochem J 382:1003–1008

35. Krippendorff BF, Neuhaus R, Lienau P, Reichel A, Huisinga W (2009) Mechanism-based inhibition: deriving K_I and k_{inact} directly from time-dependent IC_{50} values. J Biomol Screen 8:913–923
36. Morrison JF, Walsh CT (1988) The behavior and significance of slow-binding enzyme inhibitors. Adv Enzymol Relat Areas Mol Biol 61:201–301
37. Mocz G, Helms MK, Jameson DM, Gibbons IR (1998) Probing the nucleotide binding sites of axonemal dynein with the fluorescent nucleotide analogue 2′(3′)-O-(-N-Methylanthraniloyl)-adenosine 5′-triphosphate. Biochemistry 37:9862–9869
38. Yang X (2010) Dynamic ranges of detection-coupled assays and their effect on IC_{50} measurements for inhibition of enzymatic reactions. J Biomol Screen 5:556–561
39. Eccleston JF, Hutchinson JP, White HD (2001) Stopped-flow techniques. In: Harding SE, Chowdhry BZ (eds) Protein-ligand interactions: structure and spectroscopy. Oxford University Press, Oxford

Part II

Quantitation of Thermodynamics and Kinetics

Chapter 4

Isothermal Titration Calorimetry for Studying Protein–Ligand Interactions

Luminita Damian

Abstract

Isothermal titration calorimetry (ITC) is a biophysical technique that allows a thermodynamic characterization of an interactive system. It is a free in solution technique that requires no labeling, using heat as signal. ITC allows simultaneous determination of affinity K_a, stoichiometry n, enthalpy change ΔH and calculation of free energy change ΔG and entropy change ΔS in one single experiment. It is the only technique that allows direct enthalpy change measurement. By accessing the enthalpy change, we get a step closer in estimating the driving forces that characterize the interaction of a protein with a ligand, information much needed in the drug discovery process.

Key words Isothermal titration calorimetry, ITC, Thermodynamics, Affinity constant, Binding constant, Equilibrium constant, Dissociation constant, Protein interactions, Ligand binding, Enthalpy change, Entropy change, Heat capacity change, Gibbs free energy change, Stoichiometry, Competition experiments, Lead discovery, Lead optimization, Drug discovery, Hydrogen bond, Hydrophobic interactions

1 Introduction

When a protein interacts with a ligand, heat is either released or absorbed. The determination of this heat gives necessary insight into the mechanism of an interaction. ITC is the only technique that measures this heat change and therefore gives detailed thermodynamic information about the binding process. In the drug discovery process the evaluation of the compound affinity alone may not provide a clear indication for compound selection and optimization, as the active compounds may often have similar values for affinity (1). Knowing the thermodynamics of a reaction, such as fundamental parameters like the entropy change and enthalpy change, could provide valuable information for decision making in lead discovery and optimization.

Mark A. Williams and Tina Daviter (eds.), *Protein-Ligand Interactions: Methods and Applications*, Methods in Molecular Biology, vol. 1008, DOI 10.1007/978-1-62703-398-5_4, © Springer Science+Business Media New York 2013

1.1 Thermodynamics

Let us consider the following equilibrium reaction:

$$P + L \rightleftharpoons PL$$

where P is a protein and L is a small molecule (called ligand).

The change in Gibbs free energy (ΔG), under arbitrary conditions, for PL formation is related to standard Gibbs free energy change (ΔG^0) under standard conditions (e.g. 1 mol of P and 1 mol of L at pH 7 and 25 °C), by the following equation (2, 3):

$$\Delta G = \Delta G^0 + RT \ln \frac{[PL]}{[P][L]} \quad (1)$$

At equilibrium, under standard conditions, where $\Delta G = 0$, Eq. 1 becomes:

$$\Delta G^0 = -RT \ln \frac{[PL]}{[P][L]} = -RT \ln K_a = RT \ln K_d \quad (2)$$

with R = gas constant = 1.98 cal/mol/K, T = experimental temperature expressed in Kelvin, and K_a (affinity constant) = K_b (binding constant) = K_{eq} = (equilibrium constant) = $1/K_d$ (dissociation constant);

For a reaction to occur spontaneously ΔG should be negative (2). A negative ΔG indicates the equilibrium shifts toward product formation.

Consider the Gibbs Helmholtz equation which indicates that ΔG consists of two terms: enthalpy and entropy (2):

$$\Delta G = \Delta H - T\Delta S \quad (3)$$

where ΔH = enthalpy change = heat change at constant pressure, and ΔS = entropy change = heat as a measure of the molecular disorder of a system (2).

Both the enthalpy change and the entropy change are dependent on the heat capacity change based on the following equations:

$$\Delta H = \int_{T_1}^{T_2} \Delta C_p \mathrm{d}T \quad (4)$$

$$\Delta S = \int_{T_1}^{T_2} \frac{\Delta C_p}{T} \mathrm{d}T \quad (5)$$

The units of all parameters presented above are listed in **Note 1**.

1.2 Parameters Determined by ITC (4)

n = Stoichiometry

The stoichiometry gives information about the number of protein binding sites. It also reflects the purity and the functional integrity of a protein preparation if the measured and fitted stoichiometry can be compared to the known, previously determined stoichiometry (5).

K_a = Association constant

ITC allows direct measurement of dissociation constants from the mM to the nM range with the possibility of extending this range to pM by performing competition experiments. Weak interactions characterized by dissociation constants in the mM range require high concentrations of protein and ligand, while tight interactions characterized by nM to pM range require smaller concentrations.

ΔH^0 = Standard enthalpy change

ITC measures an overall heat change representing the heat contribution of several phenomena. To determine the enthalpy change ΔH^0 of the interaction itself, contributions to the overall heat from ligand dilution, from mixing of mismatched buffers and from the ionization of buffer should be determined and either subtracted or the experimental conditions changed.

The main contribution to the observed enthalpy changes arises mainly as a result of changes in hydrogen bonding interactions (3).

ΔG^0 = Standard free energy change

ΔG^0 is calculated from Eq. 2 based on the experimentally determined K_a.

A more negative ΔG implies tighter binding.

ΔS^0 = Standard entropy change

It is the heat correlated to the degree of disorder of a system. A favorable positive entropy value indicates a higher degree of disorder. Favorable entropy values are often associated with the release of water molecules from a binding interface, whereas unfavorable, negative entropy values are often linked to conformational restrictions (3).

ΔC_p = Heat capacity change

This is calculated using the following equation:

$$\Delta C_p = \frac{\Delta H_1 - \Delta H_2}{T_2 - T_1} \tag{6}$$

A thorough study implies measuring ΔH at multiple temperatures and plotting ΔH vs. T (*see* **Note 2**). The slope of the resulting line will give an accurate ΔC_p. In the majority of cases ΔC_p is negative (6). Negative (often large) values of ΔC_p, coupled with the favorable entropy changes, have been used as an indicator of hydrophobic interactions (3). Large positive heat capacity change can be attributed to extensive additional hydration, and partially to the burial of the polar groups of the interacting molecules (7).

1.3 Instrumentation

The ITC instruments currently available on the market are using the power compensation technique as detection method. The performance of a commercial ITC instrument is well described by Wiseman et al. (4) and detailed below. A reference cell and a sample cell are placed in an adiabatic jacket. In a typical experiment, the reference cell is filled with buffer (or water) and the sample cell with the protein under investigation. Both cells are heated in such a way that the temperature is almost constant, i.e. $\Delta T < 10^{-6}$ °C at all times.

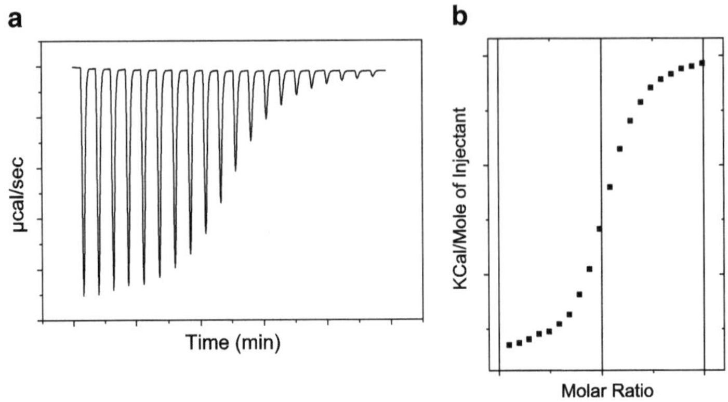

Fig. 1 (**a**) Raw ITC data showing an exothermic binding reaction. Each peak corresponds to an injection of ligand into a protein solution in the ITC cell and the area under the peak is proportional to the amount of heat released in the binding reaction. At the beginning of the titration as there is an excess of protein in the cell, all injected ligand binds and high heats are detected. When the protein becomes saturated with added ligand, the heat signal diminishes until only the background heat of dilution is observed. (**b**) Binding isotherm obtained by integrating the area of each peak. The heat released per mole of injectant is plotted against the molar ratio of the two reactants

The instrument is equipped with a syringe device that holds the ligand. This device titrates the ligand into the sample cell and also acts as a stirrer, so the binding equilibrium can be reached very quickly. During a titration the temperature in the sample cell is changing due to heat released or used in the binding event. Arising temperature differences between the reference and the sample cell are measured and compensated, calibrated to power units and saved. By monitoring the change in power (DP—differential power) applied to the sample cell compared to the reference cell as a function of time, the software generates a plot like that shown in Fig. 1a.

For the data analysis the software creates a baseline and integrates the area under each peak. The peak area is proportional to the heat released or absorbed when the ligand binds to the protein. The experimental heat change from each injection is then normalized per mole of injectant. A typical ITC binding isotherm is obtained by plotting the normalized heat vs. the molar ratio [L]/[P] (*see* Fig. 1b).

From the binding isotherm one can determine:

- n, stoichiometry—molar ratio at the inflection point of the binding isotherm.
- ΔH, enthalpy change—identified as the value at the lower plateau, if the upper plateau is reached at 0 kcal/mol; the reaction is endothermic if $\Delta H > 0$ and exothermic if $\Delta H < 0$.
- K_a, affinity constant—model dependent, cannot be read directly from the graph; the information regarding K_a is extracted from the slope at the inflection point of the binding isotherm.
- ΔG and ΔS—calculated using the Eqs. 2 and 3 above.

2 Materials

2.1 Sample Preparation

The interacting compounds have to be in solution and not involved in any other equilibrium (e.g. they do not self-aggregate or self-associate). The reactants' activity and hence the binding reaction might be pH dependent, so ensure that the experiment is performed at a meaningful pH. Due to instrument sensitivity, any heat associated with events other than binding (e.g. mixing heats due to buffer mismatch) may have a big contribution to the overall heat. In order to minimize the background heat, the protein has to be extensively dialysed and the ligand has to be dissolved in the buffer recovered from the last protein dialysis step.

Protein preparation. The protein solution should be as pure as possible and the concentration should be correctly estimated (*see* **Note 3** and Chapter 2). Presence of impurities or inactive protein will have a direct impact on the stoichiometry.

Ligand preparation. The ligand solution should be as pure as possible and the concentration accurate. Impurities or inaccurate ligand concentrations will have a direct impact on stoichiometry and enthalpy change. We recommend preparing stock solutions by weighing mg amounts using a four digits balance and using the buffer recovered from the last dialysis of the protein. We also recommend calibrating the pipettes monthly when dilution is considered.

2.2 Buffers

ITC is a free in-solution technique and in principle any buffer can be used to study protein–ligand interactions. However, if the interaction involves protonation, the intrinsic enthalpy of binding can only be determined by measurement of ΔH in buffers of different ionization enthalpy (*see* Subheading 3.4). Alternatively, buffers with low ionization enthalpy values should be considered (i.e. formate, acetate, glycine, phosphate). The observed enthalpy change will be close to the real enthalpy of reaction if the studies are performed in these buffers. The ionization enthalpy values for various buffers have been tabulated (8) or determined by calorimetry (9).

Sometimes, DMSO may be required to solubilise compounds with limited solubility in aqueous buffers. As dilution of organic solvents leads to significant heat effects, identical DMSO concentrations should be used for both protein and compound solutions. Oxidizing agents should be avoided if possible as they produce a significant baseline drift. If mandatory, 1–2 mM TCEP is recommended.

3 Methods

3.1 Experimental Design

In order to get good results from an ITC experiment one should always consider the curve shape captured in the C parameter (*see* Fig. 2):

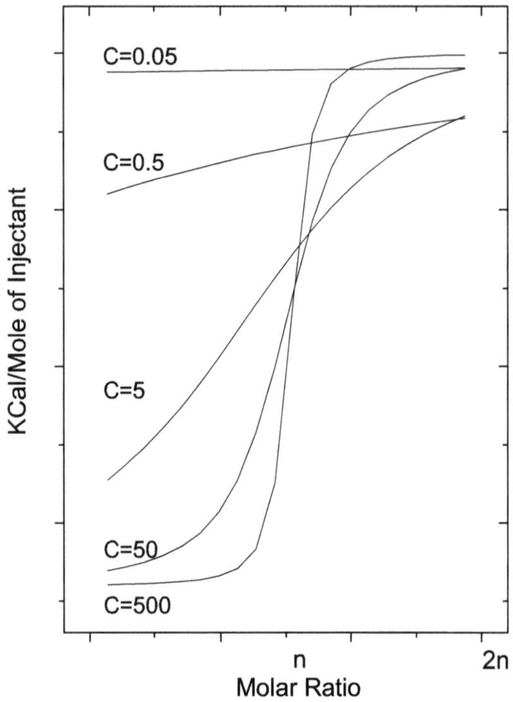

Fig. 2 Simulated binding isotherms for various values of the parameter C

$$C = n\,[\mathrm{P}]\,K_\mathrm{a} \text{ or } n\,[\mathrm{P}]/K_\mathrm{d} \quad (7)$$

The C parameter defines the sigmoidicity of the binding isotherm and depends entirely on these three parameters.

A good binding isotherm has three well-defined characteristics:

- Lower plateau—all injected ligand binds to the protein in the first few injections.
- Transition—should be covered with a good number of points to give a reliable K_a value.
- Upper plateau—protein is saturated so no more injected ligand binds; therefore the last few peaks represent the heats of dilution only.

To obtain good thermodynamic parameters, the reactant concentrations should be calculated considering C values of 5–500 (4).

(a) If K_d is known and $n = 1$, the required concentration for a successful experiment should be:

Cell: $[\mathrm{P}] \geq 5 \times K_\mathrm{d}$

Syringe: $[\mathrm{L}] \geq 10 \times [\mathrm{P}]$

(b) If K_d is unknown, we recommend

Cell: $[\mathrm{P}] = 5\ \mu\mathrm{M}$

Syringe: $[\mathrm{L}] = 50\ \mu\mathrm{M}$

If the resulting data plot is defined by a horizontal line, ($C < 1$), ten times higher reactant concentrations should be used.

For C values below 5, the binding isotherm loses the S shape curve. This type of binding isotherm is specific for weak binding (100 μM–1 mM). An alternative experimental design for this situation is described in **Note 4**.

C values higher than 500, indicate tight binding (K_d lower than 1 nM) and give accurate stoichiometry and ΔH values. Competition experiments should be used to accurately determine dissociation constants in the nM to pM range (*see* **Note 5**).

3.2 Instrument Preparation and Maintenance

For ITC, the cleaning of the cell and of the syringe is crucial (*see* **Note 6**).

If the sample cell is dirty, the instrument baseline may set at an incorrect value and may drift or may present a sudden jump during the titration. To check if the sample cell is clean, perform a water-water run every week or any time the instrument has not been used for weeks. The peaks arising from a water titration should be very small and the DP value should settle at a value within 1 μcal/s of the reference cell power.

Water (+0.02 % azide) can be added in the reference cell when the buffer is an aqueous solution. Buffers containing compounds that may change the heat capacity of the solution (glycerol, organic solvent, salt concentration higher than 150 mM) have to be used as reference instead. If binding is studied in an organic solvent, the reference cell should be filled with the same organic solvent.

Some proteins or small molecules may precipitate in the syringe. If the syringe is dirty or partially blocked, the intensity of the peaks will be irregular. It is essential not to bend the syringe needle; during stirring a bent needle will touch the surface of the cell stem creating extra heats that are sometimes much above the instrument detection limits.

3.3 Experimental Parameters

Temperature. The most common experimental temperatures are 25 °C and 37 °C. However, for sample prone to precipitate, the recommended experimental temperature is 10 °C or lower.

Reference power. A constant power applied to the reference cell during the titration; it has to be set by the user. Because the reaction might be exothermic or endothermic, choose the mid value of the dynamic range of the instrument. For example, for an instrument presenting a dynamic range of 0–12 μcal/s, a reference power of 5 μcal/s is recommended. After instrument equilibration, the expected DP value (differential power that defines the baseline during the titration) should settle at 4–5 μcal/s.

Injection volume. Depending on the instrument, the most common injection volumes are 2 μl or 10 μl.

Spacing between injections. Allow enough time between the injections. The DP value should return to the baseline prior the next injection. The most common spacing is 300 s for older instruments and 150–180 s for the most sensitive ones.

3.4 Control Experiment

The heat observed during an ITC titration is the contribution of several components:

- Dilution effects: heat released/absorbed when the ligand is titrated into the buffer, or the buffer is titrated into the protein. The ligand dilution (the injectant) shows higher heats than the protein dilution. Therefore, always perform a control experiment by titrating ligand into buffer and subtracting these values from the binding experiment, prior to data analysis.
- Background heats due to buffer mixing. Always use matched buffers (same pH, same salt concentration, etc.) in the cell and in the syringe. To minimize the heat due to buffer mismatch we recommend dialysing the protein and dissolving the ligand in the buffer from the final dialysis of the protein. The relevant control experiment would again be to titrate ligand into the dialysis buffer the protein is in.
- Heats of ionization of various buffers, $\Delta H_{\mathrm{buffer}}$. Many binding reactions occur with proton release/absorption by the protein or the ligand. In this case the reaction is pH-dependent and the binding enthalpy is dependent on the ionization enthalpy of the buffer ΔH_{buff} (10). ΔH_{buff} is specific for each buffer and it has been reported in several publications (8, 9). Repeating the experiment at the same pH in buffers with different ionization enthalpy allows to calculate the intrinsic binding enthalpy as well as the number of transferred protons using the equation below (11):

$$\Delta H_{\mathrm{obs}} = \Delta H_{\mathrm{bind}} + n_{\mathrm{H+}} \cdot \Delta H_{\mathrm{buff}} \tag{8}$$

By plotting ΔH_{obs} as a function of ΔH_{buff}, we obtain ΔH_{bind} from the intercept and the number of transferred protons $n_{\mathrm{H+}}$ from the slope (*see* Fig. 3).

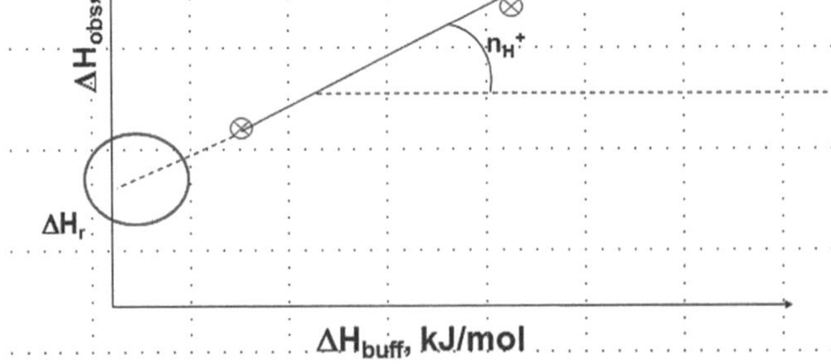

Fig. 3 ΔH_{obs} vs. ΔH_{buff}. Variation of the observed binding enthalpy with various buffers if protonation occurs upon binding

3.5 Data Analysis

Data analysis involves multiple steps such as inspection of the raw ITC data (data quality), subtraction of a control experiment (or a numerical value corresponding to the average heat change from ligand dilution), choice of a model, data fitting, assessment of a good fit, etc. Instrument software greatly helps all these steps and usually an analysis manual will explain steps in detail. In principle, for data analysis the following steps are recommended:

1. Read data: binding and control experiments.
2. Check if the integrated area of the first few peaks is higher than ±1.2 μcal (valid for a single binding event). This value is the sensitivity limit of most sensitive microcalorimeters, but should be checked with the instrument manufacturer. If the area is smaller than 1.2 μcal, the experimental data looks very noisy and the obtained thermodynamic parameters are less accurate. It is recommended to perform another experiment using higher concentrations of reactants. Consider that if there are two or more binding events, the first few peaks might be smaller than 1.2 μcal but this value may increase throughout the titration.
3. Check if the concentrations of the reactants are correct.
4. Subtract the control experiment or the numerical value corresponding to the average heats obtained from the control experiment. As an alternative option subtract a constant value that represents the average heat released/absorbed during the last 4–5 injections if saturation has been reached (the last 4–5 injections have similar heat values).
5. After the subtraction, the upper plateau should be close to 0 kcal/mol.
6. Choose the appropriate fitting model. The instrument software will have several models preprogrammed to choose from. Any others can be entered by the user. The most common binding model is 1:1 binding and it is described by the following equation:

$$Q = \frac{n[\mathrm{P_{tot}}]\Delta H V_0}{2}\left[1 + \frac{[\mathrm{L_{tot}}]}{n[\mathrm{P_{tot}}]} + \frac{1}{nK[\mathrm{P_{tot}}]} - \sqrt{\left(1 + \frac{[\mathrm{L_{tot}}]}{n[\mathrm{P_{tot}}]} + \frac{1}{nK[\mathrm{P_{tot}}]}\right)^2 - \frac{4[\mathrm{L_{tot}}]}{n[\mathrm{P_{tot}}]}}\right] \quad (9)$$

where $[\mathrm{L_{tot}}]$, $[\mathrm{P_{tot}}]$ = total concentration of the ligand (injectant) and of the macromolecule (usually the protein) inside the calorimetric cell after the *i*th injection (independent variables), Q = the heat released/absorbed after the *i*th injection (dependent variable), n (stoichiometry), K (affinity constant) and ΔH (enthalpy change) = fitting parameters, V_0 = cell volume.

The software calculates the total heat evolved after each injection. The parameter of interest for comparison with experiment is the *change* in heat content $\Delta Q(i)$ from the completion of the $i - 1$th injection $Q(i - 1)$ to completion of the ith injection $Q(i)$. Any time we inject a volume of ligand into the cell, an identical volume of the ligand–protein complex is expelled from the cell (dV_i). Therefore a correction factor corresponding to the heat contribution of the displaced volume is also considered, within the same equation.

$$\Delta Q(i) = Q(i) + \frac{dV_i}{V_0}\left[\frac{Q(i) + Q(i-1)}{2}\right] - Q(i-1) \quad (10)$$

The fitting involves initial guesses for n, K and ΔH, calculation of $\Delta Q(i)$ for each injection, comparison of these values with the measured heat for the corresponding experimental injection and improvement of initial values of n, K and ΔH by standard Marquardt methods and iteration of the above procedure until there is no further improvement in fit. To assess the correctness of the model, one would look at the fitting errors which have to be within 15 % of the actual value, the chi-square value should be minimized ("non reduced") and the parameter dependency should not be close to 1.

3.6 Applications Other than Normal Binding

- Protein activity: ITC can be used as a quality control tool to assess the protein activity, which may vary from batch to batch. This can be done by testing each batch of protein against the same ligand solution.
- Dimerization/demicellization: the K_d of a dimer or the critical micellar concentration (CMC) of a micelle together with the enthalpy change of the association event can be determined (12).
- Enzyme kinetics: If no colorimetric assay is accessible to assess the turnover of an enzyme, ITC can be a great alternative as the change in thermal power (DP value) is directly proportional to the reaction rate (13).

3.7 A Test Experiment: Carbonic Anhydrase—CBS Interaction

Note that this example has been performed on a MicroCal iTC_{200} instrument. For other microcalorimeters, the volumes and the experimental parameters may be different; however the concentrations of reactants and the results should be identical.

3.7.1 Source of Materials

PBS buffer from Sigma: cat # P5368.

Carbonic anhydrase (CA) from bovine erythrocyte from Sigma cat # C3934.

CBS from Acros Organics cat # 393170050.

3.7.2 Sample Preparation

Buffer reconstitution: Contents of one pouch is dissolved in 1 L of deionized water and will yield 0.01 M phosphate buffered saline (0.138 M NaCl and 0.0027 mM KCl), pH 7.4, at 25 °C. Add 3 % (v/v) DMSO to the final volume of 1 L.

The CA cat#C3934 is very pure so there is no need for dialysis. Dissolve CA (final concentration 50 μM) and CBS (final concentration 500 μM) into the reconstituted buffer. Check the protein concentration by performing a wavelength scan and using an extinction coefficient at A_{280} of 50070 M^{-1} cm^{-1}(14).

3.7.3 Experimental Setup

If possible, the instrument should be thermostatted at 25 °C to facilitate a faster start. A typical experimental set-up is presented in Table 1.

The control titration for this experiment is not performed here. The system reaches saturation and the small, repeatable peaks at the end of the experiment are a very good representation of control heats. If the heats at the end of the titration are not small, a control experiment by injecting CBS into buffer should be performed.

3.7.4 Sample Loading

1. Rinse the sample cell with buffer; remove the buffer as much as possible to minimize the dead volume in the cell. For low-volume microcalorimeters, if the aim is to assess the stoichiometry or to calculate the active protein concentration, the cell should be rinsed with the protein solution; this solution should be removed and the cell should be reloaded with an identical protein solution.
2. Load the cell with 300 μl CA solution and remove excess solution from the overflow reservoir.

Table 1
Experimental parameters for an ITC run using the MicroCal iTC_{200} instrument

Total # injections—20	Volume of first injection—0.4 μl
Cell temperature—25 °C	Duration of first injection—0.8 s
Reference power—5 μcal/s	Volume after first injection—2 μl
Initial delay—60 s	Duration after first injection—4 s
Syringe concentration—10X mM	Injection spacing—150 s
Cell concentration—X mM	Filter period—5 s
Stirring speed—1,000 RPM	Feedback mode/gain—High ITC equilibration options—Fast equil; auto

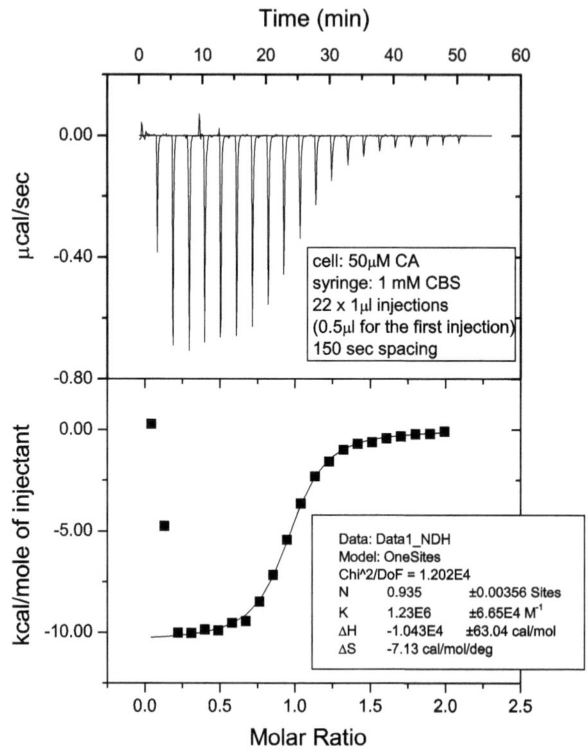

Fig. 4 Raw ITC data and binding isotherm for the interaction of CBS with CA monitored at 25 °C. The raw data shows that the first and second peak are smaller than expected, hence the two outliers in the plot of integrated heats. Their apparent randomness is due to well-known artifacts in early injections (most commonly seen on the first injection only). The data was analyzed disregarding these points as the lower plateau is clearly defined without them. If any of the plateaus are not clearly defined, this may indicate another binding event and would require further investigation. The One Set of Sites model was used for fitting and −100 cal/mol was subtracted from all points as this corresponds to the heat contribution of CBS dilution as seen from the *upper plateau*

3. Load the syringe with CBS, using the wash station module. A volume of ~65 μl CBS is required.
4. When cell reaches thermostatting stage, insert pipette and click start button.

3.7.5 Experimental Results

The results should be similar to the ones shown in Fig. 4. The obtained thermodynamic parameters are: $n = 0.94$, $K_a = 1.23E6M^{-1}$ ($K_d = 0.81$ μM), $\Delta H = -10.4$ kcal/mol. For 1 to 1 binding a stoichiometry n between 0.8 and 1.2 is acceptable (*see* **Note 7**), the reason of a deviation from 1 is inaccurate protein or ligand concentrations. The small positive peak observed on the fourth injection may be an indication of air bubble displacement. The area of the fourth

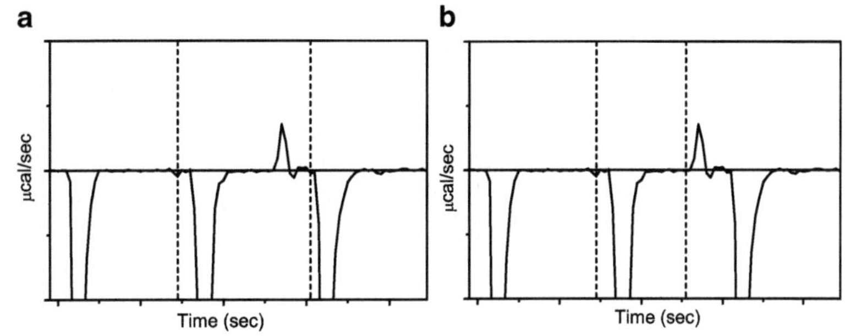

Fig. 5 Peak adjustment. During data analysis, the baseline (*horizontal line*) can be manually adjusted and, as in this example for the central peak in (**a**), the integration range can be reduced by moving the *vertical dotted line* on the *right* of the peak towards the left to any point within the injection period where the system has reached equilibrium (**b**)

peak has been calculated by manually reducing the integration area in order to disregard the positive peak (*see* Fig. 5).

4 Notes

1. *Units.* ΔH = cal/mol, ΔS = cal/mol K, $T\Delta S$ = cal/mol, ΔG = cal/mol, ΔC_p = cal/mol K, $K_a = M^{-1}$, K_d = M.
   Conversion unit 1 cal = 4.184 J
2. *Thermodynamics.* Due to the interdependence of ΔH and TΔS with ΔC_p (2), both terms may change significantly with temperature. Depending on the heat capacity, this change can be more or less significant; e.g. carbonic anhydrase has a ΔC_p close to 0 cal/mol K while TBP (TATA binding protein) has a ΔC_p of −791 cal/mol K (15). Based on this observation, one should consider that there are interactive systems with $\Delta H = 0$ kcal/mol at the chosen experimental temperature. In case the ITC plot shows no binding, although binding is expected and the correct concentrations have been used, one should perform the titration at a different temperature (±5 °C higher or lower) to avoid the T where $\Delta H = 0$ kcal/mol.
3. *Concentration determination.* The most common way to determine the protein concentration is spectrophotometry. A wavelength scan of the protein is required to ensure that the sample does not scatter light, to exclude hypo- or hyperchromicity and to check sample homogeneity. If absorption is detected at 320 nm (scattering), the protein concentration will be overestimated as those particles will also absorb at 280 nm. To determine the concentration accurately a correction needs to be applied (*see* Chapter 2).
4. *Experimental design for low C values.* For dissociation constants weaker than 100 µM, it is very difficult to meet the required

C values (5–500). We can however run ITC tests working at C values lower than 5, as long as the stoichiometry is known. At a low C value the binding isotherm loses the S shape and becomes a hyperbola. In this situation the protein concentration is much below the K_d while the ligand concentration should be high enough to saturate at least 80 % of the protein binding sites (16).

Cell: $[P] < \sim 50-100 \times K_d$

Syringe: $[L] > \sim 10 \times K_d$

For data analysis, the stoichiometry has to be fixed to the known value and the determined parameters are binding affinity and enthalpy change (consequently free energy change and entropy change).

5. *Experimental design for competition experiments* (*displacement experiments*). A competition experiment can be used to detect either weak or tight binding, i.e. dissociation constants in the mM range or in the nM to pM range. The practical approach follows an excellent protocol published by Velasquez-Campoy et al. (17) and it requires:
   - a competitor that binds the protein in the same binding site,
   - that the K_d of the weaker and tighter ligand are different by a factor of 10 or more,
   - that the binding enthalpies of the weaker and tighter ligand are different,
   - that the thermodynamic parameters of the competitor are known. To determine a weak K_d, the thermodynamic parameters of the tighter compound should be determined in a separate experiment. Likewise, to determine tight K_d the thermodynamic parameters of the weaker compound should be determined in a separate experiment. These results need to be highly accurate as they are used to calculate the K_d of interest.

   The experiment is set up in a way that the tighter binder is placed in the syringe, and the cell contains a complex of the protein with the weaker binder being in excess.
6. *Cleaning*. The sample cell should be cleaned with detergent and rinsed with water at the end of each day. However, if a precipitate is noticed at the end of a run, add detergent to the sample cell and heat up the cells at 60 °C for ~1 h. A thorough rinse with distilled water should follow.

   The reference cell does not require special cleaning but it should be rinsed several times with water on a weekly basis.

The syringe should be cleaned with distilled water after each experiment and cleaned with detergent/organic solvent at the end of the day. Sometimes, small molecules may precipitate. In this situation, the syringe should be washed with an organic solvent to solubilise the compound.

7. *Stoichiometry.*

If $n < 1$, then

- The protein concentration is lower than expected or impure or not all correctly folded.
- The ligand concentration is higher than expected.
- The fitting model is inappropriate.
- All of the above.

If $n < 1$, then

- The protein concentration is higher than expected or the protein has multiple binding sites.
- The ligand concentration is lower than expected.
- The fitting model is inappropriate.
- All of the above.

Beware that ΔH value is incorrect if the ligand (injectant) concentration is inaccurate.

References

1. Ladbury JE, Klebe G, Freire E (2010) Adding calorimetric data to decision making in lead discovery: a hot tip. Nat Rev Drug Discov 9:23–27
2. Atkins P, de Paula J (2002) Physical chemistry, 7th edn. Oxford University Press Inc., New York
3. Holdgate GA, Ward WHJ (2005) Measurements of binding thermodynamics in drug discovery. Drug Discov Today 10:1543–1550
4. Wiseman T et al (1989) Rapid measurement of binding constants and heats of binding using a new titration calorimeter. Anal Biochem 179:131–137
5. Ward WHJ, Holdgate GA (2001) Isothermal titration calorimetry in drug discovery. Prog Med Chem 38:309–376
6. Sturtevant JM (1977) Heat capacity and entropy changes in processes involving proteins. Proc Natl Acad Sci U S A 74:2236–2240
7. Niedzwiecka A et al (2002) Positive heat capacity change upon specific binding of translation initiation factor eIF4E to mRNA 5'cap. Biochemistry 41:12140–12148
8. Christensen JJ, Hansen LD, Izatt RM (1976) Handbook of proton ionization heats and related thermodynamic quantities. Wiley, New York
9. Fukada H, Takahashi K (1998) Enthalpy and heat capacity changes for the proton dissociation of various buffer components in 0.1 M potassium chloride. Proteins 33:159–166
10. Leavitt S, Freire E (2001) Direct measurement of protein binding energetics by isothermal titration calorimetry. Curr Opin Struct Biol 11:560–566
11. Jelesarov I, Bosshard HR (1999) Isothermal titration calorimetry and differential scanning calorimetry as complementary tools to investigate the energetics of biomolecular recognition. J Mol Recognit 12:3–18
12. Garidel P et al Understanding the self organisation of Association Colloids: Microcal Application Note. http://www.microcal.com/documents/colloidsappnote.pdf
13. Todd MJ, Gomez J (2001) Enzyme kinetics determined using calorimetry: a general assay for enzyme activity? Anal Biochem 296:179–187

14. Myszka DG et al (2003) The ABRF-MIRG'02 study: assembly state, thermodynamic and kinetic analysis of an enzyme/inhibitor interaction. J Biomol Tech 14:247–269
15. O'Brien R et al (1998) The effects of salt on the TATA binding protein-DNA interaction from a hyperthermophilic archaeon. J Mol Biol 279:117–125
16. Turnbull WB, Daranas AH (2003) On the value of c: can low affinity systems be studied by isothermal titration calorimetry? J Am Chem Soc 125:14859–14866
17. Velazquez-Campoy A, Freire E (2006) Isothermal titration calorimetry to determine association constants for high-affinity ligands. Nat Protoc 1:186–191

Chapter 5

Rapid Mixing Kinetic Techniques

Stephen R. Martin and Maria J. Schilstra

Abstract

Almost all of the elementary steps in a biochemical reaction scheme are either unimolecular or bimolecular processes that frequently occur on sub-second, often sub-millisecond, time scales. The traditional approach in kinetic studies is to mix two or more reagents and monitor the changes in concentrations with time. Conventional spectrophotometers cannot generally be used to study reactions that are complete within less than about 20 s, as it takes that amount of time to manually mix the reagents and activate the instrument. Rapid mixing techniques, which generally achieve mixing in less than 2 ms, overcome this limitation. This chapter is concerned with the use of these techniques in the study of reactions which reach equilibrium; the application of these methods to the study of enzyme kinetics is described in several excellent texts (Cornish-Bowden, Fundamentals of enzyme kinetics. Portland Press, 1995; Gutfreund, Kinetics for the life sciences. Receptors, transmitters and catalysis. Cambridge University Press, 1995).

There are various ways to monitor changes in concentration of reactants, intermediates and products after mixing, but the most common way is to use changes in optical signals (absorbance or fluorescence) which often accompany reactions. Although absorbance can sometimes be used, fluorescence is often preferred because of its greater sensitivity, particularly in monitoring conformational changes. Such methods are continuous with good time resolution but they seldom permit the *direct* determination of the concentrations of individual species. Alternatively, samples may be taken from the reaction volume, mixed with a chemical quenching agent to stop the reaction, and their contents assessed by techniques such as HPLC. These methods can directly determine the concentrations of different species, but are discontinuous and have a limited time resolution.

Key words Kinetics, Rate constants, Fluorescence, Stopped-flow, Data analysis, Modeling

1 Introduction

The individual steps in complex biochemical reaction schemes determine how fast systems can respond to incoming signals and adapt to new conditions (1, 2). This chapter is concerned with in vitro techniques that have been developed to study fast reactions in solution. The kinetic information obtained with these techniques is indispensable for understanding the dynamics of biochemical processes, and complements the static structural and thermodynamic information available from X-ray crystallography, NMR, and equilibrium binding studies.

Mark A. Williams and Tina Daviter (eds.), *Protein-Ligand Interactions: Methods and Applications*, Methods in Molecular Biology, vol. 1008, DOI 10.1007/978-1-62703-398-5_5, © Springer Science+Business Media New York 2013

1.1 Techniques

In all rapid mixing, or "flow" techniques, the reactant solutions are driven at high velocity into a special mixing chamber. The mixing and subsequent passage to the point of observation takes a finite amount of time, so that the mixed solution already has a certain "age" before it can be observed. The interval between the start of the mixing and the earliest possible observation time is called the instrument's *dead time*.

In *stopped-flow*, the commonest flow technique, the mixed solution rapidly flows into an observation chamber, where it is stopped and monitored by recording the change in some suitable optical signal as a function of time. The dead time of a stopped-flow instrument is typically 1–2 ms and reactions occurring on a faster time scale cannot be studied. Very slow reactions (>10 s) can be monitored but such studies may be complicated by lamp instabilities.

In *quenched-flow* the reactants are mixed, and flow down an "aging tube" at constant velocity before mixing with a "quenching agent," generally acid, that stops the reaction. The quenched reaction mixture is then analyzed using an appropriate method, such as HPLC. Because the age of the quenched sample is determined by the flow rate and the flow tube volume a series of time points is built up by doing experiments with different flow rates and/or tube volumes. Time points between ~5 ms and ~150 ms can usually be obtained in this way. Quenched-flow methods have the advantage that they can be used when no optical signal is available (3), but are much more labor intensive than stopped-flow methods.

In *continuous-flow* the reactants are mixed and an optical signal is monitored at different positions downstream from the mixer and converted into a time-dependent signal change on the basis of the known flow rate. Continuous-flow has the potential to measure reactions on a much faster time scale than stopped-flow (4) but stopped-flow is generally preferred because of its better sample economy and its ability to measure the kinetics out to longer times.

1.2 Optical Probes

Stopped-flow methods generally require that the reaction being studied is accompanied by a substantial change in an optical signal. This signal should ideally be *intrinsic* to one of the reactants. Several naturally occurring chromophoric cofactors such as NADH and pyridoxal phosphate provide useful optical signals and tryptophan is the major contributor to the optical properties of proteins (*see* **Note 1**). *Extrinsic* probes are generally introduced into proteins by covalent attachment of suitable chromophores (*see* **Note 2**). Ligands may also be modified to produce a suitable signal; for example, many ribose modified derivatives of ATP and GTP have been synthesized to study the kinetic mechanisms of ATPases and GTPases (5). Although the modified ligand is likely to exhibit the same fundamental mode of action it is unwise to assume that its kinetic properties will be the same (6). There are also some applications in which a reaction may be monitored by linking it to a second process, which must be much faster, that provides the necessary optical signal. Several useful protein based biosensors have now been developed for studying rapid reactions; for example, Webb and colleagues have produced sensors for inorganic phosphate and purine nucleoside diphosphates (7).

1.3 Reaction Kinetics and Thermodynamics

In the subsequent text, we use the symbols P and L to indicate "protein" and "ligand"—as many intracellular interactions are between proteins and small molecules—but these may be any two reactants, proteins, DNA, lipids, biomolecular assemblies, etc. PL is used to indicate a complex between P and L, and P*, L*, and PL* indicate different conformational states of these species.

The simplest reversible reaction is one where both the forward and reverse steps are unimolecular processes with first-order rate constants k_{+1} and k_{-1} (units: s^{-1}).

$$P \underset{k_{-1}}{\overset{k_{1}}{\rightleftarrows}} P^* \qquad \text{(Scheme A)}$$

The dimensionless equilibrium constant, K, for this reaction is defined as $K = k_{+1}/k_{-1}$. If the system is subjected to a change which alters the equilibrium constant the concentrations of P and P* will change until the new equilibrium position is established. If the reaction is accompanied by a change in an optical signal, S, this will change with time according to:

$$S(t) = S_{eq} - \left(S_{eq} - S_0\right) \cdot e^{-k_{OBS}t} \tag{1}$$

where S_{eq} and S_0 are the signals at equilibrium and time zero, $(S_{eq} - S_0)$ is the total signal change (or the amplitude) of the reaction, and k_{OBS} is the observed rate for the reaction (*see* **Note 3** and Fig. 1). k_{OBS} for Scheme A is equal to $(k_{-1} + k_{+1})$ and is therefore independent of concentration. Analysis of kinetic traces using Eq. 1 does not therefore give the individual rate constants but if the equilibrium constant K is known then they can be calculated using:

$$k_{-1} = \frac{k_{OBS}}{1+K} \qquad k_{+1} = \frac{K \cdot k_{OBS}}{1+K} \tag{2}$$

Reversible binding reactions, such as those in which a ligand associates with a protein, have a second-order association process, and a first-order dissociation process, and are described by:

$$P + L \underset{k_{-1}}{\overset{k_{1}}{\rightleftarrows}} PL \qquad \text{(Scheme B)}$$

Here k_{+1} is the second-order association rate constant (units: $M^{-1}\ s^{-1}$) and k_{-1} is the first-order dissociation rate constant (units: s^{-1}). The equilibrium *dissociation* constant for this reaction, K_d, is equal to k_{-1}/k_{+1} (units: M), whereas the equilibrium

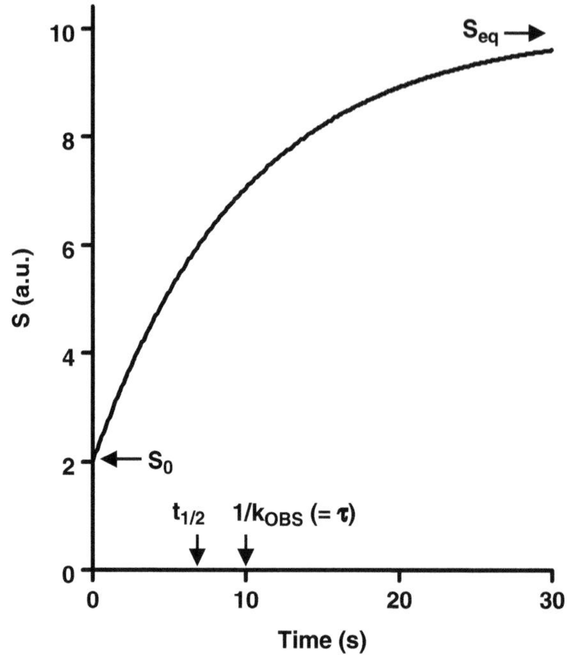

Fig. 1 A single exponential time course. A single exponential time course generated with Eq. 1 using $k_{OBS} = 0.1\ s^{-1}$, $S_0 = 2$, and $S_{eq} = 10$ (*see* text and **Note 3**)

association constant, K_a, is its reciprocal k_{+1}/k_{-1} (units: M^{-1}). There is no simple general analytical solution for the differential rate equation (*see* Subheading 3.5) that describes the change in [PL] with time. However, if one of the reactants is in large excess over the other ($[L_{tot}] \gg [P_{tot}]$ or $[P_{tot}] \gg [L_{tot}]$), the concentration of the component in large excess remains effectively constant during the reaction because $[X_{tot}] - [PL] \approx [X_{tot}]$, where $[X_{tot}]$ is the *total* concentration of the component (P or L) present in excess. The formation of PL is then said to follow pseudo first-order kinetics, and an optical signal will change with time according to Eq. 1 but with the observed rate k_{OBS} now given by:

$$k_{OBS} = k_{+1}[X_{tot}] + k_{-1} \quad (3)$$

Individual rate constants can then be extracted from the dependence of k_{OBS} on $[X_{tot}]$ (*see* Subheading 3.1). More complex schemes will often show more than a single kinetic phase. Nevertheless, under the appropriate conditions the observed time course will be the sum of two, or more, exponentials and analysis requires an appropriately extended version of Eq. 1 with two, or more, k_{OBS} values (*see* Subheading 3.4). The approach, however, remains the same: the experimental transients are analyzed to give k_{OBS} values and the rate constants are determined from the dependence of these k_{OBS} values on concentration(s).

2 Materials

2.1 Instrumentation

Instrumentation for performing rapid kinetic measurements is available from several suppliers: TgK Scientific Ltd. (Supplier of HiTech instruments: http://www.hi-techsci.com/); The KinTek Corporation (http://www.kintek-corp.com/); OLIS, Inc. (http: // olisweb.com/); Applied Photophysics (http://www.photophysics.com/); and Biologic Science Instruments (http://www.bio-logic.info/). The principal detection methods employed are fluorescence and absorbance. Fluorescence detection is widely employed because it is more sensitive than absorption and therefore allows measurements to be made at lower concentrations. Circular dichroism (CD) detection is widely employed in studies of protein unfolding, but the inherently poor signal to noise ratios of CD signals limit its use in the study of protein–ligand interactions. The available devices range from stand-alone models with different detection modes to small, hand-driven devices that can be used in conjunction with regular spectrophotometers. Hand-driven devices are relatively inexpensive and permit the study of reactions with half times as short as 10 ms (depending on the response time of the spectrometer) and are useful for studying many reactions. Most stopped-flow instruments are designed to mix equal volumes of the two reactants, but some will allow different volumes to be used. This particular technique is most widely used in studies of protein folding using chemical denaturants, where large and rapid changes in denaturant concentration are required. Devices that permit double mixing experiments are also available: two reactants are mixed and this mixed solution is then mixed with a third solution after a variable preselected time. This approach allows the study of the reactions of short-lived intermediates. Fluorescence anisotropy measurements require an instrument equipped with a polarizer filter in the excitation path which can be rotated to give either vertically or horizontally polarized light. Measurements are best made in what is known as the "T" format, with two detection photomultipliers equipped with polarizers positioned at right angles to the incident light direction for measurement of the intensity of the emitted light polarized parallel ($I_{//}$) and perpendicular ($I_{\perp}$) to the plane of polarization of the exciting light (*see* **Note 2** and Chapter 7).

2.2 Instrument Settings

As with any scientific instrument the user must understand the characteristics and limitations of the equipment being used. In the case of stopped-flow it is useful, and instructive, to demonstrate that mixing is efficient and to determine the dead time of the instrument. Detailed methods for doing this and for performing temporal calibration of a quenched-flow instrument have been given elsewhere (8, 9). The first step in designing an experiment is to choose the appropriate instrument settings and this is now discussed, with particular emphasis on fluorescence methods. Selection of the appropriate conditions is *always* facilitated by a steady-state investigation of the fluorescence changes using a conventional spectrophotometer.

2.2.1 Detection of Emission

Emitted fluorescence is generally detected by a photomultiplier after passage through cutoff and/or band-pass filters selected to pass fluorescence whilst excluding any scattered exciting light (*see* **Note 4**). It is important that the filters selected maximize the signal *change* relative to the *total* signal. Fluorescence anisotropy measurements require two photomultipliers, which will respond differently to the parallel and perpendicular light and must first be normalized. This is done by exciting the fluorophore with horizontally polarized light and adjusting the high voltage on each photomultiplier so that they give the same output signal.

2.2.2 Lamp Selection

Fluorescence measurements generally require high intensity light sources. Xenon arc lamps have a relatively smooth emission spectrum and are also ideal for collection of steady-state spectra. Mercury or xenon/mercury lamps have several intense emission bands, which can sometimes be used to advantage when doing time-based acquisition at a single wavelength. Emission from deuterium or quartz halide lamps is significantly less intense but is also less noisy and these lamps can be used in absorbance measurements and for fluorescence excitation in the visible region when the chromophore has intense fluorescence.

2.2.3 Slit Widths

A large slit width can be used to increase the light intensity for fluorophores with a large Stokes shift (the wavelength difference between the excitation and emission maxima). If the Stokes shift is small then the excitation slit width may need to be reduced to exclude scattered light from the photomultiplier. Alternatively, the wavelength of the exciting light may be set to a shorter wavelength than the excitation maximum.

2.2.4 Time Constant

The signal to noise ratio (S/N) in rapid kinetic measurements is proportional to the square root of the instrumental time constant. The time constant suppresses noise by determining how fast the instrument can respond to a changing signal (with ~95 % of any signal change taking three time constants) and must therefore be selected to be <10 % of the half time of the fastest process being observed (*see* **Note 3**). This gives the biggest noise reduction without affecting the measured rate. The S/N ratio can also be improved by averaging several individual records, or, with some instruments, by collecting data at very high sampling rates and averaging appropriate blocks of data to give the individual time points.

Once the operating conditions have been chosen the instrument should be set up to record the reaction. This should be done according to the manufacturer's instructions. In most cases the data may be collected using linear time scales (*see* **Note 5**). Analysis of kinetic transients by fitting one or more exponential terms to the curves obtained is often straightforward and the software supplied with commercially available equipment is generally adequate (*see* Subheading 3.5).

3 Methods

3.1 Simple Bimolecular Reactions

Many protein–ligand interactions are simple, reversible second-order reactions (Scheme B) and we focus first on the determination of the rate constants for such processes. Stopped-flow kinetic studies of reactions of this type are generally performed under pseudo first-order conditions with the concentration of one of the reagents (either P or L) at least tenfold higher than the other (*see* Subheading 1.3). The choice of the component to be used in excess may be dictated by availability of material; if this is not the case it should be selected so that the ratio of signal change to total background signal is maximized (*see* **Note 6**).

Initial measurements are generally made with $[X_{tot}]$ in tenfold excess. If the reaction is too fast (complete within the dead time) the concentrations will need to be reduced. The lowest usable value of $[X_{tot}]$ will be that which still maintains pseudo first-order conditions but is high enough to give good S/N. If the reaction is slow then the concentrations can be increased if this improves S/N. When a suitable signal change has been obtained it should be examined over a wide range of time scales in order to confirm that there is only a single exponential process (*see* Fig. 2). Slow changes sometimes observed in fluorescence measurements can be caused by photobleaching of the fluorophore. This is not usually a problem on time scales of <1 s and can be quantified by mixing the fluorophore with buffer and recording any decrease in fluorescence intensity. If it is a problem it can be minimized by reducing the excitation slit width.

Studies of simple bimolecular reactions performed under pseudo first-order conditions should yield single exponential transients which can be analyzed using Eq. 1 to give the observed rate (k_{OBS}). Equation 1 can also be used to analyze fluorescence anisotropy data provided that there is no change in fluorescence intensity accompanying the reaction (*see* **Note 7**). It should generally be the case that averaging five or more transients will be sufficient to give a reliable estimate of k_{OBS}. If inconsistent results are obtained from push to push it is advisable to check carefully for artifacts (*see* **Note 8**).

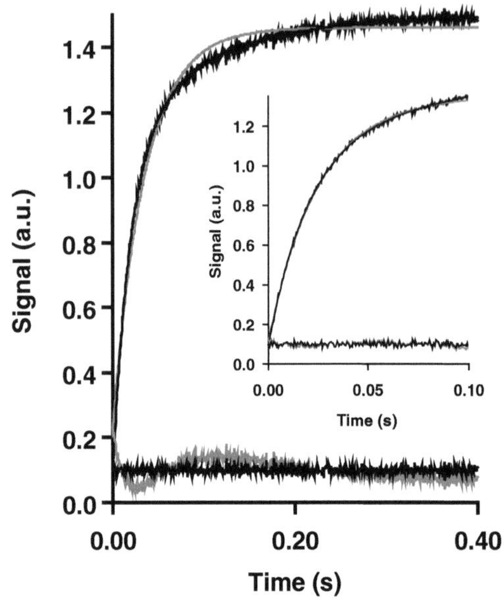

Fig. 2 Simulated double exponential time courses. The ability to detect deviations from single exponential behavior depends, of course, on the relative rates and relative amplitudes of the two phases *and* on the time range over which the transient is recorded. These transients were simulated over different time ranges with rates (and amplitudes) of 50 s^{-1} (amplitude = 1) and 10 s^{-1} (amplitude = 0.4). When analyzed over 10 ms (*Inset*) the one exponential (*grey line*) and two exponential (*black line*) fits are almost indistinguishable. When analyzed over a 40 ms time scale (*main panel*) it is clearly evident that the single exponential fit is inadequate. In both panels the residuals (observed signal minus fitted signal) are shown as *grey* and *black lines* for the one and two exponential fits respectively

The study should then be extended to higher values of $[X_{tot}]$. Although the largest possible concentration range should be studied this will not always be possible. The largest usable value of $[X_{tot}]$ will depend upon the values of the individual rate constants. Thus, for example, with an association rate constant (k_{+1}) of the order of 10^7 M^{-1} s^{-1} and an instrument dead time of 2 ms, the largest value would be of the order of 50 μM ($k_{OBS} = (k_{+1}[X_{tot}] + k_{-1}) > 500$ s^{-1}) but could be significantly lower if k_{-1} is large.

If the reaction conforms to Scheme B a plot of k_{OBS} vs. $[X_{tot}]$ will be linear with slope k_{+1} and intercept k_{-1} (*see* **Note 9**). It is important that the largest possible concentration range be covered in experiments of this type because the demonstration that k_{OBS} varies linearly over an extended concentration range is necessary to confirm that Scheme B is an adequate description of the process. Although the use of a large range of $[X_{tot}]$ should permit accurate determination of both kinetic constants this will not always be the case (*see* Fig. 3). A common problem with high affinity interactions

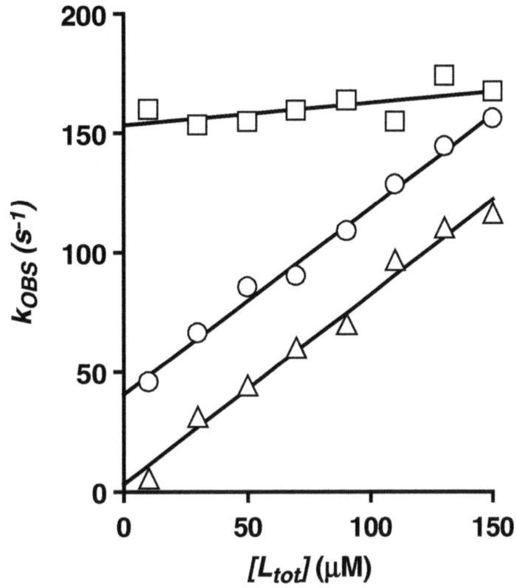

Fig. 3 Simulated plots of k_{OBS} vs. [L_{tot}]. In the ideal case (*circles*: simulated with $k_1 = 8 \times 10^5\ M^{-1}\ s^{-1}$ and $k_{-1} = 40\ s^{-1}$) both k_1 ($7.8 \pm 0.3 \times 10^5\ M^{-1}\ s^{-1}$) and k_{-1} ($40.8 \pm 2.7\ s^{-1}$) are well determined. For a high affinity interaction with small k_{-1} (*triangles*: $k_1 = 8 \times 10^5\ M^{-1}\ s^{-1}$ and $k_{-1} = 2\ s^{-1}$) k_1 is well determined ($7.8 \pm 0.3 \times 10^5\ M^{-1}\ s^{-1}$) but k_{-1} is not ($3.1 \pm 3.6\ s^{-1}$). For a low affinity interaction with high k_{-1} (*squares*: $k_1 = 1 \times 10^5\ M^{-1}\ s^{-1}$ and $k_{-1} = 150\ s^{-1}$) k_{-1} is well determined ($153 \pm 4\ s^{-1}$) but k_{+1} is not ($9.4 \pm 4.7 \times 10^5\ M^{-1}\ s^{-1}$)

is that the value of k_{-1} is too small to be accurately determined (*see* Subheading 3.3). In the case of low affinity interactions, where the dissociation rate constant (k_{-1}) is likely to be large, it may be possible to cover only a limited range of [X_{tot}] before k_{OBS} becomes too fast to measure and the association rate constant (k_{+1}) will not be accurately determined.

If both rate constants can be accurately measured they should be used to calculate a value for the equilibrium dissociation constant (K_d) for the interaction using the relationship $K_d = k_{-1}/k_{+1}$ (*see* **Note 10**). In cases where only one rate constant can be accurately determined the other can be estimated if the equilibrium dissociation constant is known from steady-state studies using $k_{+1} = k_{-1}/K_d$ or $k_{-1} = k_{+1}K_d$ as appropriate. This is, of course, will only be true if the reaction conforms to Scheme B.

3.2 Ternary Complex Formation

There are many cases where a protein binds two different ligands to form a ternary complex. Scheme C is for a protein P interacting with two ligands (X and Y) to form the ternary complex PXY.

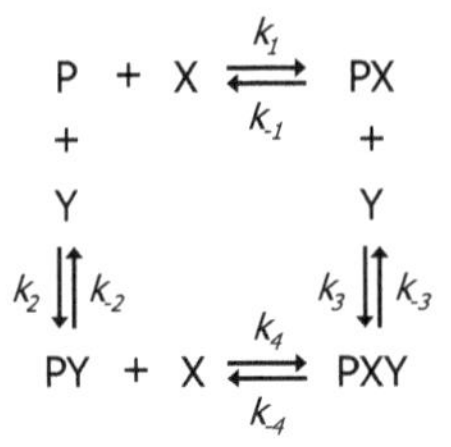

(Scheme C)

Given suitable optical signals it would be possible, at least in principle, to study all four steps of the cycle individually. The interaction of P with X and P with Y could be studied using the methods described in Subheading 3.1 for Scheme B. The interaction of Y with PX would be studied in a similar way by mixing a solution of P and X (at concentrations of X that saturate P) with a solution of Y under pseudo first-order conditions ($[Y_{tot}] \gg [P_{tot}]$), and so on. Only in ideal cases will it be possible to study all four steps individually; nevertheless it is sometimes still possible to derive useful information from the study of a subset of the steps, particularly regarding the possibility of cooperativity between binding sites (*see* **Note 11**).

3.3 Competitive Binding

It is not always possible to determine accurate dissociation rate constants from the intercept of a plot of k_{OBS} vs. $[X_{tot}]$ (*see* Subheading 3.1). Flow methods can be used to measure a more accurate value if a displacement experiment can be performed. For example, a fluorescent ligand (L) can be displaced from its complex with the protein (PL) by adding an excess of a nonfluorescent ligand (N) which is *known* to bind to the same site. For example, fluorescent analogues of ATP can be displaced by mixing with unlabelled ATP. The relevant reactions are shown in Scheme D:

(Scheme D)

The experiment involves mixing a solution of P and L (at concentrations which give a reasonable saturation of P) with a solution of N. If the concentration of N is high enough (so that when L dissociates from P it cannot re-associate before N binds) the observed rate (k_{OBS}) will be equal to k_{-1}. To confirm that this condition is met it is important to measure k_{OBS} using several different values of $[N_{tot}]$. If k_{OBS} *increases* with $[N_{tot}]$ then the condition is not met and the true value of k_{-1} will only be determined if k_{OBS} reaches a limiting (or plateau) value at high $[N_{tot}]$.

Kinetic constants for nonfluorescent ligands can be determined using similar approaches. The *dissociation rate constant* for N (k_{-2} in Scheme D) can be measured using L to induce the displacement of N from PN, although this may be technically difficult if high values of $[L_{tot}]$ are required, as the strong fluorescence will result in poor S/N. Dissociation can also be induced by mixing with an excess of a compound that reacts with N rather than P. For example, the dissociation of Ca^{2+} (N) from calcium-binding proteins such as calmodulin (P) can be studied by mixing with an excess of a fluorescent Ca^{2+} chelator (L) such as Quin 2, which forms a strongly fluorescing, high affinity 1:1 complex with Ca^{2+} (10).

The ***association rate constant*** for binding of N to P (k_{+2} in Scheme D) can also be measured using competition with L. In this case, the experiment involves mixing P with different premixed solutions of N and L. If the dissociation rate constants (k_{-1} and k_{-2}) are small and both L and N are in large excess over P then the observed first-order rate constant will be given by:

$$k_{OBS} = k_{+1}[L_{tot}] + k_{+2}[N_{tot}] \quad (4)$$

The experiment is performed using different values of $[N_{tot}]$ at a fixed value of $[L_{tot}]$. A plot of k_{OBS} vs. $[N_{tot}]$ should give a straight line with slope k_{+2} and intercept $k_{+1}L_{tot}$. More complex behavior will be observed if the dissociation rate constants are not small.

3.4 Multistep Reactions

Although many protein–ligand interactions conform to Scheme B this is by no means always the case. The most obvious indication of additional complexity is the observation of more than a single kinetic phase. When only a single kinetic phase is observed complexity is most often indicated by the observation that the variation in k_{OBS} with concentration is not linear. One of the most commonly encountered complexities is the presence of an additional step involving an isomerization. This can be a first-order isomerization of PL following an initial second-order binding event (Scheme E) or a first-order isomerization of P (or L) followed by a second-order binding event (Scheme F).

$$\mathrm{P} + \mathrm{L} \underset{k_{-1}}{\overset{k_1}{\rightleftharpoons}} \mathrm{PL} \underset{k_{-2}}{\overset{k_2}{\rightleftharpoons}} \mathrm{PL}^* \quad \text{(Scheme E)}$$

$$\mathrm{P} \underset{k_{-1}}{\overset{k_1}{\rightleftharpoons}} \mathrm{P}^* + \mathrm{L} \underset{k_{-2}}{\overset{k_2}{\rightleftharpoons}} \mathrm{PL}^* \quad \text{(Scheme F)}$$

Simple equilibrium binding measurements cannot show that different conformational states exist for P or PL as analysis of binding curves for both of these schemes will always appear to be consistent with Scheme B with the measured dissociation constants given by:

$$K_d = \frac{K_{d1}K_{d2}}{1 + K_{d2}} \quad \text{(Scheme E)} \quad (5)$$

$$K_d = K_{d2}(1 + K_{d1}) \quad \text{(Scheme F)} \quad (6)$$

The individual equilibrium dissociation constants are defined as $K_{d1} = k_{-1}/k_{+1}$ and $K_{d2} = k_{-2}/k_{+2}$ for both schemes.

The experimental approach is essentially the same as for studies of Scheme B and many of the same considerations apply. For example, if L is chosen as the component in excess then pseudo first-order conditions should be maintained ($[L_{tot}] \gg [P_{tot}]$) and the widest possible range of $[L_{tot}]$ should be covered. If the second-order binding step in Scheme E is very much faster than the isomerization step, and L is in large excess over P, then a stopped-flow record will, in ideal cases, have two kinetic phases, with the observed rate for the fast process ($k_{OBS}(F)$) varying linearly with $[L_{tot}]$ and the observed rate for the slow process ($k_{OBS}(S)$) varying hyperbolically with $[L_{tot}]$ (11):

$$k_{OBS}(F) = k_{+1}[L_{tot}] + k_{-1} \tag{7}$$

$$k_{OBS}(S) = \frac{k_{+2}[L_{tot}]}{K_{d1} + [L_{tot}]} + k_{-2} \tag{8}$$

What will actually be observed experimentally will depend on the relative contributions of the different species to the optical signal being monitored, as well as on the values of the individual rate constants (8) (*see* Fig. 4). Only in the most favorable cases where two easily resolvable kinetic events are observed over a wide range of $[L_{tot}]$ values (with the bimolecular step *always* remaining very much faster than the isomerization for all $[L_{tot}]$) will it be possible to extract all four rate constants by analysis of Eqs. 7 and 8 (*see* **Note 12** and Fig. 5). Thus, for example, although inspection of Eq. 8 shows that $k_{OBS}(S)$ should *increase* from k_{-2} when $[L_{tot}] \ll K_{d1}$ to ($k_{-2} + k_{+2}$) when $[L_{tot}] \gg K_{d1}$ this will not always be observable. If K_{d1} is low then given the nature of the typical stopped-flow experiment it is unlikely that it will be possible to work under conditions where $[L_{tot}] \ll K_{d1}$ and $k_{OBS}(S)$ will only vary significantly with $[L_{tot}]$ when $k_{-2} \ll k_{+2}$, so that k_{-2} will generally be difficult to determine. In extreme cases K_{d1} may be so low that $k_{OBS}(S)$ may well be completely independent of $[L_{tot}]$ under all attainable experimental conditions and only the sum of the rate constants for the isomerization step will be measurable.

If, on the other hand, the bimolecular step is the fast diffusion controlled formation of an encounter complex with very low overall affinity (high $K_{d1} \sim 1$ mM) then the value of K_{d2} would need to be 10^{-3} to give a typical overall equilibrium dissociation constant of 1 μM (*see* Eq. 5). A typical stopped-flow experiment could then show only a single transient process because occupancy of the intermediate PL would always be very low. The single observed rate might then vary linearly with $[L_{tot}]$ (since deviations from linearity would only be observed when $[L_{tot}]$ approached K_{d1}, *see* Eq. 8) with an apparent second-order rate constant of k_{+2}/K_{d1} and dissociation rate

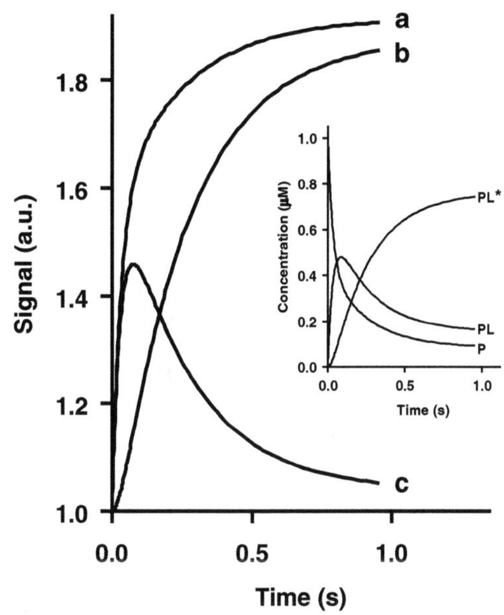

Fig. 4 Simulated double exponential time courses for Scheme E. The changes in concentrations with time (*shown in the inset*) were simulated with $k_1 = 2 \times 10^6 M^{-1} s^{-1}$, $k_{-1} = 10\ s^{-1}$, $k_2 = 5\ s^{-1}$, $k_{-2} = 1\ s^{-1}$, $[P_{tot}] = 1\ \mu M$, and $[L_{tot}] = 10\ \mu M$. Simulated time courses were produced by multiplying these concentrations by molar fluorescence coefficients (F_X) for P, PL, and PL*. *Curve a* ($F_P = 1$, $F_{PL} = 2$, $F_{PL^*} = 2$) is dominated by the fast phase with only a small contribution from the slow phase, *Curve b* ($F_P = 1$, $F_{PL} = 1$, $F_{PL^*} = 2.15$) shows only the slower phase with a small lag, and *Curve c* ($F_P = 1$, $F_{PL} = 2$, $F_{PL^*} = 0.85$) shows both phases

constant k_{-2}. When $[L_{tot}]$ does approach K_{d1}, some curvature may, of course, be observed and the *initial* slope can then be taken as equal to k_{+2}/K_{d1} (12). Association rate constants measured for protein–ligand interactions are, in fact, often significantly lower than the values predicted using theoretical calculations based on diffusion coefficients, shape, and viscosity (2). Scheme E with a high K_{d1} and a low k_{+2} is frequently invoked as an explanation for the observation of these unexpectedly low values (9, 13).

In the case of Scheme F, with the bimolecular step very much faster than the isomerization step, the expressions for experiments performed under the condition that $L_{tot} > (P_{tot} + P^*_{tot})$ are (11):

$$k_{OBS}(F) = k_{+2}[L_{tot}] + k_{-2} \tag{9}$$

$$k_{OBS}(S) = \frac{k_{-1}K_{d2}}{K_{d2} + [L_{tot}]} + k_{+1} \tag{10}$$

Scheme F can, at least in principle, be distinguished from Scheme E by the fact that the observed rate for the slow process should *decrease* from $(k_{-1} + k_{+1})$ when $[L_{tot}] \ll K_{d2}$ to k_{+1} when

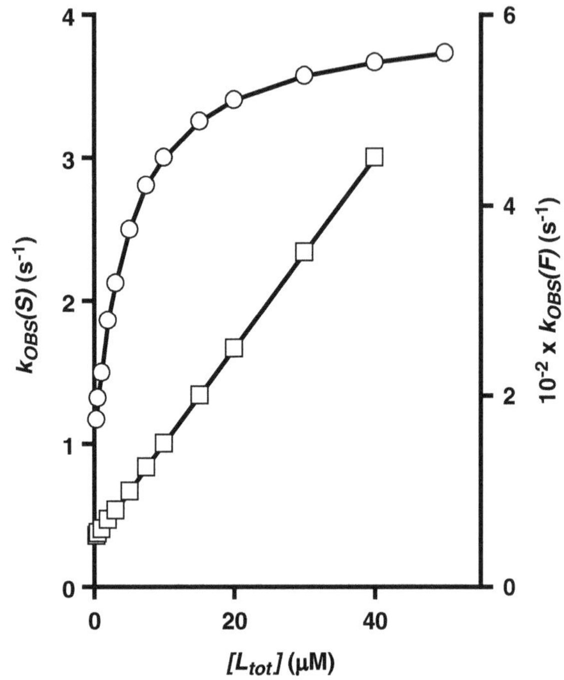

Fig. 5 Variation of $k_{OBS}(S)$ and $k_{OBS}(F)$ with $[L_{tot}]$ for Scheme E. The values were simulated using the equations for Eqs. 7 and 8 with $k_1 = 1 \times 10^7\ M^{-1}\ s^{-1}$, $k_{-1} = 50\ s^{-1}$, $k_2 = 3\ s^{-1}$, $k_{-2} = 1\ s^{-1}$, and $[P_{tot}] = 10$ nM. $k_{OBS}(F)$ (*squares, right axis*) varies linearly with $[L_{tot}]$ with slope and intercept k_{+1} and k_{-1} respectively. $k_{OBS}(S)$ (*circles, left axis*) varies hyperbolically from k_{-2} at low $[L_{tot}]$ to $(k_{-2} + k_{-2})$ at high $[L_{tot}]$. Note: This is very much an ideal case and is unlikely to be achievable in practice

$[L_{tot}] \gg K_{d2}$. However, as for Scheme E, only in the most favorable cases will it be possible to extract all four rate constants for the reaction.

Many multi-step mechanisms will consist of a series of three or more first- and second-order reactions and it is seldom, if ever, possible to derive analytical solutions for a kinetic analysis using flow methods. In this situation the most commonly used approach is to try to study the individual steps in isolation (12, 14, 15).

3.5 Data Analysis and Simulation

In the preceding discussion we have assumed that fitting the time dependence of the observed signal to one or more exponential terms will always be possible. However, the requirement for explicit analytical solutions to the rate equations places severe constraints on the experimental conditions that can be employed, and it will not always be possible to work within these constraints. For example, if it is not possible to work under pseudo first-order conditions, it will be necessary to analyze progress curves using an iterative method based on numerical integration of the appropriate differential rate equations (16).

Global analysis methods allow one to fit multiple kinetic data sets obtained under different concentration conditions (13). The simultaneous analysis of the different data sets has the potential to achieve better definition of the rate constants common to all the sets. In favorable cases it may allow the determination of kinetic constants not obtainable by traditional methods and can be used to distinguish between different kinetic models. Another strong point of global analysis is that the different data sets can be obtained using different methods, e.g., fluorescence intensity and anisotropy data, in which the kinetic constants are nevertheless the same. In such cases it is important to weight the different data sets correctly. This can be done by determining the standard deviation in the signal of a reaction that has reached equilibrium. For example, using the last 5 ms of the transient shown in Fig. 2 would give a good estimate of the standard deviation.

Having extracted rate constants by any of the methods described here it is almost always instructive to simulate the results in order to see how well the data actually fits the assumed mechanism. This is most often done at the level of simulating how k_{OBS} values depend upon the concentrations of the reagents. It can also be very instructive to simulate individual reaction traces. This can be done using any one of several freely available packages (http://sbml.org/SBML_Software_Guide/SBML_Software_Summary) that will simulate changes in concentrations with time. Although many of these methods are very sophisticated the principles are relatively easy to understand and the simplest methods can be implemented in a conventional spreadsheet. For example, Scheme E is described by the following set of coupled ordinary differential equations (ODEs):

$$\frac{d[P]}{dt} = \frac{d[L]}{dt} = -k_1[P][L] + k_{-1}[PL]$$

$$\frac{d[PL]}{dt} = k_1[P][L] - k_{-1}[PL] + k_{-2}[PL^*] - k_2[PL]$$

$$\frac{d[PL^*]}{dt} = k_2[PL] - k_{-2}[PL^*]$$

There is no analytical solution to this set of ODEs. However, if an initial set of concentrations is provided, it is possible to create a *numerical* solution. In the simplest implementation (17), a time step Δt is chosen over which none of the concentrations are expected to change by more than a very small amount. The concentration changes after Δt are calculated by multiplying the expressions for the rate by the time interval. For example:

$$\Delta[PL^*] = [PL^*]_{t+\Delta t} - [PL^*]_t \approx \Delta t(k_2[PL] - k_{-2}[PL^*])$$

where the subscripts t and $t + \Delta t$ indicate current concentration and predicted concentration after the time step Δt, respectively. The new concentrations are then calculated by adding these changes to their (known) current values. This is done for all equations in the set, and the process is repeated until a preset end time is reached (*see* **Note 13**).

Multiplying the calculated concentrations by the appropriate optical constants (such as extinction coefficients) will then generate the theoretical (noise free) transient and what might actually be observed experimentally can be created by the addition of normally distributed random noise to this theoretical curve. In Microsoft Excel, for example, one may do this using the function NORMINV by writing = NORMINV(RAND(),T,SD), where T is the theoretical value and SD is the required standard deviation on this value. Whichever fitting package or program one is using can then be tested to see how well it actually performs under a variety of different conditions.

Finally, computer simulation is also invaluable as a teaching tool and a useful aid in the design of experiments. In our experience, intuitive arguments can frequently be wrong, even in apparently simple situations.

4 Notes

1. Significant changes in tryptophan fluorescence often accompany protein–ligand interactions but this change is often of limited value if the protein contains multiple tryptophans because the signal change accompanying the interaction is often small compared with the total signal. In certain cases it may be possible to use site-directed mutagenesis to replace some of the tryptophan residues and thereby increase the size of the signal change (18). If this procedure is adopted it is, of course, important to check that the substitutions have not altered the structure and/or function of the protein.
2. Comprehensive guides to probe selection and labeling procedures are readily available (http://probes.invitrogen.com/handbook/). Labeling can be difficult if the protein contains more than a single site for the label because it may be difficult to obtain a reproducible product. Even when only a single site is available for labeling this may be far from the binding site for the reaction partner, and not therefore report on the interaction. An alternative approach is to use genetic engineering to create a protein with a single cysteine residue that can then be specifically labeled (7). For all of these approaches, it is essential that the modified protein is fully characterized and that the ratio of probe to protein should be determined. It should also

be demonstrated that the modification does not affect any biological activity of the protein. Finally, equilibrium binding measurements should be performed to determine the affinity of the modified protein for the ligand and this should be compared with that of the native protein. This can be usually done using suitable fluorescence based competition or displacement experiments (19) or by using some of the other biophysical techniques described in this book.

If the fluorescence of the labeled protein does not change upon binding it may still be possible to study the interaction using anisotropy measurements (*see* Chapter 7). In these measurements, the fluorophore is excited with vertically polarized light, and the intensity of the emitted light polarized parallel ($I_{//}$) and perpendicular ($I_{\perp}$) to the plane of polarization of the exciting light is recorded. The total fluorescence intensity is given by ($I_{//} + 2I_{\perp}$) and the anisotropy is calculated as $r = (I_{//} - I_{\perp})/(I_{//} + 2I_{\perp})$. The anisotropy is related to the fluorophore's rotational correlation time (τ_c) by the equation $r = r_o/(1 + \tau/\tau_c)$, where r_o is the limiting anisotropy of the fluorophore and τ is its excited state lifetime. Anisotropy measurements are particularly appropriate in the study of the binding of small fluorescent ligands to large macromolecules because τ_c is related to size and such reactions will therefore be accompanied by large increases in anisotropy. However, because anisotropy can be measured with high precision it is also possible to use this approach using proteins labeled with a fluorophore.

3. The reciprocal of k_{OBS} is called the *relaxation time*, or *time constant*, τ, of the system and is the time taken for the signal to change from S_0 to $(S_{eq} - (S_{eq} - S_0)/e)$. Although k_{OBS} and τ^{-1} are identical, the former is generally used to describe transients observed in flow experiments, whereas the latter is generally used to describe relaxation (or small perturbation) experiments. The *half life* $t_{1/2}$ of the reaction is defined as the time taken for the signal to change from S_0 to $(S_{eq} - (S_{eq} - S_0)/2)$, and is related to the relaxation time and observed rate through $t_{1/2} = 0.693\tau = 0.693/k_{OBS}$ (Note: $\ln(0.5) = -0.693$).
4. Scattered light arises from three sources: Rayleigh scattering of the exciting light (observed at the excitation wavelength λ_{Ex}), Rayleigh scattering of the first harmonic of the exciting light (observed at $2 \times \lambda_{Ex}$), and Raman scattering from the water. The wavelength (in nanometers) for the Raman scattering peak (λ_R) for water depends on the excitation wavelength according to $\lambda_R = \lambda_{Ex}/(1 - 0.00034\lambda_{Ex})$.
5. In more complex systems the observable processes may occur on very different time scales and it is then generally more appropriate to collect data with a logarithmic time base

which allows data to be collected at longer time intervals as the reaction proceeds. Although the time constant will need to be set to be less than the fastest process the data can sometimes be collected in the oversampling mode (collecting and averaging blocks of data) to improve the S/N for long time points.

6. Thus, for example, if the ligand (L) and the complex (PL) are fluorescent, but the protein is not, then the protein should be the component used in excess. This may not be possible in all cases, and the ligand will then have to be the component in excess. In this case it may be advantageous to use resonance energy transfer if a suitable donor/acceptor pair is available with a combination of intrinsic and/or extrinsic fluorophores. For example, the emission spectrum of tryptophan overlaps the excitation spectrum of 2′(3′)-*O*-(*N*-methylanthraniloyl)-adenine nucleotides and this has been taken advantage of in stopped-flow studies of the myosin subfragment 1 ATPase mechanism (6). By exciting the tryptophan at 280 nm and observing the methylanthraniloyl emission, the bound fluorophore is preferentially excited over free fluorophore. This allows much higher concentrations of the excess fluorophore to be used compared to the situation where the methylanthraniloyl is excited directly.

7. If there is a significant change in fluorescence intensity accompanying the reaction then the time-dependent change in anisotropy, $r(t)$, must be analyzed using (13):

$$r(t) = r_{\mathrm{PL}} + \frac{(r_{\mathrm{L}} - r_{\mathrm{PL}})}{1 - D + De^{k_{\mathrm{OBS}}t}}$$

   where r_{L} and r_{PL} are the anisotropies of L and PL and D is the fluorescence intensity of PL divided by that of L.

8. Many artifacts can occur in stopped-flow experiments and some of them can give rise to apparently perfect exponentials. One of the most common problems is the presence of air bubbles in the observation cell and it is therefore advisable to use degassed solutions for all stopped-flow measurements. Inefficient mixing, poor thermal equilibration, and small leaks in the system may all give rise to apparently real transients. Mixing solutions with very different densities may also be problematic. A suitable control experiment will usually identify problems. For example, if the reaction being studied is that of a fluorescently labeled protein with a ligand the control would be to mix the protein solution with the ligand solution, but with the ligand omitted.

9. Whenever possible, it is best to determine the variance in k_{OBS} values for each value of the independent concentration

variable, $[X_{tot}]$. The resulting sample variances may then be used to weight each k_{OBS} value by the inverse of its estimated variance. In some cases it may not be possible to obtain variances for individual samples, and it is then reasonable to assume that the *relative* error in k_{OBS} is constant. The fitting should then be done to the logarithm of k_{OBS}, since the error in $\log(k_{OBS})$ will be constant. This is particularly important in cases where k_{OBS} values vary by more than an order of magnitude.

10. A significant difference between the values may indicate that Scheme B is not an adequate description of the process. The observed variation in reaction amplitude should also be shown to be consistent with an independently measured K_d. The concentration of the protein–ligand complex formed following stopped-flow mixing is, of course, readily calculated from the total concentrations of protein and ligand present *after* mixing and the known K_d using:

$$[\mathrm{PL}] = \frac{([\mathrm{P_{tot}}] + [\mathrm{L_{tot}}] + K_\mathrm{d}) - \sqrt{([\mathrm{P_{tot}}] + [\mathrm{L_{tot}}] + K_\mathrm{d})^2 - 4[\mathrm{P_{tot}}][\mathrm{L_{tot}}]}}{2}$$

11. Reactions such as those shown in Scheme C often exhibit cooperativity in ligand binding. That is, for example, the affinity of X for P may be increased (positive cooperativity) or decreased (negative cooperativity) when Y is also bound. Changes in affinity may be caused by changes in either or both of the rate constants defining the interaction with X. Conservation of free energy for this scheme dictates that $(k_{-1}k_{-3})/(k_1k_3)$ must be equal to $(k_{-2}k_{-4})/(k_2k_4)$.

12. When analyzing rate expressions such as that given in Eq. 8 it is, in general, not good practice to transform them into linear functions, because the associated errors transform accordingly (20). There are now numerous mathematical procedures available for χ^2-minimization of nonlinear functions such as these; for example, the Levenberg–Marquardt procedure is both efficient and relatively robust (21). Fitting using the logarithms of rate and equilibrium constants is advisable because it forces them to be physically meaningful (positive) values.

13. The accumulation process described here is called numerical integration. Selecting smaller time steps will result in smaller relative changes, and in more accurate solutions, but also in an increased total simulation time. If the time steps taken are too large, the solution will not only lose accuracy, but may also become unstable. In an unstable solution the calculated values typically oscillate wildly, with amplitudes that increase with every new time step.

References

1. Cornish-Bowden A (1995) Fundamentals of enzyme kinetics. Portland Press, London, UK
2. Gutfreund H (1995) Kinetics for the life sciences. Receptors, transmitters and catalysis. Cambridge University Press, Cambridge, UK
3. Barman TE, Bellamy SR, Gutfreund H et al (2006) The identification of chemical intermediates in enzyme catalysis by the rapid quench-flow technique. Cell Mol Life Sci 63:2571–2583
4. Shastry MCR, Luck SD, Roder H (1998) A continuous-flow capillary mixer to monitor reactions on the microsecond time scale. Biophys J 74:2714–2721
5. Jameson DM, Eccleston JF (1997) Fluorescent nucleotide analogs: synthesis and applications. Methods Enzymol 278:363–390
6. Woodward SKA, Eccleston JF, Geeves MA (1991) Kinetics of the interaction of 2′(3′)-O-(N-methylanthraniloyl)-ATP with myosin subfragment 1 and actomyosin subfragment 1: characterization of two acto.S1.ADP complexes. Biochemistry 30:422–430
7. Webb MR (2007) Development of fluorescent biosensors for probing the function of motor proteins. Mol Biosyst 3:249–256
8. Eccleston JF, Martin SR, Schilstra MJ (2008) Rapid kinetic techniques. Methods Cell Biol 84:445–477
9. Eccleston JF (1987) Stopped-flow spectrophotometric techniques. In: Harris DA, Bashford CL (eds) Spectrophotometry and spectrofluorimetry: a practical approach. Oxford University Press, Oxford
10. Martin SR, Andersson-Teleman A, Bayley PM et al (1985) Kinetics of calcium dissociation from calmodulin and its tryptic fragments. A Quin 2 stopped-flow fluorescence study reveals a two-domain structure. Eur J Biochem 151:543–550
11. Halford SE (1971) Escherichia coli alkaline phosphatase. An analysis of transient kinetics. Biochem J 125:319–327
12. De La Cruz EM, Ostap EM, Sweeney HL (2001) Kinetic mechanism and regulation of myosin VI. J Biol Chem 276:32373–32381
13. Eccleston JF, Hutchinson JP, White HD (2001) Stopped-flow techniques. In: Harding SE, Chowdhry BZ (eds) Protein-ligand interactions: structure and spectroscopy. Oxford University Press, Oxford
14. De La Cruz EM, Wells AL, Rosenfeld SS et al (1999) The kinetic mechanism of myosin V. Proc Natl Acad Sci USA 96:13726–13731
15. Eccleston JF, Petrovic A, Davis CT et al (2006) The kinetic mechanism of the SufC ATPase: the cleavage step is accelerated by SufB. J Biol Chem 281:8371–8378
16. Kuzmic P (1996) Program DYNAFIT for the analysis of enzyme kinetic data: application to HIV proteinase. Anal Biochem 237:260–273
17. Schilstra MJ, Martin SR, Keating SM (2008) Methods for simulating the dynamics of complex biological processes. Methods Cell Biol 84:807–842
18. Wakelin S, Conibear PB, Wooley RJ (2003) Engineering Dictyostlium discoidum myosin II for the introduction of site-specific fluorescence probes. J Muscle Res Cell Motil 23:673–683
19. Martin SR, Bayley PM (2002) Regulatory implications of a novel mode of interaction of calmodulin with double IQ-motif target sequences from *murine* dilute myosin V. Protein Sci 11:2909–2923
20. Johnson ML (2008) Nonlinear least-squares fitting methods. Methods Cell Biol 84:781–805
21. Press WH, Teukolsky BP, Vetterling WT, Flannery BP (1990) Numerical recipes. The art of scientific computing. Cambridge University Press, Cambridge, UK

Chapter 6

Protein–Ligand Interactions Using SPR Systems

Åsa Frostell, Lena Vinterbäck, and Hans Sjöbom

Abstract

Surface plasmon resonance (SPR) biosensor technology has become an important tool for drug discovery and basic research. SPR instruments are used for a wide variety of applications including determining the binding kinetics and affinity of an interaction, specificity studies, screening, assay development as well as concentration measurements. The interacting molecules may be proteins, peptides, lipids, viruses, nucleic acids, or small organic molecules such as fragments or drug candidates. The ease with which real time information can be obtained has changed many customer workflows in both antibody and small molecule/fragment interaction analysis, from label based and affinity/IC_{50} based workflows towards a label free and kinetic based workflow. This chapter focuses on applications for drug discovery, and outlines the experimental design for screening and selection of small molecules from a focused library. Also, determination of kinetics and/or affinity constants of selected ligands, using established SPR methodology is described, together with potential issues during assay development, running of the assay, and results interpretation.

Key words Biosensor, Surface plasmon resonance, Screening, Ligand interactions, Kinetics, Affinity

1 Introduction

The interest in applying optical biosensor technology, such as Surface plasmon resonance (SPR), in drug discovery is growing (1), driven by the remarkable development of methodology and instrument technology. By measuring changes in refractive index close to a sensor surface these biosensors allow the user to study interactions between immobilized molecules and molecules in solution, in real time and without labeling. Observed binding levels and binding rates can be interpreted in different ways to provide information on the specificity, kinetics and affinity of the interaction, thermodynamics or on the concentration of a molecule.

As a result of the increased interest in SPR for interaction analysis there are now several commercially available SPR systems with different advantages when it comes to sensitivity, throughput, sample requirements, software, ease of use, and overall performance. The applied methodology is partly overlapping for several

Mark A. Williams and Tina Daviter (eds.), *Protein-Ligand Interactions: Methods and Applications*, Methods in Molecular Biology, vol. 1008, DOI 10.1007/978-1-62703-398-5_6, © Springer Science+Business Media New York 2013

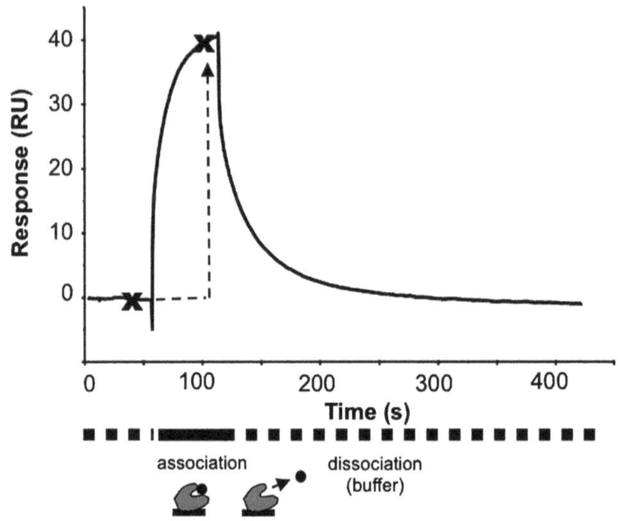

Fig. 1 Schematic illustration of a sensorgram. The bars below the sensorgram curve indicate the solutions that pass over the sensor surface. *Dotted bar* indicates running buffer, while *solid bar* indicates injection of a ligand solution resulting in binding of that ligand to a target immobilized on the chip. The shape of the sensorgram provides information about the interaction event. At time 60 s ligand is injected and starts to bind to the immobilized protein. At 120 s the ligand solution is replaced by running buffer and the ligand starts to dissociate from the target. A report point (X) records the response on a sensorgram at a specific time, typically relative to another report point. For screening and affinity analyses the report point set just before end of injection is used

systems, but our experience is mainly with Biacore™ (GE Healthcare Bio-Sciences AB) which is why the use of these systems is the focus of this chapter.

Setting up an assay with Biacore involves preparing the sensor surface by attachment of the target protein and establishment of suitable assay conditions for the ligand (*see* **Note 1**) interaction. To analyze samples, sample solution is injected over the sensor surface using automated sample handling facilities. All steps in surface preparation and analysis are monitored in a sensorgram, where changes in the molecular concentration at the sensor surface are recorded with time, Fig. 1. Experimental setup and evaluation of the results is supported by dedicated software.

Sensorgrams display the formation and dissociation of complexes over the entire course of an interaction, with the kinetics (association and dissociation rates) revealed by the shape of the binding curve. Binding responses are measured in resonance units (RU) and the response is directly proportional to the concentration of biomolecules on the surface. Report points set at specific times record obtained response levels. If the dissociation rate is rapid, the sample dissociates completely within a short time and the surface can be used directly for the next analysis. For slower dissociation

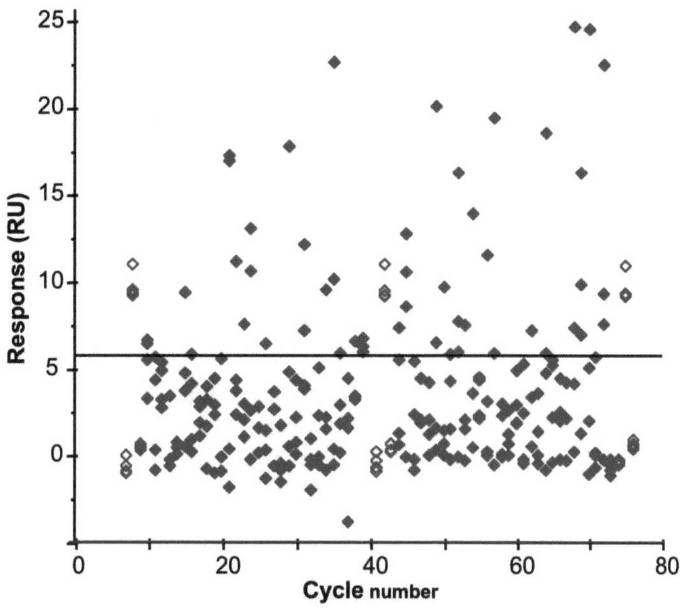

Fig. 2 Screening results from a fragment library (240 fragments with molecular weights 90–300 Da), binding to human thrombin. Screening was performed with Biacore 4000 where four different samples were analyzed in each cycle, against thrombin which was immobilized in all four flow cells. Sample concentration was 1 mM. A cut-off line was drawn based on obtaining a reasonable number of selected binders for further kinetic characterization. Controls are marked as open diamonds

rates, the surface can be washed with an injection of regeneration solution designed to remove bound ligand without affecting the protein on the sensor surface.

While true high throughput screening (HTS) with >10,000 samples is still more amenable to techniques based mainly on fluorescence (see Chapter 13), recent developments make Biacore very suitable for secondary screening of focused libraries as well as fragment screening where libraries typically contain 1,000–5,000 compounds. Figure 2 shows an example of fragment screening results.

This chapter focuses on the experimental design for screening and selection of small molecules (ligands) from a focused library, as well as the kinetic characterization of selected ligands. The protocols for characterization apply to a more basic-research style interaction study in the same way. Guidelines for assay development, running of the assays and results interpretation are given.

A quite different screening approach is the case where a ligand of interest is already at hand, and instead screening of possible target proteins is performed. Immobilization of small molecules or peptides for this approach may impose special challenges depending on the molecule. Some general guidelines for this are also given.

2 Materials

Materials for interaction studies using SPR include the protein(s) and compounds, sensor chips, immobilization reagents, buffer, and solvent correction solution. Depending on the application, regeneration and wash solutions may also be required. Materials are described along with the methods.

2.1 Selecting Buffers

Suitable running buffers for small molecule interaction studies include phosphate buffered saline (PBS) and Tris buffers. Additives such as reducing agents, cofactors necessary for target activity and glycerol may be included. It is highly recommended to include a detergent such as Surfactant P20 (GEHC) at a concentration above the critical micellar concentration (CMC) to stabilize the system; for Surfactant P20 this is typically 0.05 %. If required, dimethyl sulfoxide (DMSO) may be added to a concentration of typically 2–5 % to maintain ligand solubility, provided that the activity of the target protein is not compromised. When Sensor Chip NTA is used 50 μM EDTA should be added to the running buffer to chelate trace amounts of metal ions that otherwise may interfere with protein attachment to the sensor chip.

In general, we have found that reducing the complexity of buffer composition as far as possible often simplifies interpretation of the results. The recommendation therefore is to use the simplest buffer composition that is compatible with the requirements of the interaction. For many protein–ligand interaction studies 20 mM phosphate, 150 mM NaCl, 0.05 % Surfactant P20, 2 % DMSO, pH 7.4 (with cofactors if necessary) is a suitable buffer to try as a start. Tris buffers usually also work well.

3 Methods

Whether your aim is screening of a ligand library, kinetic characterization of ligand interactions or, the other way around, screening of protein targets over a small molecule coated surface, the first step is always preparation of your sensor surface.

3.1 Surface Preparation

3.1.1 Immobilization Level

High immobilization level is preferred for *screening* of ligands to enable detection of low level binders. For *affinity* studies one should also aim for an immobilization level that gives a clearly measurable response over the whole range of ligand concentrations so that steady state response levels can be measured with more confidence. To obtain *kinetic* data the amount of immobilized protein should be kept as low as possible (maximum ligand capacity less than 100 RU) with an acceptably measurable response. Special recommendations apply for immobilization of small molecules (see Subheading 3.1.6).

Since the SPR response is directly proportional to the mass concentration of material at the surface, the ligand binding capacity of a given target protein is related to the amount of protein immobilized as follows:

$$\text{Response}_{\text{max}} = \text{Immobilization level} \cdot \left(\text{MW}_{\text{ligand}}/\text{MW}_{\text{immob.protein}}\right) \cdot \text{Binding stoichiometry}.$$

For example, if the target molecular weight is 30,000 Da and the ligand molecular weight is 300 Da, immobilizing 5,000 RU of protein will give a theoretical ligand binding capacity of 50 RU assuming 1:1 binding and that the protein is 100 % active. In practice, the maximum observed response is also affected by the activity of the protein, the binding kinetics of ligand binding and limitations on the maximum contact time and available ligand concentration. The theoretical binding capacity can however be useful as a guide to how much target molecule to immobilize and also as a reference for assessing the activity of the surface. Measured surface activity is typically 20–80 % of theoretical maximum response.

3.1.2 Immobilization Approaches

The success of an interaction study relies in part on using an efficient and careful method for immobilization, so that the immobilized molecule will keep a high degree of activity for the duration of the experiments.

Various immobilization chemistries can be used, directed towards different functional groups on the protein (2). Here three approaches that often result in successful immobilization of proteins are described; amine coupling of target protein to the dextran matrix of Sensor chip CM5, capture of biotinylated target to a streptavidin chip, and capture of histidine-tagged target to Sensor chip NTA. In Subheading 3.1.6 some guidelines for immobilization of small molecules and peptides are given.

The running buffer described in Subheading 2.1 can often also be used during surface preparation. For immobilization procedures involving amine groups, e.g., amine coupling, running buffers containing amines (e.g., Tris and sodium azide) should be avoided. Although DMSO does not interfere with immobilization it is often omitted in the running buffer during immobilization.

3.1.3 Amine Coupling to a Dextran Matrix

Covalent immobilization of target molecules involves activation of the sensor chip followed by injection of the protein and deactivation of excess reactive groups. The most commonly used approach is amine coupling directed towards primary amine groups on the protein (3), see Fig. 3.

Amine coupling works well for many targets and the binding site for small molecules is often available after immobilization. However, for e.g., some kinases and phosphatases it can be

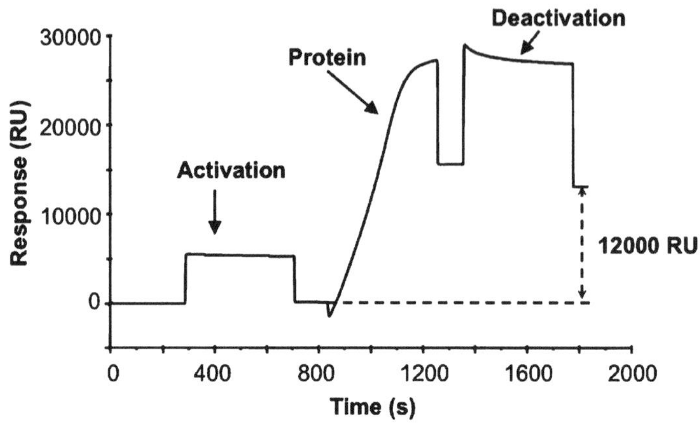

Fig. 3 Sensorgram showing amine coupling of a target protein. The first injection is activation of carboxyl groups on the dextran with a mixture of 1-ethyl-3-(3-dimethylaminopropyl)-carbodiimide (EDC) and *N*-hydroxysuccinimide (NHS). These reagents are mixed automatically in the instrument immediately before injection. Next, protein is injected and in this example binds to a level of ca 12,000 RU as measured from before the first injection to after the last injection. Finally any remaining active groups are deactivated by injection of ethanolamine, which also serves to remove loosely bound material. For low immobilization levels, below a few hundred RUs, the difference in binding between before and after the actual protein injection may better reflect the immobilization level. For very low levels injection of a binding molecule also helps estimate if a relevant immobilization level is obtained

beneficial to stabilize the protein by including a known binder with the target during immobilization (4, 5). The ligand concentration should then be about ten times above the K_D concentration for the known binder, to obtain close to saturation of the binding site.

Efficient immobilization of target molecules to a carboxymethylated dextran matrix, e.g., Sensor Chip CM5 or CM7, is possible from relatively dilute solutions (~10 μg/ml) thanks to *electrostatic preconcentration*. The attraction between the negatively charged dextran matrix on the sensor surface and positively charged protein molecules serves to concentrate the target in the matrix. Electrostatic preconcentration requires that the pH of the immobilization buffer is lower than the *p*I of the protein. In practice, suitable conditions for immobilization are quickly investigated using a procedure called *pH scouting*. This initial task of choosing buffer pH for immobilization on, e.g., Sensor Chip CM5 is a very uncomplicated experiment, and at the same time often very rewarding when starting with a new target protein, as it provides a lot of information in only a few minutes. The shape of the sensorgrams provides information about suitable pH together with indications about suitable protein concentration, immobilization time and occasional non-specific binding.

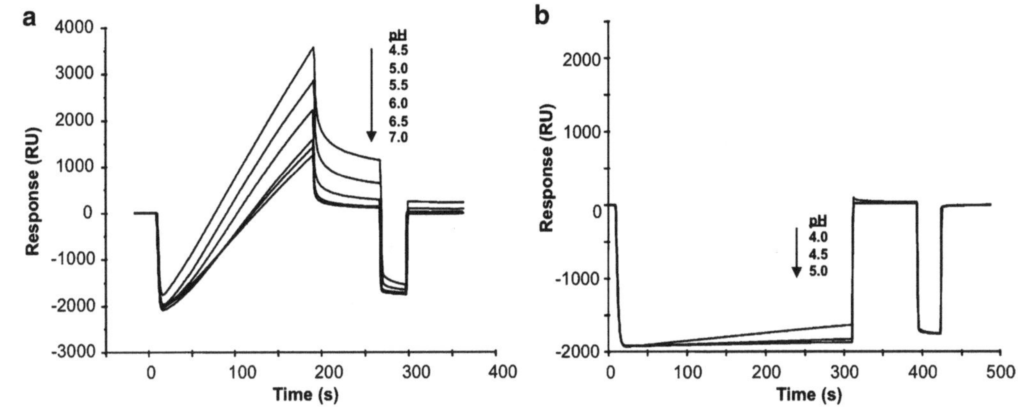

Fig. 4 Overlay plot showing results from two different pH scouting experiments. (**a**) Hemagglutinin from a human influenza subunit vaccine preparation at 5 μg/ml in buffers with indicated pH values, injected during 3 min. (**b**) A protein injected at low concentration, 0.5 μg/ml, to save material. Loosely bound protein is removed when the scouting is complete by washing with high pH (50 mM NaOH), and the surface can then be used for subsequent immobilization of protein

It is advisable to always aim for high activity of the protein and therefore to choose a buffer as mild as possible. If necessary, the time for immobilization can instead be prolonged.

For pH scouting the protein is diluted to a concentration typically in the range 1–10 μg/ml, in each of a series of low ionic strength buffers, usually in the pH range 4–7. 10 mM acetate buffers, pH 4.0–5.5 are commercially available, ready-to-use. Additional buffers are easily prepared in a "good enough" manner: 10 mM maleate, pH 6.0 and 6.5 (0.116 g maleic acid to 100 ml, pH adjusted with NaOH) and 10 mM phosphate, pH 7.0 (0.138 g $NaH_2PO_4 \cdot H_2O$ to 100 ml, pH adjusted with NaOH). The protein solutions are then injected during, e.g., 3 min over a non-activated flow cell. The flow rate is not critical, and low flow rates, e.g., 5 μl/min, may be used to save reagent, any suitable buffer can serve as running buffer. Thus, the protein is not immobilized but merely passes over the dextran matrix and the concentration of protein into the matrix is studied for different pHs. *see* **Note 2** for a few common causes of lack of preconcentration. Figure 4 shows examples of two different pH scouting experiments.

Results in Fig. 4a show that protein at all pH values displays significant preconcentration, i.e., the responses rise several 1,000 RU during 3 min, indicating that it should be possible to obtain an immobilization level of >5,000 RU for a concentration of 5 μg/ml and an immobilization time of > ca 10 min. pH 7.0 was chosen in this case. Buffers at pH 4.5, 5.0 and 5.5 all display some nonspecific binding to the dextran after the end of injection, at about 180 s. The sensorgram does not return to baseline level until after the NaOH injection at ca 270 s, when most of the remaining bound material is washed off. If possible, conditions giving significant non-specific binding should be avoided, especially conditions leading to significant remaining binding after NaOH wash. Such non-covalently bound material may later dissociate from the surface and disturb subsequent analysis. Another immobilization buffer should then be used, or another approach for immobilization should be considered.

Results in Fig. 4b show an example where only a small amount of target protein was available and still pH scouting was valuable in determining conditions for immobilization. pH 4.0 was chosen in this case and concentration was increased for the actual immobilization.

Once suitable immobilization buffer has been established the immobilization procedure simply consists of three injections; injection of activation solution, the protein and deactivation solution, see Fig. 3. Materials for amine coupling include 1-ethyl-3-(3-dimethylaminopropyl)-carbodiimide (EDC) and *N*-hydroxysuccinimide (NHS), (Amine Coupling Kit, GEHC). These are mixed (0.4 M EDC and 0.1 M NHS) in equal proportions just before injection and used for activation of carboxymethylated dextran surfaces such as Sensor chip CM5. The activity of this solution once mixed starts to decrease within minutes. 1 M ethanolamine–HCl is used for deactivation of remaining activated groups.

3.1.4 Biotinylation of the Target Molecule and Binding to Immobilized Streptavidin

The strong binding between biotin and streptavidin allows simple attachment of biotinylated molecules to streptavidin sensor chips. Although the attachment is not strictly speaking covalent, the interaction between biotin and streptavidin is of such high affinity as to be essentially irreversible. Substitution levels of around one biotin residue per protein molecule (i.e., mole equivalent) are recommended for binding to streptavidin on sensor chips (*see* **Note 3**). In general, the conditions recommended for biotinylation with commercial reagents tend to give higher substitution levels, resulting in multi-point attachment of the protein to the streptavidin chip with impairment of assay performance.

Biotinylation and subsequent binding to streptavidin is an attractive immobilization approach for sensitive targets (*see* **Note 4** for a biotinylation example). This is because neutral conditions are used both for the biotinylation step and immobilization, the latter performed by simply injecting the biotinylated target in running buffer over Sensor chip SA. 30 s to a few minutes injection at a low flow rate, e.g., 5 μl/min, is often sufficient. If only a CM5 chip is at hand, an SA chip can be prepared by amine coupling of streptavidin. This typically requires a high concentration (50–100 μg/ml) and long (20–30 min) injection time of streptavidin.

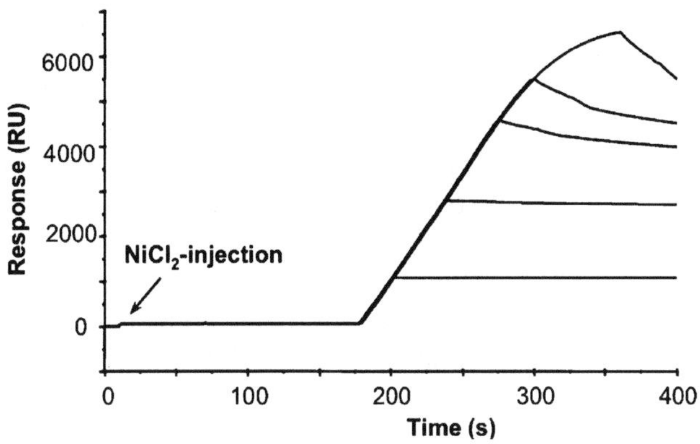

Fig. 5 Overlay plot showing binding of Histidine-tagged protein to Sensor Chip NTA. First the chip is activated by injection of 0.5 mM $NiCl_2$ (*arrow*) then histidine tagged protein is injected using five different injection times, here from 30 to 180 s. For the 30 s injection about 1,000 RU is bound to the chip and at this level binding is stable for this protein

3.1.5 Capture and Immobilization of Histidine-Tagged Proteins to Sensor Chip NTA

Poly-histidine is a widely used recombinant tag and it can chelate with Ni^{2+} ions in complex with nitrilotriacetic acid (NTA), providing a convenient approach for capturing histidine-tagged constructs on Sensor Chip NTA. This capturing approach is very attractive as it provides mild immobilization conditions for sensitive targets. It also provides homogeneous attachment points of the immobilized protein, using neutral conditions. Further, fresh histidine tagged target is used for each interaction as the Ni^{2+} is removed together with the chelated protein between each interaction cycle using a regeneration injection with EDTA. However, the latter makes capturing less attractive for screening of ligands, as more target is needed, but more attractive for kinetic characterization of hits. For *kinetics* Sensor Chip NTA is used to advantage with low capture levels.

A common mistake is to capture too much histidine tagged target; this may result in unstable binding of target to the chip. Low levels on the other hand often provide stable binding thanks to rebinding to the huge number of available NTA sites, thus may be ideal for kinetic characterization. Scouting for stable capture of histidine tagged protein to sensor chip NTA is very quick to perform. Figure 5 shows how histidine tagged tankyrase 1 (TNKS1, 5 μg/ml in running buffer) was captured during 30, 60, 90, 120, 180, or 360 s to the chip.

This protein showed stable binding level by using an injection time of 30 s, resulting in about 1,000 RU stably bound protein. With a molecular weight of 29 kDa this would give a theoretical maximum 1:1 binding level of ca 10 RU for a 300 Da compound, which proved sufficient for characterization of binding events, see Fig. 9. When capture levels are this low some further guidelines for stability may be applied (*see* **Note 5**).

Other approaches to obtain stable capture are to use more than one histidine tag on the protein if possible, or to use a tag with more than six histidines.

For *screening and affinity* analyses, a high level of immobilized oriented histidine-tagged target may be obtained on Sensor Chip NTA by first injecting $NiCl_2$ and then performing amine coupling of the target to carboxyl groups still available on the chip (*see* **Note 6**). Besides obtaining orientation of the target, another advantage compared to using, e.g., Sensor Chip CM5 is that near neutral buffers can be used as the immobilization buffer (often dilution in running buffer works well). Once immobilized, the same protein surface is used for all interactions.

3.1.6 Considerations for Immobilization of Small Molecules or Peptides

Immobilization of small molecules or peptides can be a valuable alternative to immobilizing the target protein, e.g., if the protein is sensitive to the regeneration conditions or if the binding characteristics of the protein are altered by coupling or modification reactions. Small molecules are in general more tolerant to regeneration conditions than the corresponding binding proteins (6). Some assay formats, in particular inhibition assays, involve immobilization of small molecules on the sensor surface. The general principles for immobilizing small molecules and peptides are basically the same as for proteins, but there are, however, some general guidelines that apply to both small molecules and peptides:

- There is a significant risk that immobilization will interfere with the binding properties of the molecule, either sterically or through alteration of the chemical properties. Care must be taken in selecting functional groups for coupling.
- The number and kind of functional groups available on a small molecule for coupling to the sensor surface is usually limited. Suitable groups for immobilization are primary amines, thiols, aldehydes, carboxylic groups and hydroxyls.
- Electrostatic pre-concentration is generally not useful for enhancing the efficiency of immobilization of small molecules or peptides. Coupling is instead performed at relatively high ligand concentrations (typically 1–50 mM) at pH 7–8.5.
- Immobilization of small molecules and peptides generally give rise to high response levels when binding protein. To achieve lower immobilization levels, required for kinetic determinations, ethanolamine, 1–95 mol %, can be added to the coupling solution.
- The amount of small molecule immobilized cannot usually be accurately assessed from the response levels since small molecules give inherently low responses. The immobilization level may therefore be estimated by measuring maximum binding level of binding protein.

Examples of Small Molecule Immobilizations

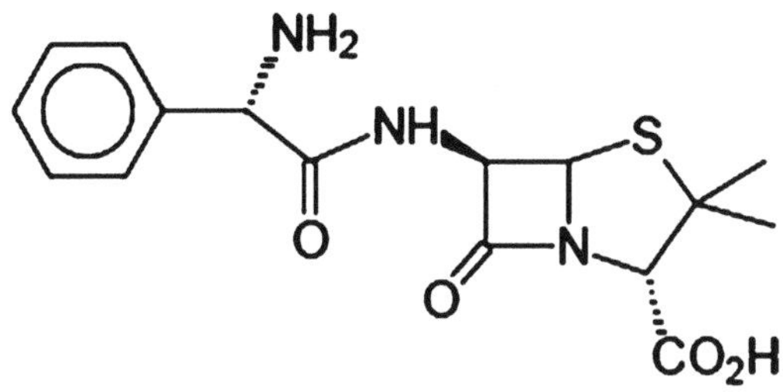

Ampicillin

Ampicillin contains a native amine group that is sufficiently reactive to allow direct immobilization by amine coupling to sensor chip CM5. The chip is first EDC/NHS-activated for 7 min followed by 7 min injection of ampicillin dissolved to 10 mM in 50 mM sodium borate pH 8.5 (coupling is efficient at high pH, and there is no electrostatic pre-concentration effect to motivate the use of low pH buffers). Finally the chip is deactivated for 7 min with 1 M ethanolamine HCl, pH 8.5.

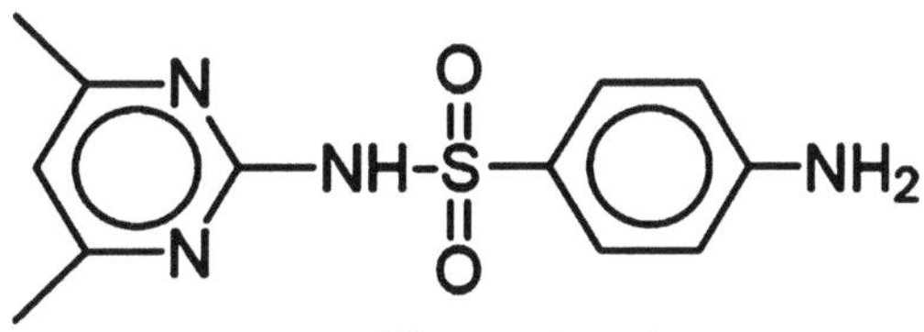

Sulfamethazine

The benzylamine group in sulfamethazine has relatively low reactivity and the compound is only sparingly soluble in aqueous buffers. Successful immobilization on sensor chip CM5 has been achieved with 7 mM sulfamethazine in 10 mM HCl pH 3.0 containing 10 % DMF (7). Immobilization was performed outside the instrument. EDC/NHS activation was performed during 18 min, sulfamethazine coupling during 3 h, and ethanolamine deactivation during 18 min.

In some cases, the molecule may need to be activated to enable immobilization. The example below illustrates activation of hydroxyl groups for immobilization on an amine surface (prepared as described in Note 7).

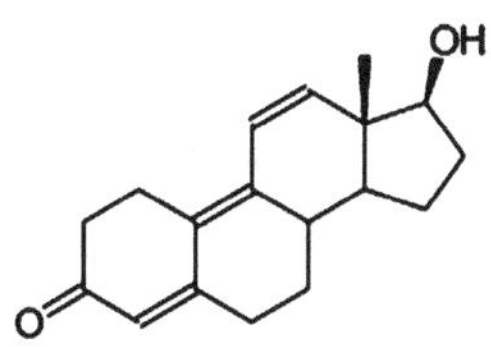

Trenbolone

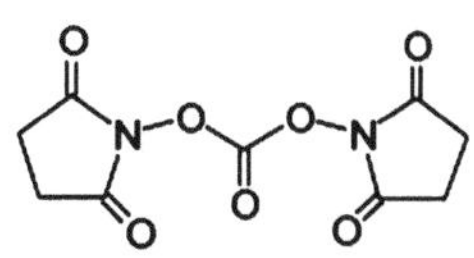

Disuccinimidyl carbonate

Modification of the hydroxyl group on trenbolone with disuccinimidyl carbonate (DSC) introduces a succinimidyl group on to the molecule which can react with amine groups on an amine-modified sensor chip surface. Modification of 10 mM trenbolone is performed outside the instrument by incubation with 40 mM DSC for 30 min in pyridine containing 40 mM 4-dimethylamin opyridine. The product is diluted 1:1 into 50 mM sodium borate buffer, pH 7 before injection for 7 min over the amine surface.

The example below shows how biotin is activated and modified with a diamine to facilitate amine coupling of biotin:

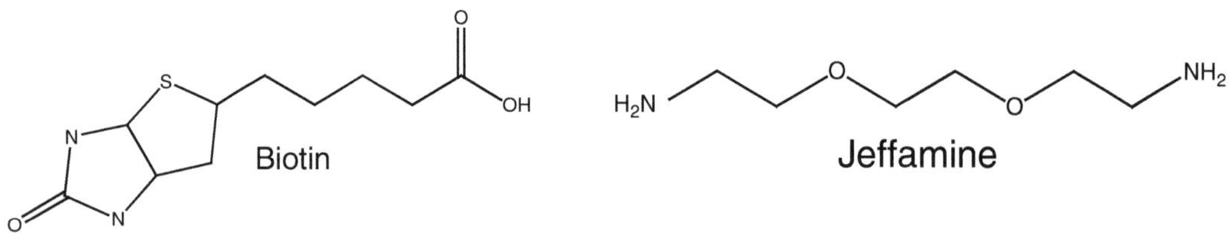

A primary amine can be introduced on to the carboxyl group of biotin by activation with EDC/NHS followed by reaction with a diamine. In this example Jeffamine (1,8-diamino-3,6-dioxaoctane) was used rather than a simple aliphatic diamine, to increase solubility of the resulting derivative. The product can be immobilized using standard amine coupling procedures (8).

Further considerations for immobilizing small molecules may be found in **Note** 7.

Examples of Peptide Immobilizations

Aniline-catalyzed oxime chemistry, through oxidation of N-terminal amino acids with $NaIO_4$, was shown to be a useful tool for achieving site-directed immobilization in ref. 9. The general methods of amine coupling, thiol linkage, and biotinylation are also applicable to peptides (*see* **Note 8** for special considerations).

3.2 Screening of Ligand Libraries

Hit selection (or secondary screening) assays are aimed at identifying lead compounds that bind to the target molecule, and are designed in the first place to provide simple "yes/no" answers from analyses of moderately large numbers of compounds (1). Normally, assays will be run with single or duplicate samples for each compound at a single concentration. The data is obtained as report points (see Fig. 1) and is used to rank compounds in terms of relative binding levels that can be interpreted as yes/no binding answers. Further, the shape of the sensorgram reveals information about binding strength as well as identification of unfavorable binding properties such as super-stoichiometric binding (more than 1:1 binding) of the ligand, precipitation, promiscuous binders and binding to the reference surface, Fig. 6.

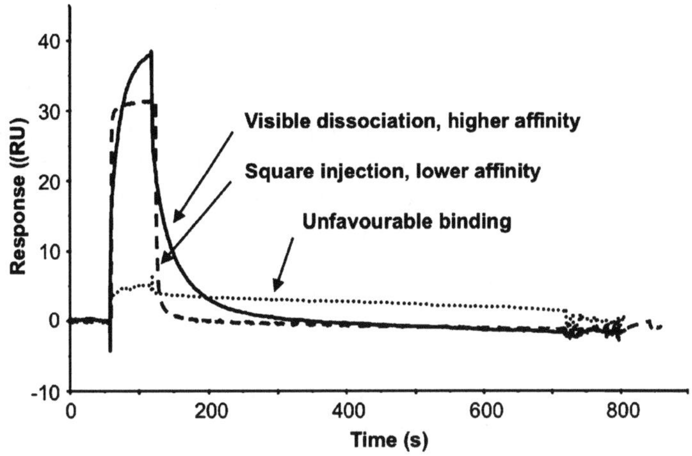

Fig. 6 Overlay plot of three sensorgrams, two displaying typical high and low affinity binding, and one sensorgram exemplifying unfavorable binding. The shape of the sensorgram provides information about the interaction event. Unfavorable binding may be displayed as a variety of different disturbed shapes

3.2.1 Before You Start

A prerequisite for obtaining high quality data is that the instrument has been well maintained using the recommended procedures. This is particularly important when low responses are expected as for small ligand interactions. Some libraries may contain "sticky" compounds that will cause carry over responses to the next cycle. If this is the case, an automatic extra wash injection (as described in the Instrument Manual) with 50 % DMSO in running buffer may be included after each compound injection. The extra wash procedure does not pass over the target molecule, but washes the rest of the flow system. Screening results with observed binding of negative controls could in rare cases also be caused by poor cleaning of laboratory glass ware. A quick rinse of glass ware with 50 mM NaOH followed by MilliQ™ water may help this.

3.2.2 Run Setup

The setup of a method for screening contains steps with the following purposes:

- *Start-up* for stabilizing the response before analyzing samples. Start-up cycles are run exactly as the samples, but with buffer as sample.
- *Solvent correction* (see Subheading 3.2.3), run at the beginning and end of the run and, e.g., every twentieth sample.
- *Samples*, injected during 30–60 s using a flow rate of, e.g., 30 μl/min, an extra wash with 50 % DMSO included if necessary.
- *Control samples*, run at the beginning and end of the run and, e.g., every twentieth sample. Run as the samples. Positive and

negative control samples should always be included when available, to provide a check on the binding activity of the surface. Use buffer as negative controls.

A reference surface is always used for the type of small molecule interactions described in this chapter. The reference serves two functions, allowing correction for bulk response changes that arise from differences in refractive index between samples and buffer, and providing a control for non-specific interaction of the sample with the reference. The choice of reference surface is to some extent determined by the design of the experiment, but most commonly a non-immobilized reference surface is used. In certain cases, immobilization of an appropriate reference protein can be very valuable in providing more stringent screening results (10).

3.2.3 Sample and Buffer Preparation

For sample preparation 1–10 mM stock solutions of compounds in DMSO are often convenient to start with. Dilution of the ligands to a concentration of 30 μM is a suitable starting concentration, if the affinity is unknown. Depending on previous knowledge of the interaction affinity, the concentration may be adjusted accordingly. In general, lower concentrations decrease the risk of artifacts such as super-stoichiometric binding, solubility issues etc. For fragment screening (fragments typically weighing <200 Da) where affinity is expected to be lower, a concentration of 200–1,000 μM is usually needed and thus also higher stock concentrations in DMSO. Which final DMSO concentration to use in the assay is dependent firstly on the DMSO sensitivity of the target protein and secondly on the solubility of the compounds, 2–5 % is typical.

It is important to match samples and running buffer as closely as possible with respect to refractive index, thus to strive for the same DMSO concentration in samples and running buffer. Even very small differences in DMSO concentration contribute significantly to the bulk refractive index (11). The need for solvent correction procedures arises because subtraction of the reference response does not exactly eliminate the contribution of bulk solution to the measured response. This is because bulk solution is excluded from the volume occupied by target molecules in the target flow cell, so the bulk contribution to the response in the target flow cell is smaller than in the reference cell. The difference is small (usually less than 10–20 RU), but it can be of the same order of magnitude as the responses expected from binding of low molecular weight compounds. In practice it is quite difficult even for a trained person to obtain a sufficiently close buffer/samples match for all samples, and a solvent correction procedure can then be applied. The solvent correction procedure simply performs a series of injections of varying DMSO concentration in running buffer; solvent correction is then applied automatically to assay results in the evaluation software, after the assay is completed, *see* **Note 9**.

Table 1
Preparation of DMSO calibration stock solutions and running buffer

	1.5 % DMSO	2.0 % DMSO	2.8 % DMSO
1.02× PBS[a]	9,800 μl	980 ml	9,800 μl
100 % DMSO	150 μl	20 ml	280 μl
Final volume	≈10 ml	1,000 ml[a]	≈10 ml

[a]Used as running buffer

Table 2
Preparation of ready-to-use DMSO calibration solutions

	1	2	3	4	5	6	7	8
DMSO concentration	2.8 %	←					→	1.5 %
1.5 % DMSO	0 μl	200 μl	400 μl	600 μl	800 μl	1,000 μl	1,200 μl	1,400 μl
2.8 % DMSO	1,400 μl	1,200 μl	1,000 μl	800 μl	600 μl	400 μl	200 μl	0 μl

Preparation of running buffer, solvent correction solutions, and samples is here exemplified for ligands diluted to 30 μM in 20 mM PBS buffer (20 mM phosphate, 150 mM NaCl, 0.05 % Surfactant P20, 2 % DMSO, pH 7.4). *See* **Note 10** for some guidelines on DMSO handling.

1. Prepare a buffer stock solution at 1.02 times the final concentration, containing all buffer components and additives except DMSO. It is important to use the same 1.02 stock buffer for samples, solvent correction solutions, and running buffer.
2. Prepare buffer containing 1.5, 2.0, and 2.8 % DMSO according to Table 1.

   The final buffer concentration will be a little higher than 1.0× in the 1.5 % DMSO solution and a little lower in the 2.8 % DMSO solution. This is intentional, since the solvent correction solutions are designed to correct for small errors in preparation of the samples, where the same variations in PBS concentration will occur.
3. Using the above two DMSO solutions, prepare a series of aliquots for the solvent correction curve, according to Table 2.

These DMSO concentrations should when injected cover a range very roughly from −500 RU to +1,000 RU relative to the baseline of the running buffer with 2 % DMSO. Use these solutions for the solvent correction cycles in the run. The solutions should be freshly prepared for each run. Note: The exact DMSO concentrations are not critical. The important factor is that the concentrations cover an adequate range, i.e., the bulk range of the samples.

To prepare 1 ml 30 μM sample solution in buffer with 2 % DMSO from a 10 mM compound stock, use, e.g., 3 μl sample stock, 17 μl DMSO, and 980 μl 1.02× buffer stock.

3.2.4 Regeneration

If the ligand dissociation is sufficiently rapid, regeneration can be achieved simply by waiting for complete dissociation. Although each sample will take a little longer to analyze this is often preferable to spending time on optimizing regeneration, a task that may be rather challenging for small molecule interactions. For kinetics, using the single cycle kinetics (SCK) format may omit the need for regeneration of high affinity binders altogether, see Subheading 3.3.1. Regeneration requires washing with a solution that favors dissociation. For *protein–protein* interactions many protein surfaces (e.g., antibodies) can be regenerated using acidic conditions such as 10 mM glycine-HCl buffers at pH 1.5–3. Other conditions that have proved useful include ethylene glycol at concentrations up to 100 %, 0.5–3 M $MgCl_2$, high pH (10–50 mM NaOH) and high ionic strength. For *regeneration of ligands* from target proteins, however, we have found that low pH rarely works, while high pH, ethylene glycol, or $MgCl_2$ may be more successful. To investigate the effect of a certain regeneration solution it is more reliable to study the repeated binding level of a known binder than to study the changes in baseline. In general, baseline changes should be viewed in relation to the immobilization level and as such are usually secondary in importance in comparison to repeated binding level of a positive control. The flow rate for regeneration is not a critical parameter; 10–30 μl/min is often used, together with an injection time of 30 s.

3.2.5 Example of Screening Results

The screening evaluation itself is based on reference subtracted data obtained as report points. However, further data adjustment is also usually applied, that is solvent correction, molecular weight adjustment, and possibly also surface activity adjustment, see Fig. 2 for a results example. Reference subtraction and solvent correction is applied first for all data, in Biacore systems this is performed automatically by the software. Molecular weight adjustment is necessary for valid comparison of responses from ligands of different molecular weights. Surface activity adjustment compensates for drift in the performance of the assay as revealed by repeated control sample measurements, and should be applied if assay drift is significant.

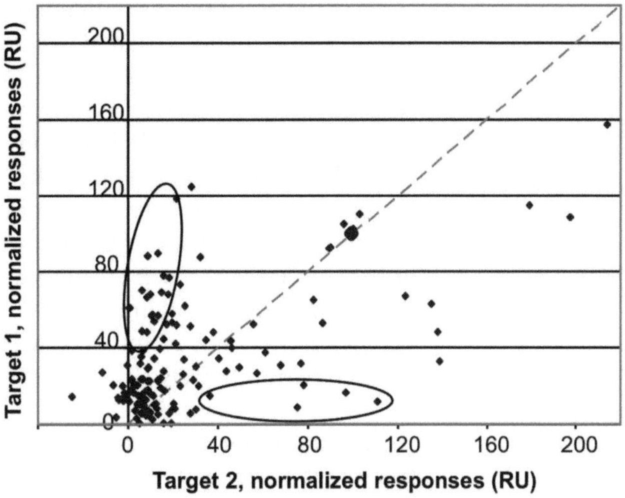

Fig. 7 Compounds screened at a single concentration, 30 μM, in 20 mM PBS, 0.05 % Surfactant P20, 2 % DMSO, pH 7.4, using Biacore T200. The compounds were injected for 60 s simultaneously over two different target proteins. Response levels were plotted against each other for visualization of selective binders. Responses were reference subtracted, solvent corrected and adjusted for differences in molecular weights. Finally data was normalized against a positive control (set to 100, *large diamond*) that was analyzed at intervals during the run. Compounds showing selective binding to respective target (*circled*) were selected

Figure 7 shows another result, this time from screening of a focused ligand library consisting of 180 compounds. The ligands were injected simultaneously over two different target proteins, immobilized in different flow cells. The histidine tagged target proteins were immobilized via capture to Sensor Chip NTA and subsequent amine coupling (5 μg/ml in 20 mM PBS, 0.05 % Surfactant P20, pH 7.4) to 5,000 RU (*see* **Note 6**).

From these results compounds showing selective binding to the respective target were chosen for further kinetic characterization. Compounds displaying visible dissociation according to Fig. 6 were also selected, showing a desired higher stability of binding.

3.3 Kinetics and Affinity Determinations

The experimental principles for determination of binding kinetics and affinity are similar. Both measurements are best approached through determination of binding characteristics over a range of ligand concentrations. The kinetic evaluation procedure determines association and dissociation constants by fitting the experimental data to a 1:1 interaction model between ligand A and target protein B (12):

$$A + B \underset{k_d}{\overset{ka}{\rightleftharpoons}} AB.$$

Kinetic constants are obtained by fitting sensorgram data to a set of rate equations that describe the 1:1 binding mechanism. Affinity constants may be calculated as the ratio of the kinetic constants, $K_D = k_d/k_a$, or determined from steady-state response levels obtained through injection of different ligand concentrations. In practice, these approaches are generally applicable in different situations that are often mutually exclusive: kinetic analysis can be used for sensorgrams that contain kinetic information, but do not necessarily reach steady state. Affinity determination from steady state measurements is most useful for interactions where the kinetics are too fast to measure with confidence but steady state is reached with relatively short injection times.

3.3.1 Kinetics Determinations

In practice to obtain kinetic information immobilization levels need to be low. In general, target immobilization levels that correspond to a maximum binding capacity for the small molecule, R_{max}, of 5–50 RU are suitable for most purposes. As the immobilization level is decreased the binding kinetics is more pronounced and becomes easier to interpret, see Fig. 8.

The range of ligand concentrations for optimal kinetic determination is related to the affinity constant of the interaction. The optimal ligand concentration series for robust kinetics should: (a) give rise to sensorgrams that for the highest concentrations approach the saturation level, (b) display curvature during injection and (c) display a significant drop of the signal during dissociation. For compounds displaying a visible dissociation phase from screening, a wide concentration range of 2.4, 12, 60, 300, and 1,500 nM may be used as a start. It is often advisable to include more than one blank injection, e.g., one blank run before, and one after the concentration series, to be able to choose the best blank for subtraction, in case minor changes to the target occur during the run. Include 1–3 start-up cycles with buffer (run exactly as the ligand samples) first in the run to stabilize the system. 1 min injection time and 10 min dissociation time may be used as a start before the need for regeneration has been explored.

An alternative approach is to inject all concentrations in sequence in the same cycle, so called *single cycle kinetics* (*SCK*) (13). The ligand does not need to dissociate between injections, rather the next concentration can be injected immediately. After the last injection the ligand is allowed to dissociate, see Fig. 9. For high affinity binders like the one in Fig. 9b this format eliminates the need of regeneration, which can be a huge advantage. Blank cycles are run exactly as the ligand cycle with the same number of injections.

The same recommendations apply as for multi-cycle kinetics regarding concentrations. To obtain high quality data it is important to include a blank cycle run exactly as the ligand cycle. Double referencing is then performed during evaluation i.e., reference flow

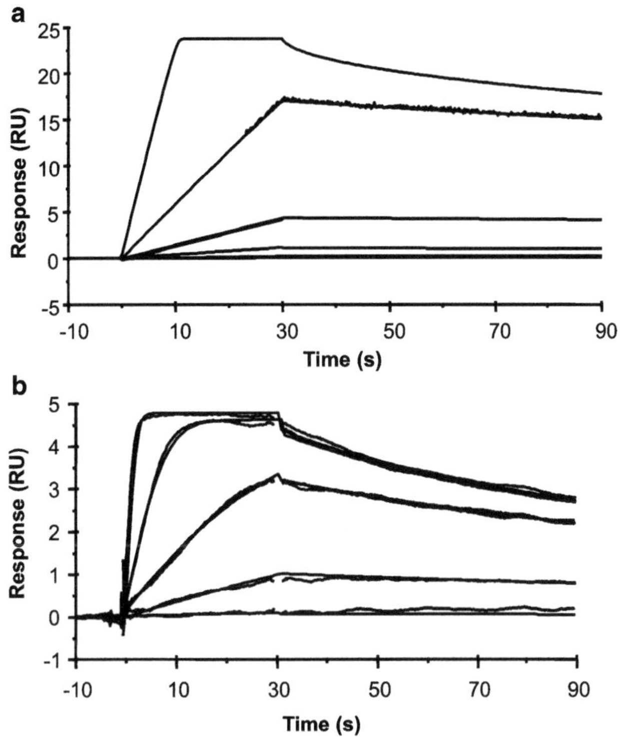

Fig. 8 The compound melagatran interacting with immobilized thrombin, data obtained in Biacore T200. A multi cycle kinetics approach was used where each concentration is injected in separate cycles. In this case the thrombin was regenerated (not shown) using 2 M $MgCl_2$ (**a**) The binding rate for this interaction is highly limited by diffusion of the molecule down to the target protein on the surface (so called *mass transfer limitation*). As soon as a molecule arrives in the matrix it is immediately bound and thus the concentration close to the surface is not the same as the injected concentration. This is seen as sensorgrams displaying mainly straight binding lines up to time 30 s (the association phase), and it is caused by too high immobilization level. These straight line sensorgrams do not contribute with any kinetic data (**b**) The immobilization level has been decreased (note the scale on the *y*-axis), diffusion is no longer limiting the binding rate, the curves during the association phase display more curvature, and kinetic data is obtained

cell data are subtracted and data from blank cycle is also subtracted. Advantages with SCK are that it saves time and the need for regeneration may be omitted altogether. Once a suitable concentration range has been established injection times may be shortened to, e.g., 30 s and for difficult interactions only 2–3 concentrations may still be sufficient to obtain kinetic data.

3.3.2 Affinity Determinations

Most commonly when injecting a small molecule, the sensorgram is displayed as a square injection, lacking a visible association and dissociation phase. The affinity is then in the μM to mM range

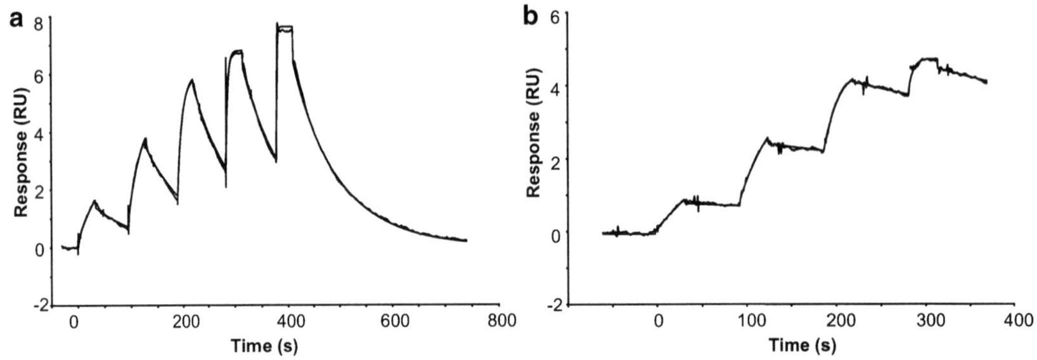

Fig. 9 Two sensorgrams showing single cycle kinetics estimation of a small compound binding to two different tankyrases. About 1,000 RU histidine tagged tankyrase was first captured stably to Sensor Chip NTA. Kinetic evaluation was then performed for the concentration series 0.06, 0.18, 0.56, 1.67, and 5 μM, injected during 30 s and data fitting well to a 1:1 interaction model after reference and blank cycle subtractions. (**a**) The two highest concentrations are visibly approaching the saturation level and the compound dissociates down to baseline after the last injection. K_D was about 70 nM with k_a 1.6 × 10^5 1/ms and k_d 0.012 1/s. (**b**) For this compound data for the highest concentration was omitted, being disturbed. Still, also here saturation is approached and the dissociation time although short is sufficient for an estimate of the dissociation rate: K_D was about 13 nM with k_a 1.1 × 10^5 1/ms and k_d 0.001 1/s. Use of the single cycle kinetic format here omitted the need for regeneration of this stably bound compound

and determination of affinity constants from steady-state response levels can be obtained by measurement over a series of ligand concentrations. The immobilization level for the target is less critical for affinity measurements, since the steady-state response is not affected by mass transport limitations. Higher immobilization levels are in fact preferable for affinity measurements since ligand response values will be higher and steady state response levels can be measured with more confidence.

As for kinetics determinations, the range of ligand concentrations for optimal affinity determination is related to the affinity constant of the interaction. The optimal ligand concentration series for robust affinity should cover a range of about 0.1–10 times the expected K_D concentration, and should give rise to sensorgrams that approach saturation level for the highest concentrations. In practice for low affinities, in particular for fragments, the used concentration is often limited by solubility issues and/or a limited amount of compound. Affinity determinations from concentration series below the K_D concentration can still be made if an R_{max} value (saturation level) can be set during evaluation. The R_{max} level may be estimated theoretically from the immobilization level (including estimated potential activity loss), or by use of a known binder injected to saturating levels. R_{max} is thus locked as a constant and a reliable fit of steady state responses may be found. Figure 10 shows an example of an affinity determination where saturation was approached using the selected concentration series and affinity was estimated through fitting to a steady state model.

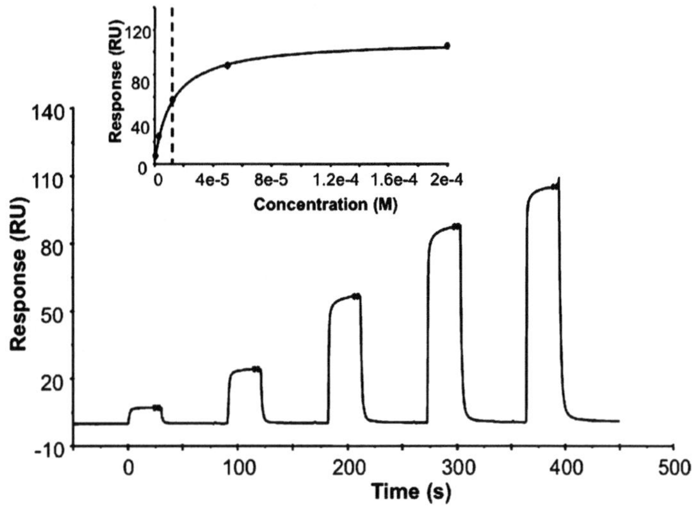

Fig. 10 Sensorgram showing single cycle affinity determination of a small compound binding to 5,000 RU immobilized target protein. Affinity was estimated for the concentration series 0.78, 3.12, 12.5, 50, and 200 μM, data fitting well to the steady state interaction model after reference and blank cycle subtractions. K_D was estimated to 12 μM (*dotted line*). The two highest concentrations are approaching saturation level which is a good sign (best seen in the insert). To confirm the affinity value one should perform several determinations. For this molecule, showing a remaining small upward slope, prolonging the injection time from 30 to 60 s is a good idea to ensure that steady state has been obtained

Finally it is worth considering the purpose of the kinetic/affinity experiment. For many purposes an estimate is sufficient to, e.g., compare different ligands of interest and the hunt for an exact number is overkill. Even just visually comparing, e.g., dissociation rates may give you the information you need about the ligand binding.

4 Notes

1. Much of the Biacore literature, including the product documentation from GE Healthcare, uses the term "ligand" for the interacting partner immobilized and "analyte" for the partner injected in solution. The current chapter adopts the accepted terminology in drug discovery and development work, where "ligand" refers to a small molecule binding to a target. In this sense, the ligand is the molecule injected in solution over a target protein immobilized on the sensor chip.
2. Lack of preconcentration effect may be caused by any of the following: (a) pH is too high or too low, (b) too much salt remains after dilution into the immobilization buffers. As a rule of thumb at least a five to tenfold dilution of a protein, from a stock solution containing physiological salt levels

(0.15 M), is required for the ionic strength to be low enough not to shield electrostatic interactions (exactly how much depends on the protein and the stock buffer content), (c) the protein concentration is too low. However, by zooming in on also very low levels of preconcentration, just visually comparing curves may still make it possible to conclude which pH is more suitable. And a slight increase in concentration and/or a prolonged immobilization time (up to ca 30 min if necessary) often gives adequate immobilization.

3. The commercially available *N*-hydroxysuccinimide (NHS)-based biotinylation reagents can differ regarding water solubility, length and characteristics of the spacer between NHS and biotin, etc. Biotinylation reagents with longer and hydrophilic spacers may result in increased capture levels on the sensor chip and improved protein–ligand availability. We have used the following reagents successfully for biotinylation and binding of active protein to streptavidin: EZ-Link™ (sulfo-)NHS-LC-biotin, EZ-Link™ (sulfo-)NHS-LC-LC-biotin (Thermo Fisher Scientific Inc), and ChromaLink™ Biotin 354S (Solulink, Inc). It is absolutely essential to remove all free biotin and for this PD Spin Trap G-25, Illustra NAP-5 (GEHC), or Protein Desalting Spin Columns (Thermo Scientific) work well.

   To achieve the substitution level distributed around 1, a molar ratio of 1.5:1 between biotin-reagent and protein is recommended and often works well. This ratio minimizes the amount of contamination from free biotinylation reagent and also minimizes the risk of influencing the binding activity of the protein. Recommended protein concentrations are 0.1–1.0 mg/ml in reaction volumes of 50–100 μl suiting a desalting micro-spin column, with a total reaction volume of 100 μl (*see* **Note 4** for a biotinylation example). The minimum amount of protein that can be biotinylated is set by the choice of the separation/desalting technique used for separation of remaining free biotin from the biotinylated protein. If small volume micro-spin columns or micro-dialysis tubes are used, it is possible to biotinylate ~10 μg protein (~100 μg/ml), depending on the protein.

   Some proteins are more difficult to biotinylate than others and some are more sensitive to covalent modification. This can sometimes motivate optimizing the degree of biotinylation. For optimization, different molar ratio of the biotin reagents (in our hands 0.5–5 mol equivalents) may be evaluated by measuring the level of binding of biotinylated protein to immobilized streptavidin.

4. Example of biotinylation of the protein Human Epidermal growth factor Receptor 2 (HER-2), molecular weight 96,800 Da:

(a) 25 μg (2.59×10^{-10} mol) HER-2 in 50 μl 0.1 M sodium-borate buffer pH 8.5 was pipetted to an Eppendorf vial (PBS, pH 7.2–7.8 also works).

(b) 1.5 molar excess of biotin, 3.88×10^{-10} mol, thus 3.9 μl of a 0.1 mM sulfo-NHS-LC-biotin solution in MilliQ water was added to the protein and incubated for 60 min at room temperature.

(c) Very careful removal of any remaining free biotinylation reagent is important, free biotin being a common cause of "no binding" of the biotinylated protein. If a column is used, it may be necessary to perform more than one buffer change. In this example, to avoid double passages and accompanying dilution of HER-2, a modified single passage through an Illustra NAP-5 column (GEHC) was used, eluting a narrower peak of the protein, perhaps losing a minor portion of the protein but avoiding free biotin. First the sample was applied to the column (54 μl), then 650 μl assay running buffer was added and finally elution of biotinylated protein was performed by addition of another 600 μl running buffer.

(d) The concentration after passage was 45 μg/ml (theoretically at 100 % recovery after passage), the protein was diluted to 30 μg/ml and injected during 3 min over Sensor Chip SA (GEHC), resulting in 2,900 RU bound HER-2.

5. For extremely low capture levels, assay stability benefits from minimizing any possible disturbances. Thus, use the same flow rate for all injections ($NiCl_2$, capture of histidine-tagged protein, the ligand and regeneration). A compromise to save histidine-tagged reagent and at the same time obtain reliable kinetic data is to use, e.g., 15 μl/min for all injections. This flow rate is sufficiently high to avoid mass transfer limitations for most small molecules. Also, avoid using Extra Wash with 50 % DMSO for kinetics determinations at these low levels, or place the extra wash after the last ligand injection for each compound.

6. Amine coupling of target protein to Sensor Chip NTA starts with conditioning of the chip. For this 0.35 M EDTA is injected for 60 s followed by Extra Wash with running buffer to get rid of remaining traces of EDTA. This EDTA solution is preferably prepared from ready-to use EDTA solution (Sigma). From solids the pH needs to be adjusted to dissolve all EDTA. Inject 500 μM $NiCl_2$ in running buffer for 60 s followed by Extra Wash with running buffer with 3 mM added EDTA, to get rid of any remaining traces of $NiCl_2$ in the flow system. Inject EDC/NHS for 7 min. Inject histidine tagged target protein, typically 5–50 μg/ml in running buffer using injection time of, e.g., 7 min. In rare cases, the protein concentration/injection time needs to be increased significantly. Deactivate

with ethanolamine for 7 min. The flow rate, as for any immobilization, is not critical for any of the injections, typically 5–10 μl/min is used.

7. Especially for small molecules the following should also be considered:
   (a) The solubility of many organic compounds is limited in aqueous buffers, so that it may be difficult to achieve sufficiently high concentrations in coupling buffer. Solutions for immobilization are often best prepared by dissolving the small molecule in a water-miscible solvent (e.g., DMSO or dimethyl formamide, DMF) and diluting the solution in immobilization buffer. Amine coupling has been performed successfully in the presence of up to 50 % DMSO, DMF or acetonitrile. Check the respective Instrument Handbook for compatibility of the Biacore flow system with organic solvents.
   (b) Small molecules are usually inherently "sticky" in their nature and may, due to the high concentration used in immobilization, be adsorbed to tubing and channels in the flow system. Residual amounts of small molecule adsorbed to the flow system may interact with injected binding protein, and diminish or in worst case totally quench the binding to the immobilized molecule on the chip. One way to overcome this is to immobilize the small molecule outside of the instrument flow system, either by dispensing molecule solution directly to the sensor chip, outside of the instrument (avoid high concentration of organic solvent, 10 % of organic solvent is in most cases sufficient to dissolve the molecule that should be immobilized), or by using the Surface Prep Unit (only available for Biacore 3000) for immobilization. If immobilization is done in the instrument, extensive wash of flow system is recommended. A useful washing solution is 50 % isopropanol in 1 M NaOH and 1 M NaCl.
   (c) To obtain immobilization to selected flow cell/s only, selected flow cells can be activated in the Biacore instrument, followed by undocking the chip. The molecule solution is then dispensed directly to the chip outside the instrument, followed by deactivating the chip by dispensing ethanolamine solution.
   (d) In some cases, the small molecule may need to be activated to enable immobilization. Activation of carboxylic groups with EDC/NHS or hydroxyls with DSC is examples of this.
   (e) Modification of the sensor surface to introduce new functional groups extends the range of chemistries that can be used for immobilization. One alternative is to immobilize

a diamine (e.g., ethylendiamine or 1,8-diamino-3,6-dioxaoctane) compound using amine coupling on Sensor Chip CM5 to create a surface carrying free amine groups. This allows, for example, immobilization via carboxyl groups on the molecule, by an "inverted" amine coupling procedure where the small molecule is activated with EDC/NHS and reacts with amine groups on the surface. Activation is first performed by injection of EDC/NHS for 7 min followed by 7 min injection of 0.1 M ethylenediamine dissolved in 50 mM sodium borate pH 8.5. Finally the chip is deactivated for 3 min with ethanolamine. Other diamines or other functional groups can be introduced using similar principles. Note that the amine surface is not stable and should be used directly after preparation.

(f) Small molecules which do not contain a native group amenable to immobilization can be modified to introduce such a group (14). While modification may require expertise in organic chemical synthesis, this approach has some inherent advantages:

- The functional group used for immobilization can be chosen to suit a selected immobilization chemistry. Side reactions with other functional groups on the molecule can in this way be reduced or eliminated.
- A spacer, such as 1,8-diamino-3,6-dioxaoctane can be introduced between the small molecule and immobilization site, which can often help to reduce steric interference in the interaction with the protein.
- The position of the molecule modification can be chosen to minimize adverse effects on the interaction with protein.

8. Especially for peptides the following should also be considered:

(a) Acidic ($pI < 3$) peptides may be electrostatically repulsed from the carboxymethylated sensor surface at neutral or slightly acidic amine coupling pH's (pH 4–5.5). Preconcentration can be facilitated by using acidic (pH < 3) coupling buffer or a coupling buffer with high ionic strength (e.g., 1 M NaCl).

(b) To facilitate amine coupling of peptides without affecting the binding properties, use of existing or introduction of ε-amino group of the C- or N terminal lysine is preferred.

(c) As an alternative, if the peptide contains lysines that participate in binding, a cysteine can be added at the C or N-terminus of the peptide. The cysteine can then be used in so called "Ligand thiol coupling" to reactive disulfides at the sensor chip.

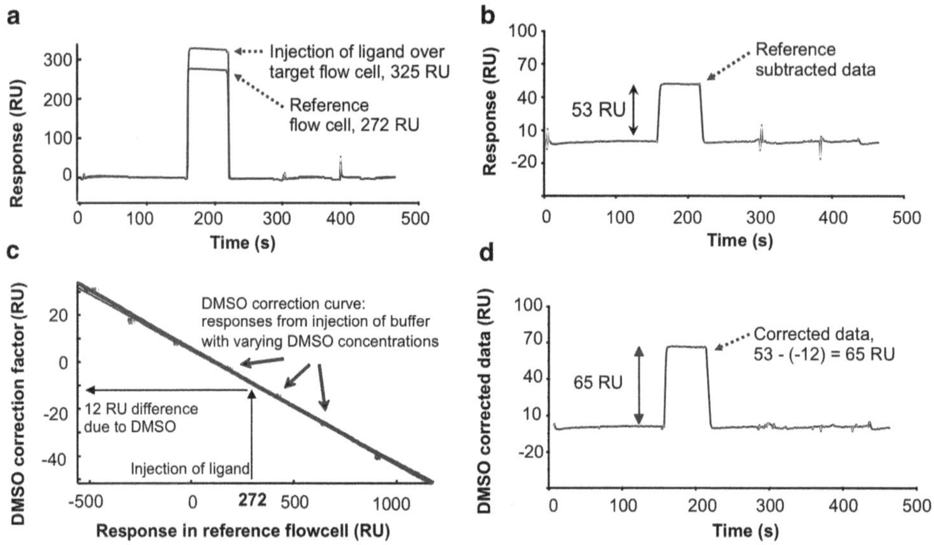

Fig. 11 (**a**) Injection of a ligand over reference and target flow cell. A high response level is obtained since the DMSO concentration of sample and assay buffer was not exactly matched in sample preparation. (**b**) Reference data subtracted from target data. (**c**) A series of eight calibration solutions (running buffer with different DMSO concentrations) were run at intervals during the experiment. The responses of the calibration solutions, obtained from the reference surface, covered a range from ca −500 to +1,000 RU relative to the baseline. A DMSO calibration curve was created by plotting difference in response between target and reference flowcell (DMSO correction factor) versus response in reference flow cell. (**d**) Ligand responses were corrected for DMSO bulk differences via the calibration curves

(d) Peptides can be also be modified with biotin to facilitate capture to streptavidin, using Sensor Chip SA or Biotin CAPture kit www.gelifesciences.com/biacore.

9. Figure 11 shows the principle for how solvent correction is applied by the software.
10. Do not store DMSO in polystyrene vessels. Make sure that all plastic equipment used in handling the solutions is resistant to DMSO. Polypropylene (used for, e.g., Eppendorf tips, vials, and micro-plates) is preferred. Prepare all solutions carefully. Keep all vials and microplates covered with correct caps and foil respectively to avoid evaporation of samples, and note that DMSO in itself is hygroscopic.

Acknowledgments

The authors thank Francis Markey, Marie Arvola, Olof Karlsson, and Robert Karlsson for helpful suggestions in the preparation of this manuscript. We also acknowledge Natalia Markova, Johan Nyberg, and Jos Buijs for the contribution of results data.

References

1. Danielson UH (2009) Integrating surface plasmon resonance biosensor-based interaction kinetic analyses into the lead discovery and optimization process. Future Med Chem 1:1399–1414
2. Johnsson B, Löfås S, Lindquist G, Edström A, Müller Hillgren RM, Hansson A (1995) Comparison of methods for immobilization to carboxymethyl dextran sensor surfaces by analysis of the specific activity of monoclonal antibodies. J Mol Recognit 8:125–131
3. Johnsson B, Lofas S, Lindquist G (1991) Immobilization of proteins to a carboxymethyl-dextran-modified gold surface for biospecific interaction analysis in surface plasmon resonance sensors. Anal Biochem 198:268–277
4. Stenlund P, Frostell-Karlsson Å, Karlsson OP (2006) Small molecule interactions with protein phosphatases using biosensor technology. Anal Biochem 353:217–225
5. Nordin H, Jungnelius M, Karlsson R, Karlsson OR (2005) Kinetic studies of small molecule interactions with protein kinases using biosensor technology. Anal Biochem 340:359–368
6. Alterman M, Sjöbom H, Säfsten P, Markgren PO, Danielson UH, Hämäläinen M, Löfås S, Hultén J, Classon B, Samuelsson B, Hallberg A (2001) P1/P1′ modified HIV protease inhibitors as tools in two new sensitive surface plasmon resonance biosensor screening assays. Eur J Pharm Sci 13:203–212
7. Sternesjö Å, Mellgren C, Björck L (1995) Determination of sulfamethazine residues in milk by a surface plasmon resonance-based biosensor assay. Anal Biochem 226:175–181
8. Wilbur DS, Pathare PM, Hamlin DK, Weerawarna SA (1997) Biotin reagents for antibody pretargeting. 2. Synthesis and in vitro evaluation of biotin dimers and trimers for cross-linking of streptavidin. Bioconjug Chem 8:819–832
9. Lempens EH, Helms BA, Merkx M, Meijer EW (2009) Efficient and chemoselective surface immobilization of proteins by using aniline-catalyzed oxime chemistry. Chembiochem 10:658–662
10. Perspicace S, Banner D, Benz J, Müller F, Schlatter D, Huber W (2009) Fragment-based screening using surface plasmon resonance technology. J Biomol Screen 14:337–349
11. Frostell-Karlsson A, Remaeus A, Roos H, Andersson K, Borg P, Hamalainen M, Karlsson R (2000) Biosensor analysis of the interaction between immobilized human serum albumin and drug compounds for prediction of human serum albumin binding levels. J Med Chem 43:1986–1992
12. Karlsson R, Fält A (1997) Experimental design for kinetic analysis of protein-protein interactions with surface plasmon resonance biosensors. J Immunol Methods 200:121–133
13. Karlsson R, Katsamba PS, Nordin H, Pol E, Myszka DG (2006) Analyzing a kinetic titration series using affinity biosensors. Anal Biochem 349:136–147
14. Mitchell JS, Wu Y, Cook CJ, Main L (2007) Direct ring conjugation of catecholamines and their immunological interactions. Bioconjug Chem 18:268–274

Part III

Spectroscopic Methods

Chapter 7

Fluorescence Techniques in Analysis of Protein–Ligand Interactions

Gabor Mocz and Justin A. Ross

Abstract

Fluorescence spectroscopy may serve as a universal tool for the study of protein–ligand interactions. Applications of fluorometry have made use of various aspects of fluorescence such as intensity, emission and excitation spectra, lifetime, quantum yield, polarization state, and anisotropy, as well as energy transfer and other electronic phenomena. An experimentalist has to consider each of these characteristics carefully, frequently in combination with each other, for the analysis of protein–ligand complexes and for the determination of binding constants. Most of the available techniques are of a rather general nature and a wealth of possibilities exists for their utilization. In this chapter we will provide a short survey of selected techniques that can be used for measuring binding constants and probing protein–ligand interactions. Basic principles and phenomena are discussed followed by experimental considerations and examples of binding constant determination. Emphasis is placed on steady-state techniques that employ the use of intrinsic protein fluorescence, labeled ligands, as well as anisotropy and resonance energy transfer.

Key words Anisotropy, Binding constant, Fluorescence, Lifetime, Ligand, Polarization, Protein, Quantum yield, Spectrum, Resonance energy transfer

1 Introduction

Interactions between proteins and ligands underlie many biochemical and biophysical processes. Fluorescence spectroscopy is one of the most convenient and successful approaches for the efficient and reliable study of such interactions. High sensitivity as well as generality and relative simplicity of common fluorescence techniques are clearly main factors in the widespread application in this and many other areas (1–4). In studies of protein–ligand complexes it may be necessary to work at very low concentrations, and fluorescence provides thousand times higher sensitivity than absorption spectroscopy. In a fluorescence experiment, it is typical to measure the signal above a low background level which usually is not the case in absorption spectroscopy where two relatively large signals are normally compared. This high sensitivity and low noise permit the

Mark A. Williams and Tina Daviter (eds.), *Protein-Ligand Interactions: Methods and Applications*, Methods in Molecular Biology, vol. 1008, DOI 10.1007/978-1-62703-398-5_7, © Springer Science+Business Media New York 2013

analysis of molecular interactions in various experimental arrangements ranging from standard fluorometry to single molecule spectroscopy (5–7).

While having an abundance of fluorescence applications in modern laboratories (8–10), our focus in this chapter will be to provide a general overview of the fundamental principles of fluorescence and to highlight the employment of steady-state techniques (11) for detection of molecular interactions and determination of binding constants. For more detailed information, the interested reader is directed to the Subheading 5 of this chapter as well as to the cited literature.

1.1 Basic Concepts

Fluorescence originates from electronic transitions in molecules with conjugated bonds (12). When a fluorophore, which is excited by a photon, returns from an excited singlet electronic state to the ground singlet electronic state, it may emit light in the form of fluorescence. An exciting light beam can induce the transition between two electronic states if the energy of the light quantum equals the difference in the energy levels between the two states. A number of non-radiative processes may compete with fluorescence to deactivate the excited state and the one with the fastest rate constant will predominate (*see* **Note 1**). Thus, fluorescence in competition with non-radiative decay can be regarded as a transition probability that depends on many factors including the local environment of the fluorophore and the solution conditions. This probability is called the fluorescence quantum yield Φ and is expressed as the ratio of radiative k_f and the sum of non-radiative k_{nr} rate constants associated with excited state deactivation (*see* **Note 2**):

$$\Phi = \frac{k_f}{k_f + \Sigma k_{nr}} = \frac{\tau}{\tau_o}. \quad (1)$$

Here $\tau = 1/(k_f + \Sigma k_{nr})$ is the effective fluorescence lifetime, and $\tau_0 = 1/k_f$ stands for the natural or intrinsic lifetime of the fluorophore. Typically fluorescence lifetimes range from picoseconds to hundreds of nanoseconds, and can be measured with a variety of techniques (*see* **Note 3**). Absolute measurements of Φ are quite difficult and tend to be a specialty. In most applications, relative quantum yields are determined by comparison of the fluorescence emission with a standard of known quantum yield (*see* **Note 4**).

The observable fluorescence intensity I_f is a wavelength dependent quantity and is related to the quantum yield and optical density. At wavelength λ (usually at the emission band maximum) it is given as

$$I_f = I_0 \Phi (1 - 10^{-\varepsilon cl}) \approx 2.303 I_0 \Phi \varepsilon cl, \quad (2)$$

where I_0 is the intensity of the excitation light, ε is the molar extinction coefficient at the excitation wavelength, c is the molar concentration, and l is the optical path length (usually 1 cm). Equation 2 follows from the Beer–Lambert law of absorption, namely, $A = \varepsilon cl$, where A is the absorption of the sample at the excitation wavelength. This linearity with concentration forms the premise for the quantitative treatment of fluorescence binding data. It implies that the observable intensity will be the sum of the contributions from all fluorescent species as long as absorption or inner filter effects can be neglected (*see* **Note 5**).

Excitation and emission spectra contain a great amount of information about a fluorophore. In this regard especially important is the wavelength-dependence of I_f since many applications to binding are based on intensity measurements. When the spectrum allows, one may selectively excite only the free ligand or the free ligand and the complex or complexes formed. The emission collected may originate from the free or the bound ligand. If the ligand is nonfluorescent, the change in the fluorescence of the protein may be followed for determination of the binding constants. In other circumstances, the emission spectra of the free ligand, the protein, and the protein–ligand complex may all overlap resulting in mixed fluorescent signals that need to be decomposed for quantitative analysis. In general, the optimal spectroscopic conditions should be selected to exploit the properties of the excitation and emission scans of the complex and its constituents.

1.2 Förster Resonance Energy Transfer (FRET)

Upon favorable conditions, non-radiative energy transfer may take place between an excited fluorophore and a neighboring fluorophore in the ground state. The intermolecular energy migration is based on a quantum-mechanical resonance process (13). If the ground-state molecule is fluorescent, it can emit the transferred energy as fluorescence. If it is not fluorescent (often called a dark acceptor), the energy is lost to the solvent as heat. The transfer probability is a function of the overlap of the emission spectrum of donor D and the absorption spectrum of acceptor A. The electrostatic interaction between a D $\rightarrow$ A pair may involve short-distance multipole- and electron exchange terms known as Dexter transfer and long-distance Coulombic terms commonly referred to as Förster transfer (14). The latter can be described in the form of dipole–dipole interactions between the excited state dipole moment of the donor and the ground state dipole of the acceptor. The rate constant of the Förster transfer may be expressed as $k_{et} \approx (R_0/R)^6/\tau_D$ where R is the distance between the D $\rightarrow$ A pair, and R_0 is the critical distance or Förster radius at which the excitation transfer probability equals the probability of all other deactivation processes, i.e., where half of the fluorescence decay is due to energy transfer (the energy transfer is 50 % efficient), and τ_D is the excited state lifetime of the donor (*see* **Note 6**). The Förster radius is typically 10–100 nm.

One of the main difficulties with FRET measurements is that multiple donors and acceptors may be present, for example multiple tryptophan residues in a protein from which the fluorescence originates. Then the calculation of efficiency becomes more complex (*see* **Note 7**). If the D → A pairs have different spectral properties, R_0 will be different for each pair and the interpretation of data can be difficult. Therefore, practical applications of FRET largely focus on cases where only a few D → A pairs or a large number of D → A pairs with assumed uniform distribution are present (15).

The study of protein–ligand complexes with FRET is mostly directed at obtaining spatial information about the location of the ligand and also at fluorescence sensing (16). However, its utilization in the measurement of binding parameters can also be productive. If energy transfer is present, it may affect the determination of binding constants by a quenching method because the fluorescence intensity will not only depend on quenching, but also on energy transfer (17, 18). However, this further decrease can also be used in the determination of the binding constant. In some cases, additional experiments may be required to distinguish between quenching due to energy transfer and decreases in the intensity due to the fluorophore environment.

We briefly note that non-FRET quenching mechanisms most commonly include dynamic and static quenching (*see* **Note 1**). Both mechanisms are governed by the Stern–Volmer relationship and may be distinguished by measuring the fluorescence life time (2, 3, 19):

$$\frac{I_0}{I_f} = \frac{\tau_0}{\tau_f} = 1 + k_q\tau_0[\mathrm{Q}] = 1 + K_{SV}[\mathrm{Q}]. \quad (3)$$

Here I_0 and I_f denote the intensities of fluorescence without and with the quencher, respectively. k_q is the quenching rate constant, K_{SV} is the Stern–Volmer constant, and [Q] stands for the quencher concentration. According to Eq. 3, quenchers reduce fluorescence intensity proportionally to their concentration. However, nonlinearity may arise when quenching by the bound quencher is incomplete or when other fluorophores are also present which are not quenched. Note also that light scattering may result in loss of fluorescent signal which can appear as quenching.

1.3 Fluorescence Polarization and Anisotropy

The light emitted by a fluorophore may be polarized to different degrees depending on the exciting light, the fluorophore, and the molecular environment (10, 20, 21). Linearly polarized light preferentially excites fluorophores with transition moments aligned parallel to the incident electric field vector in a process called photo-selection. Fluorescence polarization may be considered as a molecular stopwatch, where the clock is started at the absorption of the photon and stopped upon emission of a photon. Thus, fluorescence polarization is a measure of the fluorescence lifetime and the rotational relaxation time which in turn depends on the size of the molecule. If a small molecule binds to a large molecule, its fluorescence polarization will become dependent on the relaxation time of the large molecule. This change in polarization is direct evidence of binding (22–25) and can be used to study not only the binding of ligands to proteins but also to monitor conformational changes in proteins and to analyze the self-association of proteins and other macromolecules (26).

The polarized state of fluorescence is customarily characterized in relation to a set of reference coordinates which are defined by the direction of the excitation beam and the angle of observation, usually 90° or magic angle orientation (27, 28). When the sample is excited by vertically polarized light and the emission intensity is measured at parallel $I_{\|}$ and perpendicular $I_{\perp}$ orientations, the degree of polarization is defined in terms of the detected fluorescence intensity, $I_{\|} + I_{\perp}$, as:

$$p = \frac{I_{\|} - I_{\perp}}{I_{\|} + I_{\perp}}. \tag{4}$$

A related measure, steady-state emission anisotropy (for vertically polarized excitation) can be found from:

$$r = \frac{I_{\|} - I_{\perp}}{I_{\|} + 2I_{\perp}}, \tag{5}$$

where $I_{\|} + 2I_{\perp}$ corresponds to the total fluorescence intensity (*see* **Note 8**). The quantities r and p are connected via the relationship $r = 2p/(3 - p)$. The information content of anisotropy is slightly greater than that of polarization; however, for all practical purposes they are essentially equivalent (29). Both measures are ratiometric, i.e., they are invariant of the absolute intensity of emission light because the changes of emission proportionally affect each of the measured intensity components. Thus, polarization and anisotropy are largely unaffected by variations of the fluorophore's concentration which makes these measures particularly useful for determination of binding constants when the polarization state of either the protein or the ligand changes upon their interaction. One computational advantage of anisotropy is that r is directly additive, whereas polarization is not. This additive function can be formulated as $r = \sum I_i r_i / \sum I_i = \sum f_i r_i$ where r_i and a f_i are the anisotropy and fractional intensity of the ith molecular species, and $\sum f_i = 1$ (*see* **Note 9**).

It is of interest to mention that a variety of factors may affect the observed polarization or anisotropy. When a ligand binds to a protein, the rotational volume of the molecule and therefore its rotational correlation time (i.e., the average time it takes for the fluorophore to rotate through 55°) may display characteristic changes. τ_{rot}arises from the rotational Brownian motion reflecting

Table 1
Fluorescence lifetime τ, molar absorptivity ε, quantum yield Φ, and effective brightness *B* of the three aromatic residues in proteins. The data represent typical values from ref. 32. Individual data sets may differ to some extent, especially the quantum yield of tryptophan which varies from 0.01 to 0.35 depending on the environment and the temperature

	Lifetime	Absorption		Fluorescence			
Amino acid	τ (ns)	λ (nm)	ε (M^{-1} cm^{-1})	λ (nm)	Φ	*B* at λ_{absmax}	*B* at 280 nm
Tryptophan	0.5, 3.0–8.7	279	5,579	348	0.13	725	722
Tyrosine	3.3–3.8	274	1,405	303	0.14	197	170
Phenylalanine	~7.5	257	195	282	0.024	5	0.11

how fast the molecule can tumble in solution. The joint effects of these properties of the molecular entity and its surrounding environment are best described by the Perrin equation (10, 30):

$$\frac{1}{p}-\frac{1}{3}=\left(\frac{1}{p_0}-\frac{1}{3}\right)\left(1+\frac{\tau}{\tau_{rot}}\right), \tag{6}$$

which can be used to probe molecular parameters of protein–ligand complexes (*see* **Note 10**).

1.4 Intrinsic Protein Fluorescence and Fluorogenic Labels and Ligands

Intrinsic fluorescence of proteins can be observed in the ultraviolet (UV) from tryptophan, tyrosine, and phenylalanine (31–33). Tryptophan dominates over the other two aromatic amino acids as a relatively strong fluorescence emitter due to its higher molar absorption coefficient, higher quantum yield, and the possibility of efficient energy transfer mechanisms from tyrosine and phenylalanine to tryptophan (Table 1)

For most practical applications, the fluorescence from phenylalanine residues is overwhelmed by the fluorescence of tryptophan and tyrosine. Tyrosine fluorescence may be easily minimized through off-maximum excitation of tryptophan, for example at 300 nm, where tyrosine absorption is insignificant. However, under special circumstances tyrosine fluorescence can be unmasked by quenching the tryptophan residues. In general though, the intrinsic fluorescence of tyrosine and tryptophan containing proteins is complex and strongly depends on the environment and photophysics of these residues, especially of tryptophan (*see* **Note 11**).

A distinct class of intrinsic protein fluorescence arises from spontaneous formation of a fluorophore inside particular protein structures. These special proteins from marine invertebrates emit fluorescence in the visible region and are known as the green fluorescent protein, GFP, and its many colorful relatives out of which the blue, yellow and red variants are especially important besides the green prototype (34, 35). In GFP, from the hydrozoan medusa *Aequorea victoria*, the fluorescent chromophore forms in an autocatalytic reaction by a sequential mechanism from an internal tripeptide Ser–Tyr–Gly. The resultant chromophore is composed of a highly conjugated *p*-hydroxybenzylidene imidazolinone heterocycle, whereas in yellow and red-emitting variants further covalent modifications lead to a more extended fluorophore center. While the GFP family is extremely useful in fluorescent imaging and as a fusion protein or molecular probe, its utilization in studies of binding of small ligands to proteins is less straightforward due to the large molecular size and consequent slow rotation of the fluorescent molecule.

Common cofactors of proteins such as NAD, FAD, and porphyrins also exhibit fluorescence. Because of their abundance in bacteria, archaea, and eukaryotes, this fluorescence is less useful for identification of interactions or structures in cellular systems. However, fluorescent cofactors have many uses in in vitro applications.

Many ligands of various chemical structures exhibit fluorescence and it is not possible here to give a comprehensive account of all of these ligands. In general, the most useful fluorescent ligands are fluorescein or rhodamine derivatives as well as cyanine compounds (16). Fluorescein compounds may be coupled to a ligand via reactive amine- or thiol groups, and possess relatively high quantum yield and fluorescence intensity but only moderate photostability and marked sensitivity to pH. They may also serve as donors for FRET to rhodamines which absorb and emit light at longer wavelengths. However, at high concentrations, rhodamines may be subject to self-quenching due to formation of a ground-state dimer, called the rhodamine exciton. Cyanine compounds have high quantum yields and good photostability; however, their lifetime is usually short.

Naphthalene sulfonate derivatives, a group of environment-sensitive fluorophores deserve special attention. These solvatochromic compounds alter their fluorescent properties in response to variation in the polarity of the medium or upon binding to proteins (36). ANS (8-anilino-1-naphthalenesulfonate) as an archetypical fluorophore in this category (23) and its later derivatives such as Prodan and Nile Red are effective tools in ligand binding studies. Also of interest, ketocyanines represent a new generation of solvent sensitive fluorescent probes. For further details on fluorescent ligands, the interested reader is directed to refs. 16, 37.

1.5 Probing Protein–Ligand Interactions by Fluorescence Spectroscopy

The interaction of a particular protein P with a particular ligand L may be represented by the general scheme $m\mathrm{p} + n\mathrm{L} \rightleftharpoons \mathrm{P}_m\mathrm{L}_n$. In the state of dynamic equilibrium, fast binding and dissociation events take place continually but the concentrations of the interacting species do not change over time. In other words, ligand binding moves forward with the same rate as dissociation in the

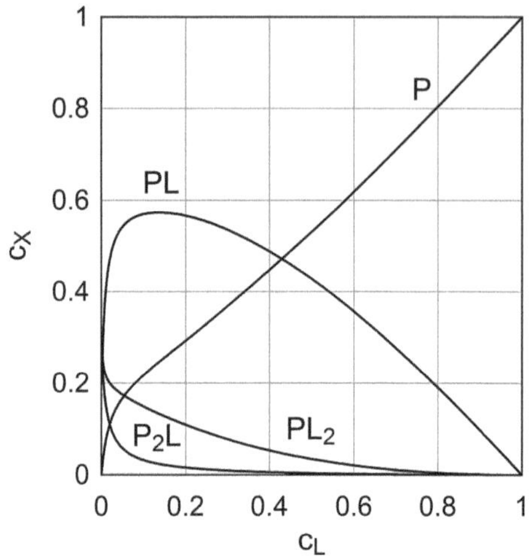

Fig. 1 Species distribution diagram for 1 μM protein and 1 μM ligand as a function of free ligand concentration [L]. [*X*] is the component concentration and the equilibria are $\mathrm{P} + \mathrm{L} \rightleftharpoons \mathrm{PL}, \mathrm{PL} + \mathrm{L} \rightleftharpoons \mathrm{PL}_2$, and $\mathrm{P} + \mathrm{PL} \rightleftharpoons \mathrm{P}_2\mathrm{L}$ with $K_{11} = 10^7$/M, $K_{12} = 10^6$/M and $K_{21} = 10^5$/M, respectively. Distribution obtained by simulation

reverse reaction. The equilibrium constant determining the stability of the complex is customarily defined either as the association (binding) constant $K_\mathrm{a} = K_\mathrm{b}$ or as its inverse function, the dissociation constant K_d (*see* **Note 12**). K_a has inverse molar units, whereas K_d is expressed directly in molar units. The simplest, and biologically most important, binding stoichiometries are PL, PL_2, and $\mathrm{P}_2\mathrm{L}$. Each complex is formed in a bimolecular step, consequently these complexes are connected by the following equilibria: $\mathrm{P} + \mathrm{L} \rightleftharpoons \mathrm{PL}$, $\mathrm{PL} + \mathrm{L} \rightleftharpoons \mathrm{PL}_2$, and $\mathrm{P} + \mathrm{PL} \rightleftharpoons \mathrm{P}_2\mathrm{L}$. Based on conservation of mass, we can write the *stepwise binding constants* as a quotient involving the concentrations of the components: $K_{11} = [\mathrm{PL}]/[\mathrm{P}][\mathrm{L}]$, $K_{12} = [\mathrm{PL}_2]/[\mathrm{PL}][\mathrm{L}]$, and $K_{21} = [\mathrm{P}_2\mathrm{L}]/[\mathrm{P}][\mathrm{PL}]$ where $[X]$ represents the equilibrium molar concentrations (*see* **Note 13**). The *overall binding constant* is given as the product of stepwise constants: $K_{\mathrm{b}(12)} = K_{11}K_{12}$ and $K_{\mathrm{b}(21)} = K_{11}K_{21}$, or in general

$$K_{\mathrm{b}(mn)} = \frac{[\mathrm{P}_m\mathrm{L}_n]}{[\mathrm{P}]^m[\mathrm{L}]^n} = \prod_{i=1}^{n} K_{mi}. \quad (7)$$

Figure 1 shows a species distribution diagram for joint formation of PL, PL_2, and $\mathrm{P}_2\mathrm{L}$. Such plots are valuable tools in visualizing the nature of the binding equilibrium that has been investigated experimentally.

The analysis of fluorescent binding data is based on the principle that the measured intensity is a sum of contributions from all components. Then, for a 1:1 complex where all species may fluoresce, we can write $K_\mathrm{b} = [\mathrm{PL}]/[\mathrm{P}][\mathrm{L}]$ and $I_\mathrm{f} = \alpha_\mathrm{P}[\mathrm{P}] + \alpha_\mathrm{L}[\mathrm{L}]$ $+\alpha_{\mathrm{PL}}[\mathrm{PL}]$

where α_x signifies the experimentally determined proportionality constants that relate the measured intensities to the concentrations of the components, i.e., α_X is the fluorescence emission intensity of species X per unit concentration. If the stoichiometry, analytical composition, or mass balance of the experimental solution is known, and if we can determine the concentration of at least one of the components in equilibrium—for instance by fluorescence intensity or anisotropy titration—then we can calculate the concentrations of all of the remaining species and hence determine the equilibrium constant. For an in-depth example illustrating the utilization of proportionality constants in fluorescence measurements and direct estimation of binding constants, *see* **Note 14**.

The most useful experimental methods for determination of K_b are based on fluorometric titration. This can be done by measuring either the intensity or the anisotropy of the fluorescent components and applying the mass balance equations. In a titration experiment one gradually changes the ratio of the protein and the ligand by varying their concentrations, usually by serial dilution, or in some cases by simple combination of the constituents. A little known special case of serial dilution, the so-called coordinates of dilution method, may also be of practical interest with samples from biological sources when the protein or the ligand cannot be easily isolated in pure form but their interaction can be monitored by the change of the emission intensity of the complex caused by varying its equilibrium concentration (*see* **Note 15** for more details).

The typical binding isotherm, i.e., the fraction of the protein molecules or binding sites to which ligand molecules are bound, is a hyperbolic function. As the concentration of the ligand increases, the binding sites attain saturation and at the half-saturation point the dissociation constant is equal to the concentration of bound ligand. In the simplest case of 1:1 binding, the binding isotherm can be written as

$$x_p = \frac{[PL]}{P_t} = \frac{K_b[L]}{1 + K_b[L]}, \qquad (8)$$

where x_p is the fraction of protein to which ligands are bound, i.e., moles of bound ligand per mole of protein. Introducing the degree of saturation $\Theta = x_p$, Eq. 8 may be also expressed as

$$\frac{\Theta}{1-\Theta} = K_b[L]. \qquad (9)$$

In the case of n identical sites, these expressions become $\Theta = x_P/n$ and $x_P = nK_b[L]/(1 + K_b[L])$.

For practicality, the isotherm is frequently rewritten in terms of experimental variables and parameters. For example if the 1:1 complex is the *only* fluorescent component then $I_f = \alpha_{PL}[PL]$ and $I_\infty = \alpha_{PL}P_t$. Here I_∞ is the fluorescence intensity measured when all the protein binds ligand and the other quantities are $\alpha_{PL} = I_f/[PL] = I_\infty/P_t$ and $[L] = L_t - [PL]$. P_t and L_t are the total concentrations of protein and ligand at a given point in the titration curve. With these substitutions Eq. 8 becomes

$$x_p = \frac{I_f}{I_\infty} = \frac{K_b(I_\infty L_t - I_f P_t)}{I_\infty + K_b(I_\infty L_t - I_f P_t)}, \tag{10}$$

which turns out to be the well-known formula first used to analyze the binding of ANS to BSA (24). From Eq. 9 K_b can be determined by nonlinear least-square fitting of the data as discussed in Subheading 1.6.

In the general case, when fluorescent P and fluorescent L are also present, we can write $I_f = \alpha_P[P] + \alpha_L[L] + \alpha_{PL}[PL]$ and $I_\infty = \alpha_L[L] + \alpha_{PL}[PL] = \alpha_L[L] + \alpha_{PL}P_t$. Then the fraction of the protein in the bound form can be obtained from

$$x_P = \frac{I_f - I_0 - \alpha_L[L]}{I_\infty - I_0 - \alpha_L[L]}, \tag{11}$$

where the intensity $I_0 = \alpha_P P_{tot} = \alpha_P[P] + \alpha_P[PL]$ is obtained at a fixed concentration of protein in the absence of ligand and subsequently K_b can easily be calculated from the binding isotherm defined by Eq. 8. Note that I_∞ is constant only if $\alpha_L \approx 0$. Accurate measurement of I_∞ requires relatively high concentration of ligand which can shift the binding equilibrium into the direction of complex formation. Under more limiting conditions when $\alpha_P = 0$ and only L and PL fluoresce, Eq. 11 reduces to

$$x_p = \frac{I_f - \alpha_L[L]}{I_\infty - \alpha_L[L]}, \tag{12}$$

and when $\alpha_L = 0$ and the fluorescence intensity is a sum of contributions from P and PL, the fraction of the bound protein is obtained from

$$x_P = \frac{I_f - I_0}{I_\infty - I_0}. \tag{13}$$

Often, the fraction of the bound ligand, rather than the bound protein can be measured more easily. For example, in fluorescence anisotropy titrations, the measured anisotropy may be related to the fraction of bound ligand as

$$x_L = \frac{[PL]}{L_t} = \frac{r - r_{free}}{r_{bound} - r_{free} + (g - 1)(r_{bound} - r)}, \tag{14}$$

where g is the quantum yield enhancement factor (38). An analogous expression can be used for polarization data:

$$x_L = \frac{(3 - p_{bound})(p - p_{free})}{(3 - p)(p_{bound} - p_{free}) + (g - 1)(3 - p_{free})(p_{bound} - p)}, \tag{15}$$

with only slightly larger complexity. This complexity is, however, only superficial since most real data analysis is done using computers and thus investigators can conveniently use either anisotropy or polarization functions (39).

Finally, we have to mention that there are important alternatives to treat fluorescence binding data. By merging Eq. 7, the mass balance on P, and the functions for I and I_0 into a single expression, we arrive at a broadly applicable generic equation that covers various types of scenarios and experimental arrangements for determination of K_b (40):

$$\frac{I_f}{I_o} = \frac{\alpha_P + \alpha_{PL} K_b[L]}{\alpha_P + \alpha_P K_b[L]} + \frac{\alpha_L[L]}{\alpha_P P_{tot}}. \quad (16)$$

The interested reader can find details on the application of Eq. 16 for three common scenarios under Subheading 5: increasing quantum yield (*see* **Note 16**), decreasing fluorescence intensity due to FRET (*see* **Note 17**), and competitive equilibrium with a quencher (*see* **Note 18**).

1.6 Nonlinear Regression Analysis for Determination of K_b

A generally applicable numerical solution for Eq. 7 and its specific derivatives such as Eqs. 8–15 may be obtained by iterative nonlinear curve fitting (41). Starting with an approximation of K_{ij} and the mass conservation equations for each components[X] in terms of the total species concentrations $T_i = \sum_{j=1}^{N} e_{ij} K_{ij} \prod_{k=1}^{i} [X]_k^{e_{ij}}$ and stoichiometric coefficient e_{ij}, we minimize S, the sum of weighted squares of the deviations between the measured and calculated concentrations of the species of interest such that $S = \sum w([X]_{obs} - [X]_{calc})^2$ where the weighting factor is $w = 1/([X]_{i+1} - [X]_{i-1})^2$. Subsequently the binding constants are incremented or decremented as needed and the computation is repeated until minimum standard deviation is found: $\sigma^2_{FIT} = S/\sum w$ from which the final values of K_{ij} are obtained. Initial estimates of K_{ij} can be obtained by assuming equivalent binding sites and estimating the binding constants from the data set as in ref. 38: $K_b = \sum x_P / \sum [L](1 - x_p)$, where x_p is the fraction of bound protein.

For obtaining accurate values of K_b or K_d, sufficient number of equilibrium measurements over a sufficient range of conditions are needed to be made. Usually it is best to compute the equilibrium constant using the entire titration curve. The best precision can be gained from fitting several experimental data sets globally. While this is generally preferable, in practice various mathematical simplifications or linear transformations are often used to aid the analysis of binding data (*see* **Notes 19 and 20**).

While very precise measurements are readily obtainable with modern instrumentation, fluorescence experiments may still be

1.7 Errors in Determination of Binding Constants by Fluorescence Techniques

affected by process and measurement uncertainty. In most cases, the uncertainty functions are additive and follow a Gaussian distribution over time or show a Poisson process at low photon counts (42). Measurement errors may especially affect the determination of binding constants as they propagate into the calculated parameters to significant degrees. For detailed information on the various properties of measurement errors and in depth analysis of error propagation in fluorescence measurements, we refer the reader to ref. 39.

2 Materials

2.1 Instrumentation

The steady state fluorescence spectrometer, also called a fluorimeter, consists of several key components regardless of the brand or make. These fundamental components are: the light source, an excitation wavelength selector, the sample, an emission wavelength selector, and the detector (Fig. 2). Other optical components such as polarizers can also be present.

The light source used in most fluorescence spectrometers is the xenon lamp. The spectral distribution of light emitted from a xenon lamp is shown in Fig. 3. It has significant intensity in the visible region of the spectrum and tapers slowly as the wavelength is decreased into the UV domain. Some xenon lamps, called *ozone free*, significantly

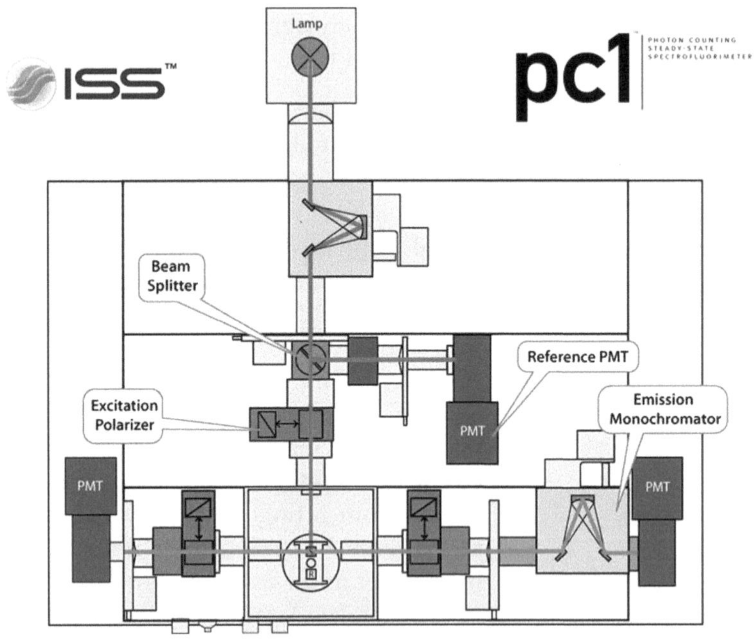

Fig. 2 Schematic of steady state fluorescence spectrometer. Used with permission from ISS Inc

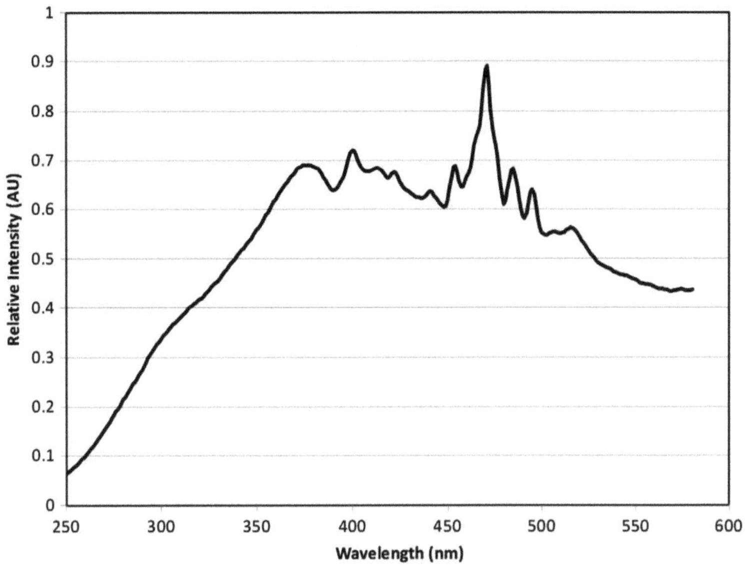

Fig. 3 Emission spectrum of a xenon arc lamp

attenuate the UV part of the spectrum which is responsible for the generation of ozone. These types of lamps are not suitable for use in intrinsic protein fluorescence studies, where one excites the sample at 260–300 nm, due to the reduced UV light output. Other types of lamps including mercury (Hg) lamps and xenon–mercury lamps are also used, but xenon lamps are the most common due to their relatively flat spectral output and UV emission. Additionally, laser and light-emitting diode (LED) sources can be employed, including fixed and tunable lasers although they are less common.

Following the light source in the optical path is an element which selects the wavelength of light which will be used to excite the sample. Typically this device is a monochromator which uses diffraction gratings to disperse the light depending on its wavelength. At the input and exit of the monochromator are the slits, which serve to specify the spectral bandwidth of the light transmitted through the monochromator. The spectral bandwidths of the slits are specified in nm, for instance 8 nm, which means that the full width at half maximum (FWHM) of the light exiting the monochromator is 8 nm (*see* **Note 21**).

Following the excitation monochromator is a beam splitter, a thin piece of quartz which reflects a small amount of the excitation light to go to a reference detector, used to measure the intensity of the excitation light. Prior to the reference detector is a cell filled with concentrated Rhodamine B or Rhodamine 101 solution which acts as a quantum counter. Due to the high optical density of the cell, it absorbs all light incident on it, regardless of the wavelength, and reemits it as fluorescence which is measured by the reference detector (*see* **Note 22**).

In fluorescence spectrometers capable of measuring fluorescence polarization and anisotropy there are two polarizers, one in the excitation path before the sample and the other in the emission path immediately after the sample. The polarizers transmit light with a particular polarization and reject the other polarization, i.e., that which is at 90° to the first. They are made from either double refracting calcite crystal prisms or absorbing thin films such as Polaroid type H sheets. While the calcite polarizers are more expensive, they have much higher transmission in the UV region and a higher extinction ratio (i.e., rejection of the unwanted polarized component).

The excitation light is then focused through a lens onto the sample with the emission collected at 90° to the excitation through another lens. The sample is usually held in a cuvette. Emission light from the sample is collected by the emission lens and spectrally filtered by either a monochromator or an optical filter. This element is critical to remove any scattered excitation light before it can reach the detector. Even very small amounts of scattered light can significantly affect fluorescence measurements and therefore it is critical that as much as possible be removed.

The most common detectors used in fluorescence spectrometers are photomultiplier tubes (PMTs) and occasionally avalanche photodiodes (APDs). The PMTs can work in one of two modes, photon counting or analogue mode. As the name suggests, in photon counting mode, each photon is individually converted into an electrical pulse and counted using electronics. In the analogue mode, it is the photocurrent which is proportional to the intensity of incident light that is measured. APDs intrinsically operate in photon counting mode but are less common due to their higher cost and smaller active area.

2.2 Instrument Calibration

Before using the fluorescence spectrometer for any type of experiment, it is recommended to verify the calibration of the monochromators. These are generally the only components within the instrument which need to be calibrated. The exact procedure will be explained in detail in the instrument manual. Usually, the emission monochromator will be calibrated first. This calibration can be checked by setting the instrument to do an emission scan (i.e., where the excitation wavelength is fixed and the emission wavelength is scanned) and blocking the excitation light either with the built in shutter or some other object. If available, one can use a calibrated standard lamp or laser source (e.g., HeNe laser), but otherwise it is possible to use the light from fluorescent lights which are present in most laboratories. These fluorescent lights contain mercury and will have several characteristic peaks in their emission. If using a portable laser source, it can be mounted above the sample compartment, with the light incident on the sample cuvette. Inside the cuvette should be a slight scattering solution, such as glycogen or even coffee creamer. The scattering solution can also be utilized if using the fluorescent room lights but often it is not necessary.

Once the light source and instrument is ready, conduct an emission scan across the applicable wavelength range. If using a HeNe laser (peak wavelength 632.8 nm) scan from ~600 to 650 nm, or if using the room lights, scan from 400 to 600 nm (Hg has peaks at 436 and 546 nm). Remember to adjust the intensity of light incident on the detector to avoid saturating it. If the measured maximum matches the correct value then move onto checking the excitation monochromator, otherwise adjust the calibration of the emission monochromator (this is usually done in using the system software).

To calibrate the excitation monochromator, use the same slightly scattering solution in a quartz 10 mm × 10 mm cuvette and place it in the sample chamber. Set the emission wavelength to 350 nm and scan the excitation wavelength from 330 to 370 nm. If the maximum is not at 350 nm, adjust the calibration of the excitation monochromator and repeat to verify that it is now correct.

2.3 Optical Filters

Optical filters are used to modify the properties of the light beam incident on them, either the wavelength, intensity or both. Filters which only reduce the intensity of light, regardless of the wavelength, are called neutral density filters. Ideally they attenuate all wavelengths of light equally. They are commonly used to decrease the intensity of the excitation beam on the sample. Neutral density filters are often silver or grey colored but can also take the form of fine wire grids which serve the purpose of reducing the transmitted intensity by scattering the light. Alternatively there are filters with optical properties that transmit some wavelengths and block others. These are named depending on whether the blocked wavelength is higher or lower (or both) than the transmitted wavelengths (Fig. 4). If a filter passes longer wavelengths and blocks shorter ones, then it is called a longpass filter. The converse is called a shortpass filter. If a filter passes a certain range of wavelengths but blocks longer and shorter wavelengths, it is called a bandpass. The opposite of a bandpass filter is the bandstop filter, which blocks a narrow band of wavelengths (*see* **Note 23**).

Although they are not generally tunable, optical filters can be utilized instead of monochromators. Their advantage is that the transmission efficiency can be much higher than that of a monochromator (*see* **Note 24**).

2.4 Cuvettes

There are many different types, shapes, and sizes of fluorescence cuvettes each with their own specific purpose, the most common

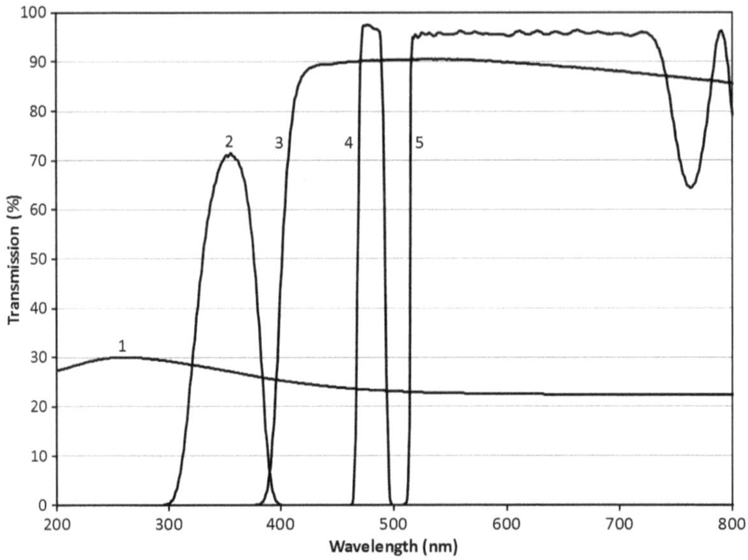

Fig. 4 Transmission spectra of optical filters. *Curve 1*: 25 % neutral density, *2*: 354 nm bandpass colored glass, *3*: KV399 longpass colored glass, *4*: 482/15 nm bandpass interference filter, *5*: 500 nm longpass interference filter

of which is the 10 mm × 10 mm square made from quartz with all four sides polished. Cuvettes with 4 and 2 mm × 10 mm compartments are also common (*see* **Note 25** and Fig. 5). Larger size cuvettes can also be used for specific purposes which require longer path lengths. One cuvette of note is the triangular cuvette, which is often used for highly concentrated or highly scattering samples. This cuvette is shaped like a right angled triangle, with the angled face at 45° or 30/60° to the excitation beam. Using this geometry, emission is collected from only the region of the sample immediately behind the angled face, thus minimizing the effect of the highly concentrated or scattering sample (*see* **Note 26**).

2.5 Techniques

From the technical point of view, the experiments performed in this section fall under three measurement types: excitation scan, emission scan or polarization assay. In an emission scan, the excitation wavelength is fixed and the emission monochromator is scanned across a certain range. Such an experiment is described in Subheading 3.1. Figure 6 shows the emission scan of 8-anilino-1-naphthalenesulfonate (ANS) in water, a ligand to be used in some of the experiments. Note that there are four peaks present but only one of them is due to fluorescence emission from the sample.

If the user is unsure whether a peak is an artifact or fluorescence, one can change the excitation wavelength. The fluorescence peak will remain stationary while the other peaks will shift. In order

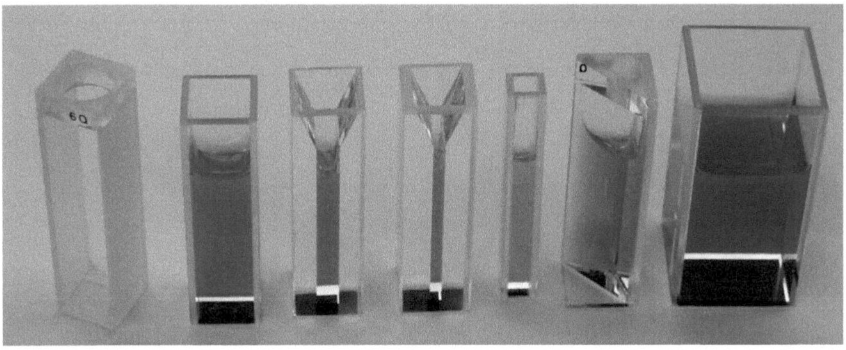

Fig. 5 Image of types of cuvettes, from *L* to *R*. 10 mm × 10 mm absorption cuvette (note frosted sides), 10 mm × 10 mm fluorescence cuvette, 4 mm × 10 mm fluorescence cuvette, 2 mm × 10 mm fluorescence cuvette, 3 mm × 3 mm fluorescence cuvette, triangular fluorescence cuvette, 20 mm × 20 mm fluorescence cuvette

to determine the true emission spectrum of a compound, the measured (also called technical) spectrum must be corrected for the wavelength and polarization dependence of the monochromator and for the wavelength dependence of the detector. The wavelength dependence of the monochromator is most severe at the Wood's anomaly (*see* **Note 27**). At this wavelength, only vertically polarized light is transmitted (10). On most instruments the effect of the Wood's anomaly can be removed by placing a vertical polarizer before the emission monochromator; however, on some instruments it is removed using the polarizer oriented horizontally. The wavelength dependence of the monochromator and detector is usually corrected using manufacturer supplied correction factors.

The excitation scan may be considered as the conjugate of the emission scan. In the excitation scan, the emission wavelength is fixed and the excitation monochromator is scanned across some range of wavelengths. The excitation scan provides the excitation spectrum of the sample. In principle the excitation spectrum should appear identical to its absorption spectrum counterpart. Recall that the intensity of the light source has a dependence on the wavelength (Fig. 3). In order to determine the true *corrected* excitation spectrum, this wavelength dependence must be accounted for. This correction can be done to a first approximation using the beam splitter and reference detector, although this method is not perfect due to the variation of the polarization of the light transmitted through a monochromator. Some manufacturers provide lamp spectra which can be used to correct the excitation spectrum although these should be checked because the lamp spectrum can change as the lamp ages. In many cases, a corrected excitation spectrum is not required, in particular if multiple samples will be compared at the same excitation wavelength

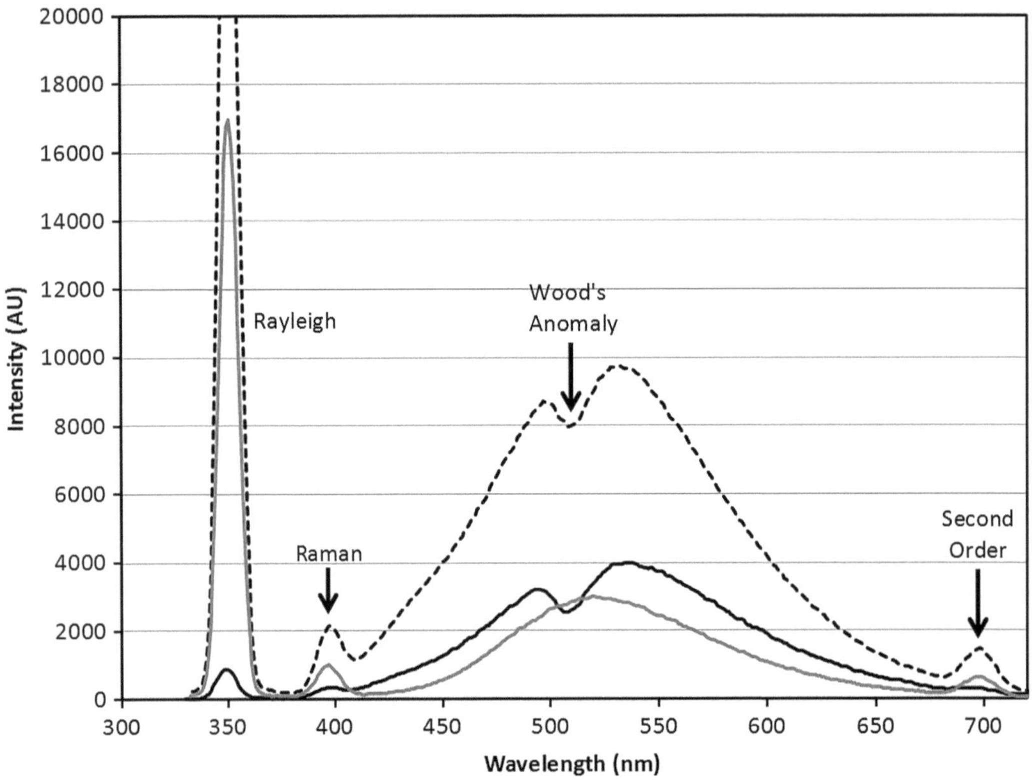

Fig. 6 Emission scan of ANS in water excited at 350 nm with different orientations of the emission polarizer showing the Rayleigh peak, Raman peak, Wood's anomaly, and second order peak. *Dashed line*—no polarizer, *solid black*—horizontal polarizer, *grey line*—vertical polarizer. Note the absence of the Wood's anomaly with the vertical polarizer. The Rayleigh peak is due to Rayleigh scattering from the sample. It always has exactly the same wavelength as the excitation light. The Raman peak is due to Raman scattering from water in the sample. It is of slightly lower energy (longer wavelength) than the excitation wavelength and is caused by inelastic scattering from the O–H bond of the water solvent. It can be calculated using the following relationship: $1/\lambda_{\text{Raman}} = 1/\lambda_{\text{Excitation}} - 0.00034/\text{nm}$. The next peak is the second order scattering peak. This peak is the result of scattered excitation light transmitted through the emission monochromator due to the second order diffraction from the grating inside the monochromator. This light is actually of the same wavelength as the excitation but appears in the emission spectrum at twice the excitation wavelength. The final peak is the fluorescence due to emission from the sample. If the same scans were conducted with a slightly different excitation wavelength, then the Rayleigh, Raman and second order peaks would be at different wavelengths while the fluorescence peak would remain unchanged

with the same intensity. Only when different excitation wavelengths will be used for different samples is this necessary (*see* **Note 28**).

The polarization or anisotropy of a sample is determined by measuring the intensity of the emission light which has the electric field oriented parallel and perpendicular relative to the laboratory or instrument axis. The measurement requires both the excitation and emission polarizers to be in the optical path. The excitation polarizer is oriented vertically, V and the intensity measured with the emission polarizer oriented vertically and horizontally, H. It is usual for the detector to have a slightly different response to light with different polarizations. This effect is the result of an intrinsic property of the detector and that there may be a slight optical effect as the polarizer is moved or rotated. A correction must be made to measure the true polarization. The correction factor is known as the *G* factor, named because it was initially a correction due to the grating. It is an instrument dependent parameter. Determining the *G* factor does not require a special sample but can be measured using the sample of interest. The excitation polarizer is set horizontally and the emission polarizer rotated to measure the intensity at each orientation. *G* is then calculated from $G = I_{\mathrm{HV}}/I_{\mathrm{HH}}$. Thus, the corrected polarization can be determined from

$$p = \frac{I_{\mathrm{VV}} - GI_{\mathrm{VH}}}{I_{\mathrm{VV}} + GI_{\mathrm{VH}}}. \tag{17}$$

Compare this equation with Eq. 4. If the emission is collected through a monochromator, then the *G* factor must be experimentally determined at each emission wavelength (*see* **Note 29**).

2.6 Proteins, Ligands, and Buffers

In the range of experiments presented under Subheading 3 you will study two common proteins: bovine serum albumin (BSA) and human serum albumin (HSA). You will explore three ligands: 8-anilino-1-naphthalenesulfonate (ANS), fluorescein, and furosemide (4-chloro-2-(furan-2-ylmethylamino)-5-sulfamoylbenzoic acid). All protein and ligand solutions are prepared in 20 mM HEPES ((2-hydroxyethyl)-1-piperazineethanesulfonic acid) buffer, pH 7.5, at concentrations given in the protocols, except for the experiment with fluorescein where a HEPES buffer of pH 8.0 is used. The ideal concentrations are typically low for any experiment, *see* **Notes 5** and **30**. The ANS and fluorescein solutions are stored in light-proof containers.

3 Methods

Most of the experiments described in this section use the same basic principle of a serial dilution to analyze protein–ligand interactions but with a different fluorescence property of the sample being measured. In a serial dilution experiment, we are changing the ratio of the two components, protein and ligand, over a large range and measuring some fluorescent property at each point. One starts, for example, with 1 ml of solution containing both protein and ligand, removes a certain volume (e.g., 0.5 ml) and then replaces that with the same volume of only ligand or only protein. In general it is best to keep the concentration of the fluorophore of interest constant, and thus if there is a change in the intensity, it is easily related back to a binding or dissociation event. The volume of sample removed each time depends on the conditions and the particular experiment but volumes in the range of 1/2 or 1/3 of the total volume usually result in a sufficient number of data points without being extremely time consuming. Each of the experiments is selected as an example to demonstrate the application of a specific fluorescence method to the particular system (*see* **Note 31**).

3.1 Change in Emission Spectrum upon Ligand Binding Analyzed by Center of Mass Method

Measuring the shift of the emission spectrum is appropriate for fluorophores whose fluorescence properties vary depending on the environment. Examples include ANS, Bis-ANS, Nile Red, and tryptophan. The experiment described here is a serial dilution to obtain data on ANS binding to BSA. If the user keeps the excitation intensity constant for each of the individual measurements, then information about the relative emission intensity of the fluorophore is obtained simultaneously. Often, these intensities are plotted over the concentration to obtain a titration curve. However, we describe the more accurate spectral center of mass method (CMM) (43).

CMM reduces the entire emission spectrum to a single number which can be monitored as a function of the protein–ligand ratio (43). The center of mass $\langle \nu_g \rangle$ is calculated as:

$$\langle \nu_g \rangle = \frac{\Sigma \nu_i I(\nu_i)}{\Sigma I(\nu_i)}, \tag{18}$$

where $I(\nu_i)$ is the fluorescence intensity per wave number interval emitted at wave number ν_i and the sum covers the entire emission spectrum. At wavelength λ, $I(\nu_i)$ is related to fluorescence intensity as $I(\nu_i) = \lambda^2 I(\lambda_i)$ and the conversion from wavelength to wave number is given by $\nu = 1/\lambda$. The advantage of CMM lies in that it is sensitive to changes in the emission profile, not only the emission maximum or intensity.

To conduct the experiment

1. Start with a 3 ml combined solution of 100 μM BSA and 1 μM ANS, placed into a 10 mm × 10 mm quartz cuvette. If using smaller cuvettes, e.g., 4 mm × 10 mm, then the volumes used can be reduced accordingly.
2. Measure the emission spectrum with excitation wavelength 350 nm and the emission scan range from 360 to 500 nm. Ensure the emission polarizer is oriented vertically and placed in front of the emission monochromator. Check the signal level at the maximum of the emission (~450 nm). It should be just below the maximum of the linear range of your detector. If the intensity is too high, add a neutral density filter to the excitation path or decrease the slit widths (*see* **Note 30**).

3. Remove 1 ml of solution and discard. Then add 1 ml of 1 μM ANS solution, mix thoroughly and measure another emission spectrum.
4. Repeat these steps until the effective BSA concentration is almost zero or the intensity gets too low to accurately measure.

To analyze the data

1. Calculate the center of mass according to Eq. 18.
2. Plot $\langle \nu_g \rangle$ as a function of BSA concentration
3. Estimate K_d from the points of inflection of the curve where the binding site on the protein is half occupied and the concentration of BSA corresponds to the dissociation constant (*see* **Note 12**). There is a significant point of inflection at ~10 nM (high affinity binding site) and another much less pronounced one at ~20–30 μM (one or more secondary weak binding site).
4. Obtain K_b by nonlinear least-squares regression. Make the substitutions $I_f \leftarrow \left(\langle \nu_g \rangle - \langle \nu_g \rangle_{min}\right)$ and $I_\infty \leftarrow \left(\langle \nu_g \rangle_{max} - \langle \nu_g \rangle_{min}\right)$ and generate the binding curve as a function of protein concentration from Eq. 10 by solving the quadratic expression in x_p for the physically realizable root. Equation 10 then becomes $\langle \nu_g \rangle = \langle \nu_g \rangle_{min} + \left(\langle \nu_g \rangle_{max} - \langle \nu_g \rangle_{min}\right) \{(\mathrm{P_t} + \mathrm{L_t} + K_d - \sqrt{(\mathrm{P_t} + \mathrm{L_t} + K_d)^2 - 4\mathrm{P_t}\mathrm{L_t}})/2\mathrm{L_t}\}$. The parameters $\langle \nu_g \rangle_{min}$ and $\langle \nu_g \rangle_{max}$ denote the asymptotic minimum and maximum of center of mass in the range of interest. For simplicity we only consider the high-affinity binding site which can be determined from part of the curve in the 1 nM to 1 μM concentration range. To recover K_b, perform nonlinear regression analysis on the binding curve using a standard software package, or adapt your own algorithm as summarized in Subheading 1.6.
5. For more rigor in the analysis, use Eq. 12 to correct for the (slight) contribution of unbound ANS to the intensity, or indeed, use Eq. 7 with appropriate substitutions to obtain K_b both for the high and low affinity sites. This goes, however, beyond the scope of the present experiment.

3.1.1 Discussion

Figure 7 shows the center of mass and maximum wavelength of ANS as a function of BSA concentration from 100 μM down to 1 nM in the serial dilution experiment. Both plots show similar features but the center of mass plot is much more sensitive to changes in the spectral properties of ANS. One can estimate the K_d values from the points of inflection of the curve where the binding site on the protein is half occupied and the concentration of BSA corresponds to the dissociation constant (*see* **Note 12**). There is a significant point of inflection at ~10 nM (high affinity

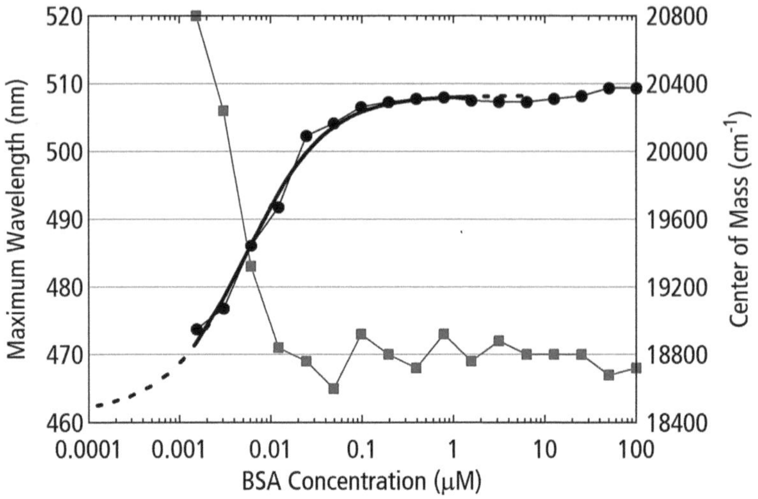

Fig. 7 Maximum wavelength (*grey squares*) and center of mass (*black circles*) plotted for the ANS–BSA complex as a function of BSA concentration. The *heavy line* shows the best-fit nonlinear regression curve for a high affinity binding site, restricted to a submicromolar range. The binding isotherm and $K_d = 5.4$ nM were obtained as described in Subheading 3.1 by computing the mean squared difference between the observed values of fraction of bound components and the corresponding values calculated from the final value of the binding constant. The *dotted line* is an extension of the fit using the final fit parameters. The best-fit set of minimum and maximum of center of mass was found as 18,460 and 20,330/cm when the low affinity sites were excluded from the binding model

binding site) and another much less pronounced one at ~20–30 μM (one or more secondary weak binding sites). In this context, the center of mass method can be used to qualitatively estimate binding constants, although its use for accurate determination can be difficult because of the contribution from numerous fluorescent species.

The center of mass method can be useful with any probe whose emission properties vary as function of environment. In particular this method is valuable when looking at changes in the intrinsic fluorescence of a protein. These changes can be the result of direct interaction of the ligand with solvent-exposed aromatic amino acids, or due to conformational or oligomeric changes in the protein with ligand binding.

3.2 Ligand Binding to Multiple Sites Determined from Energy Transfer

In the case of a fluorescent ligand, it is often possible to observe energy transfer from the aromatic amino acid residues in the protein to the ligand. The energy transfer can be manifested in several ways. Firstly through the emission spectrum, as illustrated in Fig. 8. In this experiment, the emission spectrum is measured for a solution of BSA with different concentrations of ANS while exciting at 280 nm. As the amount of bound ligand increases, we see an increase in the emission due to the ANS and a decrease in

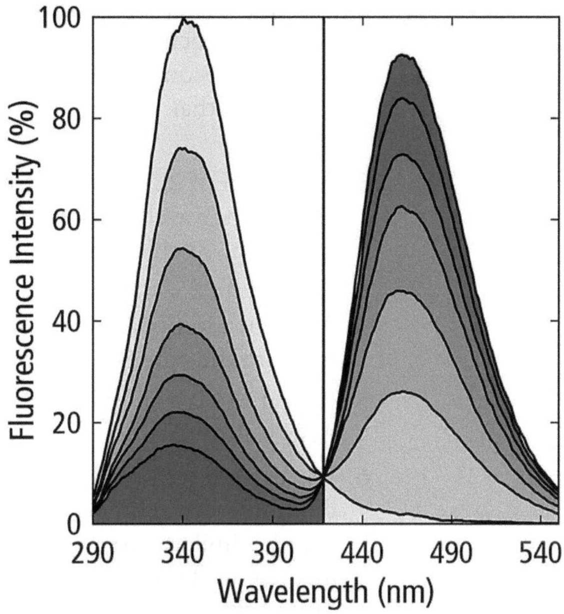

Fig. 8 Emission spectrum of ANS and BSA at increasing ratios of ligand to protein. The *grey* levels correspond to ratios of 0, 1/3, 2/3, 1, 4/3, 5/3, and 2 from *light grey* to *dark grey*. A *reference line* is drawn at the isoemissive point

the emission from the tryptophan. For more details on the binding of ANS to BSA see refs. 44–46. The presence of a single isoemissive point indicates that while multiple ligands may bind to the protein, each of the bound ligands has the same fluorescence properties.

Secondly, an equivalent experiment can be performed using the excitation spectrum, where one would monitor the emission of the ANS (acceptor) and perform excitation scans across the absorption bands of BSA and ANS. In this case an increase of the excitation spectrum at ~280 nm would be seen as the amount of bound ligand increases. A corrected excitation spectrum is not needed for comparison of the spectra but would only be required if quantification of the amount of energy transfer was required (3).

Thirdly, polarization can be used to detect the presence of energy transfer. Energy transfer is manifested in the Perrin equation (Eq. 5) as a lowering of the effective p_0. However, upon binding it is usual for the rotational correlation time and the lifetime of the donor to also be altered and great care must be taken to isolate the effect on the polarization due to each of these factors.

To conduct the experiment

1. Start with a 3 ml solution of 1 μM BSA, added to a 10 mm × 10 mm quartz cuvette.

2. Set the excitation wavelength at 280 nm and obtain the emission spectrum. Set the emission wavelength at the maximum emission point and measure the excitation spectrum. Ascertain that the excitation maximum is near 280 nm as expected.
3. Add 50 μl of 20 μM ANS solution to the 3 ml of BSA. Set the excitation wavelength to 280 nm and obtain the emission spectrum. At this wavelength both ANS and BSA absorb near maximally.
4. Add an additional 50 μl of ANS to the same cuvette and measure the emission spectrum.
5. Repeat the last step four times until the total added volume of ANS is 300 μl.
6. For each successive value of ligand to protein ratio, plot the dilution corrected fluorescence intensity as a function of wavelength similarly to Fig. 8, and interpret the decrease and increase in fluorescence intensity in the domain of both emission peaks.
7. For a quantitative evaluation of the experiment, determine $\bar{n}$, the average number of ANS molecules bound to BSA and obtain the maximum value of fluorescence under conditions when all ANS is bound. This can be done by repeating the experiment for each concentration of ANS with 150 μM (excess) BSA.

To analyze the data

1. Calculate $\bar{n}$ by the relation $\bar{n} = \mathrm{L_t} I_\mathrm{f}/\mathrm{P_t} I_\mathrm{b}$ where I_f is the fluorescence of a given solution of the ANS-BSA complex and I_b is the fluorescence of an equal concentration of ANS in 150 μM (excess) BSA. t stands for the total concentration of protein and ligand in the particular complex to which $\bar{n}$ applies. Under the conditions described, $\bar{n}$ ranges from 0 to 2.
2. Use the stoichiometric relation $[\mathrm{L}] = \mathrm{L_t} - \bar{n}\mathrm{P}$ to obtain the concentration of free ANS in the experiment with 1 μM BSA. From this the *overall* dissociation constant for each protein–ligand ratio can be estimated via the relation $\bar{n}/N = [\mathrm{L}]/([\mathrm{L}] + K_\mathrm{d})$. Here N is the maximum number of ligand molecules that can be bound by the protein under these conditions, i.e., $N \approx \bar{n} + 1$ truncated to a whole number.

3.2.1 Discussion

This experiment utilized a simple combination of BSA with ANS rather than serial dilution. Such an approach provides a quick and easy way for qualitative demonstration of the underlying physical phenomena, in this case resonance energy transfer. For quantitative evaluation, the observed fluorescence intensities have to be corrected for the dilution, or alternatively, serial dilution can also be used. In research applications, the latter would be preferred. As the concentration of ANS is increased, the amount of energy transfer increases resulting in a lowering of the emission peak at ~350 nm and an increase in the peak at ~460 nm. As mentioned previously, the presence of a single isoemissive point indicates that all of the binding sites are equivalent.

3.3 Ligand Binding Determined by Change in Polarization

In this experiment, we will follow the change in polarization due to a change in the rotational mobility of the fluorophore. As the fraction of fluorophore bound to the protein increases, so will the polarization. This type of experiment is best suited when the fluorescent ligand has a low polarization and one intends to determine the binding to a protein or other ligand which has a much larger molecular mass. The greater the difference in the molecular mass, the greater the change in polarization and the more easily the two species can be resolved.

To conduct the experiment

1. Start the dilution series with a 3 ml combined solution of 100 μM BSA and 1 μM fluorescein, placed into a 10 mm × 10 mm quartz cuvette.
2. Set the excitation wavelength to 490 nm and place an appropriate emission filter in the emission path (*see* **Note 29**). If the spectrometer does not allow the use of an emission filter, then use the emission monochromator set to 530 nm.
3. If the G factor for the spectrometer is not known, determine it by setting the excitation polarizer horizontal and calculate G from Eq. 17 by measuring the intensity with the emission polarizer set vertically and horizontally.
4. Using the G factor to correct the intensities, measure and save the polarization of the sample.
5. Remove 1 ml of solution and discard. Then replace with 1 ml of 1 μM fluorescein solution and mix thoroughly. Measure the polarization of the sample.
6. Repeat step 5 until the effective BSA concentration is almost zero or the polarization reaches a plateau.
7. Measure the polarization using the exact same instrument parameters but of only 1 μM fluorescein. This will be the polarization of the free ligand.

To analyze the data

1. Determine the binding constants for two near-identical sites starting from x_L, the fraction of the bound ligand, see Eq. 15 for calculation of this quantity. You can employ nonlinear regression as detailed below, or use linear transformation of the experimental binding profile the same way as described for intensity data (*see* **Notes 19** and **20**).

2. To carry out nonlinear least squares regression, first calculate [L] from x_L by $[L] = L_t - x_L L_t$ and rewrite x_L in terms of the fraction of the bound protein as $x_L \approx 2x_p P_t / L_t$.
3. Based on Eq. 7, the multiple stepwise equilibrium model, define the binding function as $x_p = \sum j[L]^j \prod K_j / \left(1 + \sum [L]^j \prod K_j\right)$. where j ranges from 1 to n, the number of binding sites, and K_j is the jth binding constant. For simplicity, assume $n = 2$. In general, n may vary from 1 to 5 depending on experimental conditions (44).
4. Plot x_P versus [L] and determine K_j from the binding function by nonlinear regression.
5. Once the binding constants are recovered, the experimental binding profile (polarization versus protein concentration) can be straightforwardly reconstructed: compute x_L from Eq. 7 as a function of the total protein concentration, and compute the value of polarization from Eq. 15 as a function of x_L just derived. Use the stoichiometric relations $[P] = (1 - x_p)P_t$ and $[L] = (1 - x_L)L_t$ in expressing the required quantities from Eq. 7.

When measuring polarization, one is simultaneously recording the total intensity of the emission. It is best to plot both as a function of the protein concentration to determine if there is any change in the quantum yield of the ligand upon binding. In the case of fluorescein binding to BSA, we can see about twofold quenching of the fluorescein when bound (Fig. 9). Therefore, we can use $g = 2$ for the quantum yield enhancement factor in the calculation of x_L as a starting point for the nonlinear curve fitting. If g can be measured more accurately then this parameter can be fixed. There is also a change in the polarization from >0.38 to ~0.02, but inspecting the plot we can estimate the polarization of the bound ligand to be ~0.45. These values provide valuable information for determining the starting parameters for the fitting algorithm and by setting values which are known explicitly (such as the polarization of the free ligand), one can significantly enhance the accuracy of the parameters determined from the fit.

3.4 Total Intensity Method for Ligand Binding

In the previous experiment, polarization was used to investigate the binding of fluorescein to BSA. In that case, there was about a twofold change in the quantum yield (or brightness) of the fluorescein in the bound and free states. In some instances there is a very large change in the quantum yield of the bound ligand and in this situation often it is better to use the intensity change to investigate the binding properties. This measurement can be done using two similar methods. Firstly one can just measure the polarization of each sample as it is diluted and calculate the total intensity from: $I_t = I_{\parallel} + 2GI_{\perp}$. This method has the advantage of simultaneously

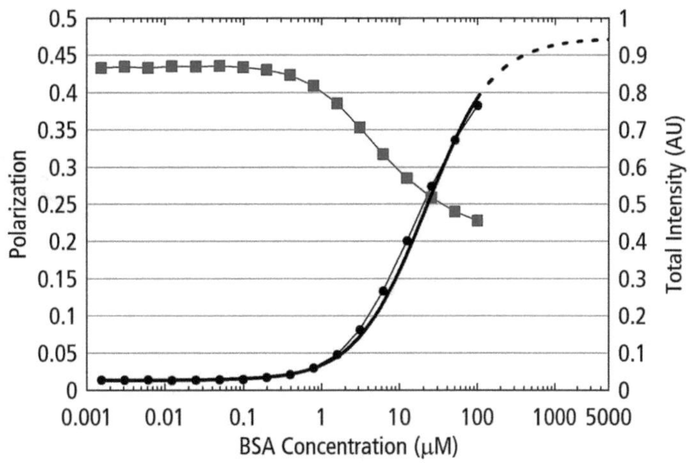

Fig. 9 Binding isotherm of fluorescein and BSA as a function of BSA concentration. *Grey squares*—total intensity, *black circles*—polarization. The *heavy line* is the nonlinear regression curve obtained according to Subheading 3.3 by assuming two binding sites on the protein. The two sites display virtually identical affinities with $K_{d1} = 21.2$ μM and $K_{d2} = 21.3$ μM, respectively. The *dotted line* is an extension of the fitted curve to illustrate that the polarization reaches a plateau at high BSA concentrations. Final polarization values provided by the fit: $p_{free} = 0.013$ and $p_{bound} = 0.47$

having the polarization and total intensity data. Alternatively, one can set the emission polarizer to 54.7° (with the excitation polarizer vertical—one of the so-called *Magic Angle* conditions (28)) and just measure the intensity through an emission filter or monochromator.

To conduct the experiment

1. Start with a 3 ml combined solution of 100 μM BSA and 1 μM ANS, added to a 10 mm × 10 mm quartz cuvette.
2. Set the excitation wavelength to 350 nm and place an appropriate emission filter in the emission path. Check for scattered light. If the spectrometer does not allow the use of an emission filter, then use the emission monochromator set to 450 nm.
3. If the *G* factor for the spectrometer is not known, determine it by setting the excitation polarizer horizontal and calculate *G* from Eq. 17 by measuring the intensity with the emission polarizer set vertically and horizontally.
4. Using the *G* factor to correct the intensities, measure and save the polarization or intensity of the sample.
5. Remove 1 ml of solution and discard. Then replace with 1 ml of 1 μM ANS solution and mix thoroughly.
6. Repeat this step until the effective BSA concentration is almost zero or the polarization or intensity reaches a plateau.

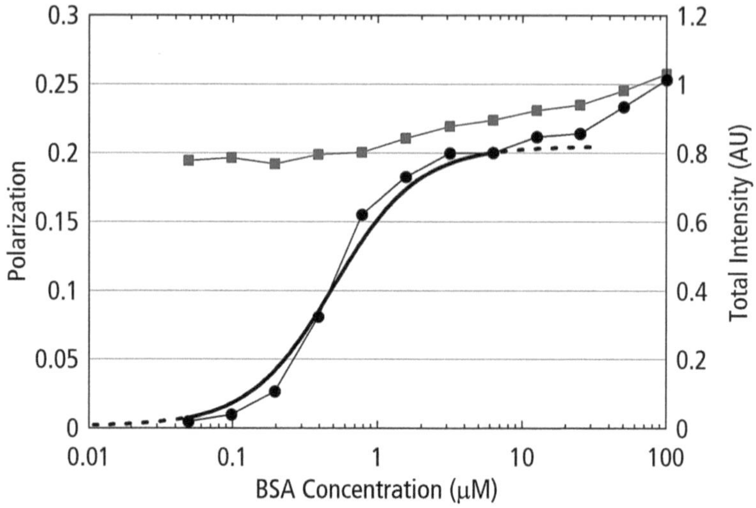

Fig. 10 Binding isotherm of ANS and BSA as a function of BSA concentration based on Subheading 3.4. *Grey squares*—polarization, *black circles*—total intensity. The nonlinear fitted curve for a moderately high affinity binding site is shown as a *heavy line* over a data range from which $K_d = 0.5$ μM is obtained. The *dotted line* is an extension of the fit beyond the data points used in the fit. Note that low affinity binding sites are not included in the fitting model. The final fitting parameters for the free and bound intensities are $I_{free} = 0.007$, $I_{bound} = 0.818$, respectively

To analyze the data

1. Calculate the overall binding constant from the binding isotherm by the application of Eqs. 7 and 10. To generate the isotherm, plot data as intensity versus total protein concentration, and fit to the derived equation using nonlinear regression: $I = I_{free} + (I_{bound} - I_{free})\left\{P_t + L_t + K_d - \sqrt{(P_t + L_t + K_d)^2 - 4P_tL_t}\right\} / 2L_t$.

   The quantities I_{free} and I_{bound} are the fluorescence intensities of the free and bound ligand, respectively. For an initial estimate of I_{bound}, use the maximum value of intensity measured in the low micromolar concentration range where the curve levels off due to saturation of the high affinity binding site.

ANS undergoes a very large enhancement of its quantum yield when bound to a protein or in a hydrophobic environment (47). In water, the quantum yield is very low, but increases about 200-fold upon binding to BSA. This change in the quantum yield makes using the emission intensity preferable to the polarization. Also, in contrast to the case of fluorescein binding to BSA in Subheading 3.3, the lifetime of ANS in water is much shorter at ~0.2 ns that that of fluorescein at 4 ns which results in a much higher polarization of only ~0.1. Figure 10 shows the total intensity and polarization of ANS as a function of BSA concentration. From the intensity, it is apparent that there are two binding regimes, one with a K_d in the sub micromolar range and one much weaker binding in the tens of micromolar range. It is much more difficult to accurately determine the binding constant in this case from the polarization due to the smaller change and very large enhancement upon binding. Thus, in this case the total intensity was used to investigate the binding. The total intensity I_t measured at each concentration is given by $I_t = f_{bound} I_{bound} + f_{free} I_{free}$ where f is the fraction of photocurrent and I is the intensity due to that species (for two species) as described in Subheading 1.5, see Eq. 12. The most rigorous approach is to employ nonlinear regression using as many experimentally determined parameters as possible, such as the intensity of the completely bound and unbound ligand.

3.5 Quenching of Intrinsic Protein Fluorescence by Ligand Binding and Stern–Volmer Relationship

In this experiment, we will use a nonfluorescent ligand, furosemide to quench the intrinsic tryptophan fluorescence of HSA (48). HSA contains only one tryptophan and 18 tyrosine residues. Again we will use the serial dilution method and measure the fluorescence emission from only the tryptophan in the HSA.

To conduct the experiment

1. Start with a 3 ml combined solution of 25 μM furosemide and 1 μM HSA, placed in a 10 mm × 10 mm quartz cuvette.
2. Set the excitation wavelength to 295 nm and place an appropriate emission filter in the emission path to collect emission only from the tryptophan. A bandpass filter that transmits from ~320 to ~380 nm is optimal. If the spectrometer does not allow the use of an emission filter, then use the emission monochromator set to 350 nm. Check for scattered light.
3. If the G factor for the spectrometer is not known, determine it by setting the excitation polarizer horizontal and calculate G from Eq. 17 by measuring the intensity with the emission polarizer set vertically and horizontally. Using the G factor to correct the intensities, measure and save the polarization or intensity of the sample.
4. Remove 1 ml of solution and discard. Replace with 1 ml of 1 μM HSA solution and mix thoroughly. Measure the polarization.
5. Repeat these steps until the effective furosemide concentration is almost zero or the polarization or intensity reaches a plateau.

To analyze the data

1. Plot Stern–Volmer graphs in the form of $I_0/I_f - 1$ versus [Q] and estimate the quenching constant K_{SV} from the slope of the linear regression line using Eq. 3.

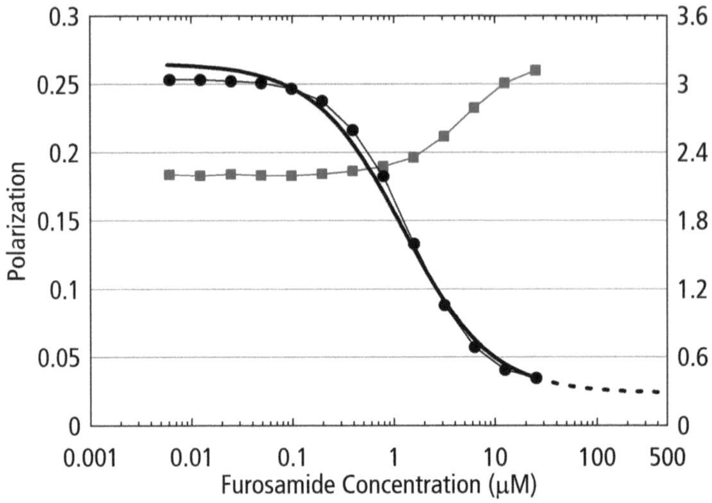

Fig. 11 Binding isotherm of Furosemide and HSA as a function of Furosemide concentration. *Grey squares*—polarization, *black circles*—total intensity. The *heavy line* represents the results of nonlinear regression analysis for the Stern–Volmer intensity quenching plot according to Subheading 3.5. The Stern–Volmer constant provides a direct measure of quenching efficiency (the higher the K_{SV}, the lower the concentration of quencher required for quenching) and its apparent value from the curve fit is $K_{SV} = 0.8\ \mu M^{-1}$. The *dotted line* shows the fitted curve beyond the concentration range of the measurement. The final values of fitted intensities are respectively, $I_{unbound} = 3.17$ and $I_{bound} = 0.28$

2. To minimize the distortion of error inherent to the linear form, it is advisable to fit the original data to the curve $I_f = I_0/(1 + K_{SV}[Q])$ using nonlinear regression.

In this experiment we measured the decrease of the intrinsic protein fluorescence due to quenching by furosemide. Once again we can see that the change of the polarization is not very large between the quenched and unquenched HSA, while there is a substantial change in the fluorescence intensity (Fig. 11). The kinetics of quenching by furosemide follows the Stern–Volmer relationship, Eq. 3 (*see* **Note 18**). While obtaining the quenching constant provides useful practical information about the binding process (such as rates of diffusion, exposure to the quencher, and location of the fluorophore on a protein) it is the ratio of the intensity of the bound and free ligand that is important for the determination of the binding constant. The calculations of K_b in this experiment are similar to those of the previous one, except that the intensity of the bound ligand–protein complex is less than in the unbound state, and thus the same methodology can be employed and Eq. 11 be used for least-squares fitting.

4 Conclusions

We have described a theoretical basis and numerous experimental procedures which employ fluorescence in the investigation of ligand binding. The choice of which experiment to use depends on the fluorescence properties of the system under investigation. In cases where there is a change in the quantum yield of the ligand or protein, it may be best to use the total intensity while a change in rotational mobility is best manifested using polarization or anisotropy. We have also included numerous practical hints and tips to illuminate the reader to common pitfalls and artifacts encountered during fluorescence measurements.

5 Notes

1. Several intra-molecular and inter-molecular deactivation pathways are available to the excited singlet state which can decrease the fluorescence lifetime and intensity (2–4). Typical competing pathways include vibrational relaxation, internal conversion (ic), and intersystem crossing (isc), described by the respective rate constants, k_x. The singlet state can interact with other molecules present in the solution, including the solvent, potentially leading to external conversion (ec), and electron transfer and energy migration between two identical or two different molecules. These quenching processes may be either dynamic or static in nature depending on diffusion constants and the quenching rate constant. In the dynamic case, the fluorophore and the quencher collide with each other while the fluorophore is in the excited state. In static quenching, the fluorophore forms a complex with the quencher preventing the fluorophore from emitting fluorescence. As a result, the overall rate constant becomes, in general, $k_f + k_{isc} + k_{ic} + k_{ec} + k_q[Q]$ where [Q] is the concentration of the quencher if present.
2. The quantum yield may also be defined as $\Phi = n_f / n_a$ where n_f and n_a represent the respective numbers of photons emitted and absorbed by the fluorophore.
3. The singlet lifetime can be determined by means of various experimental techniques, including pulse fluorometry in the time domain, phase and modulation fluorometry in the frequency domain, picosecond transient absorption spectroscopy, and single photon timing (31, 49).
4. Quantum yield standards should be selected to ensure maximum overlap of the emission between sample and reference, but also sufficient absorption at the excitation wavelength. For example, for spectral characterization of the far-red fluorescent

proteins, crystal violet can be used in methanol as a quantum yield standard with correction for the refractive index difference between the sample and reference solutions (2). A detailed description on the determination of quantum yields is given in ref. 50.

5. At low optical density (OD, i.e., absorbance) the measured fluorescence intensity is proportional to the sample concentration, but as the optical density increases, the intensity deviates from linearity. In an ideal case, the optical density is very low, less than 0.05 at the excitation wavelength. Under these conditions the fraction of light absorbed by the sample is negligible and does not result in an artifact of the measurement. At an optical density of 0.1, the intensity is underestimated by ~10 % (3). At very high optical densities, an increase in optical density can in fact result in a decrease of the measured intensity due to the inner filter effect (12). In some cases, however, it is not possible to have a low OD, for example at high concentrations of the fluorophore. Under such circumstances this problem can be alleviated through the appropriate choice of cuvette and/or sample geometry. At higher ODs, if using for example 4 or 2 mm × 10 mm cuvettes, one can rotate the cuvette such that the shorter path length side is facing the excitation beam.

6. The efficiency of energy transfer can be written in the form: $E = k_{et}/(k_{et} + k_f + \sum k_{nr}) = R_0^6/\left(R_0^6 + R^6\right)$ from which some rearrangement leads to R. Experimentally, the efficiency may be easily obtained from measurements of the fluorescence lifetime or the quantum yield of the donor in the presence and absence of the acceptor: $E = 1 - \tau_{DA}/\tau_D = 1 - \Phi_{DA}/\Phi_D$. If the acceptor is fluorescent, E also can be determined by measuring the induced fluorescence since at a given wavelength the magnitude of the excitation spectrum is a linear combination of the absorption terms: $\bar{I}_f(\lambda) = \varepsilon_A(\lambda) + E\varepsilon_D(\lambda)$.

   The Förster distance can be evaluated in terms of the normalized spectral overlap integral J of the donor emission spectrum with the acceptor absorption spectrum as: $R_0^6 = 9\Phi_D\kappa^2 J \log_e 10/\ 128\pi^5 n^4 N_A$ where κ^2 is the dipole orientation factor characterizing the mutual molecular orientation of the D → A pair, n is the refractive index of the medium (typically ~1.33), and N_A is Avogadro's number. The spectral overlap integral can be calculated using the relationship $J = \int \bar{I}_D(\lambda)\, \varepsilon_A(\lambda)\, \lambda^4 d\lambda$ where $\bar{I}_D$ is the normalized emission spectrum. For practical use, the integral may be approximated as a weighted sum of emission intensities: $J = \sum \bar{I}_D(\lambda)\varepsilon_A(\lambda)\lambda^4 \Delta\lambda/\sum \bar{I}_D(\lambda)\Delta\lambda$.

   The orientation factor accounts for the effect of the relative orientation of the excited state dipole of the donor and the absorption dipole of the acceptor. It is very difficult to experimentally determine, but its upper and lower limits can be calculated from fluorescence anisotropy measurements (51). For random orientations and rapid rotations of the D $\rightarrow$ A pair, $\kappa^2 = 2/3$ is often assumed. This assumption should only be made when the rotational rate of the fluorophore is much greater than the fluorescence lifetime of the donor, which is not the case for most fluorescent proteins. In this latter case, κ^2 falls in the range from 0 to 4, and extra experimental or structural information is required to limit the values of κ^2 (52).

7. For multiple donor–acceptor pairs, the transfer efficiency is given by $E = (1/n_D) \sum_D \left\{ \left(\sum_A (R_0/R)^6 \right) / \left(1 + \sum_A (R_0/R)^6 \right) \right\}$ where n_D is the number of donors and each term in the sum corresponds to a distinct D $\rightarrow$ A pair.

8. The theoretical limits of polarization that correspond to parallel and perpendicular transition moments in isotropic solutions are given by $-1/3_{\perp} \leq p \leq 1/2_{\parallel}$. For anisotropy, the limits are $-1/5_{\perp} \leq r \leq 2/5_{\parallel}$. In the absence of rotation, for example in highly viscous solutions such as glycerol, the average angle β between the absorption and emission dipoles of the fluorophore can be determined from $1/p_0 - 1/3 = (5/3)\{2/(3\cos^2\beta - 1)\}$ where p_0 is the polarization in the absence of rotation, also termed the limiting polarization. When the rotation of the fluorophore is very fast relative to the fluorescence lifetime, the fluorophore randomizes its orientation and loses its correlation to the excitation polarization and thus $p \approx 0$. At comparable values of the rotational period and the fluorescence lifetime, the extent of polarization attains intermediate values. Note also that p_0 is wavelength dependent. If one plots this quantity as a function of wavelength, such an excitation polarization spectrum yields information about the relative orientation of absorption bands since each transition dipole is associated with a particular direction.

9. The reciprocal of polarization can be summed in a similar fashion as the straight function of anisotropy, namely, $1/(1/p - 1/3) = \sum_i f_i/(1/p_i - 1/3)$ but this is less convenient in common calculations. Therefore, some investigators resort to direct substitution of polarization data into equations formulated for anisotropy. This practice is strictly incorrect and leads to systematic deviations in determination of binding constants described in ref. 39.

10. The rotational correlation time can be analytically defined for rigid spherical molecules or molecular complexes in terms of the hydrated volume of the molecule, V_h, the viscosity of the medium, η, and the rotational diffusion coefficient, D_{rot}, as $\tau_{rot} = \eta V_h/kT = (1/6)D_{rot}$ (Stokes equation) where k is the Boltzman constant and T is the absolute temperature. According to this relationship, depolarization and its dependence on

molecular size and environment may be studied by observing the fluorescence of molecules in solvents of various viscosities at low fluorophore concentrations. Combining the Stokes equation and the Perrin equation, the polarization of a molecular complex increases with its volume and decreases with increasing fluorescence lifetime.

11. Tryptophan in proteins tends to display multi-exponential fluorescence decay. This complexity is a result of the photo-physical properties of the indole ring and its interaction with surrounding amino acid residues. Heterogeneity of the tryptophan fluorescence lifetime can arise from different conformations and rotamers of the indole side chain in the ground state in addition to energy and electron transfer, dynamic quenching, and dipolar relaxation among other processes (32). In addition, due to the degeneracy and inter-conversion of energy levels in the excited state, non-exponential fluorescence decay may be frequently observed even in proteins with a single tryptophan residue. This is further complicated in the case of multi-tryptophan proteins where only average lifetimes can be determined. For a very detailed description of the fluorescence properties of tryptophan, see ref. 53.

12. The equilibrium constant is directly related to the Gibbs free energy of complex formation. Here the free energy change is the difference between the free energy of the complex and the sum of the free energies of its components. Combining the concentration terms into a single quotient, namely, the equilibrium constant, and expanding the free energy in terms of the related enthalpy and entropy changes, ΔG is given by $\Delta G = RT \ln K_d = -RT \ln K_a = \Delta H - T\Delta S$ where ΔH and ΔS are the change in enthalpy and entropy respectively, and T is the absolute temperature. The free energy becomes more negative during the course of complex formation and it reaches a minimum at equilibrium. Thus, ΔG determines the stability of the molecular complex and as a result K is a measure of the affinity of binding. The larger the binding constant, the more strongly bound the ligand is. The enthalpy change originates from the formation of internal interactions of the ligand with the protein and from the corresponding desolvation of the interacting atoms. The entropy term, $T\Delta S$, expresses the alteration of molecular order (through a reduction in the number of free molecular entities) and the reduction of the translational degrees of freedom on complex formation. The dissociation constant measures the propensity of the complex to disintegrate into its components. The value of the dissociation constant in molar units corresponds to the concentration of the ligand at which the binding site on the protein is half occupied. At this concentration of the ligand, the concentration of the

complex equals the concentration of protein with no ligand bound.

13. In the rigorous sense, the binding constant in real nonideal solutions is defined in the standard state of the components, i.e., in unit activity. The activity is related to the concentration by the activity coefficient such that $a = \gamma c$ and thus $K_b = K_c K_\gamma$ where K_c and K_γ are the concentration equilibrium constant and the activity quotient, respectively. K_b is then of the form $K_b = a_{P_mL_n}/\,a_P^m a_L^n = ([P_mL_n]/[P]^m[L]^n)\left(\gamma_{P_mL_n}/\gamma_P^m\gamma_L^n\right)$. Since concentrations parallel activities and fluorescence measurements are normally performed at large dilutions where the activity coefficients change very little and are not significantly different from unity, it has become standard practice to measure K_c in fluorometry and then to apply the near-equivalence identity $K_b \cong K_c$ in the range of compositions studied.

14. To address some possible strategies and utilities in the measurement of equilibrium constants, consider the interaction of the anti-inflammatory drug flurbiprofen (FBP) with HSA (54). When FBP is titrated with HSA, the fluorescence emission of FBP is quenched, while that of the HSA remains virtually the same. The protein, the free ligand, and the bound ligand all emit fluorescence in the same spectral region, i.e., the total fluorescence at emission wavelength is $I_f = I_P + I_L + I_{PL}$ where the component intensities are measured via Eq. 2 as $I_X = 2.303 I_0\, \Phi_X \varepsilon_X [c] l\ (X = P,\ L,\ PL)$. Let I_P^0 be the fluorescence intensity of the free protein, I_L^0 the fluorescence intensity of the free ligand, and I_{PL}^0 the fluorescence intensity of the protein–ligand complex which corresponds to that of the fully bound ligand in this particular system. P_t and L_t are the total concentrations of protein and ligand at a given point of the measurement. Then the proportionality constants can be calculated as $\alpha_P = I_P^0/P_t$, $\alpha_{PL} = I_{PL}^0/L_t$, and $\alpha_L = I_L^0/L_t$, and thus the fluorescence intensity becomes $I_f = \alpha_P[P] + \alpha_L[L] + \alpha_{PL}[PL]$. The equilibrium concentrations of all components can be derived in terms of the concentration of the free ligand, namely, $[PL] = L_t - [L]$, $[P] = P_t - [PL] = P_t - L_t + [L]$, and—by making the necessary substitutions—$[L] = (I_f - \alpha_P P_t - \alpha_{PL} L_t + \alpha_P L_t)/\,(\alpha_P - \alpha_{PL} + \alpha_L)$. Since I_{PL}^0 cannot be measured directly, we have to obtain α_{PL} from an auxiliary measurement. After the ligand becomes fully bound, the fluorescence intensity becomes $I_b = \alpha_P[P] + \alpha_{PL}[PL]$. Then $[PL] = L_t$, $[PL]/L_t = 1$, and $[P] = P_t - [PL] = P_t - L_t$ from which we find $\alpha_{PL} = (I_b - \alpha_P P_t + \alpha_P L_t)/L_t$. Finally, substituting the equilibrium concentrations defined above, we obtain a direct estimate of the binding constant for the HSA–FBP complex as $\widehat{K_b} = [PL]/[P][L]$. For statistical significance, average values of K_b can be calculated by $K_b = (1/n)\sum_n \widehat{K}_b$ using data taken

from at least $n = 7$ points on the titration curve. The most accurate value of K_b, however, may be obtained by nonlinear regression using equilibrium concentrations calculated in a manner similar to that described above.

15. In the case of weak to moderately strong binding, K_b can be also determined by the application of the coordinates of dilution method, a special case of serial dilution (55). In this approach the data are represented in a linearized coordinate system of dilution points. The only operation which is required is a serial dilution of the preexisting protein–ligand complex. Dilution of the samples in which both the protein and the ligand are present, will shift the dynamic equilibrium into the direction of dissociation. Measuring the concentration of the bound ligand in the sequence of dilutions permits the calculation of K_b. If d_i is the dilution factor at the ith step of the dilution series, then K_b can be defined from the species concentration [X] as the function of d_i: $K_b = [\text{PL}]/(\text{P}_t/d_i - [\text{PL}])\ (\text{L}_t/d_i - [\text{PL}])$ where d_i is given as the ratio of sample volume and the aliquot volume in each step, and the sample volume equals to the aliquot volume plus the diluent volume. In other words, the total concentration of the protein and the ligand is decreased by d_i times at each step, and a new concentration of the complex is attained. In biologically meaningful cases when $\text{L}_t/d_i \gg [\text{PL}]$, the expression reduces to $K_b = [\text{PL}]/(\text{P}_t/d_i - [\text{PL}])\ (\text{L}_t/d_i)$, and thus the equilibrium concentration of the complex depends on the dilution factor as $[\text{PL}] = \text{P}_t\text{L}_t K_b/d_i(d_i + \text{L}_t K_b)$. Similar equations can be developed when it is easier to measure the free protein than the bound ligand. If the binding is very strong, extreme dilutions are required to dissociate the complex and therefore standard fluorescent techniques may not be applicable. At very low concentrations, alternative techniques may be needed such as fluorescence correlation spectroscopy which is beyond the scope of this chapter (56, 57).

16. Upon complex formation, the quantum yield of fluorescence of either P or L increases resulting in an enhancement of fluorescence of the solution. A case in point is certain environmentally sensitive fluorescent probes, for example ANS, whose quantum yield is very low in aqueous solution but significantly increases on binding to nonpolar regions of proteins. Here $\alpha_L \approx 0$ and thus the rightmost term the numerator of Eq. 16 reduces to 0. The resultant equation has the form of a rectangular hyperbola which can be converted into a linear form to expedite data analysis. The most convenient transform has the properties of a x-reciprocal or Scatchard-type plot: $(I_f/I_0 - 1)/[\text{L}] = K_b(\alpha_{PL}/\alpha_P - I_f/I_0)$. If both $\alpha_L \approx 0$ and $\alpha_{PL} \approx 0$ then Eq. 16 further simplifies to $I_0/I_f = 1 + K_b[\text{L}]$ and K_b can be

determined from the linear plot of I_0/I_f as function of [L]. Finally, when $\alpha_L \neq 0$, Eq. 16 applies without simplification and α_P and α_L can be obtained from measuring the protein and ligand individually. To correct for the emission by the ligand, one may use the initial slope estimates of the dependent variable $I_f/I_0 - \alpha_L[L]/\alpha_P$.

17. In this arrangement, either P or L is fluorescent and the fluorescence intensity decreases upon complex formation. The quenching is due to FRET between the donor (usually P) and the acceptor (usually L). The interaction can be conveniently followed by evaluating the function $I_0/I_f = \Phi_0/\Phi_f$ or $I_0/I_f = \tau_0/\tau_f$ which then yields E and R (*see* **Note 6**). For determination of the binding constant, Eq. 16 is applicable with experimentally determined (non-zero) values of α_P, α_L, and α_{PL}.

18. It happens that we study competitive equilibria in which a nonfluorescent ligand and nonfluorescent quencher Q both bind to an intrinsically fluorescent protein resulting in a decreased probability of complex formation with the ligand. Then if the concentration of the ligand is kept constant while the concentration of the quencher is increased, the binding of the quencher effectively decreases the protein emission, for example via energy transfer or other quenching mechanisms governed by the Stern–Volmer relationship, Eq. 3 (2, 3, 19). Ideally, in diffusion (or time)-limited quenching where the quencher and the excited molecules collide with perfect effectiveness, the quenching rate constant can be calculated as $k_q = (8/3)\,RT/\eta$ with η being the viscosity of the solution, and R and T are the universal gas constant and absolute temperature, respectively. In most real applications, however, k_q has to be determined experimentally, for example from direct measurement of the fluorescence lifetime or from the slope of the equation because not all collisions lead to effective quenching.

19. Equation 10 may be readily transformed into two commonly used linear forms. A double reciprocal linear plot of $1/I_f$ versus $1/(L_tI_\infty - P_tI_f)$ may be written in the form of a Klotz plot: $1/I_f = 1/K_b \times 1/(L_tI_\infty - P_tI_f) + 1/I_\infty$ from which K_b can be determined from the slope. The other alternative may be expressed as an *x*-reciprocal plot of $I_f/(L_tI_\infty - P_tI_f)$ versus I_f/I_∞ similar to a Scatchard plot: $I_f/(L_tI_\infty - P_tI_f) = K_b - K_b \times I_f/I_\infty$. The two linear forms are not statistically equivalent as the experimental points are not equally spaced on the axes. This can be compensated for by using statistical weighting or by defining the experimental points non-equidistantly.

20. Some other common linear forms for 1:1 binding are the *y*-reciprocal plotting form and the *log–log* plot of the binding isotherm: $[L]/x_P = [L] + 1/K_b$ and $\log(x_P/(1 - x_P)) = \log[L] - \log K_b$, respectively. Still further forms of linearization are possible provided there is no interaction among the binding sites. In general, however, it is better to use nonlinear least-square fitting of the original data. Nonlinear regression may also be applied to linearly transformed data with some advantages that it is easy to obtain starting values of the parameters.
21. Alternatively, an optical filter can be utilized which only transmits a limited spectral region of the light. The advantage of a filter is that the transmission efficiency can be much higher than that of a monochromator, but the transmission through a filter cannot normally be tuned to other wavelengths.
22. The main purpose of the quantum counter is to abrogate the wavelength dependency of the reference detector as the efficiencies of PMTs are strongly dependent upon the energy of the incident light. Recall that a PMT works using the photoelectric effect. The quantum counter ensures that regardless of the excitation wavelength, the light which reaches the reference detector will only be that of the quantum counter cell emission.
23. These types of filters fall into two broader categories, depending on their composition and manufacturing process. Colored glass filters are, as the name suggests, made from a solid piece of glass or quartz impregnated with certain chemicals to give them the required optical properties. Alternatively, filters can be made by deposition of layers of chemicals on the surface of a glass or quartz substrate. These types of filters generally have a sharper transition from the transmitted to blocked region and a higher maximum transition but are also much more expensive. The use of plastic filters should generally be avoided because of very poor transmission at shorter wavelengths and inherent fluorescence from the plastic.
24. One needs to be aware of the material which makes up the filter. If conducting experiments in the UV, the filters need to be made from quartz or spectrosil otherwise they will absorb the entire light incident on them. Standard glass, e.g., BK7 absorbs almost all light below ~350 nm and is not suitable for intrinsic protein fluorescence.
25. Fluorescence cuvettes differ from absorption cuvettes as they have all four sides polished, while absorption cuvettes only have two polished sides. The choice of appropriate cuvette depends on the optical properties of the sample (i.e., optical density), sample volume and excitation beam width. It is important that the width of the cuvette be greater than the width of the excitation beam, otherwise optical artifacts may be introduced.

26. When placing the cuvette into the sample compartment, it is important to check the position of the meniscus relative to the beam height. If the beam hits the meniscus it will cause spurious results. The absolute minimum volume for 10 mm × 10 mm, 4 mm × 10 mm and 2 mm × 10 mm cuvettes is 1,500, 600 and 300 μl respectively. Using these small volumes it is generally necessary to prop up the cuvette so that the beam does not hit either the bottom of the cuvette or the meniscus. Often the lid of the cuvette is a convenient height for such a purpose.
27. The presence of the Wood's anomaly has often been incorrectly interpreted as the presence of two populations of fluorophores. The Wood's anomaly always occurs at the same wavelengths on a particular instrument but different wavelengths on different makes of instrument.
28. When measuring the excitation spectrum of a sample, a longpass filter could be used instead of the emission monochromator which results in a significantly greater amount of light detected, as the entire emission can be collected instead of just a small portion. This also allows one to use a lower excitation intensity, thus reducing photobleaching.
29. Small amounts of scattered excitation light can severely affect the polarization measurement. In order to check the amount of scattered light detected, one should place the emission filter into the excitation path and check the intensity. If it is greater than 0.5–1 % of the sample intensity then modifications to the experiment need to be done, whether it is another excitation wavelength or another emission filter. Ideally the scattered light should be at most the level of the dark counts.
30. It is crucial that the intensity always stays within the linear range of the detector. In order to achieve this, the excitation intensity should be reduced, usually by using a neutral density filter or by reducing the slit width. The filter could be placed in the emission path also but this does not provide the significant advantage of reducing the excitation intensity and thus the photobleaching of the sample.
31. Prior to obtaining a complete binding isotherm, it is recommended to perform the intended experiment only at the two extremes of protein–ligand ratio, i.e., with excess ligand or protein and with no ligand or protein, e.g., BSA + fluorescein. One would measure the polarization (and total intensity) of 100 μM BSA with 1 μM fluorescein and also of only 1 μM fluorescein. This approach also provides useful information on any fluorescent enhancement or quenching of the protein–ligand complex. Furthermore, it estimates an approximate range for the binding constant.

Acknowledgments

This publication was made possible in part by grant no. G12RR003061 from the National Center for Research Resources (NCRR), a component of the National Institutes of Health (NIH). Its contents are solely the responsibility of the author and do not necessarily represent the official view of NCRR or NIH (GM). This work was supported by National Institutes of Health grants RO1GM076665 (JAR). We thank Professor David M. Jameson for useful discussions and for proofreading the manuscript.

References

1. Brand L, Johnson ML (eds) (2008) Fluorescence spectroscopy. Methods Enzymol 450:1–358
2. Lakowicz JR (2006) Principles of fluorescence spectroscopy, 3rd edn. Springer, New York
3. Valeur B (2002) Molecular fluorescence. Principles and applications, Wiley-VCH, Weinheim
4. Sharma A, Schulman SG (1999) Introduction to fluorescence spectroscopy. Wiley-Interscience, New York
5. Valeur B, Brochon JC (eds) (2001) New trends in fluorescence spectroscopy. Applications to chemical and life sciences. Springer, Berlin
6. Kraayenhof R, Visser AJWG, Gerritsen HC (eds) (2002) Fluorescence spectroscopy, imaging and probes. New tools in chemical, physical and life sciences. Springer, Berlin
7. Gell C, Brockwell D, Smith A (2006) Handbook of single molecule spectroscopy. Oxford University Press, Oxford
8. Roehrl MHA, Wang JY, Wagner G (2004) A general framework for development and data analysis of competitive high-throughput screens for small-molecule inhibitors of protein–protein interactions by fluorescence polarization. Biochemistry 43:16056–16066
9. Eccleston JF, Hutchinson JP, Jameson DM (2005) Fluorescence-based assays. Prog Med Chem 43:19–48
10. Jameson DM, Ross JA (2010) Fluorescence polarization/anisotropy in diagnostics and imaging. Chem Rev 110:2685–2708
11. Weljie AM, Vogel HJ (2002) Steady-state fluorescence spectroscopy. Methods Mol Biol 173:231–253, Clifton, NJ
12. Jameson DM, Croney JC, Moens PD (2003) Fluorescence: basic concepts, practical aspects, and some anecdotes. Methods Enzymol 360:1–43
13. Förster T (1948) Intermolecular energy migration and fluorescence. Ann Phys 2:55–75
14. Förster T, Sinanoglu O (eds) (1996) Modern quantum chemistry. Academic Press, New York
15. Hammes G (1981) Fluorescence methods. In: Frieden C, Nichol LW (eds) Protein–protein interactions. Wiley-Interscience, New York, pp 257–287
16. Demchenko AP (2009) Introduction to fluorescence sensing. Springer Science + Business Media B V, New York
17. Xiao J, Wei X, Wang Y, Liu C (2009) Fluorescence resonance energy-transfer affects the determination of the affinity between ligand and proteins obtained by fluorescence quenching method. Spectrochim Acta A 74:977–982
18. Shaw AK, Pal SK (2008) Resonance energy transfer and ligand binding studies on pH-induced folded states of human serum albumin. J Photochem Photobiol B Biol 90:187–197
19. Permyakov EA (1993) Luminescent spectroscopy of proteins. CRC Press, Boca-Raton, FL
20. Weber G (1966) Polarization of the fluorescence of solutions. In: Hercules DM (ed) Fluorescence and phosphorescence analysis. Wiley-Interscience Publishers, New York
21. Jameson DM, Croney JC (2003) Fluorescence polarization: past, present and future. Comb Chem High Throughput Screen 6:167–173
22. Dandliker WB, Feijen GA (1961) Quantification of the antigen-antibody reaction by the polarization of fluorescence. Biochem Biophys Res Commun 5:299–304
23. Dandliker WB, de Saussure VA (1970) Fluorescence polarization in immunochemistry. Immunochemistry 7:799–828
24. Weber G, Young LB (1964) Fragmentation of bovine serum albumin by pepsin. I The origin of the acid expansion of the albumin molecule J Biol Chem 239:1415–1423
25. Levine RJ, Teller DN, Denber HC (1968) Binding of chlorpromazine and thioproperazine in vitro. 3. Fluorometric measurement of

changes in *Limulus polyphemus* (horseshoe crab) myosin B structure and enzyme activity after treatment with phenothiazine drugs. Mol Pharmacol 4:435–444

26. Jameson DM, Sawyer WH (1995) Fluorescence anisotropy applied to biomolecular interactions. Methods Enzymol 246:283–300
27. Kusbaa J, Lakowicz JR (1999) Definition and properties of the emission anisotropy in the absence of cylindrical symmetry of the emission field: application to the light quenching experiments. J Chem Phys 111:89–99
28. Fixler D, Namer Y, Yishay Y et al (2006) Influence of fluorescence anisotropy on fluorescence intensity and lifetime measurement: theory, simulations and experiments. IEEE Trans Biomed Eng 53:1141–1152
29. Mocz G (2006) Information content of fluorescence polarization and anisotropy. J Fluoresc 16:511–524
30. Perrin F (1929) La fluorescence des solutions. Ann Phys Ser 10(12):169–275
31. Ross JA, Jameson DM (2008) Time-resolved methods in biophysics. 8. Frequency domain fluorometry: applications to intrinsic protein fluorescence. Photochem Photobiol Sci 7:1301–1312
32. Callis PR, Liu T (2004) Quantitative predictions of fluorescence quantum yields for tryptophan in proteins. Phys Chem B 108:4248–4259
33. Chen Y, Barkley MD (1998) Toward understanding tryptophan fluorescence in proteins. Biochemistry 37:9976–9982
34. Mocz G (2007) Fluorescent proteins and their use in marine biosciences, biotechnology, and proteomics. Mar Biotechnol (NY) 9:305–328
35. Tsien RY (1998) The green fluorescent protein. Annu Rev Biochem 67:509–544
36. Das K, Sarkar N, Bhattacharya K (1993) Interaction of urea with fluorophores bound to protein surface J Chem Soc Faraday Trans 89:1959–1961
37. Del Castillo B, Alvarez-Builla J, Lerner DA (1991) Fluorogenic reagents and fluorescent probes. In: Baeyens WRG, De Keukeleire D, Korkidis K (eds) Luminescence techniques in chemical and biochemical analysis. Marcel-Dekker, Inc, New York, pp 73–139
38. Mocz G, Helms MK, Jameson DM et al (1998) Probing the nucleotide binding sites of axonemal dynein with the fluorescent nucleotide analogue 2′(3′)-O-(–N-methylanthraniloyl)-adenosine 5′-triphosphate. Biochemistry 37:9862–9869
39. Jameson DM, Mocz G (2005) Fluorescence polarization/anisotropy approaches to study protein–ligand interactions: effects of errors and uncertainties. Methods Mol Biol 305:301–322
40. Connors KA (1987) Binding constants. The measurement of molecular complex stability. Wiley-Interscience, New York
41. Martell AE, Motekaitis RJ (1988) The determination and use of stability constants. VCH Publishers, Inc, New York
42. Jameson DM, Weber G, Spencer RD et al (1978) Fluorescence polarization measurements with a photon-counting photometer. Rev Sci Instrum 49:510–514
43. Weber G (1989) From solution spectroscopy to image spectroscopy. In: Kohen E (ed) Cell structure and function by microspectrofluorometry. Academic Press, New York, pp 71–85
44. Daniel E, Weber G (1966) Cooperative binding by bovine serum albumin. I. The binding of 1-anilino-8-naphthalenesulfonate. Fluorimetric titrations. Biochemistry 5:1893–1899
45. Weber G, Daniel E (1966) Cooperative effects in binding by bovine serum albumin. II. The binding of 1-anilino-8-naphthalene-sulfonate. Polarization of the ligand fluorescence and quenching of the protein fluorescence. Biochemistry 5:1900–1907
46. Togashi DM, Ryder AG (2008) A fluorescence analysis of ANS bound to bovine serum albumin: binding properties revisited by using energy transfer. J Fluoresc 18:519–526
47. Laurence DJR (1952) A study of the adsorption of dyes on bovine serum albumin by the method of polarization of fluorescence. Biochem J 51:168–180
48. Voelker JR, Jameson DM, Brater DC (1989) In vitro evidence that urine composition affects the fraction of active furosemide in the nephrotic syndrome. J Pharmacol Exp Ther 250:772–778
49. Jameson DM, Ross JA, Albanesi JP (2009) Fluorescence fluctuation spectroscopy: ushering in a new age of enlightenment for cellular dynamics. Biophys Rev 1:105–118
50. Rurack K (2008) Fluorescence quantum yields: methods of determination and standards. In: Resch-Genger U (ed) Standardization and quality assurance in fluorescence measurements I: techniques, vol 5. Springer, Berlin, Heidelberg
51. Dale RE, Eisinger J, Blumberg WE (1979) The orientational freedom of molecular probes: the orientation factor in intramolecular energy transfer. Biophys J 26:161–194; Errata: (1980) 30:365
52. Cheung H (1991) Resonance energy transfer. In: Lakowicz JR (ed) Topics in fluorescence spectroscopy: principles. 2:127–176, Plenum Press, New York
53. Callis PR (2007) Exploring the electrostatic landscape of proteins with tryptophan

fluorescence. In: Geddes CD (ed) Reviews in fluorescence, 4th edn. Springer, New York
54. Takla PM, Schulman SG, Perrin JH (1985) Meaurement of flurbiprofen-human serum albumin interaction by fluorimetry. J Pharm Biomed Anal 3:41–50
55. Bobrovnik S (2005) New capabilities in determining the binding parameters for ligand–receptor interaction. J Biochem Biophys Methods 65:30–44
56. Rigler R, Elson ES (eds) Fluorescence correlation spectroscopy. Theory and applications. Springer, Berlin
57. Barbieri B, Terpetschnig E, Jameson DM (2005) Frequency-domain fluorescence spectroscopy using 280-nm and 300-nm light-emitting diodes: measurement of proteins and protein-related fluorophores. Anal Biochem 344:298–300

Chapter 8

Circular and Linear Dichroism Spectroscopy for the Study of Protein–Ligand Interactions

Tina Daviter, Nikola Chmel, and Alison Rodger

Abstract

Circular dichroism (CD) is the difference in absorption of left and right circularly polarized light, usually by a solution containing the molecules of interest. A non-zero signal for solutions is only measured for chiral molecules such as proteins whose mirror image is not superposable on the original molecule. A CD spectrum provides information about the bonds and structures responsible for the chirality. When a small molecule (or ligand) binds to a protein, it acquires an induced CD (ICD) spectrum through chiral perturbation to its structure or electron rearrangements (transitions). The wavelengths of this ICD are determined by the ligand's own absorption spectrum, and the intensity of the ICD spectrum is determined by the strength and geometry of its interaction with the protein. Thus, ICD can be used to probe the binding of ligands to proteins. This chapter contains an outline of how to perform protein CD and ICD experiments, together with some of the issues relating to experimental design and implementation. Addition of a quarter wave plate to a CD spectropolarimeter converts it to a linear dichroism (LD) spectrometer. When protein samples are aligned either in flow (as for fibers or membrane proteins in liposomes) or on surfaces the orientations of ligands with respect to the protein backbone or other subunits can be determined.

Key words Circular dichroism, Proteins, Chirality, Ligand binding, Induced circular dichroism, Linear dichroism

1 Introduction

A key feature of any biological system is its chirality or asymmetry or handedness: a chiral molecule has a mirror image that is not superposable on itself. This means the two mirror images cannot be rotated so that they look exactly the same. Macroscopic as well as smaller scale chirality is ultimately dependent on the molecular level. Since many molecules in biological systems are chiral and are present in only one enantiomeric form, the macroscopic structures they build are also chiral. Molecular chirality is perhaps most obvious with a helical molecule such as the double helical structure of B-DNA, but it is a feature of all proteins and nucleic acids.

Mark A. Williams and Tina Daviter (eds.), *Protein-Ligand Interactions: Methods and Applications*, Methods in Molecular Biology, vol. 1008, DOI 10.1007/978-1-62703-398-5_8,

Circular dichroism (CD), which is the difference in absorption of left and right circularly polarized light, is probably the simplest technique for non-destructively providing solution phase structural information about the asymmetry of molecules. Many ligands (usually small molecules that bind to a macromolecule or alternatively another macromolecule) in biological systems are also chiral, in which case they have their own CD spectrum, which will probably be perturbed upon binding to a protein. If a ligand is achiral, then it will have no intrinsic CD but will gain an induced CD (ICD) signal in its transitions when it binds to a protein. It is this ICD signal that contains the information about the asymmetry of the protein–ligand interaction. In this chapter we shall focus on how to measure CD spectra of proteins and protein–ligand complexes and how to analyze the data.

As many of the principles are the same, we shall also briefly note how linear dichroism (LD) can be used to study fibrous and membrane systems to give geometric information not obtainable from any other technique.

1.1 Protein Absorbance Spectroscopy

In order to understand CD and LD spectroscopy and use the data intelligently, it is essential that one measures the absorbance spectrum of one's sample since this shows where to expect CD or LD signals. The Beer–Lambert law for the absorption of light by a sample of concentration C is

$$A = \varepsilon C \ell, \tag{1}$$

where ℓ is the length of the sample through which the light passes, and ε is known as the molar extinction coefficient and depends on the wavelength at which the absorbance is being measured. If ℓ is measured in centimeters and C in M = mol/l = mol dm^{-3}, then ε has units of mol^{-1} dm^{3} cm^{-1} = M^{-1} cm^{-1}. For uniform samples, the Beer–Lambert law is valid as long as the spectrometer can measure the intensity of photons passing through the sample (i.e., the concentration is not so large that essentially all photons are absorbed) *and* there are no concentration-dependent intermolecular interactions. One needs to avoid inhomogeneous samples and light scattering samples to be able to apply the Beer–Lambert law (or indeed any other analysis of spectroscopic data) with confidence (1).

In the case of peptides and proteins, the spectroscopy of the amide bonds, the side chains, and any prosthetic groups (such as heme) determines the observed UV/visible absorption spectra, with their intensities and wavelengths often being affected by the local environment of the groups. UV spectra of proteins are usually divided into the "near" and "far" UV regions. The near UV in this context means 250–300 nm and is often described as the aromatic region due to the absorption of the aromatic amino acids, though transitions of disulfide bonds (cysteine–cysteine bonds) also contribute to the total absorption intensity in this region. The far

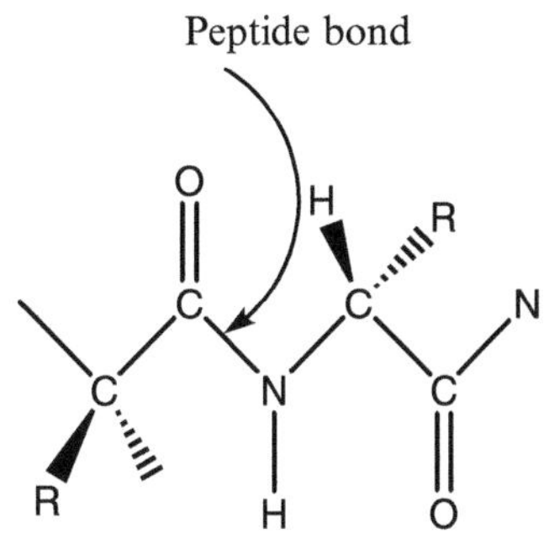

Fig. 1 L-Amino acids joined via rigid peptide bonds, indicated by the *bold lines*

UV (<250 nm) is dominated by absorption due to the peptide backbone of the protein (see Fig. 1), but transitions from some side chains also contribute. The far UV absorbance of a protein is typically of the order of a factor of 100 stronger than the near UV absorbance—though the latter can range from nothing to quite high in a peptide, such as gramicidin, with a very high percentage of aromatic residues.

It should be noted that any components of the buffer that absorb light will contribute to the absorbance even if the molecules are achiral and do not contribute to the CD signal (except to worsen the signal to noise ratio of the CD spectrum by reducing the available photons). The significance of this issue is discussed further below.

1.2 Protein Circular Dichroism

CD is defined as the difference in absorption of left and right circularly polarized light (1)

$$\mathrm{CD} = \Delta A = A_{\mathrm{l}} - A_{\mathrm{r}}, \quad (2)$$

where the subscripts indicate the polarization of light incident on the sample. The CD of macromolecules differs from that of small molecules in that the forms of the CD spectra result largely from the arrangement of the units in space and the coupling between the transitions of those units (1). The CD of proteins is most commonly used in one of two ways: (1) to probe the secondary structure of the protein itself, and (2) to probe the protein's binding interaction with small molecules (usually referred to as ligands) or other macromolecules. Although both these applications are in practice largely empirical, it is useful to have some understanding of the origin of the signal being measured.

Proteins are long chains of amino acids. If the side chain on an amino acid is not simply a hydrogen (i.e., the residue is not glycine), then the tetrahedral carbon of that amino acid is a chiral center and we should expect a CD signal in transitions of the neighboring amide groups and side chains. If, however, there is free rotation about the bonds of the main chain, the observed amide $n \rightarrow \pi^*$

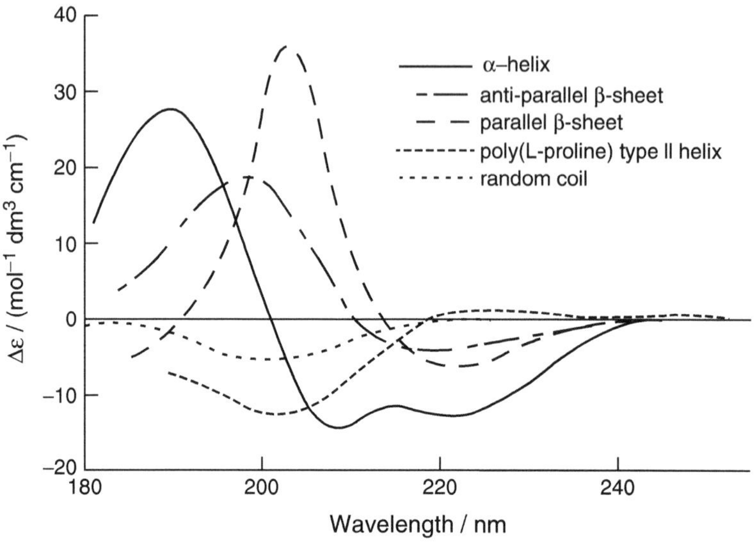

Fig. 2 Backbone (i.e., 180–250 nm) CD spectra for different protein secondary structural motifs (5, 26, 27). α-Helix (average of poly(γ-methyl-L-glutamate) in F_6iPrOH and poly(L-alanine) as a film), anti-parallel β-sheet (BOC (L-alanine)$_7$OMe as a film), parallel β-sheet (BOC (L-valine)$_7$OMe as a film), poly(L-proline) II type helix (system in 10 mM phosphate buffer at 5°C, pH 7), and random coil (collagen at 45°C in 0.01 M phosphate buffer, pH = 3.5)

(a weak transition with $\varepsilon \sim 100$ mol^{-1} dm^3 cm^{-1} occurring at 210–230 nm) and $\pi \rightarrow \pi^*$ (a stronger transition with $\varepsilon \sim 7{,}000$ mol^{-1} dm^3 cm^{-1} centered at 190–200 nm) (1–3) CD signals will be relatively small. Thus, a truly random or denatured protein will have only a small CD signal at the wavelengths accessible to most CD machines. The most stable conformation of a protein under physiological conditions is, however, not random, but composed of well-defined structures that give macro-chiral units (including the well established α-helix and β-sheet) with significant CD intensities which add together to give the observed protein spectrum. A wide range of fitting programs are available to deconvolute experimental CD data into percentages of the structural motifs (4). Some methods work in terms of the spectra of a set of identified structural motifs, others use a basis set of real protein spectra where the percentage of different structural units is known. The approach of Johnson et al. (5) uses a large basis set of such proteins, chooses a subset of these randomly until a good fit is obtained. Almost all approaches will give a good estimate of α-helix content, *if* the protein concentration is known. This is because the α-helix profile is the most distinct and also largest in magnitude as illustrated in Fig. 2. The distinctive α-helical CD spectral shape arises from the coupling of the $\pi \rightarrow \pi^*$ transition moments in each amide chromophore and results in a negative signal at about 208 nm that is not present for other motifs such as β-sheets. Good estimates of percentage content will also be obtained for proteins without α-helices. Reasonable estimates will be obtained for mixed structure proteins, though here accurate values of the amino acid residue concentration are essential.

1.3 Units of CD Spectroscopy

If equal intensities of left and right handed circularly polarized light were passed at the same time (the two actually add together to make a linearly polarized light beam) through a CD active sample the result of differential absorption would be slightly elliptically polarized light. For historical reasons many CD spectropolarimeters therefore produce a CD spectrum in units describing this degree of *ellipticity*, θ, in millidegrees (often referred to as machine units) even though machines actually measure differential absorbance (ΔA). The conversion between these two units is (1, 3)

$$\begin{aligned} \mathrm{CD} &= \Delta A \\ &= \frac{4\pi\theta(\deg)}{180 \ln 10} \\ &= \frac{\theta(\mathrm{mdeg})}{32{,}982}. \end{aligned} \tag{3}$$

The CD analogue of the Beer–Lambert law is

$$\mathrm{CD} = \Delta A = \Delta\varepsilon C\ell, \tag{4}$$

where $\Delta\varepsilon$ is the differential molar extinction coefficient (in units of $\mathrm{mol}^{-1}\ \mathrm{dm}^3\ \mathrm{cm}^{-1}$), C is concentration in M, and ℓ is the path length of sample through which the light passes, in cm. This allows conversion to molar absorbance (or if millidegrees have been retained, molar ellipticity, see below). In this context it is important to have clear in one's mind which concentration units are being used. For example, protein concentrations are sometimes in mg/mL, sometimes in moles of protein molecules per dm^3 and sometimes (as for protein structure fitting programs) in moles of amino acid residues per dm^3. Thus, for example a 5 μM solution of a 129 residue protein has a residue concentration of 645 μM or a weight/volume concentration of approximately 0.07 mg/mL.

Another commonly used unit for reporting CD intensity is molar ellipticity M_θ.

$$M_\theta = \frac{[\theta]M}{100} = \frac{100\theta}{C\ell}, \tag{5}$$

where M is molar mass in g, θ is in degrees, C is in M, ℓ is in cm and $[\theta]$ (or M_θ) is the specific ellipticity

$$[\theta] = M_\theta = \frac{\theta}{C_g d}, \tag{6}$$

where C_g is sample concentration in $g/cm^3 = g/mL$ and d is path length in dm (defined for the convenience of old-fashioned polarimeter measurements of α_D where the path length was 10 cm and the concentration reported in g/cm^3). We then write

$$\begin{aligned} M_\theta &= 3298(\varepsilon_l - \varepsilon_r) \\ &= 3298\Delta\varepsilon \end{aligned} \tag{7}$$

which has a disconcerting factor of 10 difference from Eq. 3 due to the dm to cm conversion that intruded along the way. With proteins an extremely useful empirical check is to note that a 100 % α-helical protein would have $\Delta\varepsilon = 12$ mol^{-1} cm^{-1} dm^3 and $M_\theta = 40{,}000$ degrees^{-1} cm^2 dmol^{-1}, approximately a factor of 3300 between them.

1.4 Protein LD Spectroscopy

LD is the difference in absorbance of light polarized along an orientation direction and perpendicular to it

$$\mathrm{LD} = (A_{\parallel} - A_{\perp}). \tag{8}$$

LD is the only technique that gives relative orientations of proteins and ligands (or other chromophores) in solution. However, the challenge with LD is often how to orient the samples. Most protein LD is performed by flow orienting samples, usually in a Couette flow cell or by drying them on a surface (1, 6–8). Fibers orient readily by either method and membrane proteins can be flow oriented when bound to liposomes as well as dried in bilayers (9–12).

2 Materials

2.1 Samples and Solvents

The materials for protein–ligand CD and LD spectroscopy are most obviously the protein and ligand samples of interest, the accurate concentrations of which need to be known (*see* **Note 1**). In addition, there is the solvent in which the sample is prepared, e.g., most proteins will be in a buffer, as many proteins need buffers to retain their structure. However, if the buffer of choice has an absorbance above 1 (at the concentrations to be used) anywhere in the wavelength range of interest for CD or LD (including below 200 nm where most absorbance spectrometers do not measure), then the buffer will be absorbing most of the light, and the quality of your spectrum will be poor or even nonsense (*see* **Note 2**). Phosphate is a good spectroscopy buffer; Tris is satisfactory; cacodylate is spectroscopically good but is an arsenic salt; acetate is far from ideal and may limit the accessible range to above ~215 nm; and any buffers with aromatic groups are impossible. Also beware of high concentrations of Cl^- as this can cut off reliable data as high as 215 nm. F^- does not have this problem and can often be substituted for Cl^-.

It must be remembered that the solvent and buffer can sometimes affect the actual protein structure and therefore the CD spectra obtained. Even if there are no structural changes, changing the chemical environment can affect electronic transitions, and therefore also affect the CD spectra.

2.2 Instrument Calibration

It is important to calibrate the CD spectropolarimeter for both intensity and wavelength. It is straightforward to carry out a check on the state of calibration of a CD instrument. If the intensity calibration is outside specification, it is usually easy to reset the intensity oneself, but it is more difficult to reset the wavelength, and a service engineer should be called.

Wavelength calibration (*see* **Note 3**) requires a wavelength standard. One option is a solid neodymium filter, which usually comes with the CD spectrometer. The absorbance maximum of this should be 586 ± 0.8 nm. However, wavelength accuracy at 586 nm does not guarantee it at 200 nm, which is more relevant for most proteins. Another option is a holmium oxide filter, or solution, which has a characteristic absorbance spectrum across the near UV and visible regions. Benzene vapor (if one can get benzene) is ideal in the far UV region. The absorbance maximum in a CD machine can usually be determined by determining where the gain voltage on the photomultiplier tube (often called the high tension voltage) is maximum (this may require asking the software to display the high tension voltage trace—which one should always do in any case as discussed below). Many instruments also produce an approximate absorbance spectrum from the high tension voltage trace and known properties of the instrument components. A baseline absorbance spectrum (see below) in the same cuvette should be subtracted to give the sample absorbance.

Intensity calibration (*see* **Note 4**) is conventionally carried out using aqueous ammonium d-10-camphor sulfonate (ACS), which has been related to the established hygroscopic primary standard camphor sulfonic acid (CSA) (13, 14). ACS is available from a number of suppliers (including Katayama Chemical Co. 05-1251), although care should be taken to ensure its enantiomeric as well as chemical purity. The concentration of ACS usually used is 0.06 % w/v, i.e., 60.00 mg in 100 mL or 6.000 mg in 10 mL, which should give a CD signal of 190.4 ± 1 mdeg at 290.5 nm (the CD maximum) in a 1 cm path length cuvette. (Note: CSA (which is very hygroscopic so far from an ideal standard) has a different molecular weight from ACS so a 0.06 % solution has different molar concentrations and hence different CD signals for the two standards.) It is important to weigh the solid material accurately (to 4 significant figures). For 6.000 mg this requires a 6 figure balance. The standard solution can be stored at 4 °C. However, the storage time seems to be dependent on the container in which it is stored and storage of longer than 2 weeks may be problematic. The concentration of ACS can also be determined from its absorbance (15); however, the extinction coefficient is very small and in most people's hands this is less accurate than measurement by weight. The problem with ACS is that it is used as a single wavelength calibration standard. An alternative standard with multiple wavelengths and both enantiomers will be available soon: Na[Co(EDDS)]·H_2O (EDDS = *N*, *N*-ethylenediaminedisuccinic acid) (16) (*see* **Note 4**).

2.3 Path Length Determination

In order to use and compare CD data, it is important that the path length of the cell is known—often the nominal path length of a cell, is far from the actual path length for path lengths smaller than 1 mm. One approach is to use an aqueous basic potassium chromate solution of known concentration (*see* **Note 5**). The Beer–Lambert law can then be used to calculate the path length from a measurement of absorbance. For 0.01 mm path length, a 0.2 M potassium chromate solution is required. For longer path lengths, a more dilute solution is required. To prepare a 0.2 M potassium chromate solution, accurately weigh 0.971 g potassium chromate (Aldrich 216615, 99 % ACS reagent) and transfer it to a 25 mL volumetric flask. Add approximately 20 mL water and 1 pellet of potassium hydroxide (e.g., VWR BDH AnalaR 26668.263). Make the solution up to volume and mix well. This solution may be stored indefinitely at 4 °C (as long as it is sealed tightly). If lower concentration samples are prepared by diluting the 0.2 M solution, the potassium hydroxide is also diluted, and so the solution stability is reduced.

3 Methods

3.1 Collecting CD Data

Measuring a CD spectrum is a routine procedure assuming one has access to a CD spectropolarimeter. If your sample gives a good UV–visible absorbance spectrum then it is highly likely that (if it is chiral) you will get a good CD spectrum. The essential features of a CD spectrometer are a source of (more-or-less) monochromatic left and right circularly polarized light and a means of detecting the difference in absorbance of the two polarizations of light. CD machines are much more expensive than UV machines as one is typically expecting absorbance differences of the order of 10^{-3} to 10^{-4} to be accurately measured against a background absorbance of ~1. For protein CD, the spectropolarimeters also need to be nitrogen purged, not only to prevent the high energy lamps producing ozone and therefore damaging the delicate optics within the instrument, but also to avoid having O_2 in the sample compartment absorbing the incident radiation and limiting the lowest wavelengths that can be measured. In practice this means a moderate nitrogen flow rate (3–5 cm^3/min) at all times, with an increase to 10–20 cm^3/min when collecting data below 200 nm.

The subheadings below describe the important steps you need to consider while making CD measurements.

3.1.1 Before You Start

In order to collect good quality and reliable CD data you should ensure that your instrument is properly calibrated both in terms of wavelength and intensity (*see* **Notes 3** and **4** as well as Subheading 2). You should also carefully select a sample cuvette that is suitable for CD measurements (*see* **Note 6**), and is an appropriate path length for the concentration of the sample that you are measuring (*see* **Notes 1** and 7). Since the actual path length of cells can differ significantly from the nominal values, you should always ensure that you measure the path length of the cell that you are using (*see* **Note 5** and Subheading 2). For fixed path length sealed cells, it is useful to calibrate all of your cells, and keep a calibration table handy for easy reference.

3.1.2 Sample Absorbance

Since the wavelength and absorbance ranges are the same for both normal absorption and CD it is advisable to run a normal absorption spectrum of the sample for which you wish to measure the CD spectrum first. *Always* leave the reference beam of the absorbance spectrometer empty for this experiment, since we are concerned with the total absorbance of the sample including any achiral molecules in the buffer. It is preferable to have the maximum absorbance less than 1.5 (and certainly less than 2).

3.1.3 Wavelength Range

CD spectrometers usually scan from longer wavelengths to shorter ones. Ensure you collect data for at least 20 nm to the longer wavelength side of the longest wavelength absorbance band for reasons discussed below. When the baseline spectrum (see Subheading 3.1.4) is subtracted from the sample spectrum, the region outside the absorption envelope should be flat. If it is not then this probably means either there is a very weak absorbance band that has a large disymmetry factor and hence large CD signal compared with its normal absorbance intensity, or, more probably, there is some light scattering by the sample. Sources of light scattering include dirty cuvettes (inside or outside, *see* **Note 8**), undissolved sample, condensation of samples, and particulate samples (e.g., fibers or liposomes). Given that the wavelengths of light being scattered are less than 1 μm, one does not necessarily expect to be able to see by eye the presence of such particles.

3.1.4 Measuring the Baseline and Zeroing

A machine baseline (i.e., CD spectrum of air) measured on a standard CD spectrometer will not be flat since the optical components in the instrument are birefringent. This can be overcome by storing a machine baseline in the instrument. However, the cuvette used for an experiment will also have its own CD spectrum—even CD matched cuvettes usually have slightly different intrinsic spectra. So always collect a baseline spectrum of your solvent/buffer under the same conditions as the sample spectrum using the same cuvette in the same orientation with respect to the light beam. Anything looking like a (probably weak) regular CD signal might well be contamination on the cuvette, in which case it has to be thoroughly cleaned (*see* **Note 8**). Subtract the baseline spectrum from the sample spectrum to produce the final CD plot. Although one can often have the cuvette baseline automatically subtracted by the software, this may not be a good idea for small signals or sticky samples that could be retained on the cuvette (*see* **Note 8**). It is better to see any problems in the baseline spectrum. It is often the case that even when the baseline is subtracted, the CD is not exactly zero outside the absorption envelope. However, if it is flat outside the absorption envelope, the spectrum may be zeroed by adding or subtracting a constant (either within the spectrometer software or using your chosen data plotting software).

3.1.5 Improving the Signal to Noise Ratio

The signal-to-noise ratio in a CD spectrum increases with the following: $\sqrt{n}$, where n is the number of times the spectrum is accumulated (and the data averaged); $\sqrt{\tau}$, where τ is the time over which the machine averages each data point (the response time); and $\sqrt{I}$, where I is the intensity of the light beam. The actual definition of τ depends on whether your instrument moves to a wavelength and sits there to collect data or whether it continuously scans. A way to determine the best combination of parameters (see below) for measuring a good spectrum in minimum time is to try some scans with different time constants or response times to see if the spectrum is distorted. I is usually influenced by bandwidth and also wavelength since the lamp does not have uniform intensity at all wavelengths. In fact when spectra become noisy, one of the reasons is often that the light intensity is too low, either due to age of the lamp or the mirrors or the wavelength range being used having intrinsically low flux. Most CD spectrometers have both short timescale (millisecond to minutes) and long timescale (minutes to hours) baseline variations. To average over short timescale fluctuations, collect a number of spectra or have a longer response time. Longer timescale fluctuations are usually dealt with by alternating collection of sample and baseline spectra or by assuming the fluctuation is a wavelength independent shift of the zero point. While appearing to be wishful thinking, the assumption of linear baseline drift seems to be valid especially with more recent instruments.

3.1.6 Choosing Parameter Sets

CD spectrometers give the operator considerable control over response time (τ), scan speed (s), bandwidth (b) [the wavelength range (error) of the incident light], and data interval (d). To optimize signal to noise effects (see Subheading 3.1.5) τ should be selected to be as large as possible subject to $\tau \times s$ being less than or equal to $b/2$. If τ is too long for the chosen s and b, then maxima of peaks (both positive and negative) will be cut off and their wavelengths shifted. A control scan using $\tau' = \frac{\tau}{2}$ (or $s' = 2s$) should be used to check that spectra are not being distorted by the chosen parameters. If one wishes to reduce collection time for a large number of spectra (as is often required for a binding constant study, see Subheading 3.6), then it may be worth seeing if the parameters can be abused by testing $\tau' = 2\tau$. The data interval determines the wavelength interval between data points. This may or may not have implications for the time a scan takes depending on whether the instrument operates in a stepped scan mode or continuous scan mode.

Scan speeds of 100 nm/min, $\tau = 1$ s, $b = 1$ nm, and a data step of 0.5 nm are usually a good starting point as a parameter set for most protein–ligand experiments (i.e., for samples that have the broad band shapes usually found for proteins and most of biological ligands). It is often advisable to perform a fast preliminary scan to determine whether there is any point in collecting an accurate spectrum: the most significant consideration for reliable data is whether the sample has too high an absorbance—this is summarized by the high tension voltage (or equivalent label) of the photomultiplier tube (*see* **Note 2**). It may also be necessary to select an appropriate sensitivity scale. On modern instruments this is usually only an issue for samples with a very high CD signal for their absorbance. It occurs more often when CD machines are used for LD. The need for this will be indicated by a flat straight line across peaks in the spectrum.

3.1.7 Linearity

It is good practice to check that the CD signal is proportional to sample concentration (the CD analogue of the Beer–Lambert law, Eq. 4) by running a spectrum of a sample and a diluted sample and checking that the signal scales with concentration. If the signal is not proportional to sample concentration, i.e., the signal does not increase linearly with concentration, then one has evidence of solute–solute interactions in the sample or of too high an absorbance for the instrument to be able to count the photons accurately or of inhomogeneous samples.

3.2 Protein–Ligand Interactions Probed by Circular Dichroism: Titration Experiments

When a ligand binds to a protein molecule, its transitions will acquire an ICD spectrum. If it is itself achiral then this will change its CD spectrum from zero to something; if it is chiral then its ICD is the difference between its own CD spectrum and the spectrum in the presence of the protein. The ICD in the protein regions of the spectrum are usually a combination of protein and ligand ICD as most ligands have absorption intensity below 300 nm. A few examples are given in Subheading 3.3. Figure 3, however, shows a titration series where it appears that a large ICD is being observed when NADH binds to lactate dehydrogenase. However, after subtracting the intrinsic CD signal of the same concentration of the

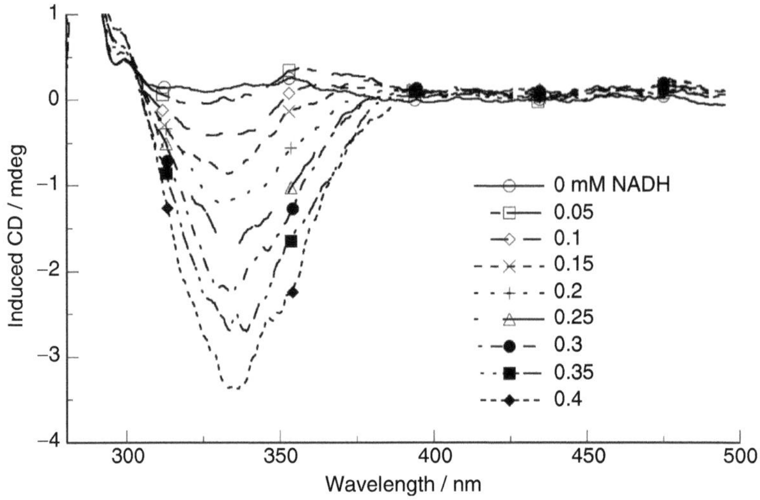

Fig. 3 CD of lactate dehydrogenase (LDH, 22.5 μM) with increasing concentrations of NADH. LDH concentration was kept constant by adding equal volumes of NADH (from a 10 mM stock solution) and LDH from a 45 μM stock solution. The measured spectrum is the sum of the intrinsic spectra of the two components, so there is no evidence of interaction by CD

chiral ligand the remaining spectrum is flat, like that of pure protein, which indicates that we are actually only measuring the ligand's intrinsic CD signal being added to the protein CD signal.

If there is just one binding mode (or equal occupancy of more than one) between protein and ligand then the magnitude of the ICD will be proportional to the concentration of protein–ligand complex (see Subheading 3.3). This is illustrated in the example of human serum albumin and quercetin discussed in Subheading 3.6. Modern computer programs help enormously with data analysis. In manual data analysis, the challenge is to determine the proportionality constant between bound ligand concentration and ICD, especially if high or low loadings cannot be achieved experimentally (see Subheading 3.6). If there is a succession of binding modes being occupied, the ICD spectral shape usually changes.

Apart from the challenges of analyzing the data once collected, there are a number of issues that must be considered for ligand binding CD experiments. One typically proceeds by titrating (or adding) one solution to another. There are many ways of doing this. Conceptually the simplest is to make up a series of independent solutions. However, this is usually the least accurate method as pipetting errors are most significant. Alternatively one can add very small volumes of a highly concentrated stock solution of ligand or protein and either ignore the dilution effect or account for it by assuming it has no structural effect on the protein. A simple way to avoid dilution effects is as follows. Consider a starting sample that has concentration *x* M of species *X*. Each time *y* μL of *Y* is added also add *y* μL of a 2*x* M solution of *X*. The concentration of *X* remains constant at *x* M. An infinite number of variations on this theme are possible, one of which is to add a solution already containing *x* M of *X*. One must always consider whether the order of addition of components or the stock solution concentrations (particularly local high concentrations during addition of the titrant) in such an experiment has any effect on the nature of the interaction. All of this assumes we have a reasonably accurate estimate of the protein concentration (*see* **Note 1**).

Whether the protein or the ligand concentration is kept constant during a titration series depends on a number of issues. If the protein CD spectrum changes as a function of its own concentration then it is usually wise to keep it constant. If it does not, by keeping the ligand concentration constant one can more easily monitor any changes in ligand ICD as a function of protein–ligand ratio. A further complication may arise here, however, in that this titration series begins/ends with low protein concentration, which may cause secondary binding sites to be occupied. Any ICD in the protein region of the spectrum is likely to be composed of both protein and ligand ICD spectra, as it is hard to find ligands with no absorbance intensity below 300 nm.

3.3 Linear Dichroism of Protein–Ligand Systems

To measure LD spectra one usually converts the alternating circularly polarized light beams of a CD spectrometer into alternating horizontal and vertical linear polarizations either by inserting a quarter wave plate or using the software to increase the voltage on the photoelastic modulator (if such an option is available) to turn it into a half wave plate. The main challenge of LD spectroscopy is usually to orient the samples. Other aspects of LD spectroscopy are the same as discussed above for CD spectroscopy.

3.4 Examples of Protein–Ligand CD Spectra

3.4.1 Protein–Solvent Interactions

Solvents can interact with and thus have an effect on the conformation of peptides and proteins. Figure 4 shows the example of a short peptide (MSLSRRQFIQASGIALCAGAVPLKASA in single letter code). A change from ~50 % α-helical (in TFE) to random coil (in water) is observed—this is an extreme change and not to be expected for most ligand binding, but it illustrates the point clearly.

3.4.2 Protein–Protein Interactions

Actin is a protein that binds to myosin during muscle contraction. Figure 5 shows the CD spectra measured for F-actin (the polymerized form of actin), a myosin molecule (S1) and the mixture of the two. The theoretical spectrum calculated from adding the actin and myosin spectra is also shown. The difference between this theoretical spectrum and the experimental mixture is small but clear and

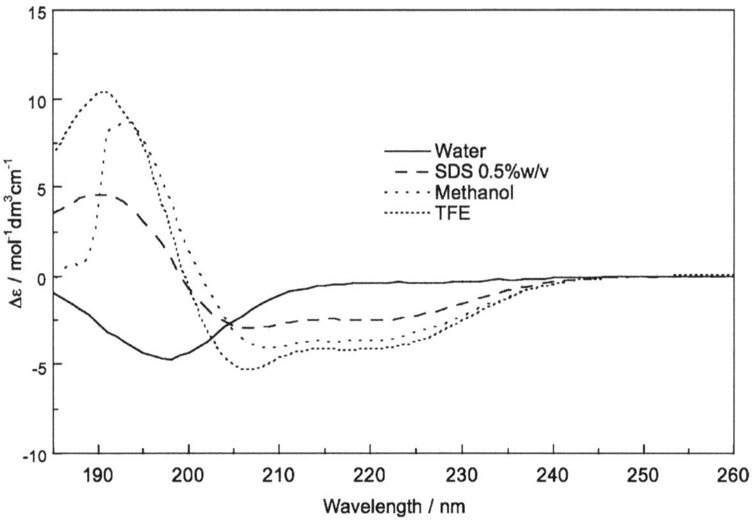

Fig. 4 CD of a peptide (MSLSRRQFIQASGIALCAGAVPLKASA) in different solvents (28). In water the peptide adopts a random coil structure. The TFE spectrum is ~50 % α-helical

reproducible, showing that either the actin or the myosin (or both) secondary structure is perturbed by their interaction or there is some electronic perturbation of their transitions. It is not possible to distinguish these possibilities.

3.4.3 ICD of a Protein–Ligand System

Bovine serum albumin binds almost anything that can be put into solution with it; however, one of its main roles is as a steroid transporter protein in blood. An example of the CD it induces in a steroid coupled to a platinum metal complex is shown in Fig. 6. The signal above 300 nm is due to perturbations to the ligand spectroscopy. That below 300 nm is a combination of protein and ligand ICD signals.

3.4.4 Peptide–Peptide Titration Series

Vancomycin (V) and ristocetin (R) are glycopeptide antibiotics that prevent cross-linking and transglycosylation during bacterial cell wall formation. Noncovalent dimerization plays a key role in their activity and CD can be used to give binding constants for such an interaction. The data in Fig. 7 are for the equilibrium

$$V + R \rightleftharpoons V - R$$

where V denotes vancomycin, R ristocetin and V–R their noncovalent complex. The ICD is assumed to be proportional to the concentration of dimer. Figure 7 shows the data collected as increasing concentrations of vancomycin are added to a solution where the concentration of the ristocetin is maintained at 25 μM. The equilibrium constant for the dimerization can then be calculated to be $K_{\text{dimerization}} = 20 \pm 5\ \text{mM}^{-1}$ as outlined in Subheading 3.6.

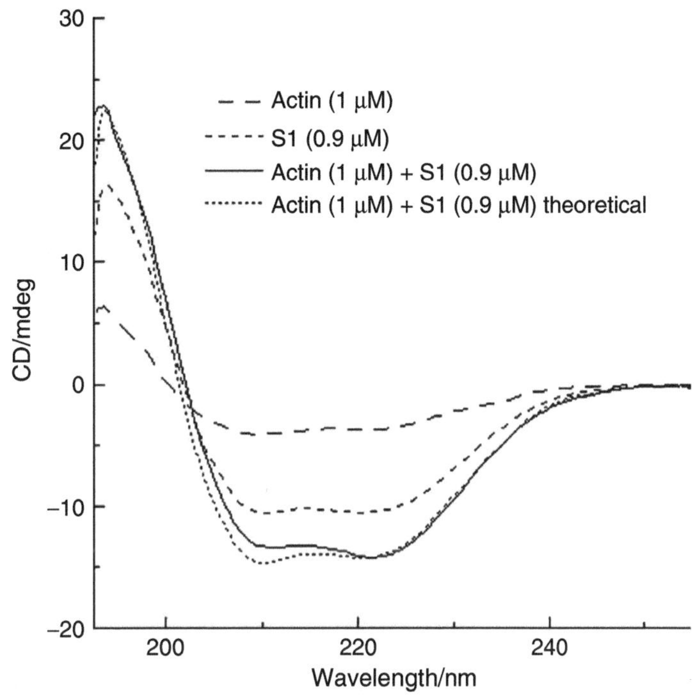

Fig. 5 Circular dichroism spectra of F-actin (1 μM actin monomer), S1 myosin (0.9 μM), the mixture of the two proteins, and the sum of the two independent spectra. All samples were made up in 5 mM Tris buffer, pH 8, 20 μM $CaCl_2$, 1 mM ATP, 1 mM $MgCl_2$, and 50 mM KCl. Path length used was 1 mm

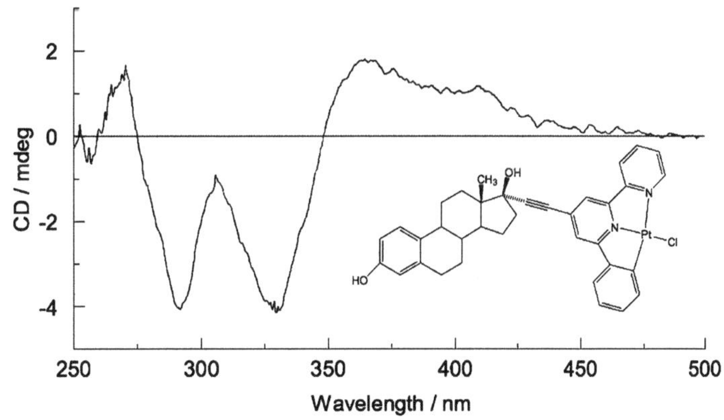

Fig. 6 ICD for PtEEtpy (30 μM, structure is shown in the figure) bound to bovine serum albumin (30 μM in 5 mM sodium cacodylate buffer). The spectra of both PtEEtpy and bovine serum albumin have been subtracted to give the ICD illustrated

3.5 Examples of Protein–Ligand LD Spectra

LD has the major advantage of only detecting molecules that are oriented. So if a fibrous protein (e.g., FtsZ or actin, see Subheading 3.4.2) is flow oriented, only the protein fibers and any ligands

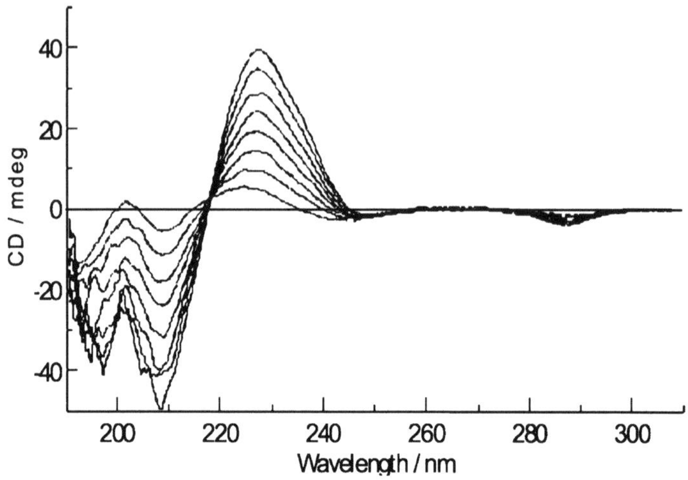

Fig. 7 ICD measured when increasing volumes of a concentrated solution of vancomycin containing 25 μM ristocetin are added to a 25 μM ristocetin solution. The intrinsic CD spectra of the peptides have been subtracted to give only the ICD. The 0 μM vancomycin spectrum is therefore a *flat line* and has been omitted. Vancomycin concentrations (increasing ICD magnitude) are as follows: 25, 50, 75, 100, 125, 150, 175, and 200 μM. The ICD signal initially increases almost linearly with the concentration of vancomycin added. This indicates that essentially all the ligand is bound, and the system is far from saturation. Data are from ref. 29

that bind to the fibers are detected. All others are invisible. This means that LD is ideal for following polymerization or depolymerization reactions as the monomers are invisible. Membrane proteins are probed by binding them to liposomes that can be flow oriented. In this case we then only see proteins and ligands that are bound to the liposomes. An example of each type of protein is given below to illustrate what is possible.

3.5.1 FtsZ Polymerization and Bundling

FtsZ is a bacterial protein that polymerizes to form a fiber that is a polymer of protein monomers called protofilaments. It has a GTP (guanine triphosphate) sandwiched between each FtsZ monomer. *Escherichia coli* FtsZ has no tryptophans to interfere with the 280 nm region of the spectrum, so the LD in that region is dominated by GTP. The guanine LD signal in the polymers is shown by the solid line of Fig. 8. Upon addition of Ca^{2+} the FtsZ protofilaments bundle together (12) and the guanine LD changes from being a broad negative band with two transitions under it into one with a low wavelength positive (260 nm) signal and a higher wavelength (280 nm) negative one. The LD is thus telling us that the bundling process results in the guanine chromophores of the

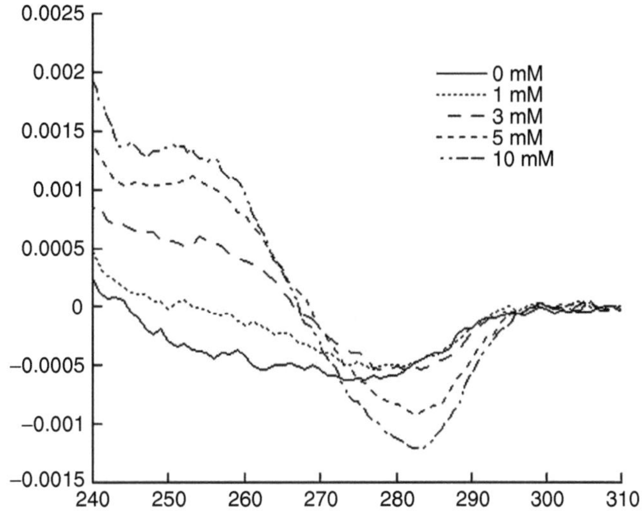

Fig. 8 LD of GTP chromophore of FstZ (11 μM) during polymerization and bundling in the presence of $MgCl_2$ (10 mM) and varying amounts of Ca^{2+} (50 mM MES buffer, pH 6.5, 50 mM KCl, 0.1 mM EDTA, and 0.2 mM GTP)

GTPs tilting so that the guanine's long axis is closer to parallel to the fiber axis than to its original perpendicular orientation (12).

3.5.2 Gramicidin Inserting into Liposomes

Many peptide antibiotics act by inserting into the bacterial cell membrane and disrupting its function in some way. One such antibiotic is gramicidin D the sequence of which is as follows: HCO-L-Val-Gly-L-Ala-D-Leu-L-Ala-D-Val-L-Val-D-Val-L-Trp-D-Leu-L-*X*-D-Leu-L-Trp-D-Leu-L-Trp-$NH(CH_2)_2OH$ where *X* is either tryptophan or tyrosine or phenylalanine (17). We can measure LD spectra of membrane peptides and proteins by mixing them with liposomes (spherical lipid bilayers) and flowing the sample in a Couette flow cell. The liposomes distort into elongated shapes and orient. Anything that inserts with its transition moments parallel to the lipids gives a negative LD signal and anything bound to liposomes with transition moment perpendicular to the lipids has a positive LD signal. The short time LD signal of gramicidin added to 1,2-dipalmitoyl-*sn*-glycero-3-phosphocholine (DPPC) liposomes at 50 °C is shown by the solid line in Fig. 9 (9). The negative 360 nm signal is due to a lipid-like probe molecule (β-DPH-HPC, 2-(3-(diphenylhexatrienyl)propanoyl)-1-hexadecanoyl-*sn*-glycero-3-phosphocholine) which inserts parallel to the lipids. The negative band at 280 nm is due to tryptophans being predominantly perpendicular to the lipids. The positive 220 nm band is due to peptide backbone transitions of the peptide. All the signals gradually decay over time as the gramicidin destroys the liposomes. It is possible to plot such data as a time course and obtain kinetic information.

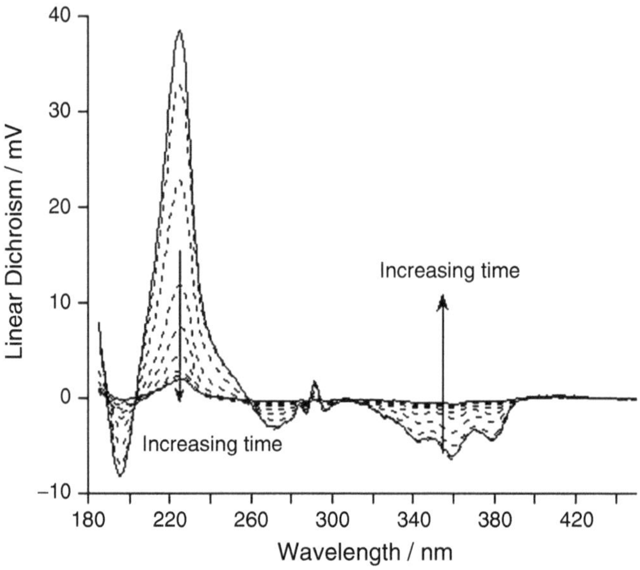

Fig. 9 LD of gramicidin D with DPPC/β-DPH-HPC (100:1 w/w) liposomes at 50 °C. 1 mg/mL lipid, 0.01 mg/mL β-DPH-HPC, and 0.1 mg/mL gramicidin in 10 % (v/v) TFE. Spectra were collected at 12 min intervals

3.6 Determination of Equilibrium Binding Constants: Quercetin Binding to HSA

3.6.1 General Considerations

When measuring protein–ligand interactions, one is frequently interested in the strength of the interaction. The simplest measure of the binding strength between a protein and a ligand is the equilibrium dissociation constant K_D,

$$K_D = \frac{[P][L]}{[PL]} \qquad (9)$$

for the equilibrium

$$\text{free ligand L} + \text{free protein P} \rightleftharpoons \text{protein} - \text{ligand complex PL}, \qquad (10)$$

where [L] is the concentration of free ligands, [P] is the concentration of free protein, and [PL] is the concentration of their complex. Note this equation and those discussed in detail below assume a 1:1 binding stoichiometry. If a protein has more than one binding site, then the total binding site concentration is $S = n[P]$, where n is the number of sites per protein. Then Eq. 9 would be $K_D = [P][L]^n/[PL_n]$ (assuming the binding sites are independent) and all subsequently discussed equations would be modified.

CD can be used to determine dissociation (or association) constants if the binding strength is of the order of μM—simply because of the concentration of the samples required to get a reasonable signal. If no ICD is observed then it is likely that the interaction is weaker than this. If the ICD is proportional to the concentration of ligand (or protein) added then the binding constant is too high to be measured by CD. Another problem often encountered while determining binding constants from CD (or any spectroscopic) data is that once "the" strongest binding site is occupied, ligands often find other sites to occupy. In this case, the signal change from a second reaction is added to the first and fitting becomes more complicated.

3.6.2 Materials and Methods

Fatty-acid and globulin free human serum albumin (A3782), quercetin hydrate (337951) and all other reagents were purchased from Sigma-Aldrich. Ringer buffer was used throughout consisting of 1.5 mM KH_2PO_4, 8.1 mM $Na_2HPO_4{\cdot}12H_2O$ pH 7.4 with 137 mM NaCl, 2.7 mM KCl, 0.8 mM $CaCl_2$, and 1.1 mM $MgCl_2$. The original quercetin stock solution was prepared by dissolving 0.8728 g in 1 ml ethanol (99.9 %, VWR) to obtain a 2.58 mM stock solution (assuming most stable dihydrate form, 338.27 g/mol). This was subsequently diluted to 1.29 mM in Ringer buffer. An HSA stock solution was prepared using 6.0 mg in 1 ml Ringer buffer, its concentration checked by absorption using an extinction coefficient of 34,445 M^{-1} cm^{-1} computed from the amino acid sequence using ProtParam (http://web.expasy.org/protparam/).

The titration was performed in a masked (black quartz) cuvette of 1 cm path length with a starting volume of 500 μl HSA at 42.6 μM. For each titration point 10 μl of quercetin solution (1.29 μM) was added, as well as 10 μl of HSA solution (85.1 μM) to keep the HSA concentration constant. Scans were recorded in a Jasco-720 CD spectrometer from 550 to 250 nm at 0.2 nm intervals with a speed of 100 nm/min, 1 s integration time, and 1 nm bandwidth at eight accumulations. The signal was reliably linear down to 285 nm, at which point the high tension voltage became too high and the data quality unreliable. The spectra were therefore cut at 285 nm (see Fig. 10a).

3.6.3 Plotting and Fitting Titration Data

Reduce the spectral noise by averaging over repeat scans collected for each set of conditions/concentrations. If the data collection software allows this, the repeats should be inspected and overlaid. Barring any outliers (scans with obvious artifacts or excessive noise), all overlaid spectra should be averaged and standard deviations at critical wavelengths obtained. Average scans for each concentration point of the titration are best shown as an overlay to give direct evidence of the data quality (see Fig. 10a). In order to obtain a titration curve, the CD signal at a chosen wavelength is plotted as a function of the changing concentration of the added component (see Fig. 10b). This CD signal can be at any fixed wavelength (302 nm in Fig. 10a), usually that of the largest change, and all background signals should have been subtracted as described in Subheading 3.1.4.

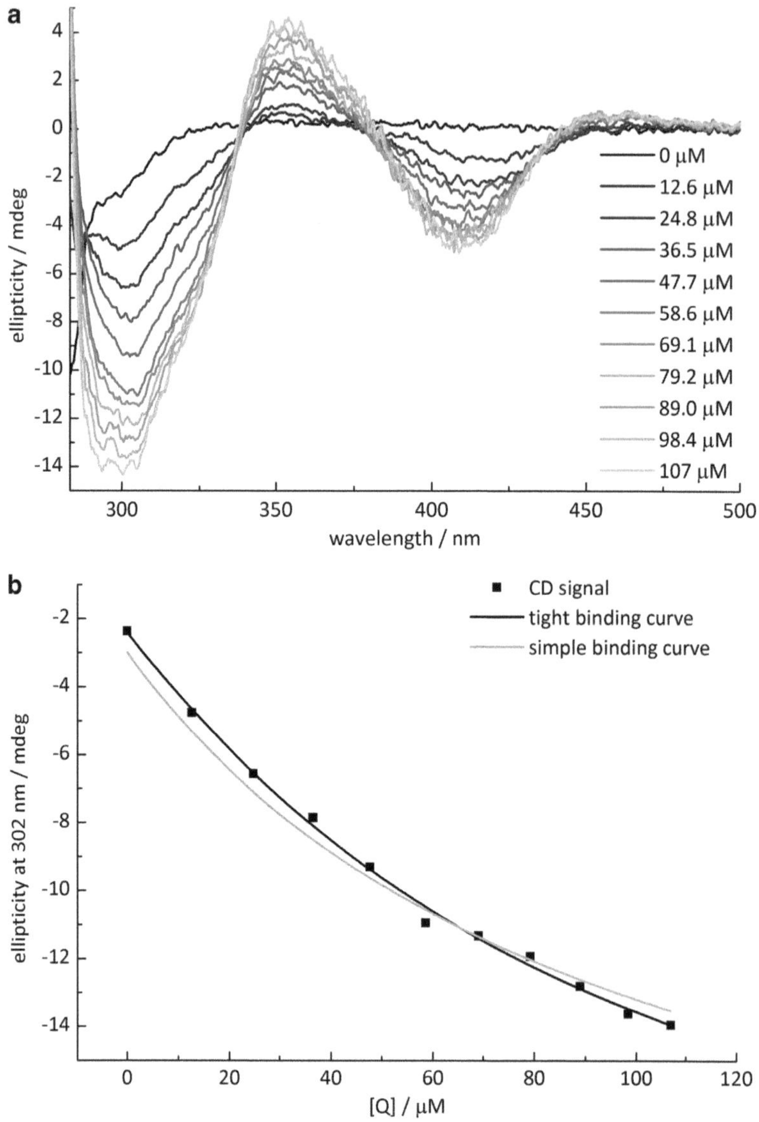

Fig. 10 (**a**) ICD spectra of quercetin upon addition to a fixed concentration of human serum albumin (HSA). (**b**) CD signal at 302 nm in (**a**) plotted against the quercetin concentration added and fitted using Eq. 12 (*grey*) and more accurate Eq. 14 (*black*)

A typical binding curve is approximately hyperbolic in shape and should reach a plateau at high ligand concentrations, i.e. show saturation of binding. The CD signals at the start and in the plateau of the titration represent the unbound and fully bound situation respectively. The signal ρ should be proportional to the fractional saturation f_b, i.e., the concentration of bound over total macromolecule with the proportionality constant α:

$$\rho = \alpha f_b = \alpha \frac{[PL]}{[P]_{Total}} \tag{11}$$

If the signal has been normalized to 1 at the maximum, the saturation of a protein P ranges from 0, when [L] is 0, to 1 (i.e., fully bound), when [L] >> K_D. If it does not reach 1, then some of the protein may be inactive, if it goes beyond one, there might be more than one binding site on the protein. The binding curve in Fig. 10a is not ideal as no plateau is reached, i.e. binding is not saturated as discussed in Subheading 3.6.4.

3.6.4 Computational Fitting of the Binding Curve

The most frequently used equation to describe binding data is derived from Eqs. 9 and 11:

$$f_b = \frac{[L]}{[L] + K_D} \tag{12}$$

(or $f_b = [P]/([P] + K_D)$ if instead the protein concentration is varied and the ligand concentration is fixed). The corresponding fitting equation implemented in data analysis software for Eq. 12, where [L] is variable and [P] is fixed, is

$$\rho = \alpha f_b = \alpha \frac{[L]}{[L] + K_D} \tag{13}$$

in accordance with Eq. 11 (this has the form: $y = A + B \cdot x/(x + C)$ where A is the starting signal for data that has not been zeroed). Note that [L] is the free ligand concentration. However, this is usually implemented by assuming $[L] = [L]_{total}$, so Eq. 12 it is only adequate when the ligand is in large excess ([L] >> [P]). To obtain sufficient signal in CD, this is rarely the case. The intrinsic method (*see* **Note 9** and (18)) provides a method of determining [L].

Linearizations based on Eq. 12, as used in the Scatchard plot, introduce systematic errors. Thus, despite their simplicity and usefulness in indicating unusual behavior if no straight line is obtained, approaches based on the simple binding equation are best avoided. The procedure described next is far superior and just as simple with modern computers.

To avoid the problem of not knowing the free ligand concentration, the binding curve obtained from changes in signals in CD spectra as a function of for example added ligand concentration should be fitted using the more universal binding/fitting function (see Eq. 19 in Chapter 1):

$$f_b = \frac{(K_D + [L]_t + [P]_t) - \sqrt{(K_D + [L]_t + [P]_t)^2 - 4[L]_t[P]_t}}{2[P]_t}, \tag{14}$$

where $[P]_t$ is the total protein concentration, for fixed protein concentration, or analogously

$$f_b = \frac{(K_D + [L]_t + [P]_t) - \sqrt{(K_D + [L]_t + [P]_t)^2 - 4[L]_t[P]_t}}{2[L]_t} \tag{15}$$

for fixed ligand concentration. In practice the difference between these equations is nonexistent as in both cases the denominators are constant and the proportionality constant α (see Eq. 11) needs to be introduced for fitting as in Eq. 13. Equation 14 does not require any approximations as the total concentration of ligand $[L]_t$ and protein $[P]_t$ are used. This equation may have to be entered as a user-defined function, if it is not in the equation library of a chosen fitting program. This would look somewhat like this for Eq. 14: $Y = A + B^*(((C + x + [P]_t) - sqrt((C + x + [P]_t)^2 - (4^*x^*[P]_t)))/(2^*[P]_t))$.

A set of data for quercetin binding to the protein HSA is shown in Fig. 10. The protein concentration is held constant and the ICD signal of quercetin is measured. The grey line in Fig. 10b is the fit from Eq. 12. As it is not true that the ligand is in excess in this titration, this only provides an upper bound on K_D. The fit to Eq. 14 is the black line in Fig. 10b and gives a K_D value of 65 ± 14 μM. Ideally, the titration should be repeated and carried to higher concentrations of quercetin to reach a better-defined plateau. Also, it could in principle be performed at a lower concentration of HSA to obtain a binding curve with a more pronounced curvature. However, this example illustrates the challenges of using CD for binding constant determinations. Increasing the quercetin concentration results in lower affinity binding modes being occupied as well as the most stable one. Decreasing the protein concentration would require significantly longer data collection times to get acceptable noise levels.

3.6.5 Manual Data Analysis: The Proportionality Constant

It is instructive to consider how to determine the equilibrium constant directly from the data. The starting point is usually to determine the proportionality constant α, which relates the ICD signal ρ to the concentration of complex or bound ligand formed, i.e.:

$$[L]_t - [L] = \alpha\rho, \tag{16}$$

where ρ is the ICD (or LD or other) signal at a chosen wavelength and α (which is a function of wavelength) is the proportionality constant over the range of binding ratios being considered. The simplest means of determining α is usually by having a large excess of macromolecule over ligand so that all the added ligand may be assumed to be bound and $[L]$ may be assumed to be zero. If this is indeed the case then a plot of ρ versus $[L]_t$ (total ligand concentration) at constant protein concentration should be a straight line with slope α. If it is not, then this indicates that the K_D (or the macromolecule concentration) is too low for this method and not all ligand is bound (*see* **Note 9**). Alternatively, the maximum ICD signal may be used to determine α if n is known. One can then calculate K using Eq. 9 for each concentration where one has measured data and take some kind of average. If neither the high, nor the low binding ratio limits can be used to determine the proportionality constant, then the intrinsic method can be used (*see* **Note 9**).

3.6.6 Other Binding Models

A glance at almost any reference dealing with titrations will make one realize that the simple equilibrium model we have assumed is quite probably invalid and also the options for data analysis are almost endless (19). In particular, while traditional approaches involve linearizing the data in some way as this enables one to see by eye whether such an approach is valid, with available computers and packages it is fairly simple to convert almost any model into a fitting equation to analyze titration data and to determine constants from the binding curve. Usually, more than one model will fit the data. It is therefore very important to obtain independent experimental evidence for the particular binding model on which the equation used is based.

3.7 Other Techniques

CD measurements of protein–ligand binding are in fact not always sensitive enough to detect ligand binding if there is little change in the protein secondary structure. The near UV CD is only perturbed if the ligand binds near an aromatic residue. In the above mentioned quercetin-HSA case we used the ligand ICD instead of the protein signal. However, that is not always possible as there may not be a ligand transition to probe or the ligand ICD signal may be very small. Thus, ligand binding may be more effectively probed by other techniques such as fluorescence or absorbance spectroscopy or LD for fibrous proteins (1, 8, 11, 12) and membrane proteins (1, 9, 10, 20) in liposomes. Measuring fluorescence polarization anisotropy (FPA) may in fact achieve some of the advantages of CD and LD spectroscopy, in that it is the difference between two polarizations of radiation so much of the background signal gets subtracted off.

CD is particularly suited to probing chiral molecules since achiral effects are cancelled out. Vibrational circular dichroism (VCD), Raman optical activity (ROA), and optical rotatory dispersion (ORD) have the same advantage. VCD and ROA are attractive as they probe the vibrations of the molecule, which are often easily identified and provide complementary information to that obtained from UV-CD and LD. There are also usually many more vibrational transitions than are found in UV-CD. Both VCD and ROA currently require much higher sample concentrations and generally much longer data accumulations than UV-CD so they are not nearly as widely used. ORD is related to CD by the Kramers–Kronig transformation so in principle contains the same information. However, the information is much harder to extract and data over the full wavelength range is at least in principle required.

4 Notes

1. Any quantitative application of CD spectroscopy to protein structure fitting or ligand binding needs a fairly accurate estimate of protein concentration. For a well-known protein the extinction coefficient at 280 nm may well be available on one of the many protein data base Web sites or in the literature. It is then a fairly simple matter to take a sample (probably of ~1 mg/mL), measure its absorbance spectrum in a 1 cm cuvette, and use the Beer–Lambert law. However, you should note that the extinction coefficient from a data base is almost certainly a theoretical one determined from adding the contributions of its aromatic amino acids and disulfide bonds, usually when they are in an aqueous environment. Thus, one should denature one's protein to use the theoretical value, then determine a native (folded) ε value for later use. Denaturing the protein is usually carried out with high concentrations of guanidinium chloride, whose purity may be sufficiently suspect that it should be determined using refractive index measurements (21). Alternatively, for a totally new protein, amino acid analysis is recommended (22).
2. The maximum absorbance of a sample (including the buffer) for CD spectroscopy should ideally be between 1 and 2. On most laboratory (i.e., not synchrotron) instruments, a rough rule of thumb is that the high tension voltage should be kept below 600 V (though if the absorbance increases sharply this may not be sufficient). A better check is to ensure that the Beer–Lambert law (Eq. 1) is obeyed when the sample is diluted or the path length reduced.

   Some protein samples are sufficiently large that they scatter incident light significantly. The photomultiplier tube does not distinguish photons that do not reach it because they are absorbed and those that are scattered. Thus, it is essential to minimize scattered light. The presence of scattering is apparent in the baseline of the spectrum if data are collected at longer wavelengths than the sample absorbs light. In practice for a sample designed to measure the protein backbone CD spectrum this means above 260 nm, whereas for concentrations appropriate for the aromatic region this means examining the data from above 300 nm. If the baseline is flat one need not worry about scattering; if it slopes upwards scattering is contributing to measured intensities.

3. The accepted wavelength calibration standard is neodymium though it is far from ideal for proteins as discussed above. Some alternative standards include holmium oxide and benzene as discussed in Subsection 2.2. After the CD machine has been on for more than 30 min run a spectrum of a neodymium filter from 610 to 560 nm with instrument parameters set for a fairly slow scan with small data pitch, e.g., 0.1 nm data pitch, 20 nm/min scan speed, 1 accumulation, 0.25 s response time, and 1.0 nm band width. Note where the maximum in the photomultiplier tube voltage occurs, as this is also the absorbance maximum; it should occur at 586 $\pm$ 0.8 nm. It is a good idea to note the wavelength accuracy in an equipment maintenance log. If the wavelength accuracy is not within specification, but the shift is constant across the wavelength range (check for the same variation with ACS, see below), then you can recalibrate the spectrum accordingly. However, it is advisable to call in an engineer. Holmium oxide or Co-EDDS are better wavelength standards for proteins than neodymium as they have bands in or near the far UV.

4. The intensity calibration of a CD machine is usually carried out by collecting a spectrum from 350 to 250 nm of 0.06 % aqueous ammonium d-10-camphor sulfonate (ACS) in a 1 cm path length cuvette. The sample needs to be weighed out with great care as instrument calibration will only be as accurate as this concentration. Unfortunately the absorbance intensity of ACS at 290 nm is very weak so using absorbance spectroscopy to determine its concentration is usually not very accurate. A typical set of instrument parameters is as follows: 0.1 nm data pitch, 50 nm/min scan speed, 1 accumulation, 1 s response time, and 1.0 nm band width. Subtract a water baseline run with the same cuvette and parameters. The wavelength and intensity of the peak should be 190.4 $\pm$ 1 mdeg at 290.5 nm. If one chooses to use d-10-camphor sulfonic acid then the magnitude must be scaled by the ratio of the molecular masses. It is a good idea to record the values in the instrument log. If the intensity is not within stated limits, use an independently made fresh ACS standard and repeat the calibration test. If the value is reproducible, all subsequent data may be scaled to bring the intensity to the correct value. Alternatively the instrument may relatively easily be recalibrated following instructions in the instrument manual.

   ACS also has a negative CD band at 191 nm. The standard 0.06% ACS solution can also be used in a 1 mm cuvette to scan from 250 to 180 nm. The ratio of this negative peak to the

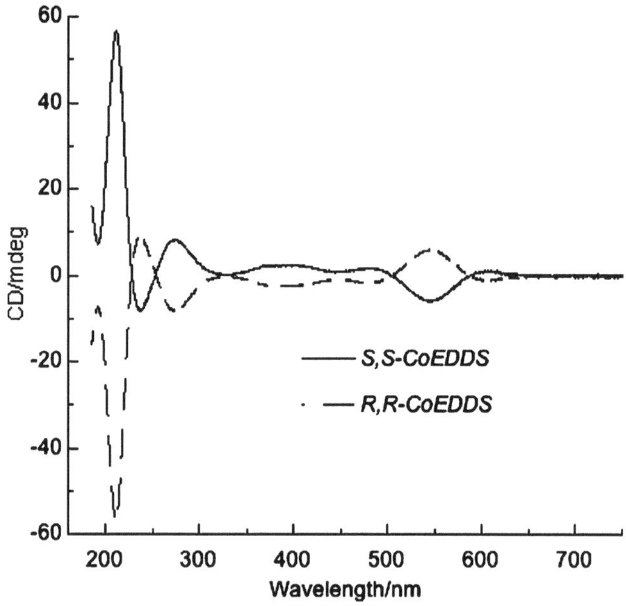

Fig. 11 *Top*: *S,S*-*N,N*-ethylenediaminedisuccinic acid (EDDS) and Co-EDDS. *Bottom*: Overlaid CD spectra of *R,R*- and *S,S*-CoEDDS (0.072 mM, 1 cm path length) illustrating the equal magnitude and opposite sign of the two enantiomers (30)

positive peak obtained at 290.5 nm at the same pathlength should be approximately 2.1. Certainly a value below 2.0 indicates poor instrument performance (23).

The ACS approach to instrument calibration is based on the assumption that a single point (or at most two) calibration is sufficient for the whole spectrum. This is transparently not the case and an alternative standard, Co-EDDS, EDDS = *N*, *N*-ethylenediaminedisuccinic acid (see Fig. 11) has been developed (24). Co-EDDS is particularly valuable for probing instrument performance from the visible region down into the far UV below 200 nm where light scattering effects become significant as the lamp intensity is reduced. Its CD spectrum is shown in Fig. 11. Co-EDDS has the added advantage of having a reasonable intensity normal absorption spectrum so its concentration can be determined spectrophotometrically.

5. It is important to determine the cuvette path length if a short path length (less than 1 mm) is used. 1 mm and longer path length cuvettes can usually be assumed to be that specified by the manufacturer (though if the path length is important this should be confirmed). 0.01 mm path lengths are almost never close to that specified. Indeed the path length of a filled demountable cuvette varies from fill to fill and user to user.

Short path lengths (0.1 mm and shorter) can be determined using interference fringes on an empty cell. However, for demountable cuvettes filling and handling by a given operator will have an effect on the actual path length. The best method we have found is for each user to fill the demountable cell with a fixed volume of the potassium chromate solution of appropriate concentration and measure the UV/Vis absorption spectrum from 600 to 350 nm. The path length is then calculated using the Beer–Lambert law ($\varepsilon = 4{,}830\ \text{mol}^{-1}\ \text{dm}^3\ \text{cm}^{-1}$ at 372 nm). It is important to always assemble the cell in the same way (mark the cuvette at one end with a pencil and note which edge is the beveled edge). Path lengths of demountable cuvettes do vary over time (as the edges of the cell get worn). It is important for at least three measurements to be performed. A new user of 10 micron cuvettes, in our experience, takes hours of reloading and remeasuring to obtain a reproducible path length. Even 0.1 mm can be challenging.

6. Either cylindrical or rectangular cuvettes may be used for CD, although a particular instrument may only take one kind. Cylindrical cells are usually deemed to have lower birefringence (baseline CD) than rectangular cuvettes, however, if UV and CD “matching” is requested when the cuvettes are purchased rectangular cuvettes seem to be equally good. Water-jacketed cylindrical cells enable the sample to be thermostatted most simply and also take the least sample volume for a given path length. With these cuvettes, you must check the configuration of your light beam and cuvette holder to ensure that the light beam passes through the sample and not the quartz walls and cooling water parts of the cuvette. Rectangular cells have a number of advantages over cylindrical cuvettes for the 1 mm and longer path length experiments: they are cheaper, may be used in standard absorption spectrophotometers (so CD and normal absorption data may be collected on exactly the same sample), and may be used for a protein–ligand serial titration experiment as ~60 % of a rectangular cell can be empty for the first spectrum and gradually filled (see below).

   If path lengths of 0.1 mm or less are required it is probably best to use demountable cuvettes where the sample is dropped onto a quartz disk or plate that is etched to a predefined depth and then another quartz disk/plate is carefully placed on top. Titrations are not possible in demountable cuvettes unless independent samples are made.

   All of the light beam incident upon the cuvette must pass through the sample and not be clipped or reflected by the walls or base of the cell or the meniscus of the solution, otherwise the measured spectrum is affected by scattered light. Thus, the narrow cells often used to minimize sample volume in normal absorption spectrophotometers *cannot* be used for CD unless the light beam is chopped or focused or is intrinsically small. While focusing of the light beam is possible, one must ensure that (1) the lenses used for the focusing are not themselves significantly birefringent (CD active), (2) the light beam does not diverge and hit the sides of the cuvette while passing through the sample, and (3) the whole light beam incident on the sample is collected by the photomultiplier tube (PMT). The light beam must not be focused too tightly on the PMT itself otherwise the PMT may be damaged. Recent work on using extruded quartz capillaries for CD spectroscopy has reduced the volume requirements down to a few microliters for a 1 mm path length experiment (25).

For UV–visible CD, high quality quartz cuvettes that transmit the full wavelength range of UV–visible light are required. In the visible region glass may be used but it is generally advisable to use quartz even here. Plastic cuvettes typically have high intrinsic birefringence so should be avoided. In any case, the need to run a baseline of each cuvette used (see below) removes the usual attraction of disposable plastic cuvettes.

7. The path length required to record a spectrum of the protein backbone region (from 260 to 190 nm) may be estimated on the basis that a 1 mm cuvette probably requires a ~0.1–0.2 mg/mL protein solution. Sometimes it is desirable to adjust concentrations to use an available cuvette, sometimes it is desirable to choose a path length to achieve certain parameters, e.g., to avoid dilution of a sample since some proteins (including monoclonal antibodies) may have a slightly concentration dependent CD spectrum or to minimize absorbance by the buffer (by using a short path length). The path length required for the aromatic region (from 300 to 250 nm) depends on the concentration of aromatic chromophores in the protein. For a protein with no aromatic groups and no disulfide bonds, there will be no aromatic region CD signal whatever the concentration or path length used. Typically, near UV measurements need 100 times more protein in the light path than far UV measurements. The path length required to measure the CD induced into ligand transitions upon binding to a protein is chosen to give an absorbance of ~1 at the wavelength of interest (usually around the absorption peaks of the ligand).

8. It is essential that the cuvette is cleaned well. Any deposit of chiral material on the quartz will have a CD spectrum. Sometimes one just hopes this subtracts off with a baseline, however, this is not good practice. To clean a cuvette one may proceed as follows. Rinse it well at least three times with high purity water

(18.2 MΩ) followed by ethanol or acetone. Dry the inside of a non-demountable cuvette with nitrogen or compressed air (but beware of oil deposits from the compressor) or a hairdryer. Dry the outside of a cuvette with a lens tissue and remove any fibers with a nitrogen line. If the cuvette shows traces of protein residue (as most easily shown by a protein CD spectrum being observed for the baseline), wash well with detergent (e.g., appropriately diluted Hellmanex), rinse with water. If the residue still remains, place the cuvette in a solution of 6 M nitric acid (beware of local safety issues, e.g., a mixture of acetone and nitric acid can be explosive) or Hellmanex (make sure the cleaning agent gets inside the cuvette) and allow it to stand for 10 min or longer before removing and rinsing thoroughly with water.

9. Sometimes the high or low binding ratio limits cannot be used to determine α. In such cases the intrinsic method can be used which we describe here. Using the fact that $[\mathrm{PL}] = [\mathrm{L}]_\mathrm{t} - [\mathrm{L}] = \alpha\rho$, Eq. 9 may be written (18)

$$K_\mathrm{D} = \frac{([\mathrm{P}]_\mathrm{t} - \alpha\rho)([\mathrm{L}]_\mathrm{t} - \alpha\rho)}{\alpha\rho} \qquad (17)$$

which can be rearranged to give

$$[\mathrm{L}]_\mathrm{t} = \frac{[\mathrm{P}]_\mathrm{t}[\mathrm{L}]_\mathrm{t}}{\alpha\rho} - [\mathrm{P}]_\mathrm{t} + \alpha\rho - K_\mathrm{D} \qquad (18)$$

For any two different total ligand concentrations L_j and L_k with the same protein site concentration,

$$\mathrm{L}_j - \mathrm{L}_k = \frac{[\mathrm{P}]_\mathrm{t}}{\alpha}\left(\frac{\mathrm{L}_j}{\rho_j} - \frac{\mathrm{L}_k}{\rho_k}\right) + \alpha(\rho_j - \rho_k) \qquad (19)$$

Thus, for several pairs of concentrations a plot of $y = (\mathrm{L}_j - \mathrm{L}_k)/(\rho_j - \rho_k)$ versus $x = (((\mathrm{L}_j/\rho_j) - (\mathrm{L}_k/\rho_k))/(\rho_j - \rho_k))$ should be a straight line with slope $[\mathrm{P}]_\mathrm{t}/\alpha$ and intercept α. In the case of non-1:1 reactions, the slope will be $n[\mathrm{P}]_\mathrm{t}/\alpha$ and thus the stoichiometry n may also be determined when $[\mathrm{P}]_\mathrm{t}$ is known.

It is sometimes convenient to perform experiments with constant ligand and varying macromolecule concentration. In this case, the equivalent expression is

$$\frac{\mathrm{P}_j - \mathrm{P}_k}{(\rho_j - \rho_k)} = \frac{[\mathrm{L}]_\mathrm{t}}{\alpha}\left(\frac{\left(\frac{\mathrm{P}_j}{\rho_j} - \frac{\mathrm{P}_k}{\rho_k}\right)}{(\rho_j - \rho_k)}\right) + \frac{\alpha}{n} \qquad (20)$$

References

1. Nordén B, Rodger A, Dafforn TR (2010) Linear dichroism and circular dichroism: a textbook on polarized spectroscopy. Royal Society of Chemistry, Cambridge
2. Nakanishi K, Berova N, Woody RW (1994) Circular dichroism: principles and applications. VCH, New York
3. Rodger A, Nordén B (1997) Circular dichroism and linear dichroism. Oxford University Press, Oxford
4. http://dichroweb.cryst.bbk.ac.uk; Dichroweb by Lee Whitmore and B.A. Wallace http://oregonstate.edu/dept/biochem/faculty/johnson.html. CDsstr by W. Curtis Johnson
5. Johnson WC (1999) Analyzing protein circular dichroism spectra for accurate secondary structures. Proteins 35:307–312
6. Marrington R, Dafforn TR, Halsall DJ, Hicks M, Rodger A (2005) Validation of new microvolume Couette flow linear dichroism cells. Analyst 130:1608–1616
7. Marrington R, Dafforn TR, Halsall DJ, Rodger A (2004) Micro volume Couette flow sample orientation for absorbance and fluorescence linear dichroism. Biophys J 87:2002–2012
8. Rodger A, Marrington R, Geeves MA, Hicks M, de Alwis L, Halsall DJ, Dafforn TR (2006) Looking at long molecules in solution: what happens when they are subjected to Couette flow? Phys Chem Chem Phys 8:3131–3171
9. Hicks MR, Damianoglou A, Rodger A, Dafforn TR (2008) Folding and membrane insertion of the pore-forming peptide gramicidin occur as a concerted process. J Mol Biol 383:358–366
10. Hicks MR, Kowalski J, Rodger A (2010) LD spectroscopy of natural and synthetic biomaterials. Chem Soc Rev 39:3380–3393
11. Marrington R, Seymour M, Rodger A (2006) A new method for fibrous protein analysis illustrated by application to tubulin microtubule polymerisation and depolymerisation. Chirality 18:680–690
12. Marrington R, Small E, Rodger A, Dafforn TR, Addinall S (2004) FtsZ fibre bundling is triggered by a calcium-induced conformational change in bound GTP. J Biol Chem 279:48821–48829
13. Chen GC, Yang JT (1977) Two-point calibration of circular dichrometer with d-10-camphorsulfonic acid. Anal Lett 10:1195–1207
14. Takakuwa T, Konno T, Meguro H (1985) A new standard substance for calibration of circular dichroism: ammonium d-10-camphorsulfonate. Anal Sci 1:215–218
15. Kelly SM, Jess TJ, Price NC (2005) How to study proteins by circular dichroism. Biochim Biophys Acta 1751:119–139
16. Chmel NP, Scott P, Rodger A (2012) Considerations of noise and measurement reproducibility of CD measurements using Na[CoIII (EDDS)]. Chirality 24:699–705
17. Hicks MR, Dafforn TR, Damianoglou A, Wormell P, Rodger A, Hoffmann SV (2009) Synchrotron radiation linear dichroism spectroscopy of the antibiotic peptide gramicidin in lipid membranes. Analyst 134:1623–1628
18. Rodger A (1993) Linear dichroism. Methods Enzymol 226:232–258
19. Polster J, Lachman H (1989) Spectrometric titrations: analysis of chemical equilibria. VCH Verlagsgesellschaft, Weinheim
20. Rodger A, Rajendra J, Marrington R, Ardhammar M, Nordén B, Hirst JD, Gilbert ATB, Dafforn TR, Halsall DJ, Woolhead CA, Robinson C, Pinheiro TJ, Kazlauskaite J, Seymour M, Perez N, Hannon MJ (2002) Flow oriented linear dichroism to probe protein orientation in membrane environments. Phys Chem Chem Phys 4:4051–4057
21. Pace CN (1986) Determination and analysis of urea and guanidine hydrochloride denaturation curves. Methods Enzymol 131:266–280
22. Gill SC, von Hippel PH (1989) Calculation of protein extinction coefficients from amino acid sequence data. Anal Biochem 182:319–326
23. Miles AJ, Wien F, Lees JG, Rodger A, Janes RW, Wallace BA (2003) Calibration and standardisation of synchrotron radiation circular dichroism and conventional circular dichroism spectrophotometers. Spectroscopy 17:653–661
24. Damianoglou A, Crust EJ, Hicks MJ, Howson SE, Knight AE, Ravi J, Scott P, Rodger A (2008) A new reference material for UV-visible circular dichroism spectroscopy. Chirality 20:1029–1038
25. Waldron DE, Marrington R, Grant MC, Hicks MR, Rodger A (2010) Capillary circular dichroism. Chirality 22:E136–E141
26. Johnson WCJ (1988) Secondary structure of proteins through circular dichroism spectroscopy. Annu Rev Biophys Biophys Chem 17:145–166
27. Johnson WCJ (1985) Circular dichroism and its empirical application to biopolymers. Methods Biochem Anal 31:61–163
28. Miguel MS, Marrington R, Rodger PM, Rodger A, Robinson C (2003) An Escherichia coli

twin-arginine signal peptide switches between helical and unstructured conformations depending on hydrophobicity of the environment. Eur J Biochem 270:3345–3352

29. Green P (1999) PhD thesis, In Chemistry, University of Warwick, Coventry

30. Damianoglou A, Rodger A, Pridmore C, Dafforn TR, Mosely JA, Sanderson JM, Hicks MR (2010) The synergistic action of melittin and phospholipase A2 with lipid membranes: development of linear dichroism for membrane-insertion kinetics. Protein Pept Lett 17:1351–1362

Chapter 9

Analyzing Protein–Ligand Interactions by Dynamic NMR Spectroscopy

Anthony Mittermaier and Erick Meneses

Abstract

Nuclear magnetic resonance (NMR) spectroscopy can provide detailed information on protein–ligand interactions that is inaccessible using other biophysical techniques. This chapter focuses on NMR-based approaches for extracting affinity and rate constants for weakly binding transient protein complexes with lifetimes of less than about a second. Several pulse sequences and analytical techniques are discussed, including line-shape simulations, spin-echo relaxation dispersion methods (CPMG), and magnetization exchange (EXSY) experiments.

Key words Protein–ligand titration, NMR line-shape analysis, CPMG relaxation dispersion, EXSY exchange spectroscopy

1 Introduction

Solution NMR spectroscopy is a powerful and versatile technique for obtaining structural and dynamic information on biological molecules and complexes (1, 2). A wide variety of NMR-based approaches can be applied to different aspects of biological interactions, including the elucidation of the three-dimensional structures (3–5) and internal motions (6, 7) of protein complexes, the characterization of transiently populated states in binding pathways (8, 9), the rapid screening of small molecule libraries (10, 11), and the determination of the thermodynamic and kinetic parameters of binding. This chapter focuses on the last application, the determination of biophysical interaction parameters by NMR.

NMR offers several advantages compared to alternative techniques for measuring the affinities and timescales of protein interactions. First, NMR spectra are generally quite sensitive to complex formation, even when interactions are weak and are not accompanied by large structural rearrangements (12). Thus, NMR is a

Mark A. Williams and Tina Daviter (eds.), *Protein-Ligand Interactions: Methods and Applications*, Methods in Molecular Biology, vol. 1008, DOI 10.1007/978-1-62703-398-5_9, © Springer Science+Business Media New York 2013

powerful diagnostic tool for determining whether two molecules interact. Second, simple analyses of the spectral changes produced by binding rapidly identify those portions of the molecules that are most involved in the interaction (13, 14). Third, NMR can accurately measure the binding kinetics for transient complexes with extremely short bound lifetimes of less than 1 ms (15). Such lifetimes are outside the range of most stopped-flow instruments, because the time required to mix the protein and ligand solutions is comparable to the time required for the binding reaction to proceed nearly to completion. Furthermore, NMR kinetics experiments are performed at equilibrium in solution, which avoids the complication of mass transport effects that are inherent to surface plasmon resonance measurements (16). Lastly, each NMR-active nucleus in the protein and ligand represents a potential probe of the binding process. Comparisons of data obtained for different nuclei can discriminate between two-state and multistate binding, and can shed light on mechanisms of allostery (8, 17, 18). The main disadvantages of using NMR to study binding kinetics are the requirement for quite concentrated (roughly 50 μM to 2 mM) isotopically labeled protein samples and restrictions on protein molecular weight (less than about 100 kDa (19), although larger systems have been successfully studied).

The following sections deal mainly with proteins and ligands that bind in a two-state manner with 1:1 stoichiometry, and in which interactions are detected via NMR signals from the protein. In these cases, the binding reaction can be described by the following simple scheme,

$$\mathrm{L} + \mathrm{P} \underset{k_{\mathrm{off}}}{\overset{k'_{\mathrm{on}}}{\rightleftharpoons}} \mathrm{PL}, \tag{1}$$

where L and P are the free ligand and protein, PL is the complex, k_{off} is the first-order dissociation rate constant such that $1/k_{\mathrm{off}}$ is the average lifetime of the bound state, and k'_{on} is a pseudo-first order association rate constant given by $k_{\mathrm{on}}[\mathrm{L}]$, where k_{on} is the second-order association rate constant and [L] is the equilibrium concentration of free ligand. The affinity of the interaction is represented by the equilibrium dissociation constant, given by the expression

$$K_{\mathrm{D}} = \frac{[\mathrm{L}][\mathrm{P}]}{[\mathrm{PL}]} = \frac{k_{\mathrm{off}}}{k_{\mathrm{on}}}. \tag{2}$$

The timescale of exchange is described by the parameter k_{ex},

$$k_{\mathrm{ex}} = k'_{\mathrm{on}} + k_{\mathrm{off}} = \frac{k_{\mathrm{off}}}{1 - f_{\mathrm{PL}}}, \tag{3}$$

where f_{PL} is the fraction of proteins in the bound state. It is evident from this relationship that the timescale of exchange is largely governed by the dissociation rate constant, k_{off}. For example, at

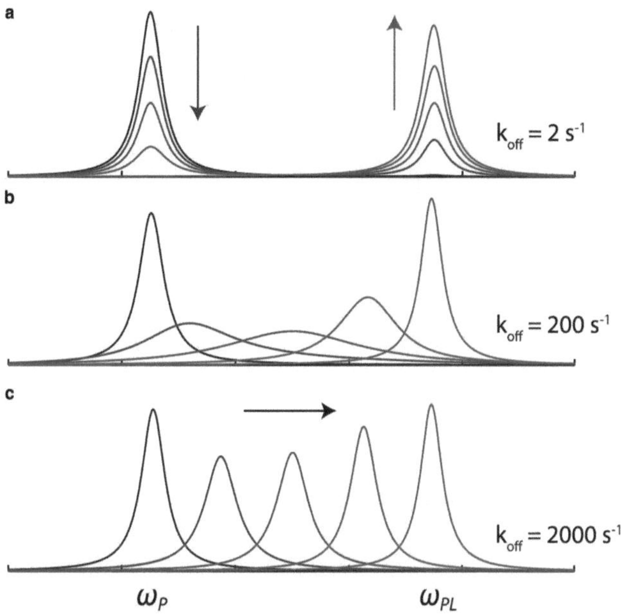

Fig. 1 Effect of ligand binding on an NMR signal. Superposition of simulated 1D NMR spectra for a nucleus with a resonant frequency ω_P in the free state and ω_{PL} in the bound state of a protein that is 1, 25, 50, 75, and 99 % ligand-bound. The spectrum of the free state is transformed into that of the bound state in a manner which depends upon the rate of exchange between the two-states. (**a**) Slow exchange regime where $k_{ex} \ll \Delta\omega$, (**b**) intermediate exchange regime where $k_{ex} \approx \Delta\omega$, (**c**) fast exchange regime where $k_{ex} \gg \Delta\omega$, where $k_{ex} = k_{off} + k_{on}[L]$ and $\Delta\omega = |\omega_{PL} - \omega_P| = 100\pi$ rad/s, in the simulations

the binding midpoint ($f_{PL} = 0.5$) of any system, the exchange rate constant is given by $k_{ex} = 2k_{off}$.

NMR detects ligand binding through changes in the resonant frequencies (chemical shifts) of NMR-active nuclei. Throughout this chapter we use ω to refer to frequencies expressed in rad/s, and ν for frequencies expressed in Hz. Suppose that a nucleus in a protein precesses at a frequency ω_P in the absence of ligand. Ligand binding alters the local electronic environment of the nucleus, resulting in a shift of the resonant frequency to ω_{PL}. The resulting NMR signal from the ensemble of nuclei in the sample depends on the fraction of binding sites that are occupied by ligand, f_{PL}, the difference in resonant frequency between the free and bound states, $\Delta\omega = |\omega_{PL} - \omega_P|$, and the exchange rate, k_{ex}. When $k_{ex} \ll \Delta\omega$, the system is in the slow exchange regime and spectra contain separate peaks at ω_P and ω_{PL}. If ligand is gradually added, the intensity (I_P) of the "free" peak at ω_P decreases while the intensity (I_{PL}) of the "bound" peak at ω_{PL} increases, according to $I_{PL} \propto f_{PL}[P]_T$ and $I_P \propto (1 - f_{PL})[P]_T$, where $[P]_T$ is the total protein concentration. This is shown for simulated 1D NMR spectra in Fig. 1a. In contrast,

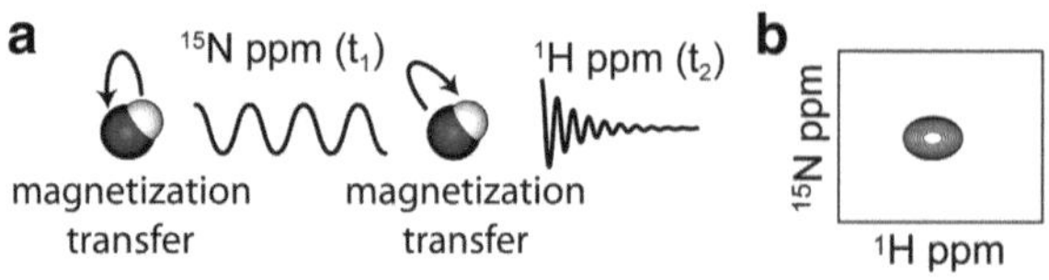

Fig. 2 Schematic representation of a 2D NMR1 correlation experiment. (**a**) Radio-frequency pulses transfer magnetization from ^{1}H to a ^{15}N nucleus to which it is covalently bonded, magnetization evolves on the ^{15}N nucleus during an incremented delay t_1, and is transferred back to ^{1}H where it is detected during a time t_2. (**b**) Fourier analysis of signal variation during t_1 and t_2 extracts the characteristic frequencies for each bonded pair of ^{15}N and ^{1}H nuclei

when $k_{ex} \gg \Delta\omega$, the system is in the fast exchange regime and the spectrum contains a single peak at the population-weighted average chemical shift. If ligand is titrated into the sample, the position of the peak gradually shifts from ω_P to ω_{PL}, according to $\omega_{obs} = \omega_P + f_{PL} \times \Delta\omega$ (Fig. 1c). Furthermore, when k_{ex} and $\Delta\omega$ are of comparable magnitudes, stochastic fluctuations in the resonant frequency caused by ligand association and dissociation lead to weak, broad peaks in NMR spectra (Fig. 1b). Thus, analyses of peak shapes, intensities and positions can yield quantitative information on the populations of different conformational states at atomic resolution.

NMR studies of proteins typically employ multidimensional techniques that resolve individual signals for tens to hundreds of chemically distinct nuclei in the molecules. A common class of experiment correlates the frequencies of directly bonded ^{15}N and ^{1}H nuclei, as depicted schematically in Fig. 2a. In this approach, the equilibrium magnetization of the ^{1}H nucleus is transferred to the directly attached ^{15}N, the ^{15}N frequency is measured indirectly via a variable delay (t_1), the magnetization is transferred back to ^{1}H, and the ^{1}H frequency is directly detected during the acquisition period (t_2) (20). The resultant 2D spectrum contains (at least) one peak for each NH pair, located at the chemical shifts of the ^{1}H and ^{15}N nuclei in the x and y dimensions (Figs. 2b and 3). Furthermore, each peak in the 2D spectrum can be assigned to the corresponding nuclei in the protein using standard methods (1). Thus, a ^{1}H/^{15}N correlation spectrum represents an atomic-resolution "fingerprint" of the protein, and forms the basis for many specialized NMR methods, as described below. Ligand titrations monitored using 2D NMR follow similar exchange behavior to that described above for 1D NMR line-shapes. As an example, a series of protein ^{1}H/^{15}N correlation spectra for a ligand titration are overlaid in Fig. 3a. The peaks shift rather than doubling as the ligand is added, indicative of association–dissociation kinetics that are fast on the NMR timescale. The residues whose peaks shift the most upon

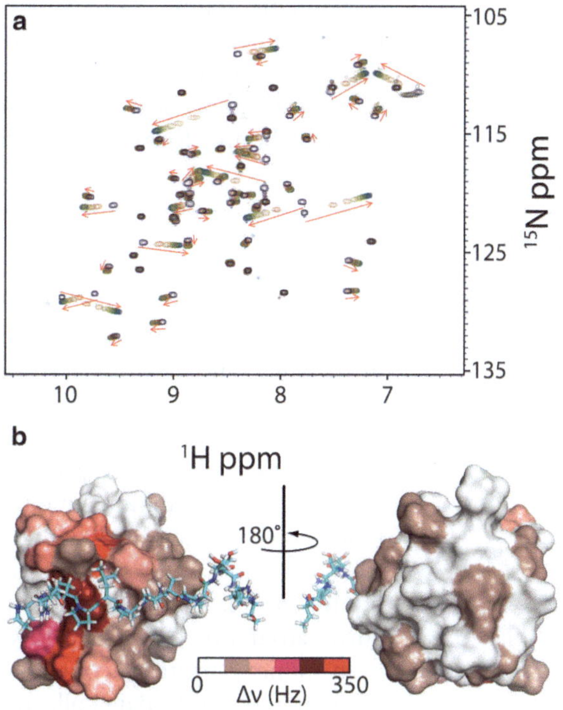

Fig. 3 Titration monitored by heteronuclear correlation spectroscopy. (**a**) Overlaid $^{1}H/^{15}N$ NMR correlation spectra obtained at 18.8 T for a titration of the SH3 domain from the Fyn tyrosine kinase with a proline-rich peptide (Ac-WSLASSPLPPPLP-NH_2). *Arrows* indicate peak displacements. (**b**) Maximum ^{1}H peak displacements ($\Delta\nu$) mapped on the surface of the Fyn SH3 domain bound to a proline-rich peptide (PDB:1AZG (69))

addition of ligand are those which experience the largest changes in local electronic environment due to binding. For this example, chemical shift displacements are mapped onto the three-dimensional structure of the protein in Fig. 3b. As expected, the largest changes are observed for residues in the binding site. Chemical shift mapping approaches, such as this, are a very powerful way to identify protein binding sites, provided that the structure of the protein is known (13, 14).

Direct inspection and quantitative analysis of NMR line-shapes and titration data can yield information on binding kinetics and affinities, as described in Subheading 3.1. In addition, we consider two specialized dynamic NMR techniques that yield even more detailed information, depending on the timescale of association–dissociation kinetics. In the case of relatively rapid dynamics (k_{ex} values of roughly 10–10^4 s^{-1}), Carr–Purcell–Meiboom–Gill (CPMG) relaxation dispersion methods can be analyzed to give the populations, interconversion rates, and chemical shift differences of exchanging states (21) (Subheading 3.2). In cases of slower binding kinetics (k_{ex} values of roughly 0.5–20 s^{-1}) NMR magnetization exchange (EXSY) experiments (22) quantify the kinetics of exchange, and help to correlate protein spectra obtained in the free and bound forms (Subheading 3.3).

2 Materials

2.1 NMR Samples

In order to study ligand binding using multidimensional heteronuclear NMR (as in Fig. 2), protein samples need to be isotopically enriched ^{15}N and/or ^{13}C. In the case of larger molecules (with molecular weights greater than about 25 kDa), additional enrichment with ^{2}H can significantly improve spectral quality (23). Isotopically labeled protein samples can be over-expressed in bacteria grown in minimal media containing $^{15}NH_4Cl$ and/or ^{13}C glucose as the sole sources of nitrogen and/or carbon, respectively (24). Protein samples are typically isolated with several stages of liquid chromatography that often conclude with size-exclusion chromatography. Sample purity is of utmost importance, since even minimal protease contamination can cause extensive protein degradation during data collection. Protein NMR samples are usually 300–600 μl in volume, contain 5–10 % D_2O (for the spectrometer frequency lock), and are usually buffered at a pH below 7.5, since base-catalyzed exchange of hydrogen atoms between protein and solvent can lead to signal loss under more alkaline conditions (25). Samples can also contain small amounts of EDTA and other protease inhibitors, as well as NaN_3 to prevent microbial contamination. The minimum protein concentration required for NMR analyses varies, depending on the spectrometer and particular experiment. In our experience, NMR titrations can be performed with protein concentrations as low as 0.1–0.2 mM, using Varian INOVA spectrometers with field strengths of 11.7–18.8 T and employing standard room-temperature probes. Analyses of dynamics experiments (CPMG and EXSY) are quite sensitive to uncertainties in peak intensity, and we find that concentrations of about 0.5–1.5 mM are preferable.

When preparing ligands for NMR binding studies, it is important to ensure that the buffer conditions match those of the protein sample as closely as possible. Otherwise, addition of the ligand may produce protein spectral changes that are due to perturbation of the solution conditions rather than binding. Buffer matching is most easily accomplished by dialyzing the protein and ligand in separate dialysis membranes in the same container of buffer. Alternatively, powdered ligands may be dissolved directly in the protein dialysis buffer. However, special care must be taken when ligands contain ionizable groups, since dissolving such molecules in the protein buffer can significantly alter the pH. Optimal ligand–protein ratios vary, depending on the experiment (*see* **Note 1**). NMR titrations should include spectra collected in the absence of ligand, spectra with ligand concentrations spanning the mid-point ($f_{PL} = 0.5$), and spectra of proteins nearly saturated with ligand (*see* **Note 2**).

CPMG and EXSY experiments require only a single sample; EXSY is most sensitive when both free and bound proteins give intense signals, and therefore, samples with $f_{PL} \approx 0.5$ are preferable. The choice of ligand–protein ratio for CPMG experiments is somewhat more complicated, since the overall exchange rate (k_{ex}) increases with increasing ligand concentration, according to Eq. 3. The extent of spectral broadening varies with both f_{PL} and k_{ex} (Eqs. 19 and 20), and thus, careful tuning of the ligand concentration can have a substantial impact on the quality of CPMG data (26). Typically, a bound fraction, f_{PL}, between about 5 and 95 % is suitable, since molecules must spend appreciable amounts of time in both the free and bound states in order to produce exchange broadening. A final important consideration is the choice of temperature (*see* **Note 3**), since this can significantly affect both the affinity and kinetics of binding. In one example, we found that binding kinetics shifted from the slow exchange regime, through the intermediate regime, and into the fast exchange regime as the temperature was raised from 10 to 50 °C (15).

2.2 Spectral Analysis

Several different software packages for processing and analyzing NMR spectra are available (for example NMRView (27), matNMR (28), FELIX (Felix NMR Inc.), CcpNmr (29), NMRLab (30), and NMRPipe-NMRDraw (31)). NMR line-shape analyses fit NMR spectra directly, as described in Subheading 3.1. This can be accomplished by outputting the spectral data in a generic format (such as text with the pipe2text command in NMRPipe) and applying freely available (32) or in-house scripts. The analysis of EXSY and CPMG data involve fitting peak intensities (volumes), which may be extracted by either automatic or manual peak-picking. In the sections below, kinetic and thermodynamic parameters are extracted from NMR data by minimizing a χ^2 target function,

$$\chi^2 = \sum \frac{\left(\Upsilon_{exp} - \Upsilon_{calc}\right)^2}{\delta \Upsilon_{exp}^2}, \quad (4)$$

in which Υ_{exp} is an experimental datum, $\delta\Upsilon_{exp}$ is the associated experimental uncertainty, Υ_{calc} is a function of the physical parameters, calculated using one of the equations below, and the sum runs over all data points analyzed. Parameter optimization can be performed using a variety of search algorithms (Simplex, Levenberg-Marquardt, etc. (33)) and software packages, for example MATLAB (Mathworks) or Mathematica (Wolfram Research).

3 Methods

3.1 Titration Analysis

3.1.1 NMR-Derived Binding Isotherms

The affinity of a protein–ligand interaction can be determined by analyzing how the fraction of proteins bound to ligand, f_{PL}, depends on the total concentration of ligand, $[L]_T$. Rearrangement of Eq. 2 yields the simple relationship,

$$f_{PL} = \frac{[L]}{[L] + K_D}, \tag{5}$$

where [L] is the concentration of free ligand and K_D is the equilibrium dissociation constant. When the concentration of protein is extremely low ($<0.01 \times K_D$), the amount of bound ligand is negligible compared to the total ligand concentration and $[L] \approx [L]_T$, yielding the familiar Langmuir binding isotherm (34). However, this situation rarely holds in NMR titrations, since the technique requires protein concentrations that are greater than about 100 μM. Instead, the quadratic form of the binding isotherm applies (35):

$$f_{PL} = \frac{1}{2}\left(1 + r_L + \frac{1}{C} - \sqrt{\left(1 + r_L + \frac{1}{C}\right)^2 - 4r_L}\right), \tag{6}$$

where $r_L = [L]_T/[P]_T$ and $c = [P]_T/K_D$. This has implications for the range of affinities that may be determined by NMR. A series of simulated binding isotherms for titrations with a total protein concentration of 100 μM and affinities ranging from 10 mM to 10 nM are shown in Fig. 4. Notably, the titration midpoints are nearly identical for $K_D \leq 10$ μM, and the curves are virtually superimposable for $K_D \leq 1$ μM. This is due to the fact that when $[P]_T$ greatly exceeds K_D, nearly all added ligand binds immediately to protein, until the binding sites are saturated. In other words,

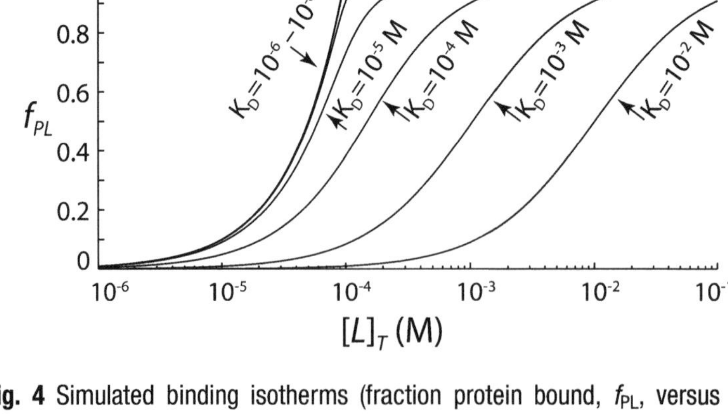

Fig. 4 Simulated binding isotherms (fraction protein bound, f_{PL}, versus total ligand concentration, $[L]_T$) for titrations of a 100 μM protein sample with ligands whose equilibrium dissociation constants range from 10^{-2} to 10^{-8} M

when $[P]_T \gg K_D$ and $[P]_T > [L]_T$, $f_{PL} \approx r_L = [L]_T/[P]_T$, independently of K_D. Thus, values of K_D that are significantly less than the protein concentration cannot be determined with accuracy using this technique. Fortunately, many interactions of biological importance have relatively weak affinities that are readily characterized using NMR.

In the case of slow association–dissociation kinetics, NMR spectra obey the expressions

$$I_{PL}\left([L]_T\right)_i = \lambda_{PL,i} \times f_{PL} \times \left(\frac{V_0}{V_0 + V_L}\right),$$

$$I_P\left([L]_T\right)_i = \lambda_{P,i} \times (1 - f_{PL}) \times \left(\frac{V_0}{V_0 + V_L}\right), \quad (7)$$

where $I_{PL}([L]_T)_i$ and $I_P([L]_T)_i$ are the peak intensities of residue i in the bound and free forms for a given ligand concentration, $[L]_T$, V_0 is the initial sample volume, V_L is the volume of added ligand solution, $\lambda_{PL,i}$ and $\lambda_{P,i}$ are normalization constants, and f_{PL} is given by Eq. 6. Peak intensities can be fitted simultaneously (using Eq. 7) to yield $\lambda_{PL,i}$ and $\lambda_{P,i}$ on a per-peak basis and a single global value of K_D. The term $V_0/(V_0 + V_L)$ accounts for dilution of the protein sample, with the assumption that peak intensity is directly proportional to protein concentration.

In the case of fast association–dissociation kinetics, the displacement of peak i from its initial location in the spectrum is given by the expression

$$\omega_{\mathrm{disp},i}\left([L]_T\right) = f_{PL} \times \Delta\omega_i, \quad (8)$$

where f_{PL} is again given by Eq. 6, $\Delta\omega_i = |\omega_{PL,i} - \omega_{P,i}|$, and $\omega_{PL,i}$, $\omega_{P,i}$ are the resonant frequencies of nucleus i in the bound and free states, respectively. When 2D NMR datasets are employed, displacements in the two frequency dimensions can be analyzed separately (Fig. 5). Peak positions can be fitted simultaneously to yield a global value of K_D, as well as $\Delta\omega_i$ values on a per-nucleus basis (usually this means one value of $\Delta\omega_i$ per peak per spectral dimension). An example of this analysis is shown in Fig. 5, for the peak displacements illustrated in Fig. 3. Note that signals may experience significant broadening during the titration, due to association–dissociation of the ligand, as discussed below. This is evident for some peaks in Fig. 3. Equation 8 is nonetheless applicable, provided that the kinetics of ligand exchange are more rapid than at coalescence, where the signals corresponding to the free and bound forms of the protein merge. At the titration midpoint ($f_P = f_{PL}$), coalescence is reached when $k_{off} = \Delta\omega_i/(2\sqrt{2})$ (36).

3.1.2 Line-Shape Analysis

Association–dissociation kinetics can lead to time-dependent fluctuations in chemical shifts that cause NMR signals to become broadened, as illustrated in Fig. 1. The effect of dynamics on

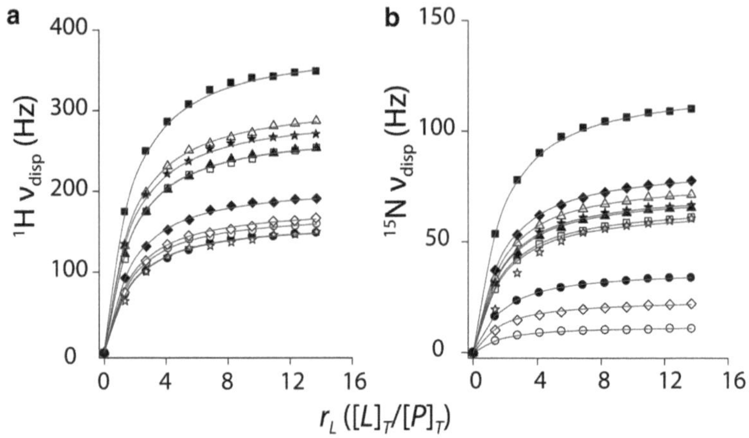

Fig. 5 Dissociation constant from a global analysis of ^{15}N and ^{1}H chemical shift displacements ($\nu_{\text{disp}} = \omega_{\text{disp}}/(2\pi)$). Displacements for amide groups of the Fyn SH3 domain, titrated with a proline-rich peptide, as illustrated in Fig. 3, plotted as a function of the mole ratio of ligand: protein, r_L. The *lines* represent the best fit ($K_D = 300\ \mu$M) according to a simultaneous analysis of all data using Eqs. 6 and 8

NMR line-shapes can be quantitatively modeled, permitting kinetic parameters to be extracted by nonlinear least-squares fitting. A compact mathematical formulation of the problem treats the precession and transverse relaxation of magnetization as rotation and decay in the complex plane, according to the expressions

$$G(t) = e^{i\omega t} \times e^{-R_2^0 t},$$
$$\frac{dG(t)}{dt} = \left(i\omega - R_2^0\right) \times G(t), \tag{9}$$

where $G(t)$ represents the time-domain NMR signal, ω is the frequency of precession, and R_2^0 is the transverse relaxation rate (37). In the case of 1:1 stoichiometry, association–dissociation kinetics are described by the coupled differential equations:

$$\frac{d}{dt}[P] = -k'_{on}[P] + k_{off}[PL],$$
$$\frac{d}{dt}[PL] = k'_{on}[P] - k_{off}[PL]. \tag{10}$$

Combining Eqs. 10 and 11 gives the matrix form of the McConnell equations (38),

$$\frac{d}{dt}\begin{bmatrix} G_P(t) \\ G_{PL}(t) \end{bmatrix} = \mathbf{A}\begin{bmatrix} G_P(t) \\ G_{PL}(t) \end{bmatrix}, \tag{11}$$

where

$$\mathbf{A} = \begin{bmatrix} i\omega_{\mathrm{P}} - R^0_{2,\mathrm{P}} - k'_{\mathrm{on}} & k_{\mathrm{off}} \\ k'_{\mathrm{on}} & i\omega_{\mathrm{PL}} - R^0_{2,\mathrm{PL}} - k_{\mathrm{off}} \end{bmatrix},$$

$$k'_{\mathrm{on}} = f_{\mathrm{PL}} k_{\mathrm{ex}}, \quad k_{\mathrm{off}} = f_{\mathrm{P}} k_{\mathrm{ex}}, \text{ and } f_{\mathrm{P}} + f_{\mathrm{PL}} = 1.$$

The total time-domain signal is then given by

$$G(t) = G_{\mathrm{P}}(t) + G_{\mathrm{PL}}(t) = [1 \quad 1] \exp\{\mathbf{A}t\} \begin{bmatrix} G_{\mathrm{P}}(0) \\ G_{\mathrm{PL}}(0) \end{bmatrix}, \tag{12}$$

where $G_{\mathrm{P}}(0) = f_{\mathrm{P}}$, $G_{\mathrm{PL}}(0) = f_{\mathrm{PL}}$ and the matrix exponential can be evaluated numerically. The 1D NMR spectrum, $g(\omega)$, is given by the Fourier transform of $G(t)$. These calculations can be extended to systems of arbitrary complexity by increasing the number of states and adding appropriate exchange terms (32). In the case of two-state exchange, $g(\omega)$ has the analytical solution (39, 40)

$$g(\omega) = \lambda \frac{\left(P\left[1 + \tau\left(f_{\mathrm{PL}} R^0_{2,\mathrm{P}} + f_{\mathrm{P}} R^0_{2,\mathrm{PL}}\right)\right] + QU\right)}{P^2 + U^2}, \tag{13}$$

where

$$P = \tau\left(R^0_{2,\mathrm{P}} R^0_{2,\mathrm{PL}} - (\bar{\omega} - \omega)^2 + \frac{1}{4}\Delta\omega^2\right) + f_{\mathrm{P}} R^0_{2,\mathrm{P}} + f_{\mathrm{PL}} R^0_{2,\mathrm{PL}},$$

$$Q = \tau\left(\bar{\omega} - \omega - \frac{1}{2}\Delta\omega(f_{\mathrm{P}} - f_{\mathrm{PL}})\right),$$

$$U = (\bar{\omega} - \omega)\left(1 + \tau\left(R^0_{2,\mathrm{P}} + R^0_{2,\mathrm{PL}}\right)\right) + \frac{1}{2}\Delta\omega\left(R^0_{2,\mathrm{PL}} - R^0_{2,\mathrm{P}} + f_{\mathrm{P}} - f_{\mathrm{PL}}\right),$$

$$\bar{\omega} = \frac{1}{2}(\omega_{\mathrm{P}} + \omega_{\mathrm{PL}}),$$

$$\tau = k_{\mathrm{ex}}^{-1} = \frac{1 - f_{\mathrm{PL}}}{k_{\mathrm{off}}},$$

and λ is a normalization constant. Although the equations above were derived for 1D NMR spectra, they can also be applied to peak cross-sections extracted from multidimensional spectra, provided that peak heights are scaled to account for broadening of the other nuclei (32). An NMR line-shape analysis of titration data comprising spectra obtained with a range of ligand concentrations can yield global values of K_{D} and k_{off}, as well as normalization constants, λ, free and bound resonant frequencies, ω_{P}, ω_{PL}, and free and bound transverse relaxation rates, $R^0_{2,\mathrm{P}}$, $R^0_{2,\mathrm{PL}}$, for each nucleus studied. When NMR data deviate significantly from the two-state model, such as when additional peaks appear during a titration, this implies that the binding mechanism includes more than two well-populated states (41).

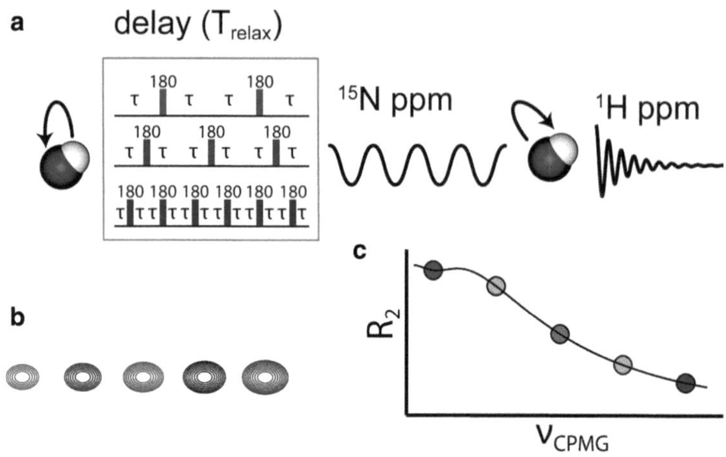

Fig. 6 Schematic representation of a ^{15}N CPMG experiment. (**a**) Equilibrium magnetization is transferred from the ^{1}H to the ^{15}N nucleus, the signal is then allowed to decay for a period T_{relax}, during which refocusing pulses are applied at times characterized by a delay τ, magnetization is returned to the attached ^{1}H, and the signal is detected. Each ^{1}H/^{15}N spin pair is characterized by two frequencies and a signal intensity which depends on the exchange dynamics during T_{relax}. (**b**) The experiment is repeated with different numbers of refocusing pulses applied during T_{relax}, yielding a series of 2D spectra with peak intensities that vary as a function of $\nu_{\mathrm{CPMG}} = 1/(4\tau)$. (**c**) The dispersion profile is calculated based on these peak intensities and Eq. 15

3.2 Carr–Purcell–Meiboom–Gill (CPMG) Experiments

3.2.1 CPMG Overview

As discussed above, association–dissociation dynamics can lead to stochastic fluctuations in nuclear precession frequencies on millisecond to microsecond timescales. This dephases transverse magnetization and results in an added contribution to transverse relaxation, such that

$$R_2 = R_2^0 + R_{ex}, \tag{14}$$

where R_2 is the observed transverse relaxation rate, R_{ex} is the exchange contribution, and R_2^0 is the relaxation rate in the absence of exchange. Line-shape analyses fit R_2, and thus R_{ex} indirectly, since peak height is proportional to $1/R_2$ and the full width at half height is equal to R_2/π. In contrast, CPMG experiments can provide detailed dynamical exchange information by studying R_{ex} more directly (42, 43). In the CPMG approach, R_2 is measured while R_{ex} is suppressed with variable numbers of refocusing pulses applied during a constant relaxation delay (Fig. 6a) (21). The dependence of R_2, and hence R_{ex}, on the pulse repetition rate can be analyzed to extract the populations, exchange rates and chemical shifts of interconverting states (44). There are other advantages to the CPMG approach. Whereas line-shape analyses require a series of spectra recorded as ligand is titrated into the sample, CPMG experiments require only a single NMR sample. Furthermore, CPMG experiments can yield detailed information on weakly populated protein states that are "invisible" in standard NMR correlation spectra (45).

The CPMG experiment is based on the Hahn spin-echo experiment, which consists of two equal delays of length τ, during which magnetization precesses (rotates) in the transverse plane, separated by a 180° refocusing pulse (τ – 180 – τ) (46). A spin's precession is completely refocused during the spin echo, provided that its precession frequency is equal in the two τ periods (Fig. 6a). If an association or dissociation event causes a change in frequency ($\omega_P \longleftrightarrow \omega_{PL}$) during the spin echo, the magnetization is incompletely refocused, which contributes to dephasing of the NMR signal and consequent loss of intensity. When pulses are applied infrequently (large τ), at least one association–dissociation event will occur during most τ delays, magnetization is largely de-phased, and the resulting NMR signal is weak. When pulses are applied rapidly, most pairs of τ delays are free of association–dissociation events, magnetization is largely refocused, and the resulting NMR signal is strong (Fig. 6b). Transverse relaxation rates (Fig. 6c) are directly related to peak intensities by the expression

$$R_2(\nu_{\mathrm{CPMG}}) = -\frac{1}{T_{\mathrm{relax}}}\ln\left(\frac{I(\nu_{\mathrm{CPMG}})}{I_0}\right), \tag{15}$$

where $\nu_{\mathrm{CPMG}} = (4\tau)^{-1}$, $I(\nu_{\mathrm{CPMG}})$ is the intensity of the peak obtained with a given pulse repetition rate, T_{relax} is the length of the relaxation delay, and I_0 is the intensity measured when $T_{\mathrm{relax}} = 0$. Optimization of the length of T_{relax} is important in obtaining good signal intensity throughout the range of ν_{CPMG} values used (*see* **Note 4**). As ν_{CPMG} is increased some cryoprobes can exhibit artifactual signal reduction, and a slight modification of the basic pulse sequence (Fig. 6a) is required to avoid this instrumental effect being mistaken for relaxation (*see* **Note 5**).

Diverse CPMG experiments have been developed to measure the exchange broadening of different types of nucleus and magnetization (47–53), and to characterize the structures and dynamics of excited protein states (54–56). In addition, $R_{1\rho}$ relaxation dispersion experiments are similar to CPMG methods, with continuous-wave radio-frequency irradiation replacing trains of refocusing pulses (57). Here we focus on CPMG experiments directed at single-quantum coherences, as illustrated schematically for an $^{15}N/^{1}H$ spin pair in Fig. 6 (21, 58).

3.2.2 CPMG Data Analysis

The shape of the CPMG relaxation dispersion profile (i.e., the CPMG frequency dependence of R_2—Fig. 6c) can be calculated numerically, analogously to Eq. 11. In the case of two-state exchange, the magnetization following n pairs of spin echoes, $n \times (\tau - 180 - \tau - \tau - 180 - \tau)$ is given by (44)

$$\begin{bmatrix} G_{\mathrm{P}}(n) \\ G_{\mathrm{PL}}(n) \end{bmatrix} = (\exp\{\mathbf{A}\tau\}\exp\{\mathbf{B}\tau\}\exp\{\mathbf{B}\tau\}\exp\{\mathbf{A}\tau\})^{n} \begin{bmatrix} G_{\mathrm{P}}(0) \\ G_{\mathrm{PL}}(0) \end{bmatrix}, \tag{16}$$

where

$$\mathbf{A} = \begin{bmatrix} -R^{0}_{2,\mathrm{P}} - k'_{\mathrm{on}} & k_{\mathrm{off}} \\ k'_{\mathrm{on}} & i\Delta\omega - R^{0}_{2,\mathrm{PL}} - k_{\mathrm{off}} \end{bmatrix},$$

$$\mathbf{B} = \begin{bmatrix} -R^{0}_{2,\mathrm{P}} - k'_{\mathrm{on}} & k_{\mathrm{off}} \\ k'_{\mathrm{on}} & -i\Delta\omega - R^{0}_{2,\mathrm{PL}} - k_{\mathrm{off}} \end{bmatrix},$$

$\Delta\omega = |\omega_{\mathrm{PL}} - \omega_{\mathrm{P}}|$, $G_{\mathrm{P}}(0) = f_{\mathrm{P}}$ and $G_{\mathrm{PL}}(0) = f_{\mathrm{PL}}$. The opposite signs of the $i\Delta\omega$ terms in **A** and **B** account for the refocusing of precession during each spin echo. The transverse relaxation rate, R_2, of the unbound peak can be obtained for any value of T_{relax} and ν_{CPMG} using the expression

$$R_2(n) = -\frac{1}{4n\tau}\ln\left(\frac{G_{\mathrm{P}}(n)}{G_{\mathrm{P}}(0)}\right), \tag{17}$$

where $\tau = 1/(4\nu_{\mathrm{CPMG}})$ and $n = \nu_{\mathrm{CPMG}} \times T_{\mathrm{relax}}$.

If separate peaks are obtained for the free and bound forms (i.e., for systems in the slow exchange regime), the R_2 value for the bound peak can be obtained by substituting G_{P} with G_{PL} in the expression above. It is assumed that $R^{0}_{2,\mathrm{P}} = R^{0}_{2,\mathrm{PL}}$ in the expressions above, since the intrinsic relaxation rates of the protein are generally not dramatically affected by ligand binding, and it is not possible, in general, to extract separate values for $R^{0}_{2,\mathrm{P}}$ and $R^{0}_{2,\mathrm{PL}}$ from CPMG data (59) (*see* **Note 6**). This approach for calculating CPMG relaxation dispersion profiles can be generalized to systems containing arbitrary numbers of states by extending the matrices **A** and **B** of Eq. 16 with additional chemical shift differences ($\Delta\omega$) and relaxation rates (R_2) along the diagonal together with exchange rate constants in the appropriate elements (60).

In the case of two-state exchange, the transverse relaxation rate of the more populated form ($R_{2,\mathrm{A}}$) is given by the analytical expression (61, 62)

$$\begin{aligned} R_{2,\mathrm{A}}(\nu_{\mathrm{CPMG}}) = \frac{1}{2}\Big(& R^{0}_{2,\mathrm{A}} + R^{0}_{2,\mathrm{B}} + k_{\mathrm{ex}} \\ & - 2\nu_{\mathrm{CPMG}}\cosh^{-1}\left(D_{+}\cosh(\eta_{+}) - D_{-}\cosh(\eta_{-})\right)\Big), \end{aligned} \tag{18}$$

where

$$D_{\pm} = \frac{1}{2}\left(\pm 1 + \frac{\psi + 2\Delta\omega^2}{\sqrt{\psi^2 + \xi^2}}\right),$$

$$\eta_{\pm} = \frac{\sqrt{2}}{4\nu_{\mathrm{CPMG}}}\sqrt{\pm\psi + \sqrt{\psi^2 + \xi^2}},$$

$$\psi = \left(R_{2,\mathrm{A}}^{0} - R_{2,\mathrm{B}}^{0} + f_{\mathrm{B}}k_{\mathrm{ex}} - f_{\mathrm{A}}k_{\mathrm{ex}}\right)^2 - \Delta\omega^2 + 4f_{\mathrm{A}}f_{\mathrm{B}}k_{\mathrm{ex}}^2,$$

$$\xi = 2\Delta\omega\left(R_{2,\mathrm{A}}^{0} - R_{2,\mathrm{B}}^{0} + f_{\mathrm{B}}k_{\mathrm{ex}} - f_{\mathrm{A}}k_{\mathrm{ex}}\right),$$

$$(\mathbf{A}, \mathbf{B}) = \begin{cases} (\mathrm{P}, \mathrm{PL}), & f_{\mathrm{PL}} \leq 0.5 \\ (\mathrm{PL}, \mathrm{P}), & f_{\mathrm{PL}} \geq 0.5 \end{cases}.$$

This expression applies to the signal from the major form of the protein (i.e., to the free state where ligand concentration is relatively low, and to the bound state close to saturation). In the limit of fast exchange ($k_{ex} \gg \Delta\omega$) this equation is well approximated by the simpler expression (63)

$$R_2(\nu_{\mathrm{CPMG}}) = f_{\mathrm{P}}R_{2,\mathrm{P}}^{0} + f_{\mathrm{PL}}R_{2,\mathrm{PL}}^{0} + \frac{\Phi_{\mathrm{ex}}}{k_{\mathrm{ex}}} \times \left(1 - \frac{4\nu_{\mathrm{CPMG}}}{k_{\mathrm{ex}}}\tanh\left(\frac{k_{\mathrm{ex}}}{4\nu_{\mathrm{CPMG}}}\right)\right), \quad (19)$$

where $\Phi_{ex} = f_P f_{PL} \Delta\omega^2$.

A typical CPMG experiment consists of series of roughly 10–20 2D correlation spectra obtained with a range of ν_{CPMG} values and yielding R_2 values for each amide, calculated using Eq. 15. For systems in intermediate timescale exchange, the analysis proceeds with nonlinear least-squares fitting of R_2 values using Eq. 18, while for systems in fast exchange Eq. 19 is employed. The exchange regime can be determined from an inspection of lineshapes throughout the titration, as discussed in Subheading 1. Series of experiments should be performed at a minimum of two different static magnetic field strengths under otherwise identical conditions (64) (*see* **Note 3**). Additional information useful in selecting the appropriate equation for analyses is provided by the dependence of exchange broadening at different static field strengths as expressed by the parameter α (65)

$$\alpha = \frac{d \ln R_{\mathrm{ex}}}{d \ln B_0}, \quad (20)$$

where B_0 is the NMR spectrometer field strength and R_{ex} is the exchange contribution to transverse relaxation. R_{ex} may be estimated as the difference between the experimental R_2 values obtained with maximal and minimal ν_{CPMG} (i.e., the reduction in R_2 with increasing frequency). In the intermediate exchange regime, $\alpha \approx 1$, and R_{ex} scales roughly linearly with field strength. In the fast exchange regime, $\alpha \approx 2$, and R_{ex} scales as the square of the field strength. Note that different peaks may be in different exchange regimes depending on their values of $\Delta\omega$, such that data

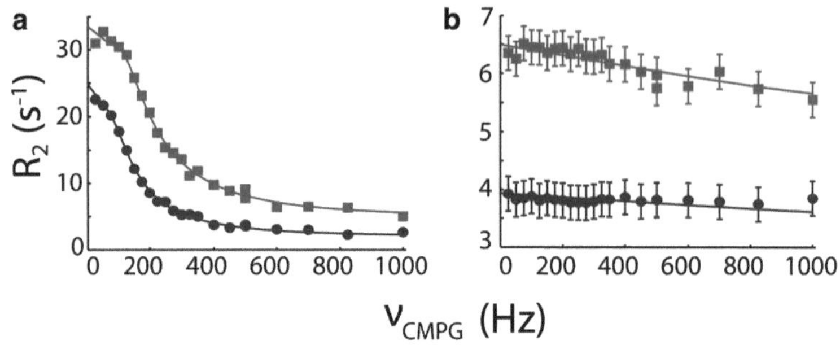

Fig. 7 Exchange rates from CPMG dispersion curves. Data for Thr14 in the Fyn SH3 domain, partially saturated with proline-rich peptides, obtained at 11.7 T (*circles*) and 18.8 T (*squares*). (**a**) Protein 7 % bound to a peptide with sequence Ac-WSLARRPLPPLP-NH_2. Exchange is in the intermediate regime, with $k_{off} \approx 200\ s^{-1}$. *Lines* correspond to best fits with Eq. 18. (**b**) Protein is 10 % bound to a peptide with sequence Ac-WSLASSPLPPPLP-NH_2. Exchange is in the fast regime, with $k_{off} \approx 2{,}000\ s^{-1}$. *Lines* correspond to best fits with Eq. 19

for peaks with smaller $\Delta\omega$ values may obey the fast-exchange Eq. 19, while peaks with larger $\Delta\omega$ values may require the intermediate exchange Eq. 18.

Fits of multiple-field data from one residue using Eq. 18 yield the parameters k_{ex}, f_{PL}, and $\Delta\omega^2$, and an R_2^0 value for each spectrometer field strength (again assuming that $R_{2,P}^0 = R_{2,PL}^0 = R_2^0$) (Fig. 7a). The dissociation rate constant is then given by $k_{off} = (1 - f_{PL})k_{ex}$ and the second-order association rate constant is calculated as $k_{on} = k_{off}/K_D$. Fits with the fast-exchange Eq. 19 yield the parameters k_{ex}, Φ_{ex}, and values of R_2^0 (Fig. 7b). Note that in the fast exchange regime, f_{PL} and $\Delta\omega^2$ cannot be determined separately, as both are subsumed into Φ_{ex}. In this case, the value of f_{PL} can be determined from observed peak displacements relative to the maximum peak displacements (ligand-free versus ligand-saturated spectra) according to Eq. 8, or from an analysis of dependence of Φ_{ex} on $\Delta\omega^2$ (*see* **Note 7**). More accurate results can be obtained by fitting dispersion profiles for all residues simultaneously, with values of k_{ex} and f_{PL} optimized globally and values of $\Delta\omega^2$ and R_2^0 extracted on a per-residue basis (66). If exchange for all residues is in the fast regime, then only k_{ex} is determined globally, while Φ_{ex} and R_2^0 are obtained on a per-residue basis. The extracted values of $\Delta\omega^2$ or Φ_{ex} can be used to validate the use of the two-state exchange model (*see* **Note 7**).

3.3 Magnetization Exchange Spectroscopy (EXSY)

3.3.1 EXSY Overview

EXSY techniques can be applied when association–dissociation kinetics are slow on the NMR chemical shift timescale. Under these conditions, separate peaks are observed for some residues in the bound and free forms of the protein mid-way through titrations with ligand (Fig. 1c). EXSY experiments are essentially simple variants of multidimensional pulse sequences with delays inserted

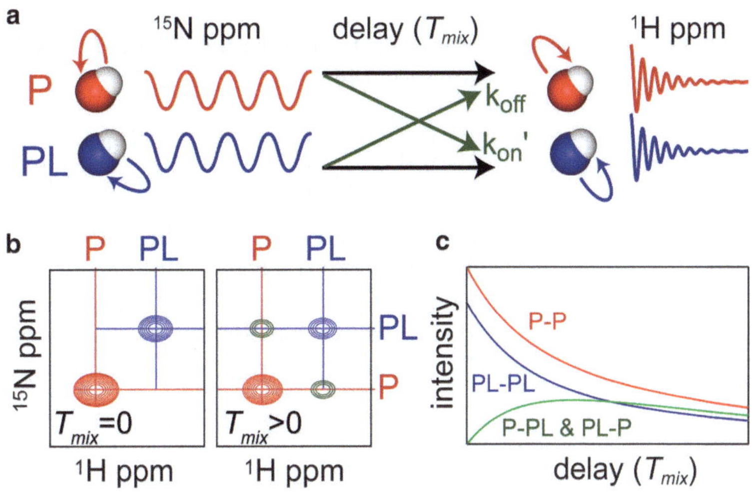

Fig. 8 Characterization of slow-timescale ligand binding kinetics by EXSY NMR spectroscopy. (**a**) Schematic representation of an EXSY pulse sequence. Equilibrium ^{1}H magnetization is transferred to the directly attached ^{15}N, the resonant frequency is measured, magnetization is aligned along the *z*-axis during T_{mix}, the magnetization is returned to the ^{1}H and the signal is detected. (**b**) Simulated spectra for a $^{1}H/^{15}N$ spin pair undergoing slow exchange. Diagonal peaks (P–P and PL–PL) correspond to the self-peaks of a residue in the free and bound states respectively, while the off-diagonal cross-peaks (P–PL and PL–P) are produced by ligand association and dissociation during T_{mix}. (**c**) Simulated intensity profiles are calculated using Eq. 23.

between the indirect and direct acquisition periods. For example, exchange rates may be quantified using a modified $^{1}H/^{15}N$ correlation experiment that includes a variable delay of length T_{mix} between the ^{15}N and ^{1}H chemical shift detection periods (22), as illustrated schematically in Fig. 8. If a free protein binds a ligand during T_{mix}, it will give rise to a cross-peak at $(\omega_P(^{15}N), \omega_{PL}(^{1}H))$. Conversely, if a bound protein releases its ligand during T_{mix}, it will give rise to a cross-peak at $((\omega_{PL}(^{15}N), \omega_P(^{1}H))$. Each exchanging amide group can therefore produce four peaks in this experiment: two self-peaks corresponding to the free and bound forms, and two cross-peaks that are produced by interconversion (*see* **Note 8**). The intensities of self-peaks and cross-peaks vary as a function of the mixing time T_{mix} and depend on the association and dissociation rate constants, k'_{on} and k_{off}, as well as the longitudinal ^{15}N relaxation rates in the two states, $R_{1,P}$ and $R_{1,PL}$. The intensities of the cross-peaks are low when $T_{mix} \approx 0$, increase due to exchange at moderate values of T_{mix}, and ultimately decay due to the dominant effect of longitudinal relaxation at large T_{mix}. Intensities of the self-peaks are initially high and decrease because of both interconversion and longitudinal relaxation. The dependences of self-peak and cross-peak intensities on T_{mix} can be analyzed to yield binding kinetic information, as described below.

3.3.2 EXSY Data Analysis

Magnetization exchange during T_{mix} is described by the differential equation (44):

$$\frac{\partial}{\partial t}\begin{bmatrix} I_P(t) \\ I_{PL}(t) \end{bmatrix} = -\begin{bmatrix} k'_{on} + R_{1,P} & -k_{off} \\ -k'_{on} & k_{off} + R_{1,PL} \end{bmatrix}\begin{bmatrix} I_P(t) \\ I_{PL}(t) \end{bmatrix}, \quad (21)$$

where $I_P(t)$ and $I_{PL}(t)$ are the instantaneous signal intensities of the free and bound states at time t, and $R_{1,P}$ and $R_{1,PL}$ are the ^{15}N longitudinal relaxation rates. Integration of Eq. 21 gives the following expression for the intensities of the self-peaks (I_{P-P} and I_{PL-PL}) and cross-peaks (I_{P-PL} and I_{PL-P}) after the T_{mix} period:

$$\begin{bmatrix} I_{P-P}(T_{mix}) & I_{PL-P}(T_{mix}) \\ I_{P-PL}(T_{mix}) & I_{PL-PL}(T_{mix}) \end{bmatrix} = \exp\left\{-T_{mix}\begin{bmatrix} k'_{on} + R_{1,p} & -k_{off} \\ -k'_{on} & k_{off} + R_{1,PL} \end{bmatrix}\right\}\begin{bmatrix} I_P(0) & 0 \\ 0 & I_{PL}(0) \end{bmatrix}, \quad (22)$$

where $I_P(0)$ and $I_{PL}(0)$ are the amounts of magnetization for the free and bound forms immediately preceding the T_{mix} delay. Solving the matrix exponential gives (44):

$$I_{P-P} = I_P^0\frac{(a_{11} - \lambda_2)e^{-\lambda_1 T_{mix}} + (\lambda_1 - a_{11})e^{-\lambda_2 T_{mix}}}{\lambda_1 - \lambda_2},$$

$$I_{PL-PL} = I_{PL}^0\frac{(a_{22} - \lambda_2)e^{-\lambda_1 T_{mix}} + (\lambda_1 - a_{22})e^{-\lambda_2 T_{mix}}}{\lambda_1 - \lambda_2},$$

$$I_{P-PL} = I_P^0\frac{\left(a_{21}e^{-\lambda_1 T_{mix}} - a_{21}e^{-\lambda_2 T_{mix}}\right)}{\lambda_1 - \lambda_2},$$

$$I_{PL-P} = I_{PL}^0\frac{\left(a_{12}e^{-\lambda_1 T_{mix}} - a_{12}e^{-\lambda_2 T_{mix}}\right)}{\lambda_1 - \lambda_2}, \quad (23)$$

where

$$\lambda_{1,2} = \frac{1}{2}(a_{11} + a_{22}) \pm \sqrt{(a_{11} + a_{22})^2 + 4k'_{on}k_{off}},$$

$$a_{11} = R_{1,P} + k'_{on}, \qquad a_{12} = -k_{off}, \qquad a_{21} = -k'_{on},$$
$$a_{22} = R_{1,PL} + k_{off}.$$

A typical EXSY dataset consists of roughly 10–20 2D correlation spectra obtained with T_{mix} varying between zero and the maximum value at which reasonably strong signals are still observable. ^{15}N longitudinal relaxation sets the upper limit for T_{mix} at about 1 s. Peak intensities are then fitted using Eq. 23 to yield k'_{on}, k_{off}, $R_{1,P}$, $R_{1,PL}$, $I_P(0)$, and $I_{PL}(0)$. As for CPMG data, more accurate values can be obtained by analyzing data for all residues simultaneously to give global values of k'_{on}, k_{off}, together with $R_{1,P}$,

$R_{1,PL}$, $I_P(0)$, and $I_{PL}(0)$ on a per-residue basis. It is assumed in Eqs. 22 and 23 that exchange during the transfer of magnetization from ^{15}N back to ^1H following T_{mix} can be neglected. When exchange is sufficiently rapid, non-negligible exchange occurs during the transfer steps, and cross-peaks are observed even with $T_{mix} = 0$. In this case, Eq. 22 can be modified to account for the observed back-transfer according to (15):

$$\begin{bmatrix} I_{P-P}(\text{obs}) & I_{PL-P}(\text{obs}) \\ I_{P-PL}(\text{obs}) & I_{PL-PL}(\text{obs}) \end{bmatrix} = \begin{bmatrix} 1 & y_{PL-P} \\ y_{P-PL} & y_{PL} \end{bmatrix} \times \begin{bmatrix} I_{P-P}(T_{mix}) & I_{PL-P}(T_{mix}) \\ I_{P-PL}(T_{mix}) & I_{PL-PL}(T_{mix}) \end{bmatrix}, \quad (24)$$

where y_{PL} accounts for differential relaxation of the self peaks, y_{PL-P} and y_{P-PL} account for dissociation and association during the back-transfer of magnetization, respectively, and all three parameters are optimized on a per-residue basis in global fits of EXSY data.

4 Notes

NMR titrations

1. An important consideration when performing NMR titration experiments is the selection of ligand concentrations. If an approximate affinity constant is already known, the appropriate ligand concentrations can be chosen using Eq. 6 so as to adequately define the saturated baseline and the curvature of the binding isotherm. The situation is more complicated if no prior information on the affinity is available. This is particularly true in the case of fast exchange, since peak displacements can only be quantitatively related to f_{PL} once the saturated baseline is obtained, i.e., once the titration is complete. In this case, it is advisable to collect several ($\geq$5) spectra over ligand–protein ratios (r_L) from 0 to about 1.5. If the interaction is of high affinity ($[P]_T > K_D$) saturation will be reached over this interval. If peak displacements vary linearly with $[L]_T$ past the stoichiometric equivalence point ($r_L = 1$), then $K_D > [P]_T$ and larger increments of $[L]_T$ can be used for additional titration points.
2. When performing titrations it is critical that the ligand and protein solutions are mixed thoroughly and that sample loss during the mixing procedure is minimized. We recommend withdrawing the sample from the NMR tube completely, using a Hamilton syringe or an NMR long-tip pipette, and

transferring it to small (≈1.5 ml) sample tube. Ligand can then be added using a standard pipettor, the sample can be thoroughly mixed, spun down, and returned to the NMR tube using the syringe. It is preferable to use the same NMR tube, sample tube, and syringe for every titration step (without rinsing) in order to minimize loss of the sample.

CPMG experiments

3. Temperature control is a major concern in studies of relaxation, particularly when performing CPMG studies, as data from more than one spectrometer are analyzed assuming that protein dynamics (and hence temperatures) are identical in all instruments. Calibration curves can be constructed for each NMR probe using perdeuterated methanol to measure the true sample temperature (67). In this case, $T(K) = -16.7467\ (\Delta\delta)^2 - 52.5130\ \Delta\delta + 419.1381$ ($\Delta\delta = \delta_{OH} - \delta_{CHD2}$), where δ is expressed in ppm. Alternatively, a nonmagnetic digital thermometer may be inserted directly into the NMR probe. Another issue is heating caused by RF pulses, particularly when proton decoupling is applied during the relaxation delay (58). This effect can be assayed by measuring the water resonance frequency (unlocked) before and after several minutes of irradiation with the pulse sequence, as the water resonance shifts by approximately −0.01 ppm/K (68). The temperature setting of the spectrometer can then be adjusted by a compensatory amount (usually by about 0.1–0.2 °C).
4. A critical parameter when performing CPMG relaxation dispersion experiments is the length of the constant relaxation delay, T_{relax}. Longer values of T_{relax} allow more time for exchange broadening to occur, whereas shorter values lead to less signal loss overall and more intense peaks. Therefore, a compromise must be reached between maximizing the sensitivity to R_{ex} and optimizing signal strength. We find that a good balance is achieved when the spectrum obtained with the maximum ν_{CPMG} is about half as intense as that obtained with $T_{relax} = 0$. This corresponds to an optimal value of $T_{relax} \approx \ln\{2\}/R_2^0$ (e.g., 40 ms for $R_2^0 = 17\,s^{-1}$). One caveat to this rule is that T_{relax} periods longer than 40 ms are generally not used, in order to limit amplifier loads, even when R_2^0 values are small.
5. Some NMR cold-probes exhibit RF pulse-dependent artifacts, such that the sensitivity decreases (and apparent R_2 increases) with increasing ν_{CPMG}. In the case of Varian cold probes, it has been found that this can be suppressed by applying CW irradiation to ^{15}N nuclei prior to each transient for a variable time period, T_{CW}, given by

$$(T_{\text{CW}})_{\text{i}} = \frac{(\upsilon_{\text{CPMG}})_{\max} - (\upsilon_{\text{CPMG}})_{\text{i}}}{(\upsilon_{\text{CPMG}})_{\max}} T_{\text{relax}},$$

where $(\nu_{\text{CPMG}})_{\max}$ is the largest value of ν_{CPMG} employed in a dispersion series, typically 1,000 Hz (15). A compensation power level 8 dB below that of the CPMG pulse train typically provides excellent artifact suppression at all fields and temperatures.

6. Alternatively, R_2 values can be measured separately for the ligand-free and ligand-saturated forms of the protein and then held fixed in analyses of CPMG data.
7. In the case of two-state exchange, $\Delta\omega^2$ values obtained from titration experiments (i.e., squared maximum peak displacements) should equal $\Delta\omega^2$ parameters extracted from CPMG data using Eq. 18, and should correlate linearly Φ_{ex} values obtained using Eq. 19 (15). If Φ_{ex} values for all residues are plotted as function of titration-derived $\Delta\omega^2$ values, the slope is equal to $f_{\text{P}}f_{\text{PL}} = (1 - f_{\text{PL}})f_{\text{PL}}$. If these correlations are not observed experimentally, it may be taken as evidence for the existence of additional millisecond-timescale exchange phenomena, i.e., dynamics involving more than two states. In this case, multistate binding parameters can be extracted by simultaneously fitting CPMG data obtained with different levels of ligand saturation (8).

EXSY experiments

8. EXSY experiments provide a convenient means for transferring spectral assignments from the free to the bound protein NMR spectrum, for systems that are in slow exchange. When association–dissociation kinetics are rapid, transferring assignments is a trivial process, since each peak gradually moves from the ligand-free frequencies to the ligand-saturated frequencies over the course of the titration. In contrast, for a slowly exchanging system, it can be challenging to associate each gradually appearing peak with its corresponding disappearing peak in the titration data. The locations of exchange cross-peaks in EXSY spectra for a partially saturated protein sample, thus, identify pairs of self peaks that are produced by the same amide group in the free and bound states (18).

References

1. Cavanagh J (1996) Protein NMR spectroscopy: principles and practice. Academic, San Diego
2. Mittermaier A, Kay LE (2006) New tools provide new insights in NMR studies of protein dynamics. Science 312:224–228
3. Ikura M, Clore GM, Gronenborn AM, Zhu G, Klee CB, Bax A (1992) Solution structure of a calmodulin-target peptide complex by multidimensional NMR. Science 256:632–638
4. Walters KJ, Ferentz AE, Hare BJ, Hidalgo P, Jasanoff A, Matsuo H, Wagner G (2001)

Characterizing protein–protein complexes and oligomers by nuclear magnetic resonance spectroscopy. NMR Biol Macromol B 339:238–258
5. Jung YS, Cai ML, Clore GM (2010) Solution structure of the IIA(chitobiose)-IIBchitobiose complex of the *N,N'*-diacetylchitobiose branch of the *Escherichia coli* phosphotransferase system. J Biol Chem 285:4173–4184
6. Finerty PJ, Mittermaier AK, Muhandiram R, Kay LE, Forman-Kay JD (2005) NMR dynamics-derived insights into the binding properties of a peptide interacting with an SH2 domain. Biochemistry 44:694–703
7. Boehr DD, McElheny D, Dyson HJ, Wright PE (2006) The dynamic energy landscape of dihydrofolate reductase catalysis. Science 313:1638–1642
8. Sugase K, Dyson HJ, Wright PE (2007) Mechanism of coupled folding and binding of an intrinsically disordered protein. Nature 447:1021–1025
9. Iwahara J, Clore GM (2006) Detecting transient intermediates in macromolecular binding by paramagnetic NMR. Nature 440:1227–1230
10. Meyer B, Peters T (2003) NMR spectroscopy techniques for screening and identifying ligand binding to protein receptors. Angew Chem Int Ed Engl 42:864–890
11. Shuker SB, Hajduk PJ, Meadows RP, Fesik SW (1996) Discovering high-affinity ligands for proteins: SAR by NMR. Science 274:1531–1534
12. Zarrine-Afsar A, Mittermaier A, Kay LE, Davidson AR (2006) Protein stabilization by specific binding of guanidinium to a functional arginine-binding surface on an SH3 domain. Protein Sci 15:162–170
13. Chen Y, Reizer J, Saier MH, Fairbrother WJ, Wright PE (1993) Mapping of the binding interfaces of the proteins of the bacterial phosphotransferase system, HPr and IIA^{glc}. Biochemistry 32:32–37
14. Liu ZH et al (2000) Structural basis for binding of Smac/DIABLO to the XIAP BIR3 domain. Nature 408:1004–1008
15. Demers JP, Mittermaier A (2009) Binding mechanism of an SH3 domain studied by NMR and ITC. J Am Chem Soc 131:4355–4367
16. Schuck P (1996) Kinetics of ligand binding to receptor immobilized in a polymer matrix, as detected with an evanescent wave biosensor.1. A computer simulation of the influence of mass transport. Biophys J 70:1230–1249
17. Tochtrop GP, Richter K, Tang C, Toner JJ, Covey DF, Cistola DP (2002) Energetics by NMR: site-specific binding in a positively cooperative system. Proc Natl Acad Sci U S A 99:1847–1852
18. Freiburger LA, Baettig OM, Sprules T, Berghuis AM, Auclair K, Mittermaier AK (2011) Competing allosteric mechanisms modulate substrate binding in a dimeric enzyme. Nat Struct Mol Biol 18:288–294
19. Tugarinov V, Choy WY, Orekhov VY, Kay LE (2005) Solution NMR-derived global fold of a monomeric 82-kDa enzyme. Proc Natl Acad Sci U S A 102:622–627
20. Bodenhausen G, Ruben DJ (1980) Natural abundance nitrogen-15 NMR by enhanced heteronuclear spectroscopy. Chem Phys Lett 69:185–189
21. Loria JP, Rance M, Palmer AG (1999) A relaxation-compensated Carr-Purcell-Meiboom-Gill sequence for characterizing chemical exchange by NMR spectroscopy. J Am Chem Soc 121:2331–2332
22. Farrow NA, Zhang O, Forman-Kay JD, Kay LE (1994) A heteronuclear correlation experiment for simultaneous determination of ^{15}N longitudinal decay and chemical exchange rates of systems in slow equilibrium. J Biomol NMR 4:727–734
23. Gardner KH, Kay LE (1998) The use of ^{2}H, ^{13}C, ^{15}N multidimensional NMR to study the structure and dynamics of proteins. Annu Rev Biophys Biomol Struct 27:357–406
24. Marley J, Lu M, Bracken C (2001) A method for efficient isotopic labeling of recombinant proteins. J Biomol NMR 20:71–75
25. Bai YW, Milne JS, Mayne L, Englander SW (1993) Primary structure effects on peptide group hydrogen exchange. Protein Struct Funct Genet 17:75–86
26. Sugase K, Lansing JC, Dyson HJ, Wright PE (2007) Tailoring relaxation dispersion experiments for fast-associating protein complexes. J Am Chem Soc 129:13406–13407
27. Johnson BA, Blevins RA (1994) NMR view – a computer program for the visualization and analysis of NMR data. J Biomol NMR 4:603–614
28. van Beek JD (2007) matNMR: a flexible toolbox for processing, analyzing and visualizing magnetic resonance data in Matlab. J Magn Reson 187:19–26
29. Vranken WF, Boucher W, Stevens TJ, Fogh RH, Pajon A, Llinas P, Ulrich EL, Markley JL, Ionides J, Laue ED (2005) The CCPN data model for NMR spectroscopy: development of a software pipeline. Protein Struct Funct Bioinf 59:687–696
30. Gunther UL, Ludwig C, Ruterjans H (2000) NMRLAB – advanced NMR data processing in MATLAB. J Magn Reson 145:201–208

31. Delaglio F, Grzesiek S, Vuister GW, Zhu G, Pfeifer J, Bax A (1995) NMRPipe: a multidimensional spectral processing system based on UNIX pipes. J Biomol NMR 6:277–293
32. Gunther UL, Schaffhausen B (2002) NMRKIN: simulating line shapes from two-dimensional spectra of proteins upon ligand binding. J Biomol NMR 22:201–209
33. Vetterling WT, Press WH, Teukolsky SA, Flannery BR (1988) Numerical recipes in C. Cambridge University Press, Cambridge
34. Langmuir I (1916) The constitution and fundamental properties of solids and liquids part I solids. J Am Chem Soc 38:2221–2295
35. Wiseman T, Williston S, Brandts JF, Lin LN (1989) Rapid measurement of binding constants and heats of binding using a new titration calorimeter. Anal Biochem 179:131–137
36. Bain AD (2003) Chemical exchange in NMR. Prog NMR Spect 43:63–103
37. Gutowsky HS, Saika A (1953) Dissociation, chemical exchange, and the proton magnetic resonance in some aqueous electrolytes. J Chem Phys 21:1688–1694
38. McConnell HM (1958) Reaction rates by nuclear magnetic resonance. J Chem Phys 28:430–431
39. Rogers MT, Woodbrey JC (1962) Proton magnetic resonance study of hindered internal rotation in some substituted N, N-dimethylamides. J Phys Chem 66:540–562
40. Gupta RK, Pitner TP, Wasylishen R (1974) Fourier transform NMR of exchanging chemical systems. J Magn Reson 13:383–385
41. Mittag T, Schaffhausen B, Gunther UL (2004) Tracing kinetic intermediates during ligand binding. J Am Chem Soc 126:9017–9023
42. Carr HY, Purcell EM (1954) Effects of diffusion on free precession in nuclear magnetic resonance experiments. Phys Rev 94:630–638
43. Meiboom S, Gill D (1958) Modified spin echo method for measuring nuclear relaxation times. Rev Sci Inst 29:688–691
44. Palmer AG, Kroenke CD, Loria JP (2001) Nuclear magnetic resonance methods for quantifying microsecond-to-millisecond motions in biological macromolecules. NMR Biol Macromol B 339:204–238
45. Korzhnev DM, Kay LE (2008) Probing invisible, low-populated states of protein molecules by relaxation dispersion NMR spectroscopy: an application to protein folding. Acc Chem Res 41:442–451
46. Hahn EL (1950) Spin echoes. Phys Rev 80:580–594
47. Esadze A, Li DW, Wang TZ, Bruschweiler R, Iwahara J (2011) Dynamics of lysine side-chain amino groups in a protein studied by heteronuclear H-1-N-15 NMR spectroscopy. J Am Chem Soc 133:909–919
48. Ishima R, Baber J, Louis JM, Torchia DA (2004) Carbonyl carbon transverse relaxation dispersion measurements and ms-mu s timescale motion in a protein hydrogen bond network. J Biomol NMR 29:187–198
49. Ishima R, Torchia DA (2003) Extending the range of amide proton relaxation dispersion experiments in proteins using a constant-time relaxation-compensated CPMG approach. J Biomol NMR 25:243–248
50. Korzhnev DM, Neudecker P, Mittermaier A, Orekhov VY, Kay LE (2005) Multiple-site exchange in proteins studied with a suite of six NMR relaxation dispersion experiments: an application to the folding of a Fyn SH3 domain mutant. J Am Chem Soc 127:15602–15611
51. Mulder FAA, Skrynnikov NR, Hon B, Dahlquist FW, Kay LE (2001) Measurement of slow (mu s-ms) time scale dynamics in protein side chains by N-15 relaxation dispersion NMR spectroscopy: application to Asn and Gln residues in a cavity mutant of T4 lysozyme. J Am Chem Soc 123:967–975
52. Skrynnikov NR, Mulder FAA, Hon B, Dahlquist FW, Kay LE (2001) Probing slow time scale dynamics at methyl-containing side chains in proteins by relaxation dispersion NMR measurements: application to methionine residues in a cavity mutant of T4 lysozyme. J Am Chem Soc 123:4556–4566
53. Lundstrom P, Lin H, Kay LE (2009) Measuring C-13(beta) chemical shifts of invisible excited states in proteins by relaxation dispersion NMR spectroscopy. J Biomol NMR 44:139–155
54. Baldwin AJ, Hansen DF, Vallurupalli P, Kay LE (2009) Measurement of methyl axis orientations in invisible, excited states of proteins by relaxation dispersion NMR spectroscopy. J Am Chem Soc 131:11939–11948
55. Hansen DF, Vallurupalli P, Kay LE (2009) Measurement of methyl group motional parameters of invisible, excited protein states by NMR spectroscopy. J Am Chem Soc 131:12745–12754
56. Vallurupalli P, Hansen DF, Stollar E, Meirovitch E, Kay LE (2007) Measurement of bond vector orientations in invisible excited states of proteins. Proc Natl Acad Sci U S A 104:18473–18477

57. Palmer AG, Massi F (2006) Characterization of the dynamics of biomacromolecules using rotating-frame spin relaxation NMR spectroscopy. Chem Rev 106:1700–1719
58. Hansen DF, Vallurupalli P, Kay LE (2008) An improved N-15 relaxation dispersion experiment for the measurement of millisecond time-scale dynamics in proteins. J Phys Chem B 112:5898–5904
59. Ishima R, Torchia DA (2006) Accuracy of optimized chemical-exchange parameters derived by fitting CPMG R_2 dispersion profiles when $R_2(0a) = R_2(0b)$. J Biomol NMR 34:209–219
60. Korzhnev DM, Salvatella X, Vendruscolo M, Di Nardo AA, Davidson AR, Dobson CM, Kay LE (2004) Low-populated folding intermediates of Fyn SH3 characterized by relaxation dispersion NMR. Nature 430:586–590
61. Carver JP, Richards RE (1972) General 2-site solution for chemical exchange produced dependence of T_2 upon Carr-Purcell pulse separation. J Magn Reson 6:89–105
62. Davis DG, Perlman ME, London RE (1994) Direct measurements of the dissociation rate constant for inhibitor enzyme complexes via the T1rho and T2(CPMG) methods. J Magn Reson B 104:266–275
63. Luz Z, Meiboom S (1963) Nuclear magnetic resonance study of protolysis of trimethylammonium ion in aqueous solution—order of reaction with respect to solvent. J Chem Phys 39:366–371
64. Kovrigin EL, Kempf JG, Grey MJ, Loria JP (2006) Faithful estimation of dynamics parameters from CPMG relaxation dispersion measurements. J Magn Reson 180:93–104
65. Millet O, Loria JP, Kroenke CD, Pons M, Palmer AG (2000) The static magnetic field dependence of chemical exchange linebroadening defines the NMR chemical shift time scale. J Am Chem Soc 122:2867–2877
66. Mulder FAA, Mittermaier A, Hon B, Dahlquist FW, Kay LE (2001) Studying excited states of proteins by NMR spectroscopy. Nat Struct Biol 8:932–935
67. Findeisen M, Brand T, Berger S (2007) A H-1-NMR thermometer suitable for cryoprobes. Magn Reson Chem 45:175–178
68. Hindman JC (1966) Proton resonance shift of water in gas and liquid states. J Chem Phys 44:4582–4593
69. Renzoni DA, Pugh DJR, Siligardi G, Das P, Morton CJ, Rossi C, Waterfield MD, Campbell ID, Ladbury JE (1996) Structural and thermodynamic characterization of the interaction of the SH3 domain from Fyn with the proline-rich binding site on the p85 subunit of PI3-kinase. Biochemistry 35:15646–15653

Chapter 10

Studying Metal Ion–Protein Interactions: Electronic Absorption, Circular Dichroism, and Electron Paramagnetic Resonance

Liliana Quintanar and Lina Rivillas-Acevedo

Abstract

Metal ions play a wide range of important functional roles in biology, and they often serve as cofactors in enzymes. Some of the metal ions that are essential for life are strongly associated with proteins, forming obligate metalloproteins, while others may bind to proteins with relatively low affinity. The spectroscopic tools presented in this chapter are suitable to study metal ion–protein interactions. Metal sites in proteins are usually low symmetry centers that differentially absorb left and right circularly polarized light. The combination of electronic absorption and circular dichroism (CD) in the UV–visible region allows the characterization of electronic transitions associated with the metal–protein complex, yielding information on the geometry and nature of the metal–ligand interactions. For paramagnetic metal centers in proteins, electron paramagnetic resonance (EPR) is a powerful tool that provides information on the chemical environment around the unpaired electron(s), as it relates to the electronic structure and geometry of the metal–protein complex. EPR can also probe interactions between the electron spin and nuclear spins in the vicinity, yielding valuable information on some metal–ligand interactions. This chapter describes each spectroscopic technique and it provides the necessary information to design and implement the study of metal ion–protein interactions by electronic absorption, CD, and EPR.

Key words Electronic absorption spectroscopy, Electronic circular dichroism, Electron paramagnetic resonance, Metalloprotein, Metal ions, Metal–protein interactions

1 Introduction

1.1 Metal Ion–Protein Interactions

Metal ions are an important part of biological systems, as living organisms use inorganic elements for many key processes. Metal ions most commonly serve as cofactors in enzymes, helping catalyze their reactions, but do take a broad range of functional roles that include the following: structural (Ca, Zn, Si), signaling (Ca), acid-base catalysis (Zn, Fe, Ni, Mn), electron transfer (Fe, Cu, Mo), and redox catalysis (Mn, V, Fe, Co, Ni, Cu, W). Some of the metal ions that are essential for life are strongly associated with proteins,

Mark A. Williams and Tina Daviter (eds.), *Protein-Ligand Interactions: Methods and Applications*, Methods in Molecular Biology, vol. 1008, DOI 10.1007/978-1-62703-398-5_10,

forming obligate metalloproteins, as in the case of the blue copper site in electron transfer proteins; while others may bind to the protein with relatively low affinity, as is the case of Ca, Mg, and Mn acting as activators of some enzymes. Common metal-donor atoms provided by protein residues include the following: nitrogen atoms of histidines, oxygen atoms of glutamate and aspartate residues, and sulfur atoms of methionines and cysteines. A comprehensive review of metal ions in biological systems can be found in (1).

The spectroscopic tools that are presented in this chapter are suitable to study transition metal ions. When a transition metal ion is coordinated by a set of ligands, as when it binds to a protein, the metal d orbitals will split in energy as a result of their interaction with the ligand orbitals (Fig. 1). The magnitude and characteristics of this ligand field effect is dependent on the geometry of the metal complex and on the strength of the metal–ligand interactions. The ligand orbitals that interact with the metal ion will be stabilized by the interaction (Fig. 1). For a more detailed description of the ligand field effect and bonding properties of transition metal complexes, the reader is referred to (2–6). When the d orbitals of a metal ion are not full (i.e., they have less than ten electrons), it is

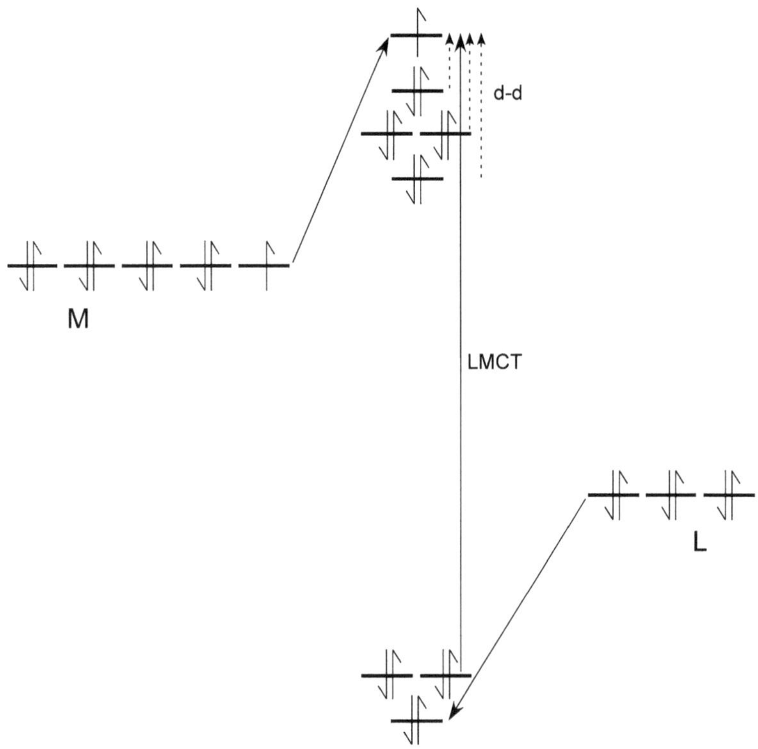

Fig. 1 Schematic molecular orbital diagram for a metal ion bound to a protein (3). Upon metal–ligand interaction, the metal d orbitals are destabilized and split in energy, while the ligand orbitals are split and stabilized. Possible ligand field (d–d) transitions are represented with *dotted arrows*, while a ligand to metal charge transfer transition (LMCT) is represented with *solid arrow*

possible to probe ligand field transitions and/or charge transfer transitions that involve ligand and metal d orbitals. Ligand field transitions refer to electronic transitions that involve orbitals with predominant metal d character, and they are often called d–d bands. Charge transfer transitions refer to electronic transitions that involve both, metal d and ligand orbitals; for example, a ligand to metal charge transfer (LMCT) is an electronic transition that originates at a ligand orbital and ends at a metal d orbital (Fig. 1). Both, ligand field and charge transfer transitions, can be probed using electronic absorption spectroscopy and circular dichroism. Finally, if the metal ion that binds to the protein of interest is paramagnetic, i.e., it has one or more unpaired electrons, then the metal–protein complex can be studied by electron paramagnetic resonance. The following sections introduce the theoretical framework for these spectroscopic tools, as inspired by their presentation in refs (7, 11).

1.2 Electronic Absorption Spectroscopy

Electronic absorption spectroscopy in the UV–visible region is a very useful, convenient and readily available technique that allows electronic transitions that occur between the ground state (Ψ_g) and an excited state (Ψ_e) of a molecule to be probed (Fig. 2); in this case the metal–protein complex. The probability of a transition due to the interaction of a photon with an electron of the metal complex is theoretically expressed through the transition moment operator $\hat{M}$. For UV–visible absorption spectroscopy, the dominant term of the transition moment operator is the electric dipole; and thus, the oscillator strength (f_{osc}) associated to an electronic transition (integrated intensity under the absorption band, Fig. 2) depends on the integral (7):

$$\int \Psi_g \hat{M}_{\text{electric dipole}} \Psi_e \, d\tau. \tag{1}$$

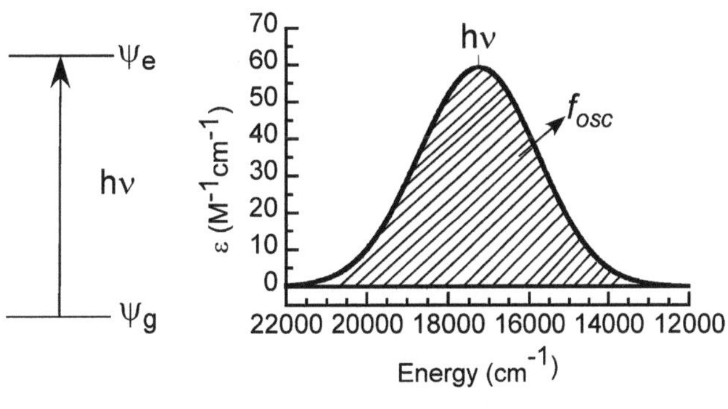

Fig. 2 Schematic diagram for an electronic transition that occurs between the ground state (Ψ_g) and an excited state (Ψ_e) of a molecule, and leads to a Gaussian-shaped absorption band (7, 11). The oscillator strength (f_{osc}) associated to the electronic transition corresponds to the integrated intensity under the absorption band, and relates to the integral in Eq. 1

If this integral is different from zero, the transition is electric dipole "allowed" and will be observed in the absorption spectrum with high intensity. If this integral is zero the electronic transition is electric dipole "forbidden"; however, this does not necessarily mean that the transition will not be observed. An electric dipole "forbidden" transition can gain intensity by several mechanisms, such as spin–orbit coupling and vibronic coupling, as discussed in refs (7–9). Thus, an electric dipole forbidden transition may be observed in the absorption spectrum, but with much lower intensity than an electric dipole allowed transition. Experimentally, an absorption band is characterized by its molar extinction coefficient (ε in M^{-1} cm^{-1}), which relates absorption intensity (A) at a given wavelength to the molar concentration (C) of the absorbing molecule through the Lambert–Beer law:

$$A = \varepsilon C l, \tag{2}$$

where l is the optical path length. The molar extinction coefficient is energy dependent with a maximum value at the energy that corresponds to the energy difference between the ground and excited states, as shown in Fig. 2.

As mentioned before, a metal complex can display ligand field (d–d) transitions and/or charge transfer transitions (Fig. 1). For metal–protein complexes, d–d transitions are usually electric dipole forbidden, and thus display small or no intensity at all. On the other hand, charge transfer transitions tend to be electric dipole allowed and they display high intensity in the absorption spectrum. From Fig. 1, it becomes evident that d–d transitions will appear at relatively lower energy, as compared to LMCT transitions. Thus, in general, in the absorption spectrum of a metal–protein complex, d–d transitions can be readily identified as they would appear in the lower energy region of the spectrum and would generally have small ε values; while LMCT transitions appear at higher energy values and have relatively high absorption intensity (*see* **Note 1**).

1.3 Electronic Circular Dichroism

Electronic circular dichroism (CD) is a widely used technique to probe the secondary structure of a protein, as discussed in Chapter 8. However, it can also be used to study electronic transitions in a metal complex that occur in the UV–visible region. CD spectroscopy uses a circularly polarized (CP) light beam, which has two components: left (L) and right (R) CP light. If a chiral molecule is exposed to CP light, its interaction with the left CP component will be different from its interaction with the right CP component, and therefore, the amount of left CP light absorbed (ε_L) by a chiral molecule will be different from its absorption of right CP light (ε_R). In a CD experiment, the difference in absorption of the two components (ε_L and ε_R) related to a given electronic transition is measured. Experimentally, a CD signal is characterized by the difference in absorption of LCP and RCP components of light

($\Delta A = A_L - A_R$); most CD instruments report this difference in ellipticity ($\theta = 32.98\ \Delta A$) in degrees. Thus, ellipticity relates to the molar concentration (C) of the absorbing chiral molecule and the difference in molar extinction coefficients for the LCP and RCP components ($\Delta\varepsilon = \varepsilon_L - \varepsilon_R$ in $M^{-1}\ cm^{-1}$) as follows:

$$\theta = 32.98\Delta\varepsilon Cl, \tag{3}$$

where l is the optical path length.

The dominant terms of the transition moment operator in UV–visible CD spectroscopy are the electric dipole ($\hat{M}_{\text{electric dipole}}$) and magnetic dipole ($\hat{M}_{\text{magnetic dipole}}$) components. The parameter that relates to the integrated area under a CD signal (Fig. 3) is the rotational strength (R), which for a given $\Psi_g \rightarrow \Psi_e$ transition, depends on the integrals (7):

$$\int \Psi_g \hat{M}_{\text{electric dipole}} \Psi_e \, d\tau \int \Psi_g \hat{M}_{\text{magnetic dipole}} \Psi_e \, d\tau. \tag{4}$$

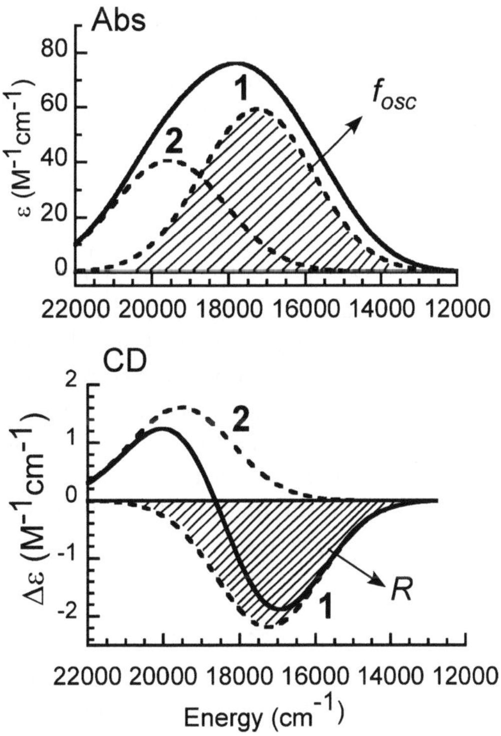

Fig. 3 Schematic diagram for two electronic transitions that are close in energy and give rise to two overlapping Gaussian-shaped absorption bands (*top*, 1 and 2). In this hypothetical case, the deconvolution of these two bands in the corresponding circular dichroism spectrum (*bottom*) is facilitated because the $\Delta\varepsilon$ for transition1 is negative, while $\Delta\varepsilon$ for transition 2 is positive. The physical parameter that relates to the integrated area under a CD signal is the rotational strength (R), which is proportional to the integral in Eq. 4 (7)

Thus, if both integrals in Eq. 4 are different from zero, the electronic transition is electric dipole and magnetic dipole allowed and it will be observed in the CD spectrum with high intensity. As in absorption, an electric dipole and/or magnetic dipole forbidden transition may be observed in the CD spectrum, as there are several mechanisms by which a forbidden transition can gain intensity (7–11).

Provided that a metal complex is chiral, as most metal–protein complexes are, CD spectroscopy can be used to probe the same electronic transitions that are probed by electronic absorption. In fact, it is quite useful to collect both absorption and CD spectra for a given metal–protein complex. Because signals observed in a CD spectrum can be positive or negative, depending on the difference $\Delta\varepsilon = \varepsilon_L - \varepsilon_R$, the overlap of two transitions in an absorption spectrum can often be more easily distinguished/deconvoluted in a CD spectrum (Fig. 3). Moreover, ligand field transitions are usually electric dipole forbidden, but magnetic dipole allowed, and thus, they gain intensity in a CD spectrum, as compared to the corresponding absorption spectrum. Thus, the ratio of the intensities displayed by a given electronic transition in CD and absorption: $\Delta\varepsilon/\varepsilon$, often called the Kuhn anisotropy factor, provides information about the type of transition. Generally, ligand field transitions display larger Kuhn anisotropy factors as compared to charge transfer transitions (7).

1.4 Electron Paramagnetic Resonance

Electron paramagnetic resonance (EPR), also called electron spin resonance (ESR), is a spectroscopic technique that allows the study of molecules with unpaired electrons. Several transition metal ions are paramagnetic, and thus, their complexes with proteins can be studied by EPR. Considering the simplest case, where a molecule has one unpaired electron, the interaction of the electron spin with the magnetic field will lead to the splitting of the spin sublevels $m_s = \pm 1/2$ of each energy state; this is called the Zeeman effect (Fig. 4). The relative energy of a particular spin sublevel m_s depends on the magnitude of the magnetic field applied to the sample (H), as given by

$$E = m_s g \beta H, \qquad (5)$$

where β is the Bohr magneton and g is the g-factor or g-value. In a typical continuous wave EPR experiment, a sample is placed in a cavity and is irradiated by microwaves with a fixed frequency (υ). At the same time, a swept external magnetic field is applied to the sample, with steadily increasing field strength. When the energy difference of the Zeeman splitting between the two spin sublevels is equal to the fixed microwave energy ($h\upsilon$), the resonance condition in Eq. 6 is met, absorption of microwaves occurs and an EPR signal is detected (12, 13).

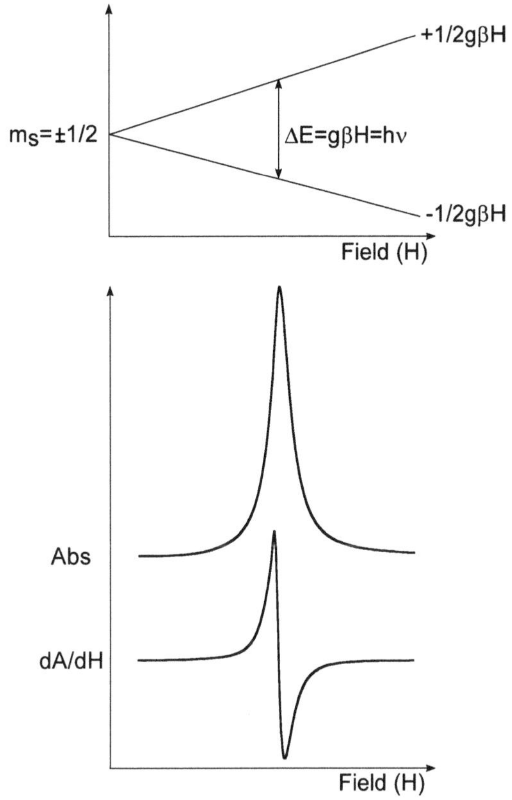

Fig. 4 Zeeman effect and the EPR experiment for a spin = 1/2 species (7, 12). The spin sublevels $m_s = \pm 1/2$ split in energy as a function of magnetic field strength (*H*) (*top*). When their energy splitting (*gβH*) is equal to the energy of the incident microwaves (*hν*), absorption of microwaves is observed. EPR spectra are displayed as the derivative of microwave absorption as a function of magnetic field (d*A*/d*H*)

$$\Delta E = g\beta H = h\nu. \tag{6}$$

EPR signals are detected by a phase sensitive detection technique, and as a result, they are usually reported as the first derivative of the microwave absorption spectrum (dA/dH) for enhanced resolution (Fig. 4). This implies that the magnetic field at which the resonance condition is achieved corresponds to the point where $dA/dH = 0$, and the peak-to-peak distance corresponds to the signal width.

Using Eq. 6, the g value associated to a paramagnetic species can be determined from the magnetic field at which the EPR signal is observed. The g-value is indicative of the chemical environment around the electron spin. For most spin = 1/2 molecules, such as free radicals, the g-value is approximately equal to 2.0023; however, for transition metal ions with one unpaired electron the g value can have small deviations from 2 due to spin–orbit coupling (*see* **Note 2**). On the other hand, transitions metal complexes with spin >1/2 can have g values that deviate greatly from 2 due to zero-field splitting (*see* **Note 2**). While metalloproteins with Cu(II), V (IV), low spin Co(II), Ni(III) or Mo(V) present g values around 2, other metal sites in proteins with Mn(II), Fe(II), Fe(III), or high spin Co(II) present EPR signals with a wider range of g values (1.5–10) (7, 11, 12). A comprehensive classification of the EPR properties of metal ions found in biological systems is presented in ref. 12.

It is important to note that, as much as the chemical environment around the unpaired electron can be anisotropic, g-values may be anisotropic. This means that the magnitude of the Zeeman splitting depends on the relative orientation of the molecule with respect to the external magnetic field (Fig. 5). Thus, a paramagnetic molecule with rhombic symmetry, i.e., three nonequivalent axes, displays different g values: $g_z \neq g_y \neq g_x$. If the relative orientation of the molecule with respect to the external magnetic field cannot be controlled (as is the case for frozen solutions of metal–protein complexes), a distribution of microwave absorptions will be observed across a range of magnetic field values that correspond to the range of g_z to g_y to g_x. The first derivative of the absorption spectrum usually allows the signals associated to the three different g values to be resolved (Fig. 5).

Additionally, any nuclear spin, of magnitude I, in the vicinity of the unpaired electron will cause a hyperfine splitting of the EPR signal into $2I + 1$ signals. The magnitude of the hyperfine interaction with the nearby nucleus (N) is given by ^{N}A, the hyperfine coupling constant, and it is of anisotropic nature. Several transition metal ions have nuclear spins that couple with their unpaired electrons; Cu^{2+} and V^{4+} display EPR signals split by metal hyperfine couplings due to their nuclear spins: $I = 3/2$ and $I = 7/2$, respectively (Fig. 6). A detailed analysis of the g values and metal hyperfine couplings can provide information of the geometry of the metal–-protein complex and the nature of the ligating residues. For example, for Cu^{2+} complexes the parallel g and ^{Cu}A values are very sensitive to the nature of the equatorial ligands, and they correlate well with the type of ligating atoms in the coordination shell (14, 15). Similar correlations between EPR parameters and the nature of the coordination environment for other types of metal centers in proteins, such as heme proteins, have been reported (12, 16–18).

Beyond the metal ion, other common hyperfine interactions with nearby nuclei in metal–protein complexes are those with nitrogens ($I = 1$) and protons ($I = 1/2$). These interactions are usually weaker than metal hyperfine interactions and lead to small hyperfine coupling constants ($10–20 \times 10^{-4}$ cm^{-1} for nitrogen and $<5 \times 10^{-4}$ cm^{-1} for proton) that can sometimes be resolved in a typical EPR spectrum; however, pulsed EPR techniques like ESEEM (Electron Spin Echo Envelope Modulation) or ENDOR

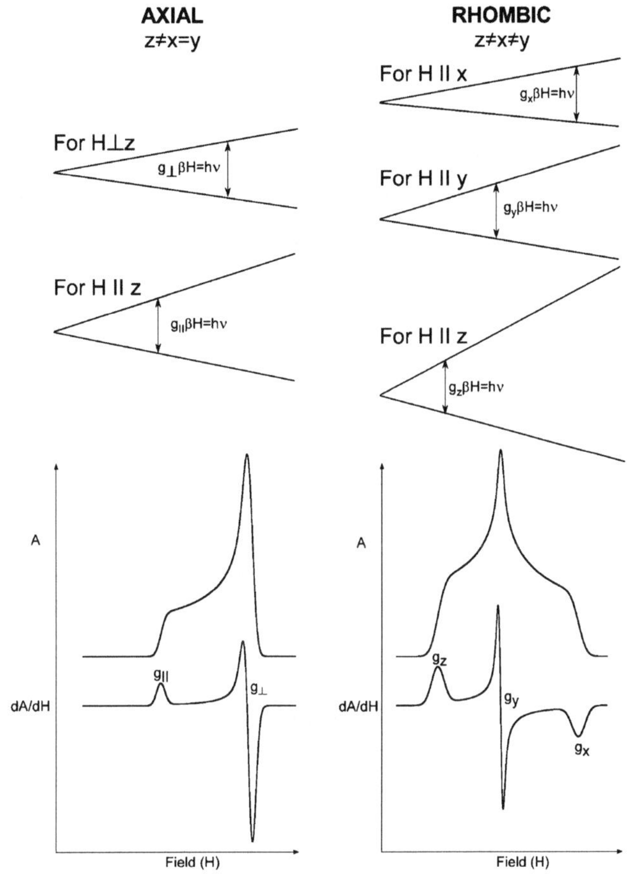

Fig. 5 Anisotropy of *g* values (7, 12). For an axial complex (*left*) with $g_z > g_y = g_x$, g_z is also called $g_{\|}$ and $g_y = g_x$ correspond to $g_{\perp}$, and the microwave absorption spectrum displays a distribution of absorptions that range from the resonance condition for $g_{\|}$ ($g_{\|}\beta H = h\nu$ when the magnetic field is parallel to *z*), to the resonance condition for $g_{\perp}$ ($g_{\perp}\beta H = h\nu$ when the magnetic field is perpendicular to *z*), including all possible intermediate orientations. Thus, the EPR spectrum (d*A*/d*H*) of an axial complex will display two signals associated to the values $g_{\|}$ and $g_{\perp}$. Similarly, for a rhombic complex (*right*) with $g_z > g_y > g_x$, a distribution of absorptions that include the three resonance conditions (for $g_z\beta H$, $g_y\beta H$, and $g_x\beta H$) will be observed

(Electron Nuclear Double Resonance) are more appropriate for an accurate measure of weak hyperfine interactions (19–21).

2 Materials

2.1 Sample Preparation

2.1.1 Water

When studying metal ion binding to proteins, it is essential to use ultra pure water with high resistivity (18 MΩ/cm). Regular distilled water may have a significant amount of metal ions that can bind to the protein under study (*see* **Note 3**).

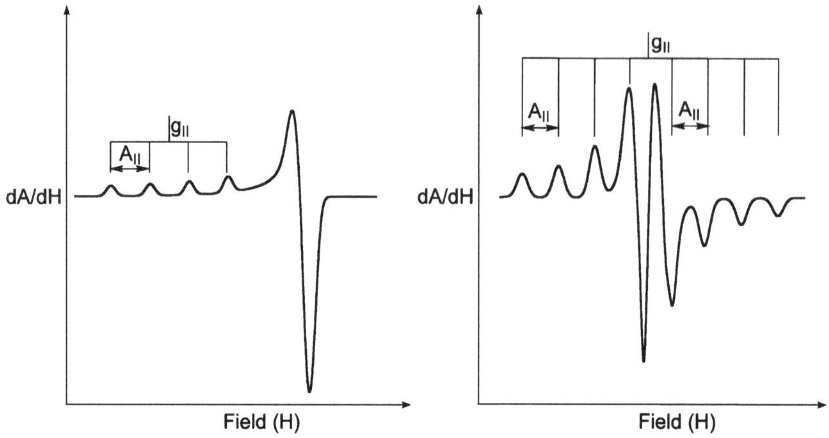

Fig. 6 EPR simulations of axial complexes of Cu^{2+} (*left*) and V^{4+} (*right*). The hyperfine interaction of the unpaired electron with the nuclear spin of ^{63}Cu and ^{65}Cu ($I = 3/2$) leads to the splitting of EPR signals into four lines, while the hyperfine interaction with the nuclear spin of ^{51}V ($I = 7/2$) leads to a splitting into eight lines. Since the metal hyperfine splitting (^{M}A where M is Cu or V) is anisotropic, and $^{M}A_{\|} >> {}^{M}A_{\perp}$ in most axial complexes, the hyperfine splitting is better resolved in the parallel region of the EPR spectra

2.1.2 Buffers

It is important to select a buffer that does not bind metal ions, such as MOPS (3-(*N*-morpholino)propanesulfonic acid) or MES (2-(*N*-morpholino)ethanesulfonic acid). If a protein is in a buffer that binds metal ions, the buffer may compete with the protein for the metal ion and shift the binding equilibrium. For a complete listing of non-complexing buffers that are appropriate to use in biological systems with metal ions see ref. 22.

2.1.3 Protein Solution

The protein concentration must be such that the absorption and CD signals (particularly the d–d transitions) show reasonable intensity and a good signal-to-noise ratio; a good starting concentration can be in the range of 0.1 and 1 mM (*see* **Note 4**). Protein or peptide concentration must be accurately known, either by absorption (using the corresponding extinction coefficient), or by BCA or Bradford assays (23, 24). When the metalloprotein is analyzed by EPR at low temperatures (4–150 K), it is advisable to prepare protein samples with a cryoprotectant such as glycerol, which also helps to achieve adequate glassing (*see* **Note 5**).

2.1.4 Metal Ion Solution

If the protein will be titrated by the metal ion, the concentration of the metal ion stock solution must be as high as possible in order to avoid diluting the protein. Some metal salts are insoluble in water; in such cases, the use of a metal ion chelator may be advisable to improve solubility. Care must be taken to choose a metal chelator with a lower metal binding affinity than that of the protein under study; additionally, data analysis must consider how the metal–protein binding equilibrium is affected by the competing chelator (25, 26).

2.2 Quartz Cells

2.2.1 Quartz Cells for Optical Measurements

For the collection of electronic absorption and CD data in the UV–Vis region, quartz cells that transmit the whole wavelength range must be used (*see* **Note 6**). Care must be taken to have the incident light beam passing through the sample, if the light is reflected or partially cut by the meniscus of the sample, the walls or base of the cell, the measured spectrum would be affected by scattered light. If the available amount of protein sample is small, semi-micro and micro quartz cells that allow data collection in very small volumes (as low as 10 μL) without decreasing the path length are commercially available. The path length of the quartz cell must be optimized in order to avoid saturation of the absorption signal (*see* **Note 4**).

2.2.2 Quartz Cells for EPR Measurements

Quartz cells must also be used for EPR experiments, as regular glass typically contains paramagnetic impurities that may interfere with the EPR signal of interest. The internal diameter of commercially available quartz tubes for EPR range from 2 to 5 mm. EPR signal intensity depends on sample concentration and on how much of the sample is inside the EPR cavity (*see* **Note 7**). Thus, the choice of internal diameter of the quartz cell depends on how much protein sample is available. Also, care must be taken to optimize the position of the quartz cell in the cavity (*see* **Note 7**).

2.3 Instrumentation

2.3.1 UV–Vis Spectrometer

There is a wide range of spectrometers commercially available, generally they are relatively inexpensive and easy-to-use. It is advisable to use one with a diode-array detector, such that the absorption spectrum can be collected from 190 to 1,100 nm at once (instead of waiting for a scan). Most UV–Vis spectrometers have two lamps: a deuterium lamp as a UV source and a tungsten lamp for the visible region. Spectrometers usually have checkup routines to make sure that the lamps are working optimally and enough light goes through the sample. If the absorption spectra are noisy, one of the first things to check is the condition of the lamps.

2.3.2 CD Spectropolarimeter

All commercially available spectropolarimeters collect CD data by scanning the wavelength and measuring CD signal at each point. Thus, it is important to check that the wavelength and CD signal intensity are calibrated (*see* **Note 8**). A flow of high purity nitrogen gas must be supplied to the lamp for at least 5 min before turning it on and as long as the lamp is on: this is to prevent the production of ozone and damage of optical components, and to displace oxygen in the sample cavity that would absorb incident light at low wavelengths. Please refer to Chapter 8 for further technical details on CD instrumentation and data collection.

2.3.3 EPR Spectrometer

The essential components of an EPR spectrometer are the following: (a) a microwave bridge that houses the microwave source and detector, (b) a resonant cavity where the sample is placed, (c) a magnet with a field controller that allows the magnetic field to be swept, and a console that performs the signal processing and controls the electronics. The microwave source can be a klystron or a Gunn diode (in newer instruments). In an EPR experiment, the microwave frequency is commonly fixed; the most widely used EPR spectrometers have a bridge that generates microwaves of ~9.5 GHz (X-band spectrometers). This frequency is useful for studying free radicals and transition metal ions using an electromagnet in the range of 0–10,000 G (Gauss = 0.0001 T). An EPR signal associated with a g value of 2 would resonate at a field of ~3,394 G at 9.5 GHz. Other microwave frequencies are commercially available for more specific applications, for example: 1 GHz (L-band), 3 GHz (S-band), 35 GHz (Q-band), and 95 GHz (W-band). EPR spectrometers with low microwave frequencies are used for in vivo detection of free radicals. Running at low microwave frequencies, such as S- and L-bands, helps resolve small hyperfine couplings; while higher frequencies (Q- and W-bands) help resolve small anisotropy of g values. It should be noted that the magnitudes of hyperfine couplings do not change when the EPR spectrum is collected at a different frequency, while the magnetic field at which the resonance condition is achieved for a given g value does change with the frequency of the microwaves used (Eq. 6).

EPR instrument calibration is important when running (or comparing) samples in different laboratories, particularly for quantitative EPR studies. The modulation amplitude must be calibrated and the precision of the magnetic field that is applied to the sample must be checked too (*see* **Note 9**).

For low temperature EPR experiments, there are commercially available cryostats to collect temperature-dependent data, using liquid nitrogen (120–200 K) or liquid helium (4–100 K). Alternatively, a commercially available quartz finger dewar containing liquid nitrogen can be used to place the sample inside the cavity to collect EPR data at 77 K.

3 Methods

3.1 Choice of Parameters for Electronic Absorption and CD Data Collection

It is common to find that the absorption spectrum of a metalloprotein is dominated in the region between 190 and 300 nm by the intense electronic transitions associated to peptide bonds and aromatic residues. These signals sometimes limit the range of wavelengths where electronic transitions associated to the metal ion (i.e., d–d and charge transfer transitions) can be observed. Additionally, when an electronic transition has a high molar extinction coefficient, the absorption can become saturated, no longer follow the Lambert–Beer law, and cause distortion of the absorption signal. For CD, high absorption can cause additional problems, as the CD signal is detected with a photo-multiplier tube (PMT), and having a very high absorption can cause high voltage and damage at the PMT. Thus, the protein concentration, optical path length, and wavelength range for collecting absorption and CD spectra must be optimized for each sample (*see* **Note 4**).

A few parameters need to be set for absorption data collection in spectrophotometers with diode arrays: wavelength range, data pitch, and integration time; collecting data with a 1 or 2 nm step and integration times of 1 s is advisable in order to get good spectral resolution. In the spectropolarimeter, the following parameters need to be set for CD data collection: bandwidth, data pitch, response time, and scanning speed. A clear discussion on how to choose these parameters to avoid CD signal distortion is presented in Chapter 8. A bandwidth of 1 nm or 2 nm, data pitch of 1 nm, response time of 1 s, and a scanning speed of 50 nm/min are suitable starting parameters for most metalloprotein samples.

For both spectroscopies, it is necessary to blank the instrument with your buffer solution. In most spectrophotometers, the blank is subtracted automatically from any sample measurement, while in spectropolarimeters, oftentimes the CD spectrum of the blank (baseline) must be collected and subtracted manually from every sample measurement.

3.2 Titration of a Protein with a Metal Ion, as Followed by Absorption and CD

An example of the titration of a prion protein fragment by Cu(II), as followed by absorption and CD spectroscopy, is shown in Fig. 7. The 0.5 mM peptide was titrated with a concentrated aqueous solution of $CuCl_2$ up to 2 equivalents of metal to peptide. The peptide solution alone has no signals in the monitored region of the spectra (12,000–42,000 cm^{-1}) (see Fig. 7a, b). In both spectra, several signals grow with the addition of Cu(II). The plots of absorption and CD signal intensities at selected wavelengths as a function of the number of equivalents of Cu(II) added (Fig. 7c, d) clearly show that there is one binding site for Cu in this peptide and that the formation of the copper-peptide complex is close to stoichiometric at these conditions. Although, it would be possible to estimate an equilibrium binding constant associated with the formation of the complex from such plots; lower concentrations of the analyte and titrations in both senses (i.e., titration of peptide with Cu and titration of Cu with peptide) would be needed to obtain a precise value (26, 27).

The data shown in Fig. 7a, b allow us to characterize the absorption and CD signals that arise upon the formation of the complex. The higher energy region of the absorption spectrum above 25,000 cm^{-1} (below 400 nm) is dominated by transitions with high ε values (~4,000 M^{-1} cm^{-1}) that can be assigned as LMCT bands (Fig. 7a). These transitions have positive signals in

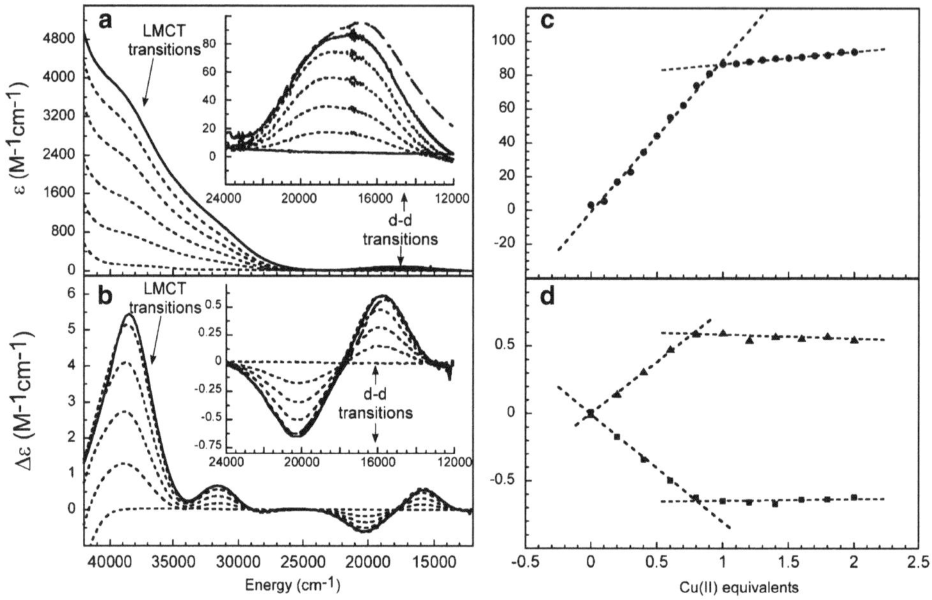

Fig. 7 Titration of the prion protein fragment KTNMKHMAGA (0.5 mM) by Cu(II), followed by absorption spectroscopy (**a**) and circular dichroism (**b**). The titration was done in 20 mM NEM buffer at pH 8.5, up to 2 equivalents of Cu(II). The spectra corresponding to titration points with <1 equivalent of Cu(II) added are shown in *dashed lines*; while spectra with 1 and 2 equivalents of metal are shown with *solid and dotted lines*, respectively. The spectral regions where d–d transitions appear are shown in the *insets*. The changes in absorption intensity at 17,760 cm^{-1} (**c**) and in CD signal intensities at 15,760 (*filled triangle*) and 20,280 cm^{-1} (*filled circle*) are shown as a function of the number of Cu(II) equivalents added (**d**). Adapted with permission from (30)

the CD spectrum (Fig. 7b). In contrast, the ligand field transitions appear at lower energy (~18,000 cm^{-1}/~550 nm) as one broad absorption signal with a very small ε value (~80 M^{-1} cm^{-1}) (Fig. 7a, inset), which is resolved in the CD spectrum as a positive band at ~16,000 cm^{-1} and a negative band at ~20,500 cm^{-1} (Fig. 7b, inset).

3.3 Absorption and CD Data Analysis

As discussed in Subheading 1.3, when absorption and CD data are available for a given metalloprotein, the electronic transitions associated with the metal complex can be better resolved and characterized. This can be done quantitatively by fitting a set of Gaussian bands to both the absorption and CD spectra, in order to deconvolve the signals that arise from each of the allowed electronic transitions of the complex, as shown in Fig. 8. The CD spectrum in Fig. 8b can be simulated with a minimum of six Gaussian bands to adequately fit all key features: four of them with a positive $\Delta\varepsilon$ and two with negative $\Delta\varepsilon$ values. The same six electronic transitions must contribute to the absorption spectrum, at the same energy as they appear in the CD spectrum, as shown in Fig. 8a. The actual height of a Gaussian band that represents a given electronic

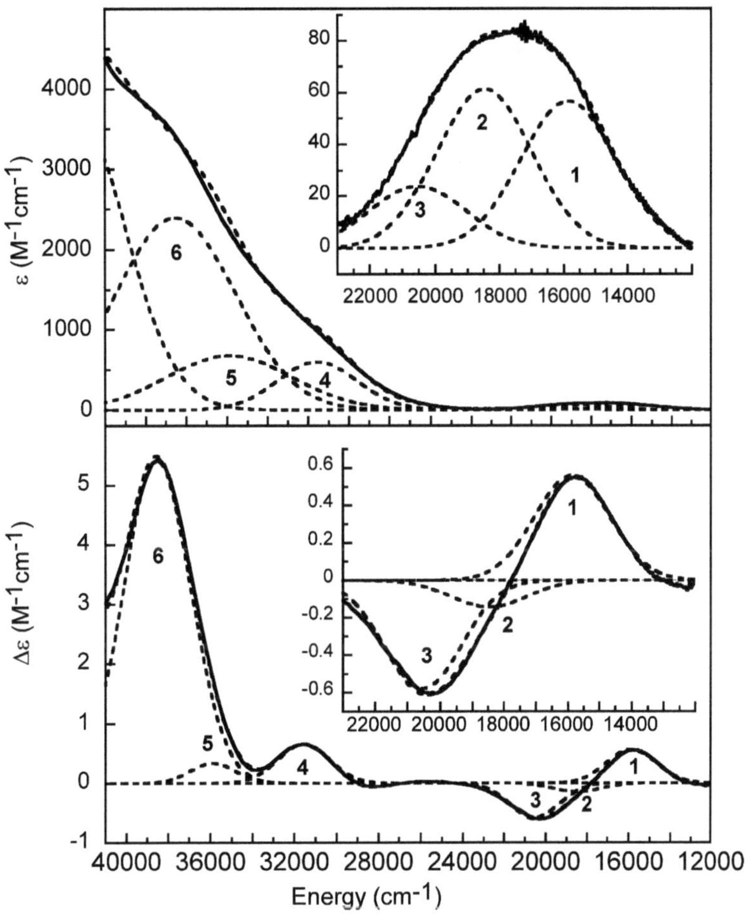

Fig. 8 Gaussian fitting/deconvolution of the UV–Vis absorption (**a**) and circular dichroism (**b**) spectra of the Cu(II)-KTNMKHMAGA complex (in 20 mM NEM buffer at pH 8.5). Experimental spectra are shown in *solid lines*, total Gaussian fit and individual Gaussian bands (with the parameters in Table 1) are shown as *dashed lines*. Adapted with permission from (30)

transition in the absorption and CD spectra corresponds to its ε and $\Delta\varepsilon$ values, respectively. The ratio of these intensities, $\Delta\varepsilon/\varepsilon$, is called the Kuhn anisotropy factor, and provides information about the nature of the transition. Gaussian bands 1–3 in Fig. 8 represent electronic transitions that can be assigned as ligand field or d–d bands, as they appear at the low energy end of the spectrum and have large Kuhn anisotropy factors (Table 1). Meanwhile, Gaussian bands 4–6 can be assigned as LMCT bands, as they appear at higher energies and have smaller Kuhn anisotropy factors (Table 1). This analysis can be carried out with any software that allows the fitting of a spectrum using the sum of several Gaussian functions.

Table 1
Parameters for the Gaussian Fits of Absorption and CD Spectra of the Cu(II)-KTNMKHMAGA complex at pH 8.5 shown in Fig. 8

Band number	Energy (cm^{-1})	ε (M^{-1} cm^{-1})	$\Delta\varepsilon$ (M^{-1} cm^{-1})	$\Delta\varepsilon/\varepsilon$ ($\times 10^{-3}$)
1 d → d	15,858	56.7	+0.561	9.90
2 d → d	18,474	61.3	−0.145	−2.36
3 d → d	20,534	23.7	−0.576	−24.3
4 N^{-}_{amide} → Cu CT	31,567	597	+0.654	1.10
5 N_{Im} π1 → Cu CT	35,891	675	+0.333	0.493
6 N_{Im} π2 → Cu CT	38,535	2,394	+5.42	2.27

Having precise information on the different electronic transitions that contribute to the absorption and CD spectra of a protein-bound metal ion can reveal the ligating residues of the metal site. For example, the LMCT bands 5 and 6 in Table 1 appear in a range of energies characteristic of imidazole binding to Cu(II) (28, 29), thus indicating that the metal ion in this complex is binding to a histidine. Similarly, band 4 falls in the range of energies associated with LMCT transitions from deprotonated amides to the Cu ion (29).

3.4 pH Titration of a Metal–Peptide Complex: Studying Protonation Equilibria by CD

The CD signals associated with a metal–protein complex are very sensitive to changes in coordination mode and/or changes that would modify the chirality of the complex. The CD spectrum of a Cu(II)-peptide complex changes drastically with pH in Fig. 9a. At the highest pH (pH = 9.5) the ligand field region of the spectrum (Fig. 9a, inset) shows a positive signal at ~16,000 cm^{-1} and a negative signal at ~20,500 cm^{-1}; while the LMCT bands have positive signals at ~31,500 and ~40,000 cm^{-1}. In contrast, at low pH (pH = 5.5) the ligand field region of the spectrum shows a weak negative signal at ~13,000 cm^{-1} and a positive signal at ~29,000 cm^{-1}; while the LMCT band at ~31,500 cm^{-1} changes sign. Figure 9b shows the changes in CD signal intensity as a function of pH at 15,800 and 20,800 cm^{-1}, energy values at which the changes are most dramatic. These data can be fit to a model that assumes a single protonation event, with the equation:

$$\Delta\varepsilon_{obs} = (\Delta\varepsilon_{low\ pH}[H^+] + \Delta\varepsilon_{high\ pH}K_a)/(K_a + [H^+]), \quad (7)$$

where $\Delta\varepsilon_{obs}$ is the observed CD signal intensity at any given pH, K_a is the equilibrium constant associated to the protonation equilibrium of the Cu(II)-peptide complex, $\Delta\varepsilon_{low\ pH}$ is proportional to the CD signal intensity of the protonated form of the complex, and $\Delta\varepsilon_{high\ pH}$ to the deprotonated form. The pKa associated with the

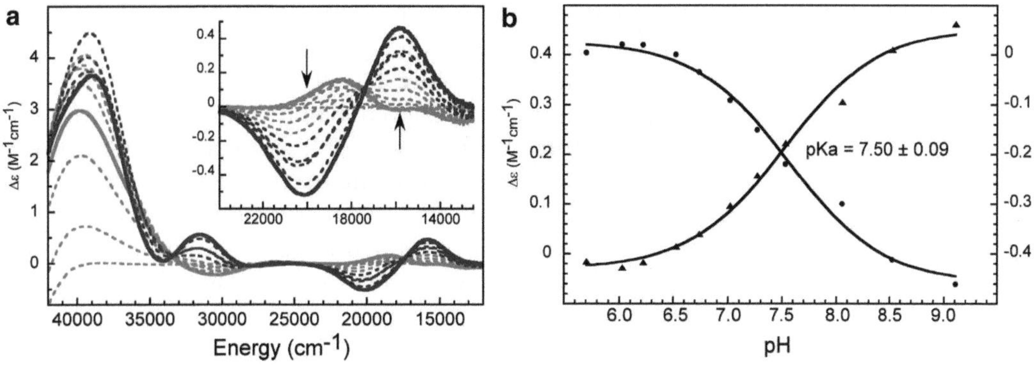

Fig. 9 pH titration of the Cu(II) complex with the prion fragment KTNMKHMAGA followed by circular dichroism. (**a**) The Cu(II)-peptide complex solutions were titrated from pH ~5.5 (*light solid line*) to pH ~9.5 (*dark solid line*); spectra for intermediate pH values are shown as *dashed lines*. The changes of CD signal intensity at 15,800 cm^{-1} (*filled triangle*) and 20,800 cm^{-1} (*filled circle*) are shown as a function of pH (**b**); the data are fit to a model (*solid lines*) that assumes a single protonation equilibrium. Reprinted with permission from (30). Copyright 2011 American Chemical Society.

protonation of the complex is 7.50 ± 0.09, as determined by the best fit of experimental data (30).

This type of analysis can be used to determine relative binding constants that relate to changes in coordination modes and/or any other effect that alters the chirality of a metal center in a protein, such as the binding of a substrate or exogenous ligand.

3.5 Getting Ready for EPR Data Collection

3.5.1 Sample Positioning in the Cavity

In an EPR spectrometer, the sample is placed in a microwave cavity, a metal box where microwaves resonate and allow the amplification of weak signals from the sample. Signal intensity depends on sample concentration and on the extent to which the microwave cavity is "filled" with the sample. For accurate quantitation of the spin concentration of the sample, the quartz cell must be filled with the sample and placed in the cavity, such that the sample occupies its full height (*see* **Note 7**).

3.5.2 Temperature

The most common EPR spectrometer uses microwaves of 9.5 GHz, corresponding to ~0.3 cm^{-1} (~0.9 cal/mol). Thus, when the resonance condition is achieved (Eq. 6) for a g value of 2 at room temperature, the Boltzmann population of spins at the lower energy spin sublevel ($m_s = -1/2$) has ~0.1 % net excess population with respect to the higher energy spin sublevel ($m_s = +1/2$). Lowering the temperature increases this difference in populations and slows down spin relaxation processes; this, in turn, increases sensitivity. Thus, the line width of the EPR spectra of most metal ions becomes sharper at low temperature. Consequently, EPR data collection on metalloproteins is usually carried out in the 4–150 K range. Reference 12 gives a general notion of optimal temperatures for the most common metal sites in proteins. It should be noted that, when the temperature is too low and the microwave power too high, saturation of the EPR signal can occur; thus, it is important to find the optimal microwave power for the selected temperature (*see* **Note 10**). Temperature stability is very important and must be achieved before tuning the cavity and collecting EPR data. When working at low temperatures, regardless of the low temperature instrumental setup in use, care must be taken to displace the air around the quartz insert (with nitrogen or helium gas) to avoid water condensation inside the cavity. Water strongly absorbs microwaves and it can cause difficulty in tuning of the cavity.

3.5.3 Microwave Bridge

The microwave radiation source and the detector are housed in the microwave bridge. As microwaves come out of the source, they pass through a variable attenuator that can precisely control their microwave power. The higher the attenuation, the lower the microwave power. EPR signal intensity generally increases with microwave power; however, at high power levels the signal saturates, i.e., it decreases and broadens with increasing microwave power levels. The optimal microwave power for EPR data collection will depend on the nature of the sample and the chosen temperature (*see* **Note 10**). The attenuated microwaves are directed to the cavity containing the sample and are then reflected back to the microwave bridge for detection (most EPR spectrometers are reflection spectrometers in which absorption corresponds to reduction in reflected microwave intensity). Detection of reflected microwaves is generally performed with a Schottky barrier diode, which converts the microwave power into an electrical current (31). In order to optimize sensitivity and simplify quantitation, the diode should operate in its linear region (where the diode current is proportional to the square root of the microwave power); this is achieved at a corresponding optimal diode current (the actual value may vary for each microwave bridge). A small amount of microwave power, "bias," can be sent directly to the detector to ensure that total power (direct + reflected) at the detector is such that it operates at the optimal diode current. The reference microwaves are phase shifted so they are in phase with the reflected microwaves from the cavity as they combine at the detector. When tuning the cavity, the "bias" and reference phase shift must be adjusted to optimize the diode current for a range of microwave powers (range of attenuation values) that are appropriate for a specific sample (*see* **Note 10**).

3.5.4 Tuning the EPR Cavity

Microwaves are directed from the microwave bridge to the cavity via a waveguide, and they are coupled into the cavity via a hole, or "iris," which controls the amount of microwaves entering and reflecting from the cavity. Microwaves must resonate in the EPR cavity; this means that the cavity stores the microwave energy. At the resonance frequency of the cavity (*see* **Note 11**), microwaves remain inside the cavity and they are not reflected back to the bridge. The efficiency of the cavity is indicated by a quality factor (Q), which is proportional to the ratio of microwave energy stored and microwave energy dissipated from the cavity. The larger the Q value of a cavity, the greater the sensitivity of the spectrometer (31). The process of tuning the cavity optimizes its Q value with a given sample inside. The first step to tune a cavity is to find its resonance frequency at a low microwave power (generally in the range of 2–200 nW for X-band spectrometers). Next, the reference microwave power ("bias") and phase shift must be adjusted in order to get the optimal diode current at the detector. Finally, the size of the "iris" must be adjusted such that the optimal diode current is maintained throughout the range of microwave powers that will be used for a given sample. When the magnetic field reaches the resonance condition (Eq. 6) and it drives the absorption of microwaves by the sample, the amount of microwaves reflected to the bridge change and an EPR signal is observed.

3.5.5 Setting Detection Parameters

The spectrometer console contains the electronics for phase sensitive detection, which enhances the sensitivity of an EPR spectrometer. It consists in a sinusoidal modulation of the magnetic field strength at a modulation frequency. When an EPR signal is observed, microwaves that are reflected from the cavity are amplitude modulated at the same frequency. The detector is only sensitive to signals with the same modulation frequency and phase as the field modulation, suppressing signals from noise or electrical interference. Care must be taken to choose the appropriate modulation amplitude, modulation frequency, and time constant to avoid signal distortion (31). Generally, the best approach is to use an amplitude for the field modulation approximately equal to the width of the EPR signals (intensity increases with the modulation amplitude, but if the latter is larger than the line-width, the signal becomes distorted). For the resolution of the splitting between two EPR signals (a hyperfine splitting for example), it is best to use a modulation amplitude that is smaller than the splitting. The modulation frequency should not be close to any resonant frequency in the sample. In the detection electronics there is also an electronic filter that "damps" the response to variation in the input signal, filtering out high-frequency noise. The longer the time constant for this filter, the lower the noise levels in a spectrum. However, if the time constant is too long for the rate of the magnetic field scan, the EPR signal can get distorted or even filtered out. Long time constants can be used to see weak signals, but the scan rate must be adjusted accordingly to assure that the time it takes to scan a single EPR signal is greater than the time constant. A modulation frequency of 100 kHz, modulation amplitude in the range of 2–5 G and a time constant in the range of 100–400 ms are good starting parameters for metalloproteins.

3.5.6 Magnetic Field Parameters

The magnetic field must be swept in a range such that the resonance condition (Eq. 6) is achieved and the EPR signal of interest is centered in the spectrum. Free radicals appear around a g value of 2, their signals are very sharp, and a sweep of only 50 G centered around the field corresponding to $g = 2$ (~3,394 G at X-band frequency), is appropriate. In contrast, paramagnetic metal ions can have a wide range of g values (*see* **Note 2**) and their signals are broader due to spin relaxation, so wider ranges of magnetic field may be needed (~500–6,000 G). The magnetic field scan rate is determined by the conversion time, which is the amount of time that the instrument takes to integrate the signal at a given field position. Thus, the number of data points in the field scan multiplied by the conversion time yields the total sweep time. As mentioned above, the magnetic field scan rate must be adjusted according to the optimal time constant for the EPR signal of interest. For further details on selection of EPR instrument parameters, please refer to (31).

3.6 EPR Data Analysis

3.6.1 Baseline Correction

There are several sources for background EPR signals that are not associated with a sample: signals due to oxygen from air in the cavity, dirt and dust impurities that contaminate the cavity, baseline drift due to electro-mechanical effects in the resonator, small amounts of paramagnetic impurities contained in the quartz cells used to contain the sample or the quartz inserts used in low temperature experimental set-ups. Thus, it is highly recommended to run a baseline EPR spectrum, having in the cavity everything but your sample (for example you can run the buffer solution in the same kind of quartz cell as you run your sample). The subtraction of this baseline from the sample EPR spectrum will help eliminate any background EPR signals that may appear in the spectrum but are not associated to the sample. Further digital baseline correction may be needed in order to correct for baseline drift and to have the baseline of the spectrum centered at zero in the regions where no EPR signals are observed (*see* **Note 12**); this is particularly important for comparison of experimental and simulated EPR spectra.

3.6.2 Extraction of EPR Parameters from Spectra

Often times, it is relatively easy to extract from an EPR spectrum approximate g values and even metal hyperfine splittings. Figure 10 shows the EPR spectrum of a Cu(II) complex with a prion protein fragment. It becomes evident from first inspection and comparison to the copper EPR spectrum shown in Fig. 6 that, the splitting between the lowest field positive signals corresponds to the metal hyperfine coupling associated to the parallel region of the spectrum (${}^{Cu}A_{\|}$). In contrast to the EPR simulation shown in Fig. 6, in Fig. 10 the signals associated to $g_{\perp}$ overlap with those associated with $g_{\|}$, and thus, only three of the four ${}^{Cu}A_{\|}$ lines can be distinguished. From Fig. 10, one can estimate that the signal associated with $g_{\|}$ is centered at ~3,070 G, corresponding to a $g_{\|} \sim 2.195$

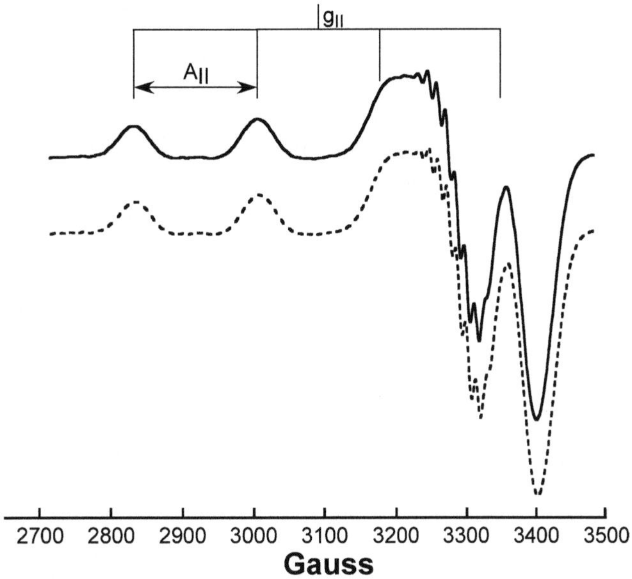

Fig. 10 Experimental X-band EPR spectrum of the Cu(II)-KTNAKHAAGA complex (*solid line*) and its EPR simulation (*dashed line*). Peptide concentration was 0.5 mM with 0.9 equivalents of Cu(II), in 20 mM NEM buffer with 50 % glycerol, at pH 8.5. The spectrum was collected in a Bruker EMX spectrometer at 150 K; microwave power = 10 mW; modulation amplitude = 5 G; modulation frequency = 100 kHz; time constant = 327 ms; conversion time = 82 ms; and microwave frequency = 9.4312 GHz. The EPR simulation was performed with the program XSophe (32), using the parameters in Table 2. Adapted with permission from (30)

(*see* **Note 11**), and that $^{Cu}A_{||}$ is ~200 G, corresponding to ~205 × 10^{-4} cm^{-1} (*see* **Note 13**). This combination of $g_{||}$ and $^{Cu}A_{||}$ values fall in the range of parameters found for tetragonal Cu(II) complexes with four nitrogen ligands in the equatorial coordination shell (14, 15). The estimation of the $g_{\perp}$ value and the metal hyperfine in this region ($^{Cu}A_{\perp}$) is much harder to obtain due to the signal overlap and the fact that $^{Cu}A_{\perp}$ is relatively small. Additionally, the fine structure observed in the perpendicular region of the spectrum at around 3,300 G is evidence for nitrogen hyperfine coupling, indicating that the metal ion has strong coupling with its nitrogen ligands. One of the nitrogen ligands is the histidine residue, while the other three nitrogens are deprotonated amides from the backbone of the prion protein fragment, which tend to yield highly covalent Cu–N bonds, and thus, strong ^{N}A hyperfine couplings (30). Note that the participation of nitrogen donor ligands (histidine residue and deprotonated amides) can be confirmed by the presence of corresponding LMCT bands observed in absorption and CD (Subheading 3.3). This is an example of how different spectroscopic techniques can provide complementary information that help elucidate the structural features of a metal binding site in a protein.

3.6.3 EPR Spectrum Simulation

Although a lot of information can be extracted from a visual inspection of an EPR spectrum as discussed above, performing EPR simulations is needed in order to determine g values and hyperfine coupling constants more accurately, particularly when signals overlap, as is the case for the EPR spectrum presented in Fig. 10. EPR simulation programs allow the user to compare the experimental spectrum with several simulated ones with different trial parameters, until the best match is found. The best approach is to first estimate the g values and any large hyperfine couplings that can be readily identified by visual inspection. For example, in the spectrum shown in Fig. 10 the $g_{\|}$ and ${}^{Cu}A_{\|}$ values can be easily estimated and readily optimized in the simulation, while finding the optimal parameters for the perpendicular region and for the fine structure due to small nitrogen hyperfine couplings may take many trial and error rounds. Additionally, line widths are an important aspect of EPR signals that need to be simulated, they tend to be larger for transition metal ions as compared to organic free radicals, and they can also be anisotropic because line broadening can vary for different parts of the spectrum. Thus, the anisotropic line widths are three more parameters (L_x, L_y, L_z in Table 2) that one would like to extract from the experimental spectrum. It is recommended to start with as few parameters as possible, and progressively build more complex scenarios if needed.

Figure 10 shows the best EPR simulation attained for the experimental spectrum, using the EPR simulation program XSophe (32) and the parameters listed in Table 2. The simulation assumes that the four nitrogen ligands are nonequivalent and it considers rhombic anisotropy for all parameters, such that, at least 12 parameters contribute to the simulation of the perpendicular region of the spectrum. This level of complexity was needed in order to adequately simulate the fine structure observed in this region of the spectrum. Note that the $g_{\|}$ and ${}^{Cu}A_{\|}$ values determined from visual inspection of the EPR spectrum are not widely different from those found through the simulation; however, perpendicular g and metal hyperfine couplings, the number of nitrogen nuclei and their corresponding hyperfine couplings that contribute to the fine structure in the perpendicular region could not have been determined without the simulation.

3.6.4 Spin Quantitation

EPR spin quantitation can be a useful tool to determine how many metal binding sites a protein has for a particular paramagnetic metal ion, or to determine metal ion loading in a metallo-enzyme sample. EPR signal intensity increases with spin concentration; however, it can also change with other factors (amount of sample in the cavity, EPR cell diameter, microwave power, etc.). Thus, the collection of EPR data on a reference standard of a known spin concentration under the same conditions as the sample is absolutely necessary to

Table 2
EPR simulation parameters for the spectrum shown in Fig. 10

g_x	2.038
g_y	2.048
g_z	2.193
A_x	17
A_y	17
A_z	195
$^{N1}A_x$	14
$^{N1}A_y$	12
$^{N1}A_z$	12
$^{N2}A_x$	12
$^{N2}A_y$	14
$^{N2}A_z$	12
$^{N3}A_x$	19
$^{N3}A_y$	17
$^{N3}A_z$	17
$^{N4}A_x$	12
$^{N4}A_y$	14
$^{N4}A_z$	12
L_x	6
L_y	5
L_z	24

Hyperfine splittings and line-widths are given in $\times 10^{-4}$ cm^{-1}. The line shape of the signal was modeled as Gaussian, as implemented in the EPR simulation program XSophe (32)

quantitate the spin concentration. The reference must be chosen based on the nature of the sample to be spin quantitated, making sure that its concentration can be accurately determined and that it has EPR properties similar to the sample; this is particularly important for transition metal ions (31). To quantify EPR signal intensity precisely the area under the EPR absorption curve must be calculated and compared to that of the reference. It is very important to make sure that the EPR spectrum is baseline-corrected before the double integration (*see* **Note 12**), as any baseline drift would contribute to the total integral. The accuracy of EPR spin quantitation relies on controlling all the different sample and instrumental factors that affect EPR signal intensity, as discussed in detail in (12, 31).

3.7 EPR as a Probe of Metal Coordination Properties

As mentioned above, the EPR parameters associated with a protein-bound paramagnetic metal ion can provide valuable information on the coordination properties of the metal–protein complex. Thus, EPR is a powerful tool to evaluate how these coordination properties change upon a given event, such as substrate binding or a change in pH. Figure 11a shows the comparison of the EPR spectra of Cu(II) bound to a prion protein fragment at pH 8.5 and 6.5 (this is the same complex that was shown in Figure 10, and it has a protonation equilibrium with a p*K*a of 7.6 by CD) (30). The drastic changes observed by CD suggest that the structure of the complex changes upon protonation; these structural changes can be probed by EPR. From Fig. 11a, it becomes evident that the $g_{\|}$ and ${}^{Cu}A_{\|}$ values change significantly upon protonation of the complex: $g_{\|}$ increases from 2.193 to 2.227, while ${}^{Cu}A_{\|}$ decreases from $195 \times 10^{-4}\ cm^{-1}$ to $167 \times 10^{-4}\ cm^{-1}$, indicating that the coordination shell of the Cu-peptide complex has changed. Specifically, the $g_{\|}$ and ${}^{Cu}A_{\|}$ values at pH 6.5 fall in the range of parameters found for tetragonal Cu(II) complexes with three nitrogens and one oxygen (3N1O) in the equatorial coordination shell (14, 15). This means that the coordination mode of the complex changes from 4N (having one His and three amide groups bound to the metal as described in Subheading 3.6.2) to 3N1O with one His, two amide groups and one carbonyl oxygen bound to Cu (30).

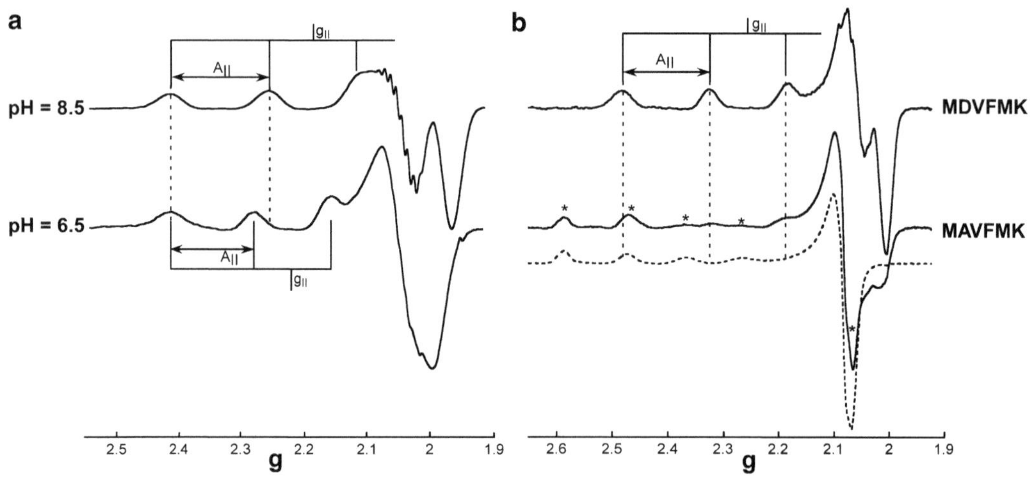

Fig. 11 Comparison of X-band EPR spectra of the Cu(II)-KTNAKHAAGA complex at pH 8.5 and 6.5 (**a**); peptide concentration was 0.5 mM with 0.9 equivalents of Cu, in 20 mM NEM and 20 mM MES buffer with 50 % glycerol. Comparison of EPR spectra of the Cu(II) complex with MDVFMK and MAVFMK peptide fragments (**b**); peptide concentration was 0.5 mM with 0.9 equivalents of Cu, in 20 mM MOPS buffer with 50 % glycerol, at pH 6.5. The EPR spectrum of free Cu(II) in the same buffer is shown as the *dashed line*, and the *asterisk* marks indicate the signals in the spectrum of the Cu(II)-MAVFMK complex that are associated with free Cu(II) in solution. Spectra were collected using the same instrument settings as in Fig. 10. Adapted with permission from (30, 33)

Additionally, the spectrum in the perpendicular region at pH 6.5 appears to be less rhombic and has less pronounced fine structure associated with nitrogen hyperfine couplings; the latter is consistent with having less nitrogen donors in the coordination shell. Combining this EPR analysis with other spectroscopic tools (absorption and CD) and electronic structure studies, models of how Cu(II) coordinates to this region of the protein can be proposed (30).

Another example for the use of EPR to probe changes in coordination properties is shown in Fig. 11b, where the EPR spectrum of Cu(II) bound to an N-terminal fragment of α-synuclein with sequence MDVFMK is compared to that of the Cu(II) complex with the peptide variant MAVFMK, to evaluate the role of the Asp2 residue in metal ion coordination. The spectrum of the Cu(II)-MDVFMK complex shows an EPR signal with $g_{\parallel} = 2.250$ and ${}^{Cu}A_{\parallel} = 189 \times 10^{-4}$ cm^{-1}, which are indicative of an 2N2O equatorial coordination with two nitrogens and two oxygen donors (14, 15). In contrast, the EPR spectrum of the Cu (II)-MAVFMK complex is quite different: it shows weak signals similar to those associated with Cu(II)-MDVFMK, but the spectrum is dominated by signals that correspond to free Cu(II) in solution (marked with * in Fig. 11b). In contrast to CD, where free aqueous Cu(II) ions form non-chiral complexes with water molecules and do not yield any CD signal, in EPR aqueous Cu (II) ions give rise to intense signals (dotted spectrum in Fig. 11b). The appearance of these signals in the spectrum of the Cu(II)-MAVFMK complex indicates that the affinity of Cu(II) for the peptide decreases upon the D2A substitution, and thus, under the same stoichiometric conditions, not all the metal ion is bound to the peptide. This example highlights the importance of keeping in mind that free paramagnetic metal ions (not bound to the protein) can also contribute to the EPR spectrum. These EPR results also indicate that Asp2 is one of the oxygen-based ligands and plays an important role in Cu(II) binding to this protein (33).

4 Notes

1. Exceptions to this general rule are observed in highly covalent metal sites in proteins, such as the blue copper center (34). Additionally, when the ligands have low-lying antibonding π molecular orbitals (such as N_2, CO or NO) metal to ligand charge transfer transitions (MLCT) may occur (9). Heme sites in proteins are also distinguished for their characteristic electronic spectral features, as they have low-lying empty π molecular orbitals and their highest occupied orbitals are π bonding, such that intense $\pi \rightarrow \pi^*$ transitions (called Q bands) are observed (35). When studying a metalloprotein, it is advisable

to review the relevant literature to understand the particular spectroscopic features associated to its metal ion.

2. Metal sites in proteins can be classified according to their EPR properties. Metal ions with spin $S = 1/2$, such as Cu(II), Ni (III), Ni(I) and Mo(V), present g values around 2 with small deviations (0.4 maximum) due to spin–orbit coupling, which allows for the mixing of excited ligand field states into the ground state (7, 11). The resulting spectrum can be axial or rhombic depending on the geometry of the complex. Low spin Fe(III) sites in heme proteins are $S = 1/2$ and have g values with larger deviations; a detailed analysis of their g values is a useful tool to characterize the nature of the axial ligand (12). Metal active sites in proteins with more than one paramagnetic metal ion bridged by one or more ligands can also have a total spin of 1/2, and they are described as coupled spin systems. Examples of such sites are found in iron-sulfur proteins and binuclear non-heme iron enzymes (e.g., ribonucleotide reductase or purple acid phosphatase). For example, a binuclear non-heme iron site with one Fe(II) ion (d^6) coupled to a Fe(III) center (d^5) would have a total of 11 d electrons that antiferromagnetically couple to yield a total spin of 1/2. This type of active site typically has rhombic EPR spectra with g values in the range of 1.6–2. Metal sites with $S > 1/2$, such as Mn(II), Fe (II), Fe(III), and high-spin Co(II), present zero-field splitting (zfs), which involves the energy splitting of the different m_s spin sublevels in the absence of the magnetic field, yielding g values that can greatly deviate from 2. For example, for a half-integer spin system with a total $S = 5/2$ (high-spin Fe(III) or Mn(II)), zfs will lead to an energy difference between the m_s levels $\pm 1/2$, $\pm 3/2$ and $\pm 5/2$ in the absence of the magnetic field; and each of these so called Kramers doublets will split further when the magnetic field is turned on due to the Zeeman effect. For non-Kramers or integer spin systems with an even number of unpaired electrons, the $\pm m_s$ doublets (e.g., ± 2 and ± 1 for a $S = 2$ center) are further split by zfs effects, as described in (12).

3. If the purity or high resistivity of water is questionable, sometimes it is advisable to add a chelating resin such as Chelex (a resin with high affinity for metal ions) to the buffer, in order to eliminate trace metal ions. After treatment with Chelex overnight, the resin can be easily filtered from the solution.

4. The metalloprotein sample must be concentrated enough to give a good signal-to-noise ratio, but not too concentrated that your absorption signals are saturated; absorption intensity should be ideally between 0.1 and 1.0 absorbance units, so it falls in a linear range and obeys the Lambert–Beer law (Eq. 10.2). This is particularly important for LMCT or MLCT bands, which tend to have high ε. The path length of

the cell that contains the sample will also affect the signal intensity in both, absorption and CD spectroscopy. The best choice of path length and sample concentration depends on the nature and availability of sample, and on the type of experiment; for example, you may prefer not to dilute your sample and use a cell with a shorter path length, or if you are determining binding constants, it might be desirable to keep protein concentrations low and use larger path lengths. At the optimal concentrations to observe signals associated with metal sites in metalloproteins (0.1–1 mM), the absorption of peptide bonds (between 200 and 230 nm) and aromatic residues (around 280 nm) are usually very intense, they may saturate and make it difficult or impossible to observe the signals associated with the metal site in this energy region. The choice of path length may depend on the absorption signal of interest: d–d transitions associated with a Cu(II)-protein complex are usually very weak and are best observed with a cell path length of 1 cm for metalloprotein solutions at concentration in the millimolar range; in contrast, LMCT bands that fall close to the region where the protein absorbs (200–280 nm) are best observed at shorter path lengths to avoid saturation. Usually, what works best for absorption in terms of sample concentration and cell path length would work well for CD spectroscopy. However, it is important to check in the spectropolarimeter that the voltage at the PMT detector (which is inversely proportional to the amount of light that it receives) is not too high in the selected wavelength range sweep, as high voltages may damage the PMT detector.

5. The use of glycerol for EPR sample preparation has a dual function: it serves as a cryoprotectant for the protein, and helps the sample form a disordered glass-state upon freezing. EPR cells are very fragile and tend to break easily upon freeze and thaw cycles; this effect can be minimized if the aqueous sample forms a glass. Glassing is achieved by adding 50 % glycerol to the buffer solution, and by freezing the sample very slowly. Glycerol is the most common cryoprotectant used with proteins, but it can bind metal ions; consequently, CD and absorption spectra in the presence and absence of glycerol should be compared to rule out any possible effect on the metal–protein complex. For more information on the preparation of samples for low temperature EPR experiments, including air-sensitive samples, refer to (36).

6. Commercial optical glass cells have a usable range of 334–2,500 nm. If electronic transitions below 350 nm need to be studied (typically charge transfer transitions), Spectrosil® quartz cells with a usable range of 170–2,500 nm must be used.

7. Quartz tubes for EPR measurements are available in different inner diameters (2–5 mm). The intensity of the EPR signal increases with the amount of sample that is placed in the cavity; thus, larger diameter tubes give larger signals. However, because liquid water absorbs microwaves strongly, it is advisable to use capillaries or quartz tubes with a small inner diameter with aqueous samples at room temperature. It is advisable that the EPR tube is placed far enough into the cavity, such that the sample occupies the full height of the cavity, in order to maximize the EPR signal intensity. The latter is also a requirement if you need to quantify the total spin concentration in your sample. In X-band spectrometers, the height of the cavity is 2.5 cm; thus, a volume of ~120 μL of sample in a 3 mm inner diameter quartz cell is enough to collect an EPR spectrum in optimal conditions; this volume increases to ~250 μL for cells with 4 mm inner diameter.
8. The wavelength calibration of a spectropolarimeter is performed using a neodymium crystal, which absorbs in the region from 610 to 560 nm. The data is collected with low scan speed (20 nm/min), small data pitch (0.1 nm), 0.25 s response time, and 1.0 nm band width. The maximum photomultiplier voltage must occur at 586 ± 0.8 nm. The calibration of the CD intensity is performed using an aqueous solution of 0.06 % ammonium d-10 camphor sulfonate in a 1 cm path length cell. The spectrum is collected from 350 to 250 nm with the parameters: 0.1 nm data pitch, 50 nm/min scan speed, 1.0 s response time, and 1.0 nm band width; the maximum intensity must be 190.4 ± 1 mdeg at 290.5 nm. For further details in CD instrument calibration, the reader is referred to the instrument manuals and Chapter 8.
9. Calibration of the modulation amplitude for the EPR cavity is essential in order to quantitatively compare EPR spectra run in different laboratories. This is done using a standard sample of a crystal of 2,2-diphenyl-1-picrylhydrazyl (DPPH), which gives a very narrow EPR signal. EPR spectra of the standard are collected at different modulation amplitudes; signal distortion is observed as the modulation amplitude is increased. For such a very narrow signal, the peak-to-peak width of the first derivative EPR signal is approximately equal to the peak-to-peak modulation amplitude (31). The DPPH standard is also used to check on the accuracy of the magnetic field as the signal should appear at a g value of 2.0037 ± 0.002. Although most EPR spectrometers have a probe that measures the magnetic field, the probe is usually not at the same place as the sample, and thus, there may be a small offset between the two (quartz inserts and cryostats can increase this offset). The DPPH standard can be used to calibrate the probe or DPPH may also be

used as an internal standard to determine precisely the field at the sample. Further details on EPR instrument calibration can be found in (31, 37).

10. Choice of microwave power for a given sample and setup. At low power levels, EPR signal intensity grows as the square root of the microwave power. At higher power levels, the effect of saturation occurs: the EPR signal broadens and its intensity decreases with increasing microwave power. EPR signal saturation can make it difficult to accurately measure line-widths and small hyperfine splittings. Also, to quantitate the spin concentration in a sample, both the EPR spectra of the sample and the reference, must be collected under non-saturating conditions. EPR signal saturation is most commonly observed at low temperatures. Low temperature is used to slow down spin relaxation processes, increase the net excess population of the low energy spin sublevel, and increase sensitivity; however, the combination of low temperature and high microwave power can yield an excess population of the higher energy spin sublevel, and thus to a decrease in signal intensity. The nature of spin relaxation processes depends on the nature of the sample, and EPR signal intensity depends on several factors including sample concentration, type of cavity, how much sample is placed inside the cavity, temperature and microwave power. Thus, for each sample and experimental setup that you use, a microwave power sweep must be performed in order to determine the optimal microwave power that would prevent saturation of the EPR signal. The range of optimal microwave powers is that where the signal intensity changes linearly with respect to the square root of the microwave power.

11. The microwave frequency in an X-band spectrometer is fixed at ~9.5 GHz. However, the exact value of ν at which the cavity is tuned with a sample inside may vary with its contents (sample tube, a quartz accessory for variable temperature studies, etc.). Thus, it is very important to register the exact frequency, ν, at which an EPR spectrum is run, in order to determine precisely the g value associated to a given signal. The g value associated with an EPR signal that is centered at a given magnetic field (H in gauss) can be calculated using the equation: $g = 714.46\ \nu/H$ (from substituting the Bohr magneton value and unit conversion factors in Eq. 6).

12. Sometimes, even after subtracting an EPR baseline spectrum from your sample spectrum, not all baseline drifts are corrected. Additional digital baseline correction can be carried out with many data analysis programs, using linear or polynomial functions to fit and subtract the drifting baseline; however, care must be taken not to filter out, nor "create" any EPR signals.

13. Hyperfine splittings are usually extracted from EPR spectra in magnetic field units. The conversion of A values to frequency units (cm^{-1}) as they are normally reported in the literature is given by: A (in cm^{-1}) = A (in Gauss) × g × 4.6686 × 10^{-5}, where g is the g value associated to the EPR signal that has the hyperfine splitting. A values may also be reported in MHz, the conversion between these two units is straightforward: 1 cm^{-1} = 2.9979 × 10^4 MHz.

Acknowledgments

L.Q. gratefully acknowledges the important scientific influence of Prof. Edward I. Solomon (Stanford University) that is clearly evidenced in this paper. This work was funded by Consejo Nacional de Ciencia y Tecnología (CONACYT grant #128255). L.R.A. has been a recipient of postdoctoral fellowships from CONACYT (grant #060366-Q and the program "Estancias Posdoctorales Vinculadas al Fortalecimiento de la Calidad del Posgrado Nacional").

References

1. Bertini I, Gray HR, Stiefel EI, Valentine JS (eds) (2007) Biological inorganic chemistry: structure and reactivity. University Science, Sausalito, CA
2. Huheey JE, Keiter EA, Keiter RL (1993) Inorganic chemistry: principles of structure and reactivity, 4th edn. HarperCollins College Publishers, New York
3. Lever ABP, Solomon EI (1999) Ligand field theory and the properties of transition metal complexes. In: Solomon EI, Lever ABP (eds) Inorganic electronic structure and spectroscopy. Wiley, New York
4. Figgis BN (1996) Introduction to ligand fields. Interscience, New York
5. Ballhausen CJ (1962) Introduction to ligand field theory. McGraw-Hill, New York
6. Ballhausen CJ, Gray HR (1964) Molecular orbital theory. Benjamin/Cummings, Reading, MA
7. Solomon EI, Hanson MA (1999) Bioinorganic spectroscopy. In: Solomon EI, Lever ABP (eds) Inorganic electronic structure and spectroscopy. Wiley, New York, pp 1–129
8. Harris DC, Bertolucci MD (1978) Symmetry and spectroscopy: an introduction to vibrational and electronic spectroscopy. Dover, New York
9. McMillin DR (2000) Electronic absorption spectroscopy. In: Que LJ (ed) Physical methods in bioinorganic chemistry. University Science, Sausalito, CA
10. Johnson MK (2000) Circular dichroism and magnetic circular dichroism. In: Que LJ (ed) Physical methods in bioinorganic chemistry. University Science, Sausalito, CA
11. Solomon EI (1984) Inorganic spectroscopy—an overview. Comments Inorg Chem 3:227–320
12. Palmer G (2000) Electron paramagnetic resonance of metalloproteins. In: Que LJ (ed) Physical methods in bioinorganic chemistry. University Science, Sausalito, CA
13. Bencini A (1999) Electron paramagnetic resonance spectroscopy. In: Solomon EI, Lever ABP (eds) Inorganic electronic structure and spectroscopy. Wiley, New York, pp 93–159
14. Peisach J, Blumberg WE (1974) Structural implications derived from the analysis of EPR spectra of natural and artificial copper proteins. Arch Biochem Biophys 165:691–708
15. Sakaguchi U, Addison AW (1979) Spectroscopic and redox studies of some copper(II) complexes with biomimetic donor atoms: implications for protein copper centers. J Chem Soc Dalton Trans 600–608
16. Cammack R, Cooper CE (1993) Electron paramagnetic resonance spectroscopy of iron complexes and iron-containing proteins. Methods Enzymol 227:353–384

17. Parish RV (1990) NMR, NQR, EPR and Mossbauer spectroscopy in inorganic chemistry. Ellis Harwood, Chichester
18. Palmer G (1979) Electron paramagnetic resonance of hemoproteins. In: Dolphin D (ed) The porphyrins. Academic, New York
19. Chasteen ND, Snetsinger PA (2000) ESEEM and ENDOR spectroscopy. In: Que LJ (ed) Physical methods in bioinorganic chemistry. University Science, Sausalito, CA
20. Deligiannakis Y, Louloudi M, Hadjiliadis N (2000) ESEEM spectroscopy as a tool to investigate the coordination environment of metal centers. Coord Chem Rev 204:1–112
21. Hoffman BM (2003) ENDOR of metalloenzymes. Acc Chem Res 36:522–529
22. Yu Q, Kandegedara A, Xu Y, Rorabacher DB (1997) Avoiding interferences from Good's buffers: a contiguous series of noncomplexing tertiary amine buffers covering the entire range of pH 3–11. Anal Biochem 253:50–56
23. Bradford MM (1976) A rapid and sensitive method for the quantitation of microgram quantities of protein utilizing the principle of protein-dye binding. Anal Biochem 72:248–254
24. Smith PK, Krohn RI, Hermanson GT, Mallia AK, Gartner FH, Provenzano MD, Fujimoto EK, Goeke NM, Olson BJ, Klenk DC (1985) Measurement of protein using bicinchoninic acid. Anal Biochem 150:76–85
25. Grossoehme NE, Spuches AM, Wilcox DE (2010) Application of isothermal titration calorimetry in bioinorganic chemistry. J Biol Inorg Chem 15:1183–1191
26. Xiao Z, Wedd AG (2010) The challenges of determining metal-protein affinities. Nat Prod Rep 27:768–789
27. Fersht A (1999) Structure and mechanism in protein science: a guide to enzyme catalysis and protein folding. WH Freeman and Company, New York
28. Bernanducci EE, Schwindinger WF, Hughey JL, Krogh-Jespersen K, Schugar HJ (1981) Electronic spectra of copper(II)-imidazole and copper(II)-pyrazole chromophores. J Am Chem Soc 103:1686–1691
29. Daniele PG, Prenesti E, Ostacoli G (1996) Ultraviolet-circular dichroism spectra for structural analysis of copper(II) complexes with aliphatic and aromatic ligands in aqueous solution. J Chem Soc Dalton Trans 3269–3275
30. Rivillas-Acevedo L, Grande-Aztatzi R, Lomelí I, García JE, Barrios E, Teloxa S, Vela A, Quintanar L (2011) Spectroscopic and electronic structure studies of copper(II) binding to His111 in the human prion protein fragment 106–115: evaluating the role of protons and methionine residues. Inorg Chem 50:1956–1972
31. Eaton GR, Eaton SS, Barr DP, Weber RT (2010) Quantitative EPR. Springer/Wien, New York
32. Hanson GR, Gates KE, Noble CJ, Griffin M, Mitchell A, Benson S (2004) XSophe-Sophe-XeprView. A computer simulation software suite (v.1.1.3) for the analysis of continuous wave EPR spectra. J Inorg Biochem 98:903–916
33. Binolfi A, Rodriguez EE, Valensin D, D'Amelio N, Ippoliti E, Obal G, Duran R, Magistrato A, Pritsch O, Zweckstetter M, Valensin G, Carloni P, Quintanar L, Griesinger C, Fernandez CO (2010) Bioinorganic chemistry of Parkinson's disease: structural determinants for the copper-mediated amyloid formation of alpha-synuclein. Inorg Chem 49:10668–10679
34. Solomon EI, Szilagyi RK, DeBeer George S, Basumallick L (2004) Electronic structures of metal sites in proteins and models: contributions to function in blue copper proteins. Chem Rev 104:419–458
35. Loew G (1999) Electronic structure of heme sites. In: Solomon EI, Lever ABP (eds) Inorganic electronic structure and spectroscopy. Wiley, New York
36. Beinert H, Orme-Johnson WH, Palmer G (1978) Special techniques for the preparation of samples for low-temperature EPR spectroscopy. Methods Enzymol 54:111–132
37. Poole CP (1983) Electron spin resonance: a comprehensive treatise on experimental techniques. Wiley, New York

Chapter 11

Monitoring Protein–Ligand Interactions by Time-Resolved FTIR Difference Spectroscopy

Carsten Kötting and Klaus Gerwert

Abstract

Time-resolved FTIR difference spectroscopy is a valuable tool to monitor the dynamics and exact molecular details of protein–ligand interactions. FTIR difference spectroscopy selects, out of the background absorbance of the whole sample, the absorbance bands of the protein groups and of the ligands that are involved in the protein reaction. The absorbance changes can be monitored with time-resolutions down to nanoseconds and followed for time periods ranging over nine orders of magnitude even in membrane proteins with a size of 100,000 Da. Here, we discuss the various experimental setups. The rapid scan technique allows a time resolution in the millisecond regime, whereas the step scan technique allows nanosecond time resolution. We show appropriate sample cells and how to trigger a reaction within these cells. The kinetic analysis of the data is discussed. A crucial step in the data analysis is the reliable assignment of bands to chemical groups of the protein and the ligand. This is done either by site directed mutagenesis, where the absorbance bands of the exchanged amino acids disappear or by isotopically labeling, where the band of the labelled group is frequency shifted.

Key words Infrared, Time-resolved, Difference spectroscopy, Rapid scan, Step scan, Bacteriorhodopsin, Retinal, GTPases, GTP, Caged-substances, Isotopic labeling, Band assignment, ATR, Global fitting

1 Introduction

The generation of difference spectra between, e.g., the inactive and active states of proteins selects out of the background absorbance of the whole sample the absorbance bands of the protein groups and of the ligands that are involved in the protein activation (1). The absorbance changes can be monitored with time-resolutions down to nanoseconds. Tr-FTIR difference spectroscopy was established in investigations of the light-driven proton pump bacteriorhodopsin (bR) (2). By this model system for light-driven proton transport (Fig. 1), the roles of the aspartate residue D96 as the internal proton donor of the retinal Schiff base and aspartate D85 as the primary proton acceptor of the Schiff base proton were shown (3, 4). In further studies, the importance of protein bound internal

Mark A. Williams and Tina Daviter (eds.), *Protein-Ligand Interactions: Methods and Applications*, Methods in Molecular Biology, vol. 1008, DOI 10.1007/978-1-62703-398-5_11, © Springer Science+Business Media New York 2013

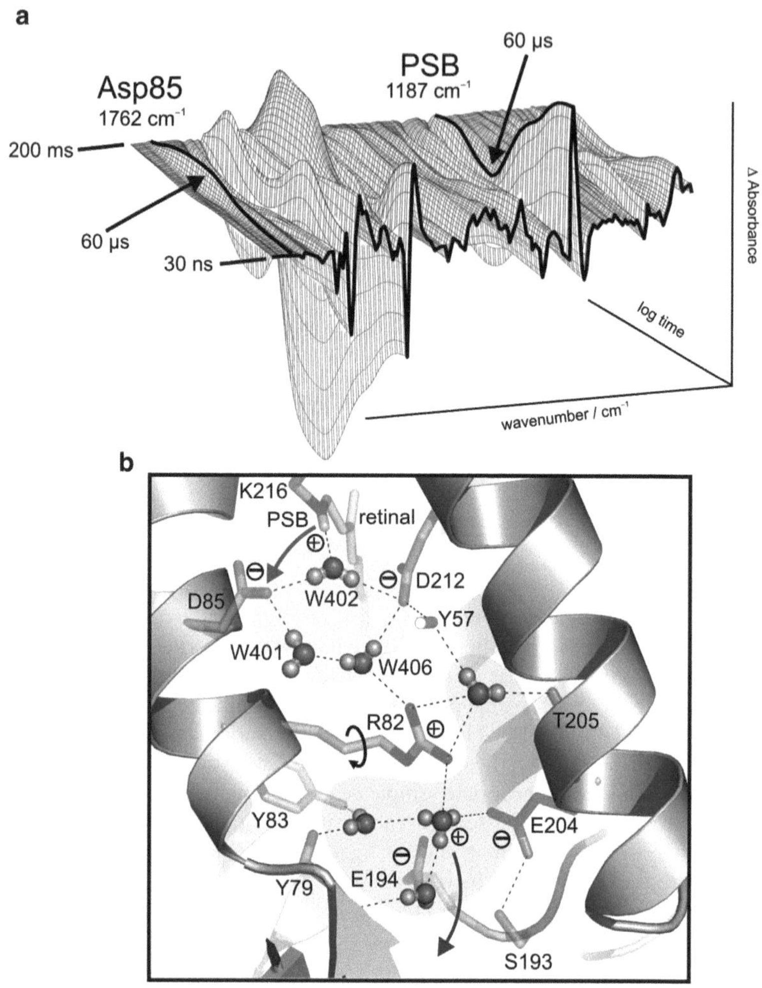

Fig. 1 (**a**) The IR absorbance changes during the photocycle of bacteriorhodopsin (bR). One band of the protein at 1,762 cm^{-1} (protonated Asp85) and one of the protein–ligand interface at 1,187 cm^{-1} (Protonated Schiff Base, PSB) are marked. In the L $\rightarrow$ M reaction (completed after 60 μs), a proton is transferred from the PSB to Asp85, as seen in the decay at 1,187 cm^{-1} and the increase at 1,762 cm^{-1}. (**b**) Model of the proton pump mechanism of bR, to which many groups have contributed (for references, see main text). After the light-induced all-*trans* to 13-*cis* retinal isomerization in the BR-to-K transition, the Schiff base proton is transferred to Asp 85 in the L-to-M transition. Deprotonation of the protonated Schiff base can be followed at 1,187 cm^{-1} in (**a**) and protonation of Asp85 at 1,762 cm^{-1} in (**a**). Due to light-induced isomerization of the chromophore retinal, the strong H-bond of water 402 is broken and approximately half of the energy is stored in the protein. After isomerization, the free OH-group of the dangling water 401 is H-bonded and can no longer stabilize the charge of Asp85 (5). The proton is transferred from the PSB to Asp85. Due to the neutralization of Asp85, a downward movement of the salt bridged Arg82 is induced. This movement of the positive charge destabilizes the protonated water complex near the protein surface. The protonated water cluster (*lower shaded area*) stores a proton, probably in an asymmetric Eigen-complex $(H^+(H_2O)_3)^6$. This destabilizes the second hydration shell. In contrast to the random Grotthuss-proton transfer in water, in the protein the water complex is deprotonated by a directed movement of Arg82. The proton is stabilized in the second hydration shell by amino acids instead of water molecules

water molecules for the directional proton transfer via a Grotthuss like mechanism was demonstrated (5, 6).

By comparing the FTIR difference spectra with FT-Resonance Raman spectra, the bands of photoactive ligands can be readily identified. In the FT Resonance Raman spectra, only the chromophore bands appear, whereas in the FTIR difference spectra, ligand and protein bands appear. Thereby, for example, the chromophore bands in photoactive yellow protein were assigned (7). Moreover, light-induced redox reactions of ligands can be monitored in these photosynthetic proteins. That the light-induced electron transfer within the prosthetic groups and the proton uptake via amino acid side chains are coupled in a ping-pong mechanism to electron transfer in photosynthetic reaction centers has been elucidated by time resolved FTIR (8, 9).

By using photolabile trigger compounds, protein–ligand interactions can be studied in proteins without chromophores. The redox driven proton transport in cytochrome C oxidase is studied using the electron donor riboflavin as a source of caged electrons (10). In a similar manner, the interactions of ATPases and GTPases with their ATP and GTP ligands can be studied using caged ATP or caged GTP as photolabile trigger (11, 12). By this approach the interaction of GTP with Ras has been studied, and it has been shown that binding of GTP to Ras induces a specific charge distribution in GTP, which reduces the free activation energy (13). Thereby, GTP hydrolysis is catalyzed. In addition, protein–protein interactions and their influence on the GTP ligand can be studied. The influence of GAP on the conformation and charge distribution of GTP bound to Ras is studied in detail. This reveals why binding of GAP to Ras catalyzes the reaction by five orders of magnitude (14, 15). In oncogenic Ras this activation is inhibited and this is one major event in the transformation of a cell into a cancer cell. FTIR spectroscopic assays also enable label free screening of drug– protein interactions (16).

In summary, the above studies demonstrate that time-resolved FTIR difference spectroscopy is a powerful tool to monitor protein–ligand or protein–protein interactions. Complementary to X-ray or NMR providing three dimensional structural models of proteins, FTIR delivers information on H-bonding, protonation state, charge distribution and time dependence of the protein–ligand interaction.

2 Materials

The following equipment is state of the art and successfully used for time-resolved FTIR difference spectroscopy in our laboratory. They have to be used in suitable environment, including stable temperature (± 0.5 K) and vibration isolation, to ensure best results. Of course, less demanding measurements are also possible using instruments with lower specifications.

Table 1
Frequently used IR-window materials

Material	Transmission range (cm^{-1})
Silicon (Si)	1000–10,000
Calcium fluoride (CaF_2)	950–66,000
Barium fluoride (BaF_2)	890–50,000
Zinc selenide (ZnSe)	450–20,000
Silver chloride (AgCl)	400–23,000

2.1 Spectrometer

Vertex 80v (Bruker Optics, Karlsruhe, Germany) with KBr-Beamsplitter.

CaF_2 windows (Korth, Kiel, Germany). This material is transparent in the infrared and the visible regions and does not react with most samples. Other material is shown in Table 1.

MCT-detector KMPV11-1-J1 (Kolmar, Newburyport, MA, USA).

Dry-Air purge gas generator Balston 75-62 (ParkerBalston, NJ, USA).

2.2 Laser

Excimer-Laser LPX Pro (Coherent, Santa Clara, CA, USA) with XeCl (308 nm).

Dye-laser NarrowScan (Radiant Dyes, Wermelskirchen, Germany) with Coumarin 153 (540 nm).

Nd:YAG-Laser Quanta-Ray GCR-170 (Spectra Physics, Mt. View, CA, USA).

2.3 Step Scan

MCT-detector KV100-1-B-7/190, cutoff 850 cm^{-1} (Kolmar, Newburyport, MA, USA).

Jäger Adwin-Pro II 18 bit/500 kHz transient recorder (Jäger, Lorsch, Germany).

Digital Oscilloscope LeCroyWavePro 715 Zi (LeCroy, Chestnut Ridge, NY, USA).

2.4 Special Equipment

Hyperion 3000 FTIR imaging microscope (Bruker, Karlsruhe, Germany).

BaF_2 windows (Korth, Kiel, Germany).

Diamond-μ-ATR-cell (Resultec, Garbsen, Germany).

Flow Cell (Biolytics, Freiburg, Germany).

Vertical ATR multireflection unit with a 52 × 20 × 2 mm trapezoidal germanium ATR plate (aperture angle 45°; 25 internal reflections) (Specac, Orpington, UK).

3 Methods

3.1 General Setup

A typical setup for a time resolved FTIR experiment is shown in Fig. 2a. The light source is a Globar (SiC heated at 1,800 K), which is a black body radiator. Its infrared light passes an aperture (0.25–12 mm) before entering a Michelson interferometer, consisting of a beamsplitter (KBr for the mid infrared), a fixed and a movable mirror. Subsequently the light passes the sample chamber, equipped, e.g., with a thermostatic transmission cell. Here, the cell can additionally be irradiated by a UV/visible light laser. Finally the infrared light reaches the liquid nitrogen cooled MCT (mercury, cadmium, telluride, HgCdTe)-detector.

FTIR-spectrometers have crucial advantages compared to dispersive spectrometers. By means of the Michelson interferometer (Fig. 2b), all wavelengths can be measured in parallel (multiplex advantage): At the beamsplitter, one half of the infrared light is reflected on a fixed mirror, the other half is transmitted to a moving mirror. Both parts are recombined at the beamsplitter. Depending on the position of the moving mirror, a path difference (ΔX) is created between the beam reflected by the fixed mirror and the beam reflected by the moving mirror. For example, a path difference between two monochromatic waves of zero wavelengths (mirror in position X_1) leads to constructive interference, whereas a path difference of half a wavelength (mirror in position X_2) leads to extinction. By variation of the path difference, the interference pattern of polychromatic light after recombination leads to an interferogram, where the intensity is plotted against the mirror position. After Fourier transformation, the intensity I as a function of the wavelength is obtained (Single Channel Transmittance spectrum, Fig. 2c). An absorbance spectrum is obtained by comparison of two single channel transmittance spectra, one with sample and one without sample.

With modern FTIR spectrometers, a complete spectrum can be obtained within 10 ms. Further advantages of FTIR-spectrometers are the absence of dispersive elements (i.e., there are no slits combined with prisms or gratings which attenuate the signal intensity; Jaquinot advantage), and the high accuracy of the wavelength (Connes advantage).

In Fig. 3 few sample cells are shown. The most common cell is a simple transmission cell with IR-transparent windows, e.g., CaF_2 (*see* **Note 1** and Fig. 3a). Due to the high absorptivity of water in the mid-infrared spectral region, meaningful spectra of hydrated proteins are obtained by transmission measurements only through very thin (2–10 μm) films. This involves placing a drop of a protein suspension or solution onto an IR transparent window and then carefully concentrating it under a nitrogen stream or under vacuum. Alternatively, a protein suspension of a membrane bound

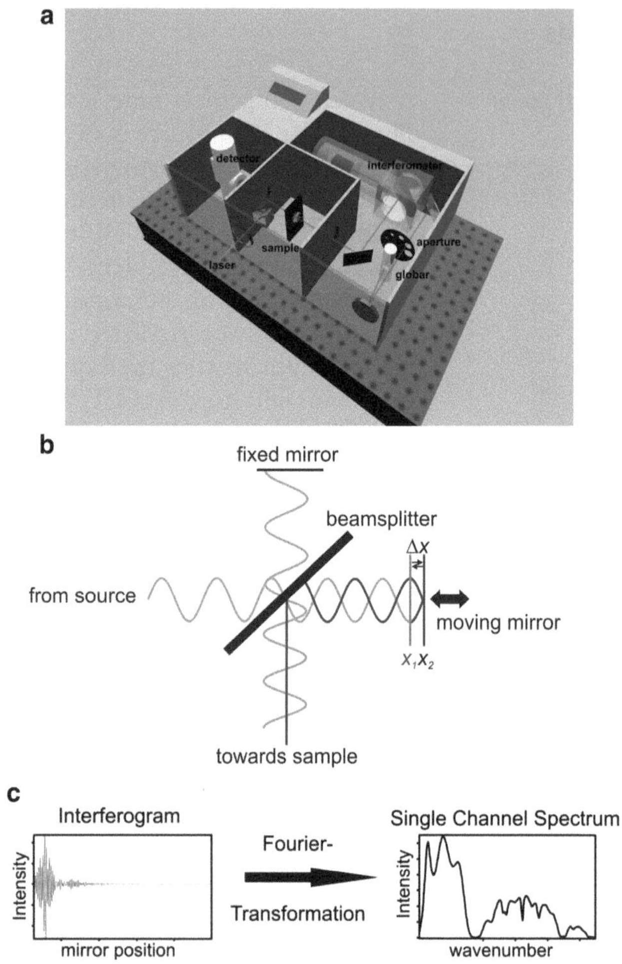

Fig. 2 (**a**) Scheme of an FTIR spectrometer on a vibration isolation table (adapted from ref. 58). (**b**) Schematic representation of a Michelson interferometer. An electromagnetic wave is split at the beamsplitter; one half is reflected at a fixed mirror, the other half is transmitted on to a moving mirror. Both parts are recombined at the beamsplitter. For mirror position X_1, the path difference, ΔX, is zero and the wave interferes constructively, whereas for mirror position X_2, the path difference, ΔX, is half a wavelength and thus the inference is destructive. (**c**) The result of the measurement is an interferogram, where the intensity is plotted against the mirror position. After Fourier transformation, the intensity I is obtained as a function of the wavelength (single channel spectrum)

protein is centrifuged and the pellet is squeezed between two IR transparent windows. A typical measurement requires about 100–200 μg protein. The required concentration of the protein in the film is 2–10 mM. The sample chamber is closed by a second IR-window, which is separated from the first by a mylar-spacer of a few micrometer thickness.

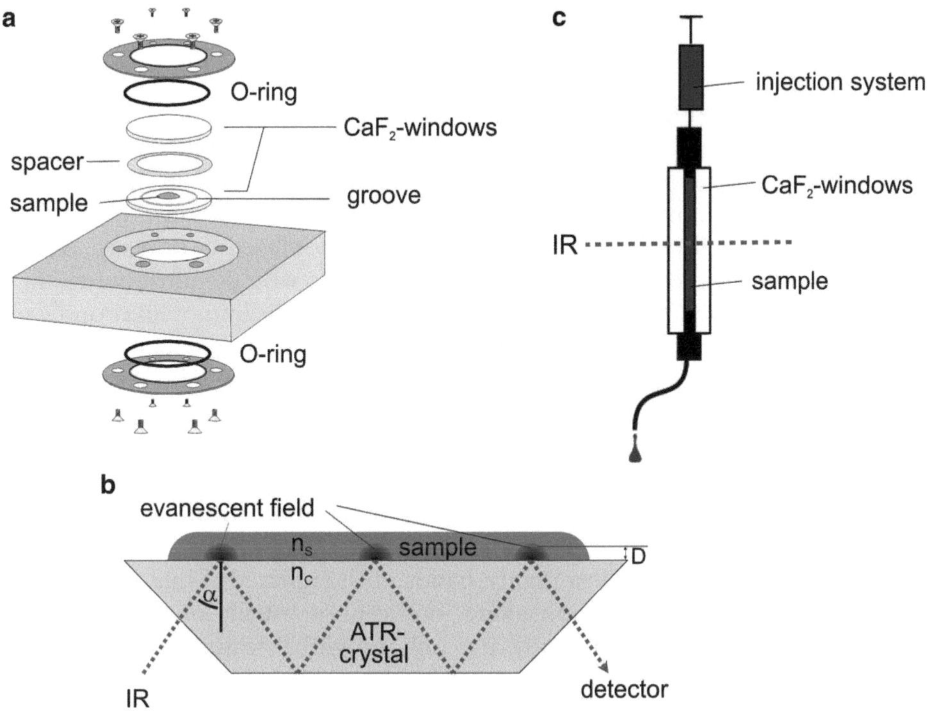

Fig. 3 Various sample cells for the investigations of protein–ligand interaction. (**a**) Transmission cell. (**b**) ATR-cell. (**c**) Flow cell

Instead of transmission cells, ATR (attenuated total reflection) cells (Fig. 3b) can be used (17, 18). The IR light is reflected at the interface of a crystal (most common are diamond, ZnSe, Si, and Ge) and the sample. In this process, an infrared evanescent wave penetrates into the sample and absorption by that part of the sample which lies within a small distance of the interface can be measured. The depth *D*, where the intensity of the wave is reduced to 13.5 %, can be calculated by

$$D = \frac{10^4}{2\pi\nu n_C\sqrt{\sin^2\alpha - \frac{n_S}{n_C}}}, \tag{1}$$

where ν is the wavenumber, n_S is the refractive index of the sample, n_C the refractive index of the crystal, and α is the angle of incidence. In most cases, *D* is in the low micrometer regime and those parts of the sample proximal to the interface participate much more strongly in the absorption. Thus, monolayers (e.g., of lipids or of hydrophobic proteins that bind to the crystal surface) can be investigated. Signal can be enhanced by the SEIRA-technique, where the ATR-cell is coated by thin film of metal (usually, gold or silver) that creates a locally increased, polarized electric field at the solution interface leading to greater IR absorption by the nearby molecules (19–21). One can increase the signal by using multiple reflections. ATR-cells with 1–25 reflections are available.

Both, transmission cells and ATR-cells can be used as flowcells (Fig. 3c). Here the sample can be easily exchanged by means of a tubing system. This can increase the quality of difference spectra enormously, because the whole setup (sample thickness, window position, etc.) can be maintained exactly the same. In a flow cell, the protein of interest has to be immobilized in order to avoid loss of protein during an experiment that would contribute to the difference spectrum. Various techniques for immobilization of proteins are possible; among them is covalent attachment via cysteine residues (19) or anchoring of the protein by lipid anchors to model membranes (22, 23). Immobilized proteins provide an excellent platform for the investigation of protein–ligand interactions as a succession of different ligands can be flowed over them.

Figure 4 shows the absorbance spectrum of a protein. Even a small protein of 20 kDa, such as shown here, has about 10^4 overlapping absorbance bands in the infrared. Thus, from the absorption spectrum alone, one cannot usually obtain information on individual bands, but only on global features of the protein. The spectrum is dominated by the amide I (C=O stretch) and amide II (NH bend coupled with C–N stretch) bands, where every amino acid contributes. From these backbone absorptions information

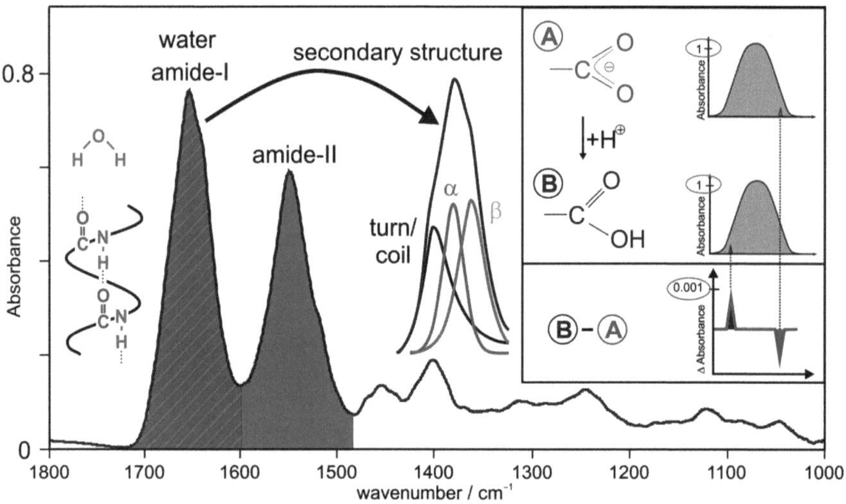

Fig. 4 Typical IR absorbance spectrum of a protein in solution (Ras). The major components are the amide I band (C = O stretching vibration), water (bending vibration) that overlaps with amide I, and the amide II band (a combination of NH bending and CN stretching). The amide I mode of amino acids of different secondary structure have different frequencies and deconvolution of the amide I band can reveal information on secondary structure. In the *inset*, two absorption spectra of a protein, which differ only in the protonation of a carboxyl group, are shown schematically. In the *lower part*, a difference spectrum of these two states is shown. The background absorptions of the unchanged part of the protein are cancelled out and the absorptions of the reacting group are now distinctly resolved

can be gained on the secondary structure of the protein (23, 24). Water absorption (O–H bend) is found in the same region as the amide I.

For a FTIR difference spectrum of a reaction A → B, one calculates the absorbance spectrum of B minus the absorbance spectrum of A. Thus, the vibrations from groups which are not changed during the reaction cancel each other out and only the absorbance changes during the reaction are selected (Fig. 4, inset). Now individual residues can be resolved. It is important to maintain accurately the same conditions during the reactions, otherwise the background, with a ~10^3 stronger absorbance, will obscure the difference spectrum. To monitor such small changes, highly sensitive instrumentation is required. FTIR spectroscopy is able to reliably detect such small changes due to the multiplex and the Jacquinot advantages, which lead to the crucial increase in the signal to noise ratio.

3.2 Equilibrium Techniques

Difference spectra can be obtained by time-resolved techniques (see below) or by shifting the equilibrium of a reaction by light, electrochemistry, a temperature-jump or any other controlled disturbance. In the investigations of protein–ligand interactions, a monolayer of an immobilized protein at an ATR crystal can be allowed to react with solutions containing different concentrations of a ligand. These measurements result in dissociation constants in the same way as SPR (surface plasmon resonance) measurements, but additionally provide the chemical information that can be obtained from a FTIR spectrum (25).

3.3 Trigger-Techniques for tr-FTIR

Since the absorbance changes are several orders smaller than the background absorbance of the protein, the sample has to be activated without removing it from the sample chamber. The time-resolved difference technique requires a sharp initiation (triggering) of the protein reaction. This can be achieved by photoexcitation or by fast mixing.

3.3.1 Photobiological Systems

Ideally suited to time-resolved study are light induced reactions in photobiological systems like bacteriorhodopsin (bR) (2, 26) and the photosynthetic reaction centers (RCs) (8), which carry intrinsic chromophores. In these systems, the chromophore can be directly activated by a laser flash, which induces isomerization or redox reactions of the prosthetic groups.

3.3.2 Caged Compounds

A much broader range of applications can be achieved using caged compounds, in which biologically active molecules are released from inactive photolabile precursors. This allows the initiation of a protein reaction with a nanosecond UV laser flash. Caged phosphate, caged GTP, caged ATP, and caged calcium have been established as particularly suitable trigger compounds (27–29). Further, production and use of caged glutamic acid (30) and caged protons (31)

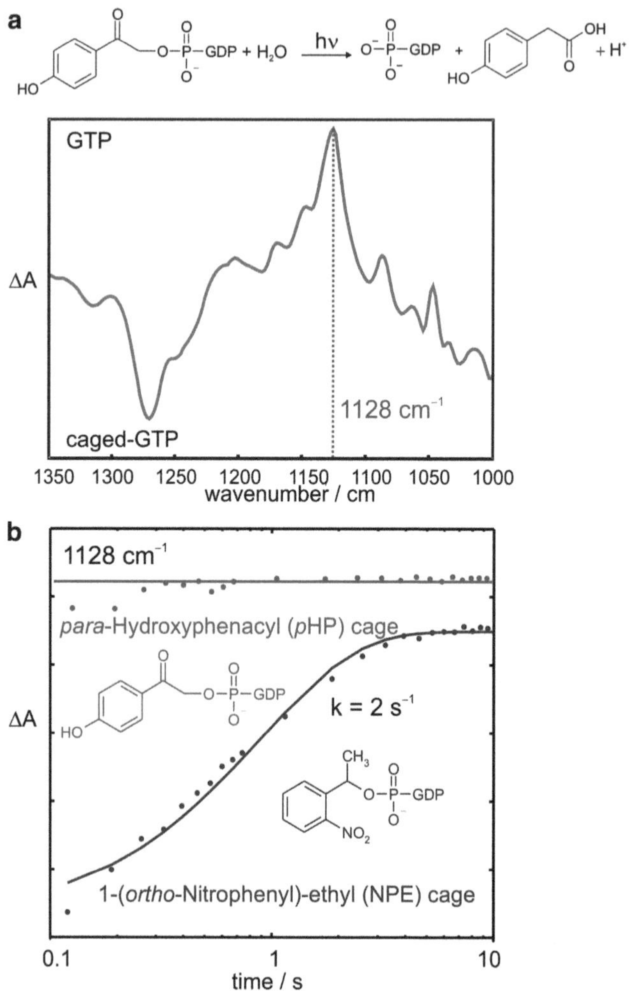

Fig. 5 (**a**) Photolysis difference spectra of pHP-caged GTP (10 mM caged-phosphate, 200 mM HEPES (pH 7.6), 20 mM $MgCl_2$). Bands facing upwards belong to the GTP. (**b**) Time dependent absorbance changes of the marker band for GTP in water at 1,128 cm^{-1}. The decaging of NPE-GTP gives an aci-nitro anion and GTP is formed with a rate of 2 s^{-1} at 260 K. The pHP cage cleaves at an excited state, producing GTP in 0.5 ns

have been reported. The most popular caging groups are *ortho*-alkylated nitrophenyl compounds (32); however, their photoreactions involve several intermediates, which limit the time resolution of the measurement of the protein reaction. The *para*-hydroxyphenacyl-cage is faster, as the photoreaction proceeds on the excited state in the subnanosecond time regime (Fig. 5) (33).

3.3.3 Caged GTP

The use of caged GTP is an instructive example of a time-resolved FT-IR study of caged protein ligands. A typical FT-IR difference spectrum of this photolysis reaction is shown in Fig. 5a (29).

A spectrum of the caged phosphate is measured prior to the photolysis as a reference, then after photolysis further spectra are recorded and the absorbance difference spectra, ΔA, are calculated. Only those vibrational modes that have undergone reaction-induced absorbance changes give rise to bands in the difference spectra. Negative bands in the difference spectrum are due to loss of the caged GTP, whereas positive bands are due to the photolysis products. The 1-(2-nitrophenyl) ethyl (NPE) moiety is frequently used to protect phosphate, nucleotides, and nucleotide analogs. The mechanism of photolysis of compounds containing the 2-nitrobenzyl group has been the topic of several investigations (32, 34). After photolysis, the caged compound decays to GTP and the by-product 2-nitrosoacetophenone. One can resolve an intermediate, the aci-nitro anion. At pH 6 and 260 K the formation of GTP is complete after 3,000 ms (Fig. 5b). A faster reaction of 300 ms is observed at pH 7.6. However, the best time resolution can be obtained with the *para*-hydroxyphenyl (pHP) caged group. Here, the cleavage takes place on an excited state and is complete within 0.5 ps (33). Again the photolysis reaction of caged GTP has been investigated in detail by FTIR spectroscopy, and the bands appearing in difference spectra have been assigned by the use of ^{18}O phosphate labeling (29).

3.3.4 Micro-Mixing-Cells

Since silicon is transparent in the midinfrared, micro-machined silicon components offer great potential for establishing FTIR spectroscopy as a method for studying microsecond mixing experiments. It has been shown that micro-scale-mixing devices can decrease the characteristic mixing time from milliseconds down to 10 μs. A continuous-flow-mixing chip (Fig. 6), has been designed for FTIR microscopy (35). The protein solution in the center and two streams of mixing buffer enter through 80 μm deep inlet channels, which intersect with the 8 μm deep observation channel. Because the three inlet channels are a factor of 10 deeper than the observation channel, at the merger there is almost equal pressure over the whole width of the observation channel. Due to the viscosity determined laminar flow, no turbulence is induced when the buffer channels merge, and a central layer of protein solution is formed between two buffer layers over the whole width of the observation channel; a fluid dynamics simulation was performed to verify this flow pattern. Because the protein layer is thin, diffusing reactant molecules (e.g., ligands) diffuse rapidly from the buffer solution into the protein solution. Thus, mixing is very fast. The time resolution is achieved by scanning along the observation channel with the focused beam of a FTIR microscope. This approach provides a time resolution of 400 μs, which is about 1,000 times faster than recent IR stopped-flow setups. Additionally, the miniaturization reduces the sample consumption by an even greater magnitude.

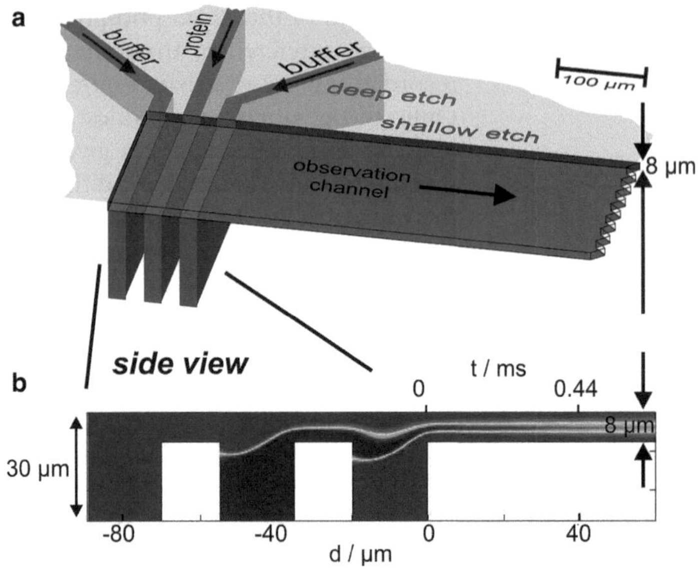

Fig. 6 Schematic of a microfabricated flow-cell (adapted from ref. 35). (**a**) *Top view* of the chip. The 80-μm-deep inlet channels (*top*) converge and combine in laminar manner in the 8-μm-deep outlet channel. (**b**) Two-dimensional fluid dynamics simulation with false color representation. A jet of the center (protein) solution between two layers of buffer solution is formed and buffer components rapidly diffuse into the protein solution

3.4 Rapid Scan

The principle of the rapid scan FT-IR mode of time-resolved spectroscopy is simple: after taking a reference spectrum of the protein in its ground state, one activates the protein (e.g., by a laser flash) and records interferograms in much shorter times than the half-lives of the reactions (26). The time pattern of such an experiment is shown schematically in Fig. 7a–c. The reference (R) interferograms represent the unreactive ground state (I). By a laser flash, the protein or the ligand becomes activated, e.g., by removing a cage group. Now the reaction can be monitored by taking interferograms during the complete reaction pathway (II to III to IV). Thus, the first interferogram after the flash will mainly represent the spectrum of state II and in the following the ratio of III will increase according to the kinetics of the reaction. Global exponential fitting or multivariate curve resolution (MCR) analysis (see below) enable the extraction of difference spectra for each reaction step (Fig. 7c). The Fourier transformation technique permits observation of processes whose half-lives are on the order of the scan time (*see* **Note 2**) for first-order reactions or even below. If the half-life of the observed process is shorter than the duration of the scan, the intensity of the interferogram is convolved with the absorption change of the sample. In case of first-order reactions, this leads to an interferogram that is convolved with an exponential

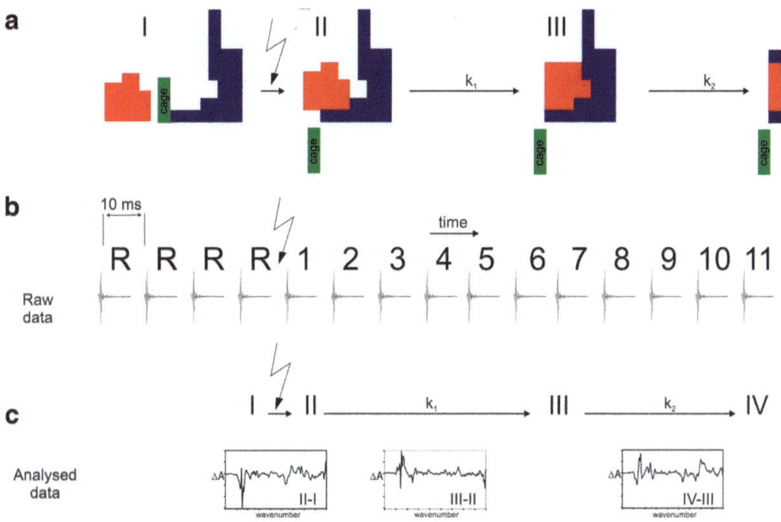

Fig. 7 (**a**) Time course for a rapid scan FTIR experiment. (I) The protein and caged ligand are in solution, (II) The laser flash removes the cage group (*green*), which allows for the interaction of the protein (*blue*) with the ligand (*red*). This leads to the formation of the final protein–ligand complex (IV) via the intermediate (III). (**b**) Time course of data acquisition: First, reference spectra (R) are taken. The laser flash initiates a reaction. During the reaction via III towards IV interferograms are taken. (**c**) Analysis of the time resolved data allows the calculation of difference spectra which show the bands of the groups involved in the subsequent reaction steps

function, which results only in Lorentzian line shape broadening in the spectrum after Fourier transformation.

3.5 Step Scan

In the step scan mode, the interferometer moving mirror may be visualized as being held stationary at the interferogram data position x_n (Fig. 8); the protein activity is initiated, for example, by a laser flash and the time dependence of the intensity change at this interferogram position x_n is measured. Then the interferometer “steps” to the next interferogram data position x_{n+1} and the reaction is repeated and measured again. This process is continued at each sampling position of the interferogram. The position of the interferometer mirror must be kept accurate to about 1–2 nm at each x_n while the intensity change of the interferogram during the time of the reaction is measured. Therefore the method is very sensitive to external disturbances (e.g., sound). The time resolution is usually determined by the response time of the detector, which is a few nanoseconds. After the measurement the data is rearranged to yield time-dependent interferograms $I(t_i)$. Using intense pulsed IR sources instead of the conventional Globar the time resolution is determined by the time duration of the probe pulse. This can give in principle femtosecond resolution with broad-band femtosecond lasers or synchrotron radiation. For more details on the step scan technique see the literature (36–40).

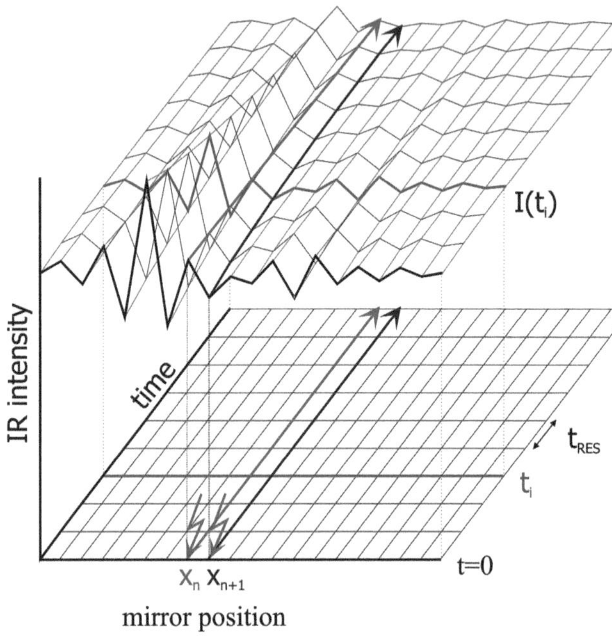

Fig. 8 Principle of the step scan technique. The interferometer mirror is stepped to a sampling position x_n, the reaction is then initiated and the time dependence of the IR intensity is measured (in direction of *arrow*). The detector limits the time resolution. After relaxation the interferometer is stepped to the next position, x_{n+1}, and the data recording process is repeated. After the measurement at all interferometer positions the data is rearranged to yield a set of interferograms $I(t_i)$, which after Fourier transformation yield the spectra at each time point

A typical step scan experimental setup is shown schematically in Fig. 9 (40). Except for the sample chamber, the FT-IR apparatus is evacuated below 3 mbar during the measurement. This increases the stability and reduces the sound-sensitivity of the movable mirror. Furthermore, the whole setup is positioned on a vibration isolation table within a temperature-controlled laboratory. With such a setup, we have determined the residual spatial fluctuations of the movable mirror to be ~0.5 nm. The sample chamber is purged with dry air (dew point −70 °C). The IR absorbance changes are detected by a photovoltaic HgCdTe detector. The detector's signal is amplified in a home-built two-stage preamplifier with an AC- and a DC-coupled output. The bandwidth of the DC part is limited to 400 kHz, whereas the bandwidth of the AC part is 200 MHz. Controls ensure that the output-signal of the preamplifier depends linearly on the IR intensity. The DC-coupled output of the preamplifier is digitized by a 18-bit, 500 kHz transient recorder connected to a PC. The offset of the input signal can be compensated to zero. This allows subsequent amplification of the signal and use of the full dynamic range of the transient recorder. In order to

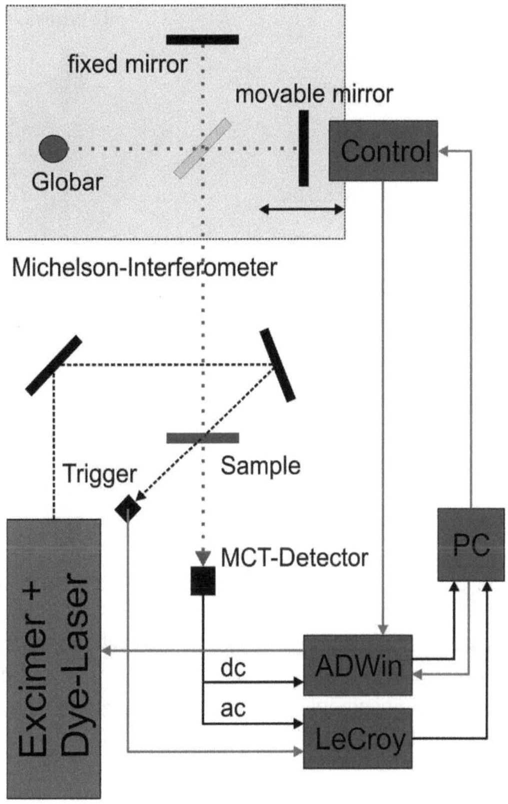

Fig. 9 A typical step scan setup. Synchronization connections for the measurement between the Excimer pumped dye laser, the control of the movable mirror, the slow AD conversion by the ADWin card, the fast AD conversion of the LeCroy digital oscilloscope are shown in *grey*. Data wires are shown in *black*, the excitation light in *dashed line*, and the infrared light in *dotted lines*

reduce the huge amount of data that would be generated in the nanosecond to the millisecond time domain, data is time-averaged at longer times. The AC-coupled output of the preamplifier is recorded by a digital oscilloscope.

At every sampling position of the interferogram the correct positioning of the movable mirror is checked before data acquisition starts. A transistor-to-transistor (TTL) output-signal is used to trigger the excimer-laser to initiate the reaction. A fast photodiode (rise-time 10 ns) is triggered by the dye-laser flash and starts data acquisition.

The spectral range can be limited below for example 1,975 cm^{-1} by a low pass filter to reduce the number of sampling points of the interferogram. This filter also shields the IR detector from both scattered light of the dye laser and the heat emitted by the sample (the dye laser's pulse causes a small warming of the sample). The resulting interferogram contains 780 data points. It is multiplied by the Norton–Beer-weak apodization function and

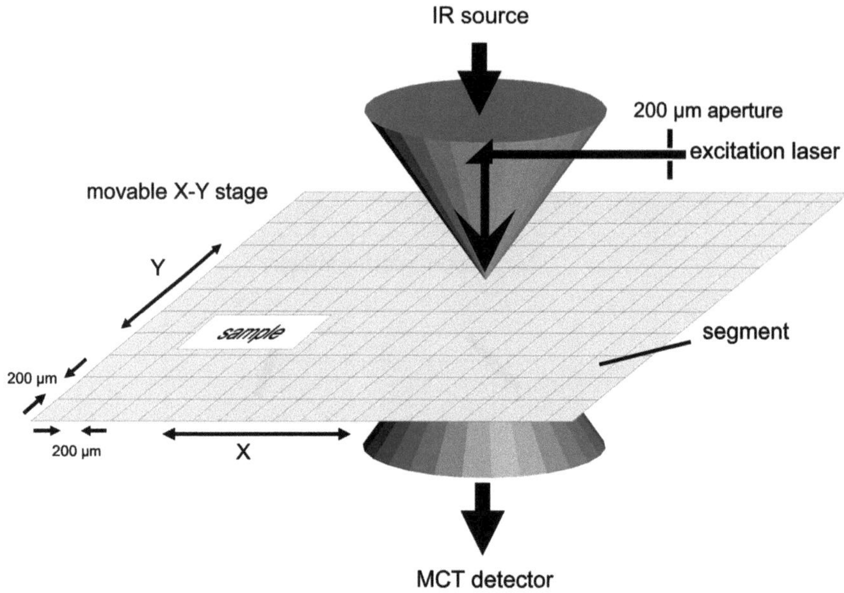

Fig. 10 Representation of a step scan setup for noncyclic reactions. Both, IR-beam and excitation laser are focused to a diameter of 200 μm. The sample is mounted on a motorized *x–y* stage and divided into several thousand 200 × 200 μm segments. The sample is renewed by consecutively moving the next segment into the focused spot (patent DP 19804279)

then zero-filled by a factor of 2. The phase-spectrum $\varphi(\nu)$ is calculated with a resolution of 50 cm^{-1}, whereas the difference spectra are recorded with a spectral resolution of approximately 3 cm^{-1} (*see* **Note 3** for further details).

3.5.1 Step Scan FT-IR of Noncyclic Reactions

The step scan technique cannot be applied directly to noncyclic reactions, because the investigated process has to be repeatedly initiated, typically at ~1,000 sampling positions to create the interferogram. Consequently, to investigate irreversible systems the sample has to be renewed at every sampling position; historically this has only been practicable in a flow-cell. We need to use 4 μm thin film cells to depress the water background absorption sufficiently to perform difference spectroscopy of hydrated biological samples.

In a novel approach, the IR beam and the excitation laser-beam are focused to a very small diameter of 200 μm (Fig. 10). Thereby, only a small segment of the sample, which has a diameter of 15 mm, is excited and probed. By moving the sample, which is mounted on a movable *x–y* stage, so that different nonexcited segments are at the laser focus each time the reaction can be repeated until a complete interferogram data set has been recorded. The technique was successfully applied to the noncyclic reaction of the photolysis of caged ATP (41). By this technique, the transiently formed aci-nitro anion complex is also measureable. This successful demonstration of the study of an irreversible reaction with 20 μs time-resolution now opens the way for many new applications of step scan FT-IR measurements to noncyclic reactions. Continuous flow through chips (35) can also be used to rapidly renew reactive sample at the detection site and thus enable step scan measurements of noncyclic reactions (42).

3.6 Global Fit

For time-resolved data, adequate kinetic analysis is important. A so-called "global-fit" analysis yields the apparent rate constants of the analyzed processes (43). The global-fit analysis does not only fit the absorbance change at a specific wavenumber, but fits the changes across the complete spectrum simultaneously. All reactions are assumed to be first order and can therefore be described by a sum of exponentials. The absorbance changes ΔA as a function of frequency/wavenumber and time (t) are modeled as sums of n_r exponentials with apparent rate constants k_l and amplitudes a_l:

$$\Delta A(\nu, t) = \sum_{l=1}^{n_r} a_l(\nu)\mathrm{e}^{-k_l t} + a_\infty(\nu). \tag{2}$$

The fit procedure minimizes the difference between the measured data $\Delta A_{\text{measured}}$ and the theoretical description ΔA, weighted according to the noise w_{ij} at the respective wavenumbers (i). In this analysis, the weighted sum of squared differences f between the fit with n_r apparent rate constants k_l and data points at n_w measured wavenumbers ν_i and n_t times t_j is minimized:

$$f = \sum_{i=1}^{n_w}\sum_{j=1}^{n_t}(w_{ij})^2(\Delta A_{\text{measured}}(\nu_i, t_j) - \sum_{l=1}^{n_r} a_l(\nu_i)\mathrm{e}^{-k_l t_j} - a_\infty(\nu_i))^2. \tag{3}$$

For unidirectional forward reactions the determined apparent rate constants are directly related to the respective intrinsic rate constants describing the respective reaction steps (44, 45). An example for a kinetic fit at a single wavenumber is shown in Fig. 11. The result of the global fit can be found in Fig. 1a. If in addition to the forward reaction, significant back-reactions occur, the analysis becomes more complicated (46). Then the reaction has to be modeled by guessing the intrinsic rate constants until they fulfill the experimentally observed time-course described by the apparent rate constants. Because the number of the intrinsic rate constants in the model is larger than the number of the experimentally observed apparent rate constants, the problem is experimentally underdetermined and the solution is not unequivocal.

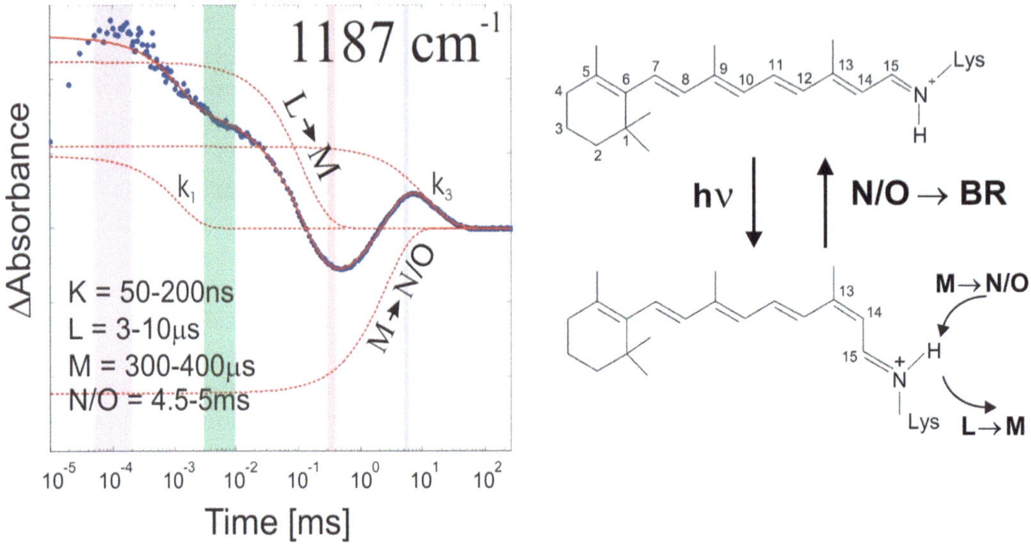

Fig. 11 Time dependent absorbance changes of the marker band for the C14–C15 stretching vibration of retinal. Due to the photoexcitation 13-*cis* retinal is formed. Deprotonation of the PSB (L → M), reprotonation (M → N/O), and reisomerization (N/O → BR) can be monitored. The *solid line* shows a kinetic fit by four exponential functions. The *dotted lines* represent the amplitudes of the four apparent reaction steps. Note that the N → O reaction cannot be resolved, because it is faster than the preceding M → N reaction. Difference spectra of almost pure intermediates can be obtained by averaging the spectra measured during each of the intervals shown as *vertical shaded bands*

An alternative method of analysis, is multivariate curve resolution (MCR) (47). Thereby, the difference spectra of the pure intermediates and their concentration profiles are calculated from all difference spectra measured. This procedure allows the determination of transient spectra independent of specific kinetic models and independent of the temporal overlap (43).

3.7 Band Assignments

In order to derive information on the mechanism of a protein–ligand interaction, the infrared bands have to be assigned to individual groups of the protein or the ligand. The frequency range in which a band appears allows a tentative assignment. For example, the retinal vibrations are expected in the fingerprint region between 1,300 and 1,100 cm^{-1}, the carbonyl vibrations of aspartic or glutamic acids between 1,700 and 1,770 cm^{-1}.

Unambiguous band assignment can be performed by using isotopically labelled proteins or by amino acid exchange via site directed mutagenesis. Isotopic labeling shifts the stretching frequency, ν, of the labelled group due to the change in mass (m) of the atoms

$$\nu = \frac{1}{2\pi}\sqrt{\frac{k}{\mu}}, \quad (4)$$

where k is the force constant, and the reduced mass $\mu = (m_1 \times m_2)/(m_1 + m_2)$. Isotopic labeling can be performed on prosthetic groups like retinal (48) or nucleotides like GTP (29, 49, 50) by chemical synthesis. As an example of the use of site-specific isotopic labeling, we consider again caged GTP (see Fig. 13). Isotopic labeling of all amino acids of one kind can be achieved by biosynthetic incorporation of isotopically labelled amino acids (51, 52). Further, site-directed exchange of an amino acid by mutagenesis eliminates the absorption band of the exchanged group. The principle of band assignment by site directed mutagenesis is demonstrated in Fig. 12a, while Fig. 12b shows the difference spectra, between ground-state BR and the N-intermediate state of its photocycle, for the wild type (wt) and two asparagine mutants as an example of a real measurement (26). Absorbance changes in the spectral range between 1,500 and 1,000 cm^{-1} are highly reproducible for wild-type and mutant proteins, indicating that these specific mutants are noninvasive and do not disturb the protein structure. The carbonyl band at 1,742 cm^{-1} is absent in the spectrum of the Asp96Asn mutant and can be assigned to protonated Asp96. Since the band is negative, Asp96 is protonated in the ground state but deprotonated in the N intermediate. On the other hand, the double band around 1,735 cm^{-1} vanishes in the Asp115Asn mutant and identifies both bands as originating from Asp115 and implies that Asp115 undergoes environmental change between the two states.

Mutation of an amino acid is easy to achieve by site-directed mutagenesis, a standard molecular biology method (53), but changes the structure of the protein to a greater or lesser degree. Isotopic labeling has the advantage of marking the molecular group without perturbation. An example for band assignment by isotopic labeling of a ligand is given in Fig. 13. Here $\gamma^{18}O_3$-labelled GTP is used in the Ras catalyzed hydrolysis reaction of GTP (14). In the difference spectrum (Δ) of the hydrolysis the negative bands correspond to the Ras-GTP state. The band at 1,143 cm^{-1} is shifted by the γ-label and can thus be assigned to an absorption of the γ-GTP group. On the other hand the α-band at 1,263 cm^{-1} (assigned by α-labelled GTP, not shown) remains in the same position after γ-labeling. Similarly one can assign positive bands (product state) at 1,078 and 992 cm^{-1} as absorptions of the appearing P_i after hydrolysis. Often, a double difference spectrum ($\Delta\Delta$), the difference of the labelled and the unlabelled difference spectra, gives distinct band positions, because only absorptions which shift due to the labeling appear.

Whereas many ligands can be synthesized with site specific isotope labels, and common cofactors are commercially available, site-directed isotopic labeling amino acids is an expensive molecular biology method and only very rarely successfully applied. Today chemical synthesis of a protein allows site-directed isotopic labeling

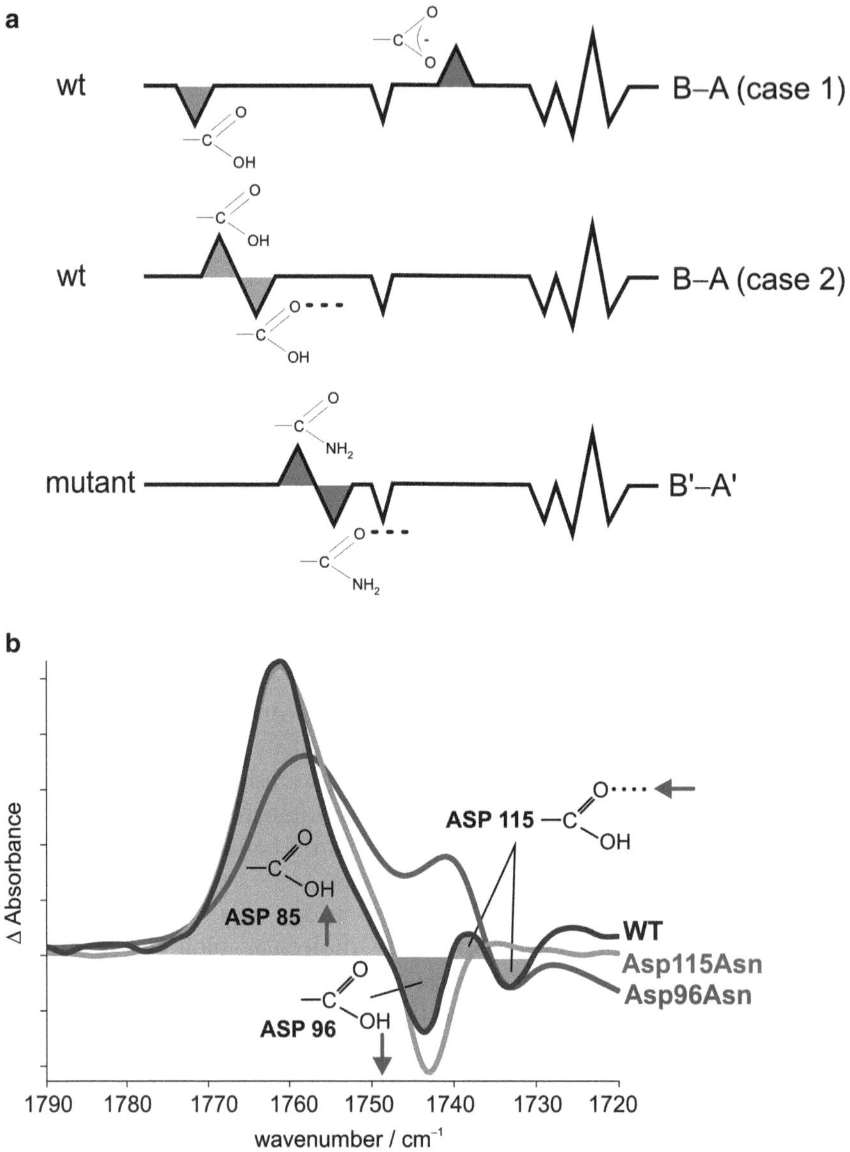

Fig. 12 (**a**) Schematic representation of the expected absorbance changes in the IR difference spectra. If in the transition to an intermediate a carboxylic acid is deprotonated (case 1) a negative carbonyl band and a positive carboxylate band should appear in the difference spectrum. If a hydrogen bond of a protonated carboxylic acid is broken (case 2), a difference band is expected due to the frequency upshift of the carbonyl vibration. If the amino acid giving rise to the absorbance changes is exchanged and the respective mutant protein is measured, these carbonyl bands should disappear as compared to the wt difference spectrum. In addition, in the mutant difference spectrum, a new carbonyl band might appear for Asp (Glu) to Asn (Gln) mutation. It is important to notice that all other bands remain the same (as indicated), showing that most of the structure is not affected and that the mutation is noninvasive. (**b**) Spectra, demonstrating the principle of band assignment by site-directed mutagenesis. Spectra show the difference between the BR ground state and the intermediate N state of wild-type and mutant bacteriorhodopsin. Bands of the N state are facing *upwards*, and vanishing bands of the ground state are facing *downwards*. The spectrum of wt shows by a negative band the deprotonation of Asp96 (*shaded*). In the corresponding spectrum of the mutant Asp96Asn, this band is absent. Additionally, the difference bands of Asp115 indicating a hydrogen bond change between the two states are now seen more clearly. This band vanishes in the Asp115Asn mutant

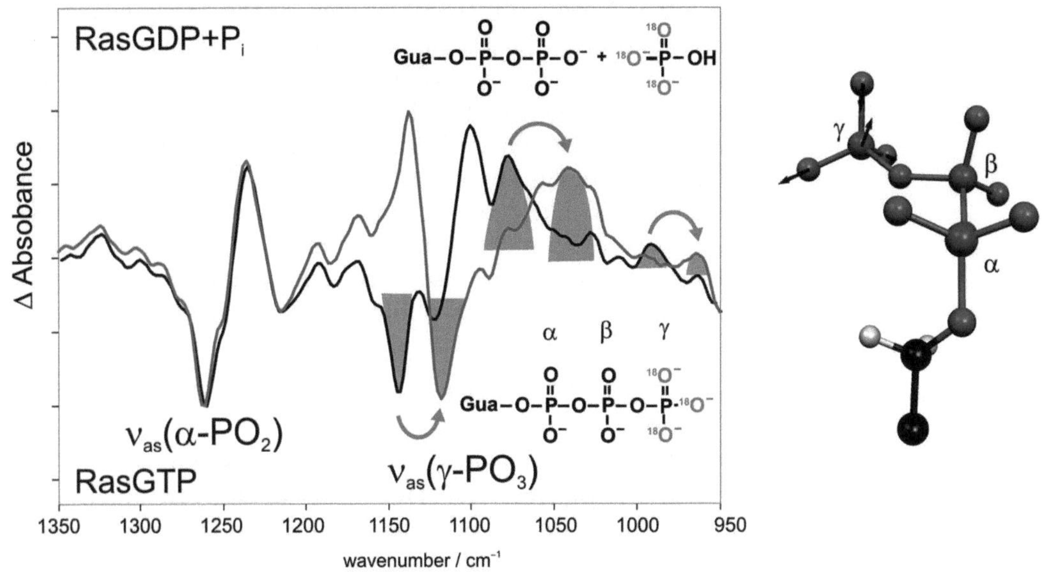

Fig. 13 Spectra, demonstrating the principle of band assignment by isotopic labeling. In *black* the difference spectrum of RasGTP hydrolysed to RasGDP is shown, in *grey* the same spectrum with $\gamma^{18}O_3$-labelled GTP. *Arrows* indicate the changes upon isotope labeling. Due to the labeling the band at 1,143 cm^{-1} is shifted and can be assigned to an γ-vibration, whereas the band at 1,263 cm^{-1} (the α-band) is not affected. Right, the calculated normal modes of the ν_{as}(γ-GTP) vibration are shown as *arrows* on the molecular structure

and may become the method of the future (54–56). Another method for site-directed modification is the use of nonnatural amino acids in the amber codon suppression technology (57).

4 Notes

1. *General setup.* For FTIR measurements numerous types of window material are available. Table 1 (see Subheading 2) gives a short list of the most frequently used materials for FTIR measurements of hydrated protein samples. Most frequently used is CaF_2, because it is completely water insoluble and transparent in the mid-IR and UV region. However, for investigation at lower wavelength than 950 cm^{-1} other materials are necessary (Table 1).

   A table that includes further materials is found at www.korth.de.

2. *Rapid Scan.* The velocity of the interferometer moving mirror, V_{max}, and the desired spectral resolution, $\Delta\nu$, determine the scan duration Δt and thereby the time resolution. Today, the fastest commercially available FTIR spectrometers are capable of yielding a time resolution of 10 ms at 4 cm^{-1} spectral resolution. Improvements in current interferometer designs that yield a significant increase in mirror scan velocity are unlikely to occur for interferometers with a reciprocating

motion. There are practical reasons for this, e.g., the extreme acceleration of the scanner at the turning points. If a rapidly rotating, rather than a reciprocating, mirror is used, the scan speed can be increased to the point that the time resolution for a 4 cm^{-1} resolution can be decreased to 1 ms. Without a radical change of design, however, the time resolution for the rapid scan is still likely to be limited to the millisecond range. A typical experimental setup for rapid scan measurements is illustrated in Fig. 2. A laser flash activates the sample. Simultaneously, a conventional photolysis setup measures the absorbance change in the visible region.

3. *Step Scan*. The time course of a step scan experiment is the following:
   (a) The mirror moves to the first acquisition point of the interferogram. The DC value of the IR intensity is measured. Afterwards the offset of the DC signal is set to zero.
   (b) A laser flash starts the reaction. Both transient recorders simultaneously measure the time-dependent IR intensity changes at the sampling position. The 200 MHz transient recorder measures the time domain from 30 ns to 20 μs, and a digital oscilloscope monitors from 5 μs until the end of the reaction in the millisecond time range. At each sampling position of the interferogram the reaction is repeated several times to improve the signal-to-noise. After measuring the time traces of all interferogram sampling positions the data are rearranged to yield time dependent difference interferograms $\Delta I(t_j)$. Because these difference interferograms contain positive as well as negative spectral features, usual Mertz phase correction cannot be directly applied. Therefore the stored phase $\varphi(\nu)$ from the first measurement in step (a) of Note 3 is used. The phase does not change between both measurements, because the movable mirror stops exactly at the same sampling points and only small absorbance changes take place. Possible errors due to transient heating of the sample by the laser flash, baseline distortions and nonlinearity of the IR detectors are discussed in detail in Rammelsberg et al. (40). For details on Fourier transformation procedures see e.g., ref. 59.

References

1. Gerwert K (1993) Molecular reaction mechanisms of proteins as monitored by time-resolved FTIR spectroscopy. Curr Opin Struct Biol 3:769–773
2. Gerwert K, Hess B, Soppa J, Oesterhelt D (1989) Role of aspartate-96 in proton translocation by bacteriorhodopsin. Proc Natl Acad Sci U S A 86:4943–4947

3. Gerwert K, Siebert F (1986) Evidence for light-induced 13-cis, 14-s-cis, isomerization in bacteriorhodopsin obtained by FTIR difference spectroscopy using isotopically labelled retinals. EMBO J 5:805–811
4. Gerwert K, Souvignier G, Hess B (1990) Simultaneous monitoring of light-induced changes in protein side-group protonation, chromophore isomerization, and backbone motion of bacteriorhodopsin by time-resolved Fourier-transform infrared spectroscopy. Proc Natl Acad Sci U S A 87:9774–9778
5. Garczarek F, Gerwert K (2006) Functional waters in intraprotein proton transfer monitored by FTIR difference spectroscopy. Nature 439:109–112
6. Wolf S, Freier E, Potschies M, Hofmann E, Gerwert K (2010) Directional proton transfer in membrane proteins achieved through protonated protein-bound water molecules: a proton diode. Angew Chem Int Ed 49:6889–6893
7. Brudler R, Rammelsberg R, Woo TT, Getzoff ED, Gerwert K (2001) Structure of the I1 early intermediate of photoactive yellow protein by FTIR spectroscopy. Nat Struct Biol 8:265–270
8. Remy A, Gerwert K (2003) Coupling of light-induced electron transfer to proton uptake in photosynthesis. Nat Struct Biol 10:637–644
9. Onidas D, Stachnik JM, Brucker S, Krätzig S, Gerwert K (2010) Coupling of light-induced electron transfer to proton-uptake in photosynthesis. Eur J Cell Biol 89:983–989
10. Lübben M, Gerwert K (1996) Redox FTIR difference spectroscopy using caged electrons reveals contributions of carboxyl groups to the catalytic mechanism of heme-copper oxidases. FEBS Lett 397:303–307
11. Kötting C, Kallenbach A, Suveyzdis Y, Eichholz C, Gerwert K (2007) Surface change of Ras enabling effector binding monitored in real time at atomic resolution. Chembiochem 8:781–787
12. Völlmecke C, Kötting C, Gerwert K, Lübben M (2009) Spectroscopic investigation of the reaction mechanism of CopB-B, the catalytic fragment from an archaeal thermophilic ATP-driven heavy metal transporter. FEBS J 276:6172–6186
13. Kötting C, Gerwert K (2004) Time-resolved FTIR studies provide activation free energy, activation enthalpy and activation entropy for GTPase reactions. Chem Phys 307:227–232
14. Kötting C, Blessenohl M, Suveyzdis Y, Goody R, Wittinghofer A, Gerwert K (2006) A phosphoryl transfer intermediate in the GTPase reaction of Ras in complex with its GTPase-activating protein. Proc Natl Acad Sci U S A 103:13911–13916
15. Kötting C, Kallenbach A, Suveyzdis Y, Wittinghofer A, Gerwert K (2008) The GAP arginine finger movement into the catalytic site of Ras increases the activation entropy. Proc Natl Acad Sci U S A 105:6260–6265
16. Kötting C, Suveyzdis Y, Bojja RS, Metzler-Nolte N, Gerwert K (2010) Label-free screening of drug-protein interactions by time-resolved Fourier transform infrared spectroscopic assays exemplified by Ras interactions. Appl Spectrosc 64:967–972
17. Fringeli UP, Baurecht D, Siam M, Reiter G, Schwarzott M, Burgi T, Bruesch P (2002) ATR spectroscopy of thin films. Handbook of Thin Film Materials 2:191–229
18. Harrick NJ (1987) Nanosampling via internal reflection spectroscopy. Appl Spectrosc 41:1–2
19. Ataka K, Heberle J (2007) Biochemical applications of surface-enhanced infrared absorption spectroscopy. Anal Bioanal Chem 388:47–54
20. Ataka K, Kottke T, Heberle J (2010) Thinner, smaller, faster: IR techniques to probe the functionality of biological and biomimetic systems. Angew Chem Int Ed 49:5416–5424
21. Osawa M (2001) Surface-enhanced infrared absorption. Top Appl Phys 81:163–187
22. Elfrink K, Ollesch J, Stohr J, Willbold D, Riesner D, Gerwert K (2008) Structural changes of membrane-anchored native PrPC. Proc Natl Acad Sci U S A 105:10815–10819
23. Güldenhaupt J, Adigüzel Y, Kuhlmann J, Waldmann H, Kötting C, Gerwert K (2008) Secondary structure of lipidated Ras bound to a lipid bilayer. FEBS J 275:5910–5918
24. Goormaghtigh E, Gasper R, Benard A, Goldsztein A, Raussens V (2009) Protein secondary structure content in solution, films and tissues: redundancy and complementarity of the information content in circular dichroism, transmission and ATR FTIR spectra. Biochim Biophys Acta 1794:1332–1343
25. Kötting C, Güldenhaupt J, Gerwert K (2012) Time-resolved FTIR spectroscopy for monitoring protein dynamics exemplified by functional studies of Ras protein bound to a lipid bilayer. Chem Phys 396:72–83
26. Gerwert K (1988) Intramolecular protein dynamics study with time-resolved Fourier-transform IR-difference spectroscopy. Berichte der Bunsen-Gesellschaft 92:978–982
27. Pelliccioli AP, Wirz J (2002) Photoremovable protecting groups: reaction mechanisms and applications. Photochem Photobiol Sci 1:441–458

28. McCray JA, Trentham DR (1989) Properties and uses of photoreactive caged compounds. Annu Rev Biophys Biophys Chem 18:239–270
29. Cepus V, Ulbrich C, Allin C, Troullier A, Gerwert K (1998) Fourier transform infrared photolysis studies of caged compounds. Methods Enzymol 291:223–245
30. Cheng Q, Steinmetz MG, Jayaraman V (2002) Photolysis of gamma-(alpha-carboxy-2-nitrobenzyl)-L-glutamic acid investigated in the microsecond time scale by time-resolved FTIR. J Am Chem Soc 124:7676–7677
31. Barth A, Corrie JET (2002) Characterization of a new caged proton capable of inducing large pH jumps. Biophys J 83:2864–2871
32. Walker J, Reid GP, McCray JA, Trentham DR (1988) Photolabile 1-(2-nitrophenyl)ethyl phosphate esters of adenine nucleotide analogs. Synthesis and mechanism of photolysis. J Am Chem Soc 110:7170–7177
33. Park C-H, Givens RS (1997) New photoactivated protecting groups. 6. p-Hydroxyphenacyl: a phototrigger for chemical and biochemical probes. J Am Chem Soc 119:2453–2463
34. Corrie JET, Trentham DR (1993) Caged nucleotides and neurotransmitters. Bioorg Photochem 2:243–305, 1plate
35. Kauffmann E, Darnton NC, Austin RH, Batt C, Gerwert K (2001) Lifetimes of intermediates in the beta-sheet to alpha-helix transition of beta-lactoglobulin by using a diffusional IR mixer. Proc Natl Acad Sci U S A 98:6646–6649
36. Palmer RA, Chao JL, Dittmar RM, Gregoriou VG, Plunkett SE (1993) Investigation of time-dependent phenomena by use of step-scan Ft-Ir. Appl Spectrosc 47:1297–1310
37. Palmer RA, Manning CJ, Chao JL, Noda I, Dowrey AE, Marcott C (1991) Application of step-scan interferometry to two-dimensional Fourier transform infrared (2D FT-IR) correlation spectroscopy. Appl Spectrosc 45:12–17
38. Weidlich O, Siebert F (1993) Time-resolved step-scan Ft-Ir investigations of the transition from Kl to L in the bacteriorhodopsin photocycle—identification of chromophore twists by assigning hydrogen-out-of-plane (hoop) bending vibrations. Appl Spectrosc 47:1394–1400
39. Uhmann W, Becker A, Taran C, Siebert F (1991) Time-resolved FT-IR absorption spectroscopy using a step-scan interferometer. Appl Spectrosc 45:390–397
40. Hessling B, Herbst J, Rammelsberg R, Gerwert K (1997) Fourier transform infrared double-flash experiments resolve bacteriorhodopsin's M-1 to M-2 transition. Biophys J 73:2071–2080
41. Rammelsberg R, Huhn G, Lubben M, Gerwert K (1998) Bacteriorhodopsin's intramolecular proton-release pathway consists of a hydrogen-bonded network. Biochemistry 37:5001–5009
42. Schleeger M, Wagner C, Vellekoop MJ, Lendl B, Heberle J (2009) Time-resolved flow-flash FT-IR difference spectroscopy: the kinetics of CO photodissociation from myoglobin revisited. Anal Bioanal Chem 394:1869–1877
43. Hessling B, Souvignier G, Gerwert K (1993) A model-independent approach to assigning bacteriorhodopsin's intramolecular reactions to photocycle intermediates. Biophys J 65:1929–1941
44. Cantor CR, Schimmel PR (1980) Biophysical chemistry, Pt. 1: the conformation of biological macromolecules. W. H. Freeman, New York
45. Fersht A (1983) Enzymes: structures and reaction mechanisms. W. H. Freeman, New York
46. Steinfeld JI, Francisco JS, Hase WL (1999) Chemical kinetics dynamics, 2nd edn. Prentice Hall, Upper Saddle River, NJ
47. Blanchet L, Ruckebusch C, Mezzetti A, Huvenne JP, de Juan A (2009) Monitoring and interpretation of photoinduced biochemical processes by rapid-scan FTIR difference spectroscopy and hybrid hard and soft modeling. J Phys Chem B 113:6031–6040
48. Lugtenburg J, Mathies RA, Griffin RG, Herzfeld J (1988) Structure and function of rhodopsins from solid state NMR and resonance Raman spectroscopy of isotopic retinal derivatives. Trends Biochem Sci 13:388–393
49. Allin C, Ahmadian MR, Wittinghofer A, Gerwert K (2001) Monitoring the GAP catalyzed H-Ras GTPase reaction at atomic resolution in real time. Proc Natl Acad Sci U S A 98:7754–7759
50. Allin C, Gerwert K (2001) Ras catalyzes GTP hydrolysis by shifting negative charges from gamma- to beta-phosphate as revealed by time-resolved FTIR difference spectroscopy. Biochemistry 40:3037–3046
51. Engelhard M, Gerwert K, Hess B, Siebert F (1985) Light-driven protonation changes of internal aspartic acids of bacteriorhodopsin: an investigation of static and time-resolved infrared difference spectroscopy using [4-13C]aspartic acid labeled purple membrane. Biochemistry 24:400–407
52. Warscheid B, Brucker S, Kallenbach A, Meyer HE, Gerwert K, Kötting C (2008) Systematic approach to group-specific isotopic labeling of proteins for vibrational spectroscopy. Vib Spectrosc 48:28–36

53. Stryer L (1995) Biochemistry, 4th edn. W.H. Freeman, New York
54. Fischer WB, Sonar S, Marti T, Khorana HG, Rothschild KJ (1994) Detection of a water molecule in the active-site of bacteriorhodopsin - hydrogen-bonding changes during the primary photoreaction. Biochemistry 33:12757–12762
55. Becker CFW, Hunter CL, Seidel R, Kent SBH, Goody RS, Engelhard M (2003) Total chemical synthesis of a functional interacting protein pair: the protooncogene H-Ras and the Ras-binding domain of its effector c-Raf1. Proc Natl Acad Sci U S A 100:5075–5080
56. Tremmel S, Beyermann M, Oschkinat H, Bienert M, Naumann D, Fabian H (2005) C-13-labeled tyrosine residues as local IR probes for monitoring conformational changes in peptides and proteins. Angew Chem Int Ed 44:4631–4635
57. Ye SX, Zaitseva E, Caltabiano G, Schertler GFX, Sakmar TP, Deupi X, Vogel R (2010) Tracking G-protein-coupled receptor activation using genetically encoded infrared probes. Nature 464:1386
58. Kolano C (2003). PhD thesis, Ruhr-Universität Bochum
59. Griffiths PR, de Haseth JA (2007) Fourier transform infrared spectrometry, 2nd edn. Wiley, Hoboken, NJ

Part IV

Ligand Discovery

Chapter 12

Biophysical Methods in Drug Discovery from Small Molecule to Pharmaceutical

Geoffrey Holdgate, Stefan Geschwindner, Alex Breeze, Gareth Davies, Nicola Colclough, David Temesi, and Lara Ward

Abstract

Biophysical methods have become established in many areas of drug discovery. Application of these methods was once restricted to a relatively small number of scientists using specialized, low throughput technologies and methods. Now, automated high-throughput instruments are to be found in a growing number of laboratories. Many biophysical methods are capable of measuring the equilibrium binding constants between pairs of molecules crucial for molecular recognition processes, encompassing protein–protein, protein–small molecule, and protein–nucleic acid interactions, and several can be used to measure the kinetic or thermodynamic components controlling these biological processes. For a full characterization of a binding process, determinations of stoichiometry, binding mode, and any conformational changes associated with such interactions are also required. The suite of biophysical methods that are now available represents a powerful toolbox of techniques which can effectively deliver this full characterization.

The aim of this chapter is to provide the reader with an overview of the drug discovery process and how biophysical methods, such as surface plasmon resonance (SPR), isothermal titration calorimetry (ITC), nuclear magnetic resonance, mass spectrometry (MS), and thermal unfolding methods can answer specific questions in order to influence project progression and outcomes. The selection of these examples is based upon the experiences of the authors at AstraZeneca, and relevant approaches are highlighted where they have utility in a particular drug discovery scenario.

Key words Orthogonal screening, SAR by NMR, Calorimetry, Surface plasmon resonance, Thermal shift, Thermodynamics, Kinetics, Plasma protein binding

1 Introduction

Drug discovery is a long, complex and expensive process. It takes a multidisciplinary approach encompassing basic and clinical sciences, up to 15 years and around $1.5 billion to bring a new drug to market. Biophysical methods have an important part to play in this process, being employed at several stages of the Drug Program Operating Model (Fig. 1). However, most biophysical

Mark A. Williams and Tina Daviter (eds.), *Protein-Ligand Interactions: Methods and Applications*, Methods in Molecular Biology, vol. 1008, DOI 10.1007/978-1-62703-398-5_12, © Springer Science+Business Media New York 2013

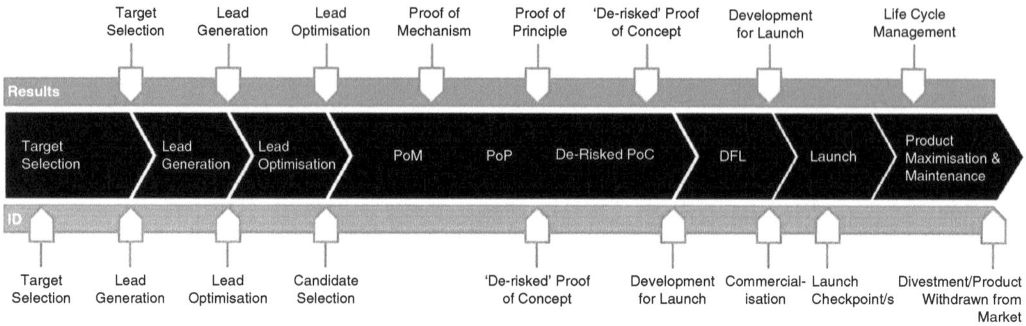

Fig. 1 Schematic representation of the Drug Program Operating Model. The drug discovery and early development process is divided into several stages, separated by investment decisions (ID) concerning whether or not to progress. Moving from one phase to the next depends upon meeting specific criteria, agreed at the start of the project. Decisions to proceed or stop are taken based upon the combination of results obtained and the resource required by the particular investment decision. In AstraZeneca, the Innovative Medicines unit is usually accountable for the drug-hunting project until a de-risked proof of concept (PoC) has been established, following which responsibility passes to the Global Medicines Development unit. Outcomes of biophysical analysis are significant in all decisions up to the de-risked PoC stage

input is made in the early stages of drug discovery, here a variety of biophysical approaches are employed in the discovery and creation of new lead compounds which will subsequently be optimized into potential candidate drugs (CDs). This chapter provides an overview of drug discovery and the biophysical methods exploited at each of the early stages of target selection and validation, and lead discovery and optimization through to proof of mechanism.

2 Discovery and Optimization

2.1 Target Selection

Of all the causes of attrition in clinical development, lack of efficacy now dominates, and can be broken down into two main components: poor molecule and poor target (1). It is essential to identify the right target in order to appropriately modify a particular disease condition, and the cost of prosecuting the "wrong" target can be enormous. Unfortunately, target identification is a challenging activity influenced by a huge number of variables.

Assuming an established clinical link with disease (by no means always the case with novel targets, at least at the outset), it is then important to understand how the target protein functions in both normal physiology and disease pathology, and secondly, whether it is capable of binding drug-like small molecules tightly enough for its function to be modulated at physiologically achievable drug concentrations. Together with other approaches, biophysical tools can be used to build up a picture that answers both questions and starts to establish some boundaries for the eventual pharmacodynamic profile of any drug directed at that target.

Information on target enzyme kinetics, concerning the identity and abundance of enzyme forms populated during the catalytic cycle, can be of considerable value in both assessing druggability in the disease tissue setting and in influencing the mechanism by which one chooses to attempt to design a drug against the target. For example, the knowledge that a cofactor binds with submicromolar K_d and is present at 100 μM in the cytosol would tend to discourage attempts to target the cofactor binding site with competitive inhibitors. For targets subject to modulation by endogenous protein or small-molecule ligands, knowledge of mechanism and equilibrium constants is also an important component of appraising tractability. Biophysical tools including ITC, NMR, and SPR can play a pivotal role in such early-stage target characterization, alongside traditional mechanistic enzymology approaches.

Increasingly prominent among the factors contributing to selection of a "good" target is its intrinsic potential for small molecule modulation or "druggability" (2): how feasible is it to design a molecule (large or small) with a good chance of achieving the desired extent of modulation of the target within the disease context? Early identification of those targets that are unlikely to be amenable to functional modulation by small molecule ligands with drug-like physicochemical properties would enable considerable resource savings (or focus on more promising targets). Conversely, early data that hinted at a high probability of successful small molecule lead generation might privilege a target over others with equally strong biological rationale.

A number of ways have been proposed for assessing druggability or the related property of "ligandability" (3). Ligandability refers to the potential for the target to bind small molecules with high affinity, but does not seek to predict the likelihood that these small molecules can become drugs, since this is dependent on many other factors, arguably beyond the scope of early target selection. Ligandability assessments fall into one of two categories—computational and experimental. Computational assessments rely on prior knowledge: structural information about the target or close homologues must be available, and/or there must be information about compounds known to bind to the target or its homologues. Armed with this knowledge, ligandability can be predicted either by assessing the match between the compound and the target binding pocket using various scoring functions, or by identifying ligandable sites on the protein surface (4–7). Computational ligandability assessments have met with mixed success in predicting the outcomes of lead generation programs. A number of simplifying assumptions are made in any computational modelling, and the reliability of the resulting predictions will depend on the degree to which these assumptions actually hold in the real situation. One of the most obvious drawbacks of most current methods is the difficulty of accounting for target and, to a lesser extent, ligand flexibility in pocket analysis.

In creating an experimental procedure to assess ligandability there are two key challenges: to provide information that could not be obtained by simply running a high-throughput screen (HTS) of the million compounds typically available in a screening bank, and to provide this information quickly enough to avoid running an expensive HTS that is unlikely to generate worthwhile hits. To address these challenges, Hajduk and co-workers (6, 8) have proposed that experimental fragment screening can be used to predict successful hit-to-lead outcomes. Fragments are molecules of small size (typically <300 Da i.e., one-third to one-half the size of drug candidates) and low intrinsic complexity. A typical drug candidate can be considered to consist of a combination of fragments of distinct chemistry. A chemically diverse set of around 1,000 fragments can be used to probe a target protein's ability to interact with molecules from a large region of chemical space and effectively search an enormous number of possible fragment combinations orders of magnitude more efficiently than a HTS of drug- and lead-sized compounds (9, 10). However, the low binding affinities typical for ligands of small size and low complexity present a challenge for robust detection, and here biophysical techniques have an advantage over conventional biochemical assays. NMR has traditionally been the method of choice for fragment screening owing to its high reliability and information content. Other techniques are increasingly used where they overcome some of the limitations of NMR, namely low sensitivity (it requires relatively large amounts of the target protein, which can often be difficult to obtain) and modest throughput. Surface plasmon resonance (SPR) in a direct-binding assay format in which the protein target is coupled to a chip surface is gaining increasing acceptance, as are so-called "thermal shift" (TS) techniques that measure the change in thermal unfolding temperature of the protein caused by ligand binding (via a fluorescent dye that preferentially binds the unfolded state). As the temperature of the sample is increased and the protein unfolds, the dye binds to the more exposed hydrophobic regions of the protein, and an increase in fluorescence signal is detected. The stability (T_m) of the protein can be measured from the midpoint of this thermal unfolding curve. Ligand binding to native protein results in an increase in stability and a measurable increase in T_m. Both SPR and thermal shift approaches tend to be intrinsically noisier than NMR and are less able to detect very weak binding ($K_d > 1$ mM) reliably, but use much less target protein. Another emerging technology for fragment screening is native nano-electrospray mass spectrometry (MS), which has low reagent consumption and potentially high throughput, although affinity detection limits and reliability remain to be determined (11).

At AstraZeneca, we have compared historical fragment screening outcomes from 36 in-house programs with success rates of targets in reaching the hit-to-lead phase. Based on a simple classification of the target into categories of low, medium, and high ligandability, on the grounds of hit rate, hit affinity range, and chemical diversity in NMR fragment screens, we observed a strong correlation with hit generation success through HTS. One particular trend was very striking: low ligandability in fragment screening has a strong correlation with failure to make the transition into the hit-to-lead phase. Medium and high ligandability correlated well with progressively increased success in entering hit-to-lead (3). We are now starting to apply fragment-based ligandability screening more routinely in AstraZeneca as we find it to be an important indication of chemical tractability for novel targets.

2.2 Hit Identification

High-throughput screening is still the most widely used approach for the identification of lead compounds (12), in which typically millions of compounds are tested for activity in miniaturized enzyme or ligand binding assays against a range of target proteins. Biophysical, affinity-based, methods have been traditionally employed, and had highest impact, during the validation and characterization of the HTS outputs. However, biophysical methods are increasingly considered as a primary approach for hit finding. In combination with the establishment of fragment-based drug discovery as a new paradigm for lead generation, affinity-based and in particular label-free methods have been further developed for screening fragments. A variety of technologies capable of dealing with the 1,000–10,000 compounds in a fragment library can be used for the primary fragment screening or during the evaluation of fragment screening hit lists: NMR, SPR, MS, TS, and optical waveguide grating (OWG) based assays. Until recently, NMR has been the primary method of choice for fragment screening (13, 14), because of its particular strength in combining, within one technique and under one set of conditions, single-atom spatial resolution information on structure and conformational dynamics with binding energetics. Furthermore, this rich repertoire of information can be obtained from the endogenous nuclear spins, which are direct reporters of the local electronic (chemical) environment, without recourse to labels other than non-perturbative stable isotope enrichment, or to immobilization of the protein to a surface. The major challenges in realizing the potential of NMR in practical applications arise from its intrinsically low mass sensitivity. However, significant improvements in sensitivity have been and continue to be made through advances in instrument technology and the development of novel experimental schemes.

Initial enthusiasm for building NMR capabilities within pharmaceutical discovery organizations stemmed from its emergence as a method for determining protein structures in the late 1980s and early 1990s (15). The promise of access to target structures in the more physiological free solution state (in contrast to the crystalline state captured by X-ray diffraction) was a strong initial lure. However, although structural analysis of ligand binding remains a strength of NMR (that is not further covered here), it is in other areas of application, adventitiously facilitated by the methods developed for structure determination, that NMR has seen its most widespread impact in hit identification. Indeed, it has become the *de facto* "gold standard" method for identifying and confirming protein–ligand interactions—including the very weak (K_d > mM) binding of fragments.

The interaction between a protein target and its putative ligands can in principle be interrogated through observation of perturbations to the NMR spectra of either component. In so-called "protein-observe" methods the protein target itself is monitored, typically by means of 2D HSQC-type experiments involving the use of stable-isotope (usually ^{15}N, but also sometimes ^{13}C)-labelled protein. Because protein-observe methods encode information from specific backbone or side chain nuclei, they carry spatial information that can identify the binding site of a small molecule ligand. This spatial encoding renders protein-observe experiments highly robust since specific binding events are readily distinguished from nonspecific effects of global physical perturbations of the protein properties (aggregation, unfolding) or arising from the ligand itself (aggregation-related or so-called "promiscuous" behavior) (16). The protein resonance shifts as a function of ligand concentration can also be used to directly determine stoichiometry and/or affinities. However, despite advances in sensitivity and experimental methodology, protein-observe methods are still restricted to targets <60 kDa that can be produced in an isotope-labelled form in the many milligram quantities needed.

"Ligand-observe" methods evade these restrictions by monitoring effects on the resonances of the small molecule ligand that accrue when the ligand transiently occupies its cognate binding site on the target protein (which does not need to be isotope-labelled). These effects arise because the properties of the ligand resonances differ substantially between its bound and free states. Various "modes" of ligand-observe experiment including waterLOGSY (17), saturation-transfer difference (STD) (18), and relaxation (T_2, $T_1\rho$)-filtered spectroscopies (19) are widely used, often in parallel, since they are sensitive to different dissociation time-constants and binding-site characteristics. In contrast to protein-observe approaches, target size is not limiting and considerably less protein is required. On the other hand, binding cannot be detected directly for potent ligands with slow dissociation kinetics and in some cases it is not possible a priori to discriminate between specific association with a target binding site and other events, e.g., self-aggregation. These limitations can be mitigated by demonstrating competition of the compound in question with a known ligand.

The pioneering work of the Abbott group on experimental fragment screening (SAR-by-NMR®) was based on 2D HSQC-based protein-observe methods (20, 21), and this remains the approach of choice in many labs in cases where it is economically viable to generate large amounts of the isotope-labelled target. However, the 1D ligand-observe techniques have proved to be both highly sensitive and very versatile for fragment screening applications. Their benefits include lower protein consumption compared with protein-observe techniques, and, crucially, their ability to detect binding at ligand concentrations substantially below the K_d of the interaction. Fragment libraries ranging in size from 1,000 to 10,000 compounds are typically screened in mixtures of 6–12 compounds. Hits in mixtures can be directly identified from ligand-observe experiments, but must be found by deconvolution when protein-observe approaches are used. The protein consumption of ligand-observe NMR can be lowered and, to an extent, sensitivity enhanced by the use of "target-immobilized NMR spectroscopy" or TINS (22). In TINS, the target protein is immobilized to an agarose resin packed in a flow-cell NMR probe. Mixtures of compounds are flowed over the target protein and binding detected by STD.

Whereas most fragment screening experiments use ^{1}H NMR, it is also possible to exploit the high sensitivity of the fluorine nucleus by detecting binding via the displacement of a ^{19}F reporter ligand (23), and the recent availability of ^{19}F-capable cryoprobes has further enhanced the attractiveness of this approach.

Lastly, weak, "second-site" fragments that bind adjacent to a known primary site can be detected by a number of NMR methods. Introducing a paramagnetic nitroxide spin label either into the protein (SLAPSTIC approach (24)) or into a potent first-site ligand (25) results in strong relaxation from the spin label, active over 10 Å or more, that is effective at identifying even very low affinity binders to a neighboring pockets. Inter-ligand NOEs have also been used for second-site fragment screening (26, 27). Further details on NMR-based screening for protein–ligand interactions can be found in Chapter 14.

The early success and practical limitations of fragment-based NMR screening has mobilized researchers to adopt alternative, affinity-based, biophysical methods. Thermal-shift assays or mass-spectrometry, optical biosensors, and SPR technologies, in particular, have been recently successfully employed (28). Optical biosensors monitor changes in certain characteristics of light by exploiting the evanescent-wave phenomenon. The penetration of a light wave a small distance beyond a reflecting interface means that the angle or wavelength dependence of reflection at the interface is affected by the refractive index of material near the interface. Binding of a ligand at a protein coated surface can be monitored via changes that results from the small change in refractive index that occur if ligand binds. SPR based instruments (from providers including, but not limited to Biacore, Bio-Rad, Reichert, and ICx Nomadics), which detect the change in angle at which there is a minimum intensity of reflected light at a coated gold sensor surface (*see* Chapter 6 for further details), are the most utilized optical biosensors that make use of this phenomenon to enable the real-time measurement of the formation of protein–ligand interactions.

These kinetic measurements potentially allow the association, equilibrium binding and dissociation phases of a test compound interaction with the target protein to be characterized in a single experiment (29). Key to all SPR assays is the successful tethering of (usually) the target protein to the biosensor surface without compromising the activity and functionality of the protein. This is typically achieved via accessible primary amines on the protein surface (e.g., lysine residues). The major advantage of protein capture is that this surface may then be used for multiple ligand binding experiments (in some cases up to a 1,000-times or more) in a flow-based system. However, there is a requirement to establish suitable conditions that allow for the reproducible regeneration of the sensor surface without losing binding activity, which can be challenging.

The regeneration and throughput restrictions of sequential flow systems have driven the development and application of optical waveguide-based biosensors, which are also based on the evanescent field phenomenon (30). Platforms like EPIC® from Corning (31), BIND® from SRU Biosystems (32), and Octet® from ForteBio offer the possibility for increased sample throughput by using a plate-based approach optical waveguides are incorporated within each well of a microplate and changes in the refractive index at the interface changes the peak wavelength of the reflected light.

A wide range of living cells can be attached to the OWG biosensors enabling the establishment of assays in living cells. Ligand binding to GPCRs at the cell surface typically leads to trafficking of molecules and protein complexes as well as receptor internalization and translocation. This movement, referred to as dynamic mass redistribution, can be followed using optical waveguides and facilitates study of cell signalling as well as the screening of compounds against endogenous or recombinantly expressed receptors (33). In contrast to cell-based assays, biochemical assays that employ biosensors coated with recombinant proteins with the aim to monitor the ligand binding to the target protein directly may be less straightforward as extra care is required to develop an assay which reliably distinguishes specific from nonspecific or promiscuous binding. The use of OWG platforms has facilitated identification of aggregation-based inhibitors (34), as these tend to display super-stoichiometric binding, that is easily monitored using these mass-based detection systems. There are emerging examples of application of these methods during the high-throughput screening for modulators of protein–protein interaction (35) and protein–DNA interactions (36). Nevertheless it is important to keep in mind that even a specific binding event has no predictive power for its functional relevance until there is further characterization of this interaction, for example, in functional assays, via competition experiments with known ligands, or using structural information.

A sometimes overlooked application for biophysical approaches is in the characterization of protein reagents and the development of a valid biochemical assay that will be employed during the hit discovery process. Biophysical methods can serve as rapid and reliable checks for the quality and validity of the protein reagents as well as the assay conditions. This is particularly useful if well-characterized reference compounds are available. Confirmation of the protein identity using MS methods, such as electrospray ionization (ESI-MS) (37), followed by a quantitation of function, e.g., through the determination of the ligand binding capacity using ITC, can prove extremely valuable. Comparison of the obtained stoichiometry with the known number of binding sites allows for an estimate of the proportion of functional protein, which can be used to normalize assay results between batches. Another calorimetric approach, differential scanning calorimetry (DSC) (38) or fluorescence-based thermal shift assays can measure the thermodynamic stability under a variety of conditions to enable optimal protein formulation (39).

MS methods allow detection of binding compounds following initial separation from non-binding compounds by rapid size exclusion chromatography employing centrifugation, liquid chromatography, ultrafiltration, equilibrium dialysis, or frontal affinity chromatography (40). After dissociation from the protein they are usually detected by ESI-MS, and the possibility to do this in larger mixtures makes it feasible for full-deck HTS (41). Through the recent enhancement of ESI allowing folded proteins or multi-subunit assemblies to be ionized without disrupting their tertiary or quaternary structure (42), it is also possible to monitor the binding of the compounds directly, which is sometimes referred to as native MS (43). Further advances in automated sample handling and nano-electrospray sources have enabled application in fragment screening as well as focused screening (44).

2.3 Hit Evaluation

Hit evaluation describes the process by which hits identified during primary screening are evaluated, often using a range of approaches, including biophysical methods, to determine whether they are truly confirmed hits, or false positives, caused by an assay artifact (e.g., interference with the detection system) or another unwanted mechanism, such as compound aggregation (45, 46). However carefully the HTS assay is designed, and the assay executed, there will always be the potential for producing false positives (47). The need for miniaturization and high levels of signal amplification in HTS creates the potential for assay interference from a number of sources, as does the presence of compounds within the screening sets that can display anomalous behaviors (16) that artifactually produce the signal expected for a bone fide active compound. This fact has led to an investment in biophysical approaches to improve the quality of hits.

The initial process of hit evaluation may often include retesting in the primary assay (including from solid stock if the original screen used liquid stocks) and concentration responses to allow IC_{50} determination. As part of the process, the chemical structure of the compound will be confirmed, as well as its purity. Typically, this characterization will involve biophysical techniques such as liquid chromatography, mass spectral analysis, and/or NMR studies of the ligand to confirm structure and purity. These tests will be accompanied by orthogonal screening with further biochemical and biophysical assays, hit ranking and clustering, assessments of chemical amenability and intellectual property position, and the start of secondary screening as part of the design–make–test cycle.

Biophysical methods (48) are viewed as complementing, but not replacing, traditional assay technologies during hit evaluation. A number of established biophysical methods, as well as some new higher-throughput techniques, are now used to augment traditional HTS in order to improve hit quality because they are typically both more inherently robust to interference from compound-related artifacts and provide additional information content that allows the discrimination of specific binding from artifacts. Methods must be carefully and systematically selected for their suitability to efficiently answer specific questions during the early stages of a drug discovery program. This combination of approaches should enable a more rigorous assessment of early hits, in order to ensure that only true hits, exhibiting the desired mechanisms of action are progressed for chemistry and later stage biological characterization.

2.3.1 Orthogonal Screening

Valid enzyme inhibitors should demonstrate an ability to reduce the rate of catalysis irrespective of the method of detection, and must also achieve this by binding to the target protein. Similarly, hits in a biological activity screen against other types of protein target, e.g., receptors or regulators require further validation. Orthogonal screening refers to those follow up studies to evaluate the outputs of the primary screen that employ an alternative methodology (49). Orthogonal screens may include alternative biochemical assays run under similar experimental conditions, e.g., with a different read-out of the catalytic rate. When the primary hit finding approach is a biochemical or biological assay, biophysical technologies provide a natural orthogonality. Biophysical methods may be employed for two principal reasons: to establish that initial hits really bind to the target protein and to identify behavior that may indicate a promiscuous or nonspecific mechanism generating activity in the primary screen. Ideally, orthogonal biophysical screens should have the capacity to deal with the typical hit rates emerging from HTS of between 0.1 and 1 % of compounds screened (50). Lower throughput biophysical approaches are more suited to affinity measurement and mechanistic analysis of hits. An important aspect of the orthogonal screening component of hit evaluation is to be able to choose from a range of well-behaved assays that will provide confidence that the identified hits operate via the desired mechanism. Achieving this situation requires that the lead generation strategy be agreed, and the desired assays validated before the primary screen is initiated.

The choice of orthogonal assay depends upon the level of detail, and the speed or throughput required at each stage of the hit evaluation process, and may be tailored for individual projects and targets. Direct methods that measure changes associated with the protein often provide more information than either direct methods monitoring the ligand or those employing indirect measures of affinity. In such protein-based methods, greater information is obtained by studying the interactions of test compounds individually. In contrast, ligand-based methods often allow mixtures of compounds screened together with little or no information loss, leading to the potential for higher throughput.

2.3.2 Direct Protein Detection Methods

There are a number of biophysical techniques that directly monitor changes occurring upon binding of test compounds to target proteins, with optical based biosensors being the most rapidly growing. Use of optical sensor-based methods (e.g., SPR or OWG platforms) subsequent to primary screening has enormous value in determining whether compounds identified as active in the primary screen truly bind to the target protein. The throughput of SPR methods, although increased over the last few years, still represents a limitation in characterizing all of the activities identified in the primary assay, and typically only a subset (usually chosen by the medicinal and computational chemistry teams based on the structure–activity relationships seen in the primary screen) can be practicably followed up. Higher-throughput methods can circumvent the necessity to subset compounds at this stage, and OWG instruments offer the possibility of full orthogonal screening (and even primary screening). However, the plate-based nature of these instruments, and thus the absence of controlled liquid flow to introduce ligand at specific times, means that it may be difficult to measure kinetics on a meaningful timescale. Hence, the powerful combination of both SPR and OWG instruments is desirable for any biophysical laboratory involved in lead generation.

NMR is also a powerful method for hit confirmation. The choice between the protein-observe and ligand-observe methodologies discussed previously in the context of fragment-based target evaluation will tend to be dictated by the characteristics of the target protein. Protein-observe experiments, usually using ^{15}N-labelled protein and HSQC-type spectra, derive their power to discriminate between genuine specific binding and nonspecific effects from the information encoded in the backbone amide shift perturbations resulting from addition of ligand. Additional effects on the observed resonance linewidths and intensities as a function of added compound concentration can highlight undesirable compound behavior, including the formation of mixed compound/protein aggregates (e.g., micelles) or denaturation and unfolding of the protein.

2.3.3 Indirect Protein Detection Methods

An alternative configuration of the SPR and OWG methods uses a compound, known to bind the protein, immobilized to the sensor surface, and this can be useful when it may be difficult to immobilize the target protein itself. The chosen compound, which usually binds at a known site, is termed a target definition compound (TDC). The TDC must be captured at high density on the surface. This situation deliberately introduces mass transport limited binding into flow systems used in SPR, so that the binding signal reflects the diffusion of protein to the surface, which is controlled by the concentration of the protein in the bulk solution. Thus, test compounds from the initial screen that bind to the target protein at the same site as the TDC compete with it for binding and reduce the effective free protein concentration, which therefore decreases the amount of bound protein and the signal in a proportional manner (*see* **Note 1**). Because binding to the target protein is monitored only indirectly, compounds of low molecular weight or with affinity values outside those typically detectable by SPR or OWG in direct binding assays can be followed.

Fluorescence-based thermal unfolding methods giving rise to a thermal shift in T_m can also be used to indirectly measure compounds binding to target protein at this stage. However, one important consideration of thermal shift as a method to measure ligand binding is that compounds with a more entropic binding component tend to give larger T_m shifts, and these tend towards lipophilic/hydrophobic compounds whilst the enthalpically driven compounds, potentially with more specific polar interactions will give smaller T_m shifts. This tends to bias the results to those entropically driven compounds (51). For these reasons, it is probably most useful to use thermal shift assays for hit validation rather than for initial screens of novel chemical entities, in order to avoid false negatives.

2.3.4 Bound Ligand Detection Methods

In addition to ligand-observe NMR methods described earlier, MS finds wide use in screening mixtures of compounds for binding to any protein target with the hit compounds being confirmed with high confidence by measurement of their molecular mass. Novartis have developed a plate based affinity screening approach, called "Speedscreen" (52). The target protein is pre-incubated with mixtures of large numbers of test compounds (400 was typically found to be optimal). These mixtures are then applied to a 96-well plate of mini size exclusion columns (a variety of separation methods, including both liquid and centrifugation based size exclusion chromatography and ultracentrifugation could be employed (40)). The bound ligands which are separated with the target protein are then dissociated from the protein, and identified by ESI-MS.

2.3.5 Identifying Promiscuous Compounds

Compounds that act through a promiscuous or nonspecific mechanism often arise as false positives in traditional HTS assays. Such mechanisms include self-aggregation, aggregation of or with the target protein, compounds which (partially) denature the target protein, or which act as chaotropic or kosmotropic agents.

No one test alone will be able to conclusively identify promiscuous compounds, but the apparent affinity of such compounds typically shows a strong dependence on assay conditions. Consequently, it is useful to explore the ratios of affinities (or IC_{50} values) found with and without a pre-incubation step and at elevated protein concentrations. Higher affinities (lower IC_{50}s) after pre-incubation or at higher protein concentration are indicative of an aggregation mechanism. Addition of the suspect compound to one or more reference assays based on proteins unconnected to the target can be used to directly check for promiscuous inhibition (*see* **Note 2**).

Compounds with other nonspecific mechanisms, for example redox reactive compounds, may also show time dependent inhibition in the presence of a reducing agent such as dithiothreitol (DTT). These can be identified using UV spectrometry. Compounds are tested with and without reducing agent to identify changes in spectra that can be attributed to redox reactivity.

Compounds that act by unfolding the target protein can be identified using biophysical methods that measure protein stability such as DSC or fluorescence-based thermal shift measurements (see above). As these compounds act by binding to unfolded protein and stabilizing the denatured state, they should be detectable as a decrease in protein stability in the presence of compound.

Ligand binding should be of the expected stoichiometry of the interaction and the ligand concentration response measured by SPR, OWG, ITC, or NMR should cover at a wide dynamic range around the expected K_d for the interaction. A saturable binding signal for the test compound should be seen for a well-behaved compound, whereas many nonspecific interactions are characterized by super-stoichiometric ligand binding signals, which increase disproportionately with compound concentration (53) (*see* **Note 3**).

2.4 Lead Optimization

Although lead optimization is often regarded as a distinct phase in the drug discovery process, it is important to note that each phase is significantly influenced by those preceding and following. Most of the foundations for high quality and speed of lead optimization are already laid out during the lead generation phase and even back as far as an effective screening campaign. The requirements for early development also need to be taken into account during lead optimization, and place restrictions the compounds within a chemical series that can move forward to be considered as CD. Lead optimization comprises the synthesis of lead compounds and analogs with improved biochemical efficacy by establishing SAR, increasing potency, reducing off-target activity and improving physicochemical and pharmacokinetic properties. During lead optimization, one measures and compares a multitude of parameters for, and properties of, various lead compounds in order to guide the selection of those with the greatest potential to be developed into safe and efficacious medicines. Often studies are conducted both in living organisms (in vivo), and with a range of affinity-based biophysical methods in vitro using recombinant target protein. Biophysical methods are primarily used to guide the "maturation" of the affinity of compounds, but the wealth of information than can be produced by a combination of methods may also influence other aspects of lead optimization.

ITC (54, 55) and SPR (29) are regarded by many researchers as "gold standard" methods for the determination of affinities as both measure the direct interaction of a compound with the target protein without any modification. Whereas both methods can deliver accurate and relevant affinity values, nevertheless great care has to be taken during the establishment of conditions for the experiment, as one might unintentionally create a situation that is not a valid surrogate for that in vivo. The use of recombinant protein constructs that lack, for example, regulatory domains, or the absence of cofactors that influence the binding of the natural ligand(s) will mislead. It is always important to follow in vitro SAR with suitable in vivo assays. If in vivo potency/efficacy is much higher (drop-on) or lower (drop-off) than suggested by the affinity, it may be indicative of limitations of the assay system and/or conditions.

Another important aspect is the correct determination of the mode of action of the leads during optimization. Establishing the mode of action not only requires demonstration of specific interaction with the target and identification of the binding site. Some preliminary indications of the compound's binding pose and major specificity-determining interactions are also of great value at this stage. A variety of NMR experiments can provide information on ligand binding mode without necessitating a full structure determination of the target–ligand complex. Isotope-filtered methods can be used to selectively measure nuclear Overhauser enhancements (NOEs) between protons on an unlabelled ligand and a protein labelled with ^{13}C and/or ^{15}N (56); selective labelling of leucine, valine, and isoleucine. Methyl groups with ^{13}C and ^{1}H in a ^{2}H-labelled background can furnish NOE distance restraints for even quite large (>50 kDa) protein targets (57). For systems where protein labelling is not possible, various ligand-observe techniques can provide binding epitope-related information. STD-based "epitope mapping" experiments observe differential signal enhancements for ligand protons buried in the binding interface compared with those exposed to solvent; this information can be compared between ligands to build a picture of the common surface implicated in molecular recognition of the target binding site. Transferred "inter-ligand" NOEs (ILOES), between fragments that bind simultaneously at adjacent subsites, can be used to estimate the relative orientation of the ligands, facilitating linking strategies (27, 58). In many cases, ILOES are effectively mediated through spin-diffusion via protons in the target binding site itself (59) and the INPHARMA (Inter-ligand NOEs for PHARmacophore MApping) technique (60) uses this phenomenon to transfer magnetization between protons of a pair of weakly binding ligands that compete for the same binding site to create an apparent inter-ligand NOE, even though the ligands never bind simultaneously to the target. By comparing the NOE intensities and build-up rates across different protons of ligand pairs, a pharmacophore map of the binding pocket can be constructed. This approach has been used to impressive effect to generate structure-based design hypotheses for a GPCR target, GPR40, in the absence of X-ray structural information (61).

For enzyme targets, characterization of reaction kinetics can provide information which can influence the correct experimental setup in subsequent biophysical experiments. For example, observation of mixed noncompetitive inhibition, where inhibitors bind to the free enzyme and the enzyme–substrate complex with different affinities, could require the inclusion of the substrate (at a concentration equal to its K_d to ensure a balance between free enzyme and enzyme–substrate complex) during such studies in order to avoid lead optimization and SAR development against an irrelevant form of the enzyme. This becomes even more important for those methods that allow for the dissection of the affinity into either its thermodynamic or kinetic components, which adds not only another dimension into medicinal chemistry design but also further sources for errors.

2.4.1 Thermodynamic Fingerprinting

The affinity of a protein–test compound interaction is determined by the change in the free binding energy which occurs when the molecular interactions of the free state are exchanged for those of the bound state, (Fig. 2). Affinity is commonly used as the main driver for lead optimization and it seems plausible that determination of the two components of the free binding energy, namely the enthalpy and the entropy, could lead to better informed decision making in medicinal chemistry design, as two thermodynamic parameters instead of one can be used in SAR and similar analyses to guide the lead optimization process (62). Isothermal titration calorimetry (ITC) is the only direct approach to experimentally determine both the free energy and the enthalpy change upon binding (the entropy is then determined from the Gibbs–Helmholtz equation). ITC thus provides a full thermodynamic description of the binding interaction that has been suggested to be a useful tool for lead optimization (63, 64). In particular, Freire has been a strong advocate for the use of thermodynamic data in lead optimization and he underpins this with a retrospective analysis for some drug target systems (65), where it is seen that there is an improvement of the enthalpy during drug evolution; with "first in class" compounds being mainly entropically driven, but with "best in class" compounds having a dominant enthalpic contribution. If enthalpy is good for selectivity and specificity, a focus should be on rational engineering of such interactions that contribute positively to the enthalpy, but this is not trivial.

There is an increasing body of evidence that there is intrinsic value in gathering thermodynamic data with the aim of using

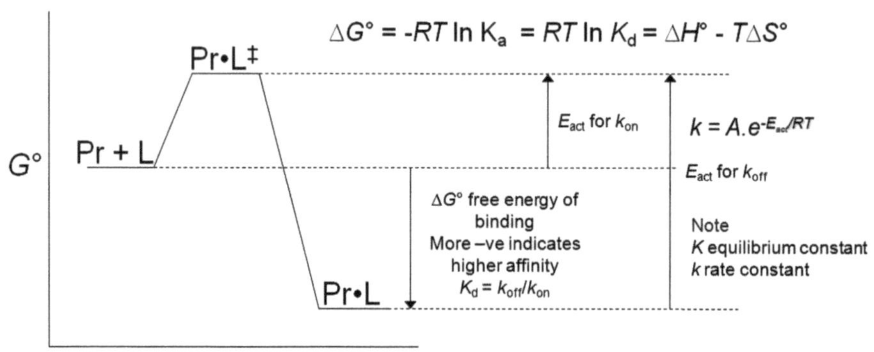

Fig. 2 Free energy diagram and its relationship to kinetic parameters for a single-step binding interaction. For binding of a ligand (L) to a protein (Pr) to be spontaneous, the change in free energy on moving from the reactants to the products must be negative. Increasing affinity is achieved by lowering the free energy of the protein–ligand complex (Pr·L). The dissection of the binding free energy into enthalpic and entropic components is described by the Gibbs–Helmholtz equation (*top right*). The rate constants for association (k_{on}) and dissociation (k_{off}) are related to the activation energy barriers for association and dissociation respectively, by the Arrhenius equation, and increasing residence time may be achieved by stabilizing the Pr·L complex, and/or destabilizing the transition state (Pr·L‡)

enthalpic as well as free energy optimization as a criterion in compound selection for further optimization (66). A sound strategy to at least monitor those thermodynamic signature changes during lead optimization is to simply map the localization and type of modifications that maximize the enthalpic and entropic contributions. It is envisaged that a combination of those modifications will help to accelerate affinity optimization in subsequent rounds. However, the accumulated knowledge and experience currently does not provide a reliable means of rationally applying the data in medicinal chemistry design, only the ability to establish correlations between the thermodynamic data and the structural elements will eventually allow a rational and prospective use of such data in lead optimization and drug design. As we gather a deeper understanding of the requirements for effective application of thermodynamic information in lead optimization, approaches such as the use of matched molecular pairs (two compounds that differ from each other by just one atom or a small, localized group of atoms) or thermodynamic double-mutant cycles (67) will help to improve our capability to use that information in drug design. Comparing the binding of closely related compounds that only have seen subtle changes in their structure simplifies the analysis of thermodynamic correlation and helps to guide the next round of optimization. Complementing those methods with structural data is currently indispensable to build confidence in this approach and will also provide a rich dataset for a future thermodynamic parameterization that may unleash its full potential.

2.4.2 Kinetic Fingerprinting

An emerging approach, which is gaining in momentum within the pharmaceutical industry, is the monitoring of binding kinetics during lead optimization. There is strong likelihood of a link between drug–target residence time and critical factors, such as drug efficacy, safety, and toxicity, as a result of the pharmacodynamics of drug action (68, 69). For a given affinity, slower onset inhibitors will spend longer times bound to their targets (c.f. rapid, reversible inhibitors) and will remain bound even when free drug concentrations are declining. The concept of using the drug–target residence time as a criterion for appropriate drug efficacy was introduced by Copeland (70, 71) and has attracted a lot of attention as a selection tool to reduce attrition rates. In an analogous manner to the thermodynamic component approach, the affinity is a result of two rate constants for association and dissociation of the drug–target complex (Fig. 2), each of which could be used as a parameter for lead optimization. The residence time is the reciprocal of the dissociation rate constant for the drug–target complex, and SPR is often ideal for the accurate determination of this parameter. Another valid but more approximate approach is the measurement of enzyme recovery rates after incubation with the inhibitor at a concentration that is tenfold above the measured IC_{50} followed by a rapid dilution of the preformed complex by a 100-fold (72). This allows the regain of enzyme activity to be monitored, and although this is strictly an approach to a new equilibrium position, it is a useful approximation to the dissociation rate constant of a reversible inhibitor. Slow or no regain of activity can suggest a slowly dissociating or irreversible mechanism.

Presently, examples detailing rational improvement of the kinetic behavior of compounds by establishing structure–kinetic relationships are quite sparse and kinetic data have mostly been used retrospectively during the affinity maturation process. Current efforts within the pharmaceutical industry are focusing on the prospective use of such data in order to design compounds with desirable kinetic properties, and in particular long residence times. A recent example has been the development of a highly efficacious polypeptide derived from compstatin, a 13-residue peptide that inhibits the activation of the complement system. SPR was used in conjunction with other approaches to determine the effects on the activity of compstatin upon modifications of some key interacting residues (73). The kinetic data identified relationships with ligand modification not apparent from the affinity, which, together with structural data, revealed essential structural features that were responsible for increasing stability of the complex and eventually supported the development of analogs with improved efficacy. That the residence time of an inhibitor has indeed predictive power of in vivo efficacy has been shown in the case of FAS-II enoyl reductase, where the residence time has been a much better predictor of the in vivo activity than affinity alone (74). These studies have broad implications for drug discovery programs in general and lead optimization in particular, and should prompt revision of current lead optimization strategies to further expand the biophysical characterization that will be key to eventually facilitating the design of compounds with prolonged residence times.

3 Physicochemical Property Determination and ADMET

3.1 Compound Availability

Off-target protein binding by a compound is monitored throughout early development and lead generation, where its evaluation plays a key role in aiding series selection. Off-target binding has an important role in the determination of in vivo free drug levels which, when combined with potency and efficacy data, allow pharmacokinetic–pharmacodynamic relationships to be established and plays an important role in determining safety margins for compounds entering development. It is essential that highly accurate methods are used to determine values for equilibrium and kinetic binding parameters. Moreover, there is a need to generate this data in a medium to high-throughput manner to fit with discovery project timelines.

3.1.1 Plasma Protein Binding

Whereas numerous methods are available for measuring the affinities of compounds for plasma protein, equilibrium dialysis remains the gold standard because of its ability to generate accurate data for a diverse range of compound types (75). The method is based on a dialysis cell which is separated into two chambers by a semipermeable membrane which does not allow the movement of albumin protein across it. Plasma is placed into one half of the cell and buffer into the other half. Compound is then added to the dialysis cell at a concentration below that of the albumin, dialysis proceeds, typically at 37 °C, until equilibrium between the compound and plasma proteins is achieved. The solutions on each side of the membrane are subsequently removed and analyzed (typically by LCMS/MS) to determine the amount of compound that is bound to the plasma protein. Because both sides of the dialysis chamber are sampled, any compound loss to the cell walls or membrane does not affect the binding figure (*see* **Note 4**).

Recent developments have focused on improving the measurement throughput of the dialysis technique with the introduction of a number of commercial devices amenable to automation. These include 96-well devices from HTDialysis LLC, from Thermo Scientific (the "Rapid Equilibrium Device"), from Harvard Bioscience (the "Equilibrium Dialyser") and a 24-well serum binding system from BD Biosciences. Together, improvements in sample analysis via use of LC MS/MS and using sample mixtures has meant that measurement of many hundreds of compounds every week is routine.

Ultrafiltration is an alternative high-throughput technique used to measured protein binding (76). Ultrafiltration is simpler and quicker to run than dialysis. Typically, compound is spiked into plasma and allowed to equilibrate with the plasma proteins prior to sampling to determine the total drug concentration. The plasma solution is then added to an ultrafiltration tube containing a semipermeable membrane and the free drug solution pushed through the semipermeable membrane using centrifugation. This solution is then sampled to determine the percentage of free drug. High-throughput formats, such as the 96-well ultrafiltration plates from Millipore are available. The main limitation of this technique is that some compounds are absorbed onto the filter membrane or device walls and give rise to over estimation of protein binding.

Less commonly, ultracentrifugation is used to separate free drug from drug bound to protein, but this technique is quite expensive in terms of equipment cost and the problem of compound adsorption to the device walls affecting results remains (*see* **Note 5**).

3.1.2 Albumin Binding

Albumin is the protein present in highest concentration (~600 μM) in plasma and usually dominates the plasma protein binding. As a result, measurement of binding to human serum albumin (HSA) alone is often used as proxy for extent of plasma protein binding. This allows for more straightforward control and standardization of binding assays. Numerous techniques have been established where albumin is immobilized onto a solid support to facilitate the binding measurement. It is assumed that in the chemically bonded albumin its binding sites remain intact and that there is no significant non-specific binding to the support (controls are necessary to establish this). In SPR-based measurements, albumin is immobilized on a sensor chip using covalent coupling via amine groups on the protein. Drug in solution is flowed over the chip and binding to albumin is detected, in the same way as for the specific protein–ligand interactions covered earlier. Recent instrumentation has greatly improved sensitivity such that detection of small and weak binding compounds is less of an issue, particularly for highly bound compound. SPR has been suggested as a suitable approach for screening large numbers of compounds to obtain binding constants (77). However, "readout" of SPR results is not necessarily straightforward as not all compounds fit a simple binding model and data analysis may be complex. Also for weaker binders, higher concentrations of drug are required and solubility may be a problem.

HPLC-based methods, in which albumin is chemically bonded to column matrix, have also proved to be useful for studying protein binding. Originally, retention time under constant buffer and flow conditions was used to estimate the albumin binding constant. However, retention times of high affinity (low off-rate) compounds could be impracticably long. Recently faster gradient-based methods have been adopted (78). Typically the logarithm of the retention time in a constantly increasing gradient designed to reduce binding (e.g., of organic solvent in water) is plotted against the log(K_b) for a set of reference compounds. This linear plot serves as a calibration for estimating unknown log(K_b) from the measured retention times. The gradient HPLC method gives a reasonable correlation with literature plasma binding log$K$ data, and where there have been differences in values this has been suggested to be due to the system being based on HSA alone and so missing other plasma binding proteins such as alpha glycoproteins. The gradient method has been shown to be particularly useful for spotting high affinity and/or stoichiometry compounds quickly.

3.1.3 Albumin Binding Site Identification

Human albumin exhibits two primary drug binding sites; site I, the primary binding site for drugs such as Warfarin and Azapropazone, and site II, which binds drugs such as Diazepam and arylpropionic acids (also referred to as the Indole-Benzodiazepine site). There are also seven fatty acid binding sites on albumin. To facilitate molecular design it is helpful to identify the primary binding site on albumin to which a compound binds. A competition assay is set up where the drug is mixed with reference compound of known binding site, which will reduce compound binding in fluorescence spectroscopy, or SPR-based assays. Together with crystal structures of ligand-HSA complexes, results of competition assays can give useful insight into ways to modify compounds to diminish off-target binding.

3.2 ADMET: Absorption, Distribution, Metabolism, Excretion, Toxicity

Efforts to reduce the overall timescale of drug discovery have led to a tremendous growth in demand for earlier ADMET information on increasing numbers of compounds (79). Consequently, a focus on maximum throughput for a "decision making" assay specification has long since replaced bespoke analysis as the main goal for research scientists supporting early discovery. The solution for many has been the deployment of generic high-throughput screening assays employing automated sample preparation and dedicated analytical platforms. Benchmark information for a range of properties including in vitro metabolic stability, intestinal absorption, drug blood levels, liability to drug–drug interaction, and potential to form reactive metabolites are obtained in this manner (80). Data are used to guide further structural refinements to produce potent compounds that meet the criteria for an effective drug.

Versatile laboratory robots are used prepare batches of samples in parallel using industry standard 96 (or more recently 384) well plate formats. The resulting plates and discrete sample coordinates are then transferred for HPLC/MS analysis operated in Multiple Reaction Monitoring (MRM) detection mode. Electrospray ionization predominates as the MS method of choice, providing unrivalled sensitivity and selectivity for small "drug like" molecules (150–800 Da) in biological matrices (81) (*see* **Note 6**).

For orally administered compounds, absorption from the gut can limit systemic exposure to a drug. Passive permeability across the intestinal epithelium, as well as active transport (uptake and efflux), are key determinants of absorption (82). In culture, Caco-2 cells form confluent monolayers and polarize, thus replicating the differential distribution of transporters and formation of tight junctions that is observed in vivo. Compound is added to either the apical or basolateral side of the monolayer and transfer from one side to the other quantified by LC-MS/MS. The apparent permeability can be used to predict the fraction absorbed and screen out compounds likely to have poor absorption in vivo. Greater movement of compound in the basolateral to apical direction is indicative of transporter mediated efflux. This may limit absorption in vivo and specific inhibitors and/or cell lines expressing individual transporters used to identify those involved (83).

Drug metabolism is one of the most important processes affecting the extent and duration of exposure to a drug in humans. The liver is the main site of metabolism, and many in vitro approaches utilizing liver microsomes (subcellular fractions containing cytochrome P450, UDP glucuronyl transferases, flavin monooxygenases) or whole hepatocytes have been developed to help model drug metabolism (84). Use of primary hepatocytes, which in contrast to microsomes possess the full complement of metabolizing enzymes, for routine screening and estimation of human metabolic clearance has grown due to improvements in human hepatocyte isolation and cryopreservation, and is now seen as the gold standard (85). Early in drug discovery, little, if anything, is known about the metabolic fate of a compound. Consequently, most methods monitor parent compound depletion from incubations with hepatocytes or microsomes using LC-MS/MS detection. The intrinsic clearance values derived from these experiments can be "scaled up" to give a predicted in vivo clearance. Other factors that affect exposure of the liver to the compound in vivo, such as binding to albumin and blood/plasma partitioning, can be factored in to assays or predictions (86). Samples from incubations are also used to identify the metabolites formed using high resolution MS, MS/MS experiments, and NMR (85). This information can be used to drop compounds from further development or to design out sites of metabolic weakness, identify pharmacologically active or potential reactive metabolites, and also give an indication of the enzymes involved in clearance.

Inhibition of cytochrome (CYP) P450s is an undesirable feature of drugs due to the potential to alter normal metabolism or the pharmacokinetics of co-administered medicines (drug–drug interaction) and cause toxicity (87). Techniques using radiolabelled probe substrates, cofactor consumption or oxygen production have been used to measure inhibition of P450 (88). However, high-throughput assays utilizing recombinant protein and fluorogenic or chemiluminescent probe substrates are now more commonly used to classify the inhibitory potency of compounds early in discovery (89). More accurate assessment of inhibition is usually carried out in human liver microsomes using a drug-like probe substrate. By using MS/MS to measure product formation and with careful selection of incubation conditions to ensure the probe substrates are specific for particular CYP isoforms, these incubations can be cocktailed to increase throughput (90). Enzymes such as CYP450 3A4 have multiple binding sites and the use of different probe substrates can provide information on how the compound is binding (91). Time dependent inhibition of P450s is of particular concern since a drug–drug interaction may persist even when drug levels fall in the plasma (92). Modifications of the standard assays (pre-incubation in the presence or absence of cofactor) are used to assess whether inhibition is time dependent (93). Dialysis of the incubation mixture or treatment with ferricyanide can then be used to determine whether this is due to covalent modification of the protein (irreversible) or complexation with the heme (quasi-irreversible, metabolite inhibitor complex) (94).

Where inhibition cannot be avoided, mathematical models are used to put the inhibition data into context with the predicted clinical exposure and assess the risk of drug–drug interactions (95).

4 Concluding Remarks

Biophysical methods can form an integral part of a sophisticated suite of techniques, a "toolbox," that enables coherent and powerful strategies for the design of novel drugs. We are entering a new era, where these biophysical methods are increasingly employed alongside, and sometimes even instead of, traditional biochemical methods for target selection, hit identification, and hit evaluation. Biophysical methods have also found increasing application in the later stages of drug discovery, where physicochemical characterization and ADME properties of lead compounds are required. This breadth and depth of application has been brought about by access to large amounts of purified protein made available through advances in molecular biology and protein science. The use of these methods is often highly automated and underpinned by customized laboratory IT systems that manage all activities from work request to sample generation, sample analysis, data processing, and uploading final results to a corporate database. This emphasizes the value of a multidisciplinary approach to drug discovery. Within the pharmaceutical industry, there has been a tendency to increasing throughput, both in terms of the number of projects and in the numbers of compounds screened. The use of biophysical methods have helped to reiterate the importance of quality within multidisciplinary drug discovery programs, by helping to identify the right target and the right compounds with the right properties to be taken forward. This approach will increase the success rate in drug discovery projects.

5 Notes

1. Care must be given to the choice of target definition compound (TDC) and certain characteristics are desirable such as a relatively high affinity, typically a sub-micromolar K_d. Ideally, the kinetics of the interaction should also be favorable (having a large association rate constant) to enhance the mass transport effect, allowing effective concentration measurement, although it may be possible to take advantage of the law of mass action, as the concentration of the TDC (immobilization density at the surface) may be increased to help promote the protein binding step. A low dissociation rate constant will help to minimize dissociation during the binding phase. Once the choice of TDC has been made, the assay is validated by ensuring that there is a linear

response between protein concentration and binding signal, and that at the chosen protein concentration, TDC added in solution competes off the protein from the surface bound TDC. The assay can then be used to identify binding compounds by incubating the test compound with the chosen protein concentration until equilibrium is attained, and then introducing this equilibrated mixture into the flow cell. The instrument detects the free protein concentration by the reduced level of binding to the TDC-coated surface. Concentration responses, obtained as the free protein concentration is reduced in the presence of increasing concentrations of test compound, can then be completed in order to measure the affinity of the interaction.

2. Nonspecific inhibition may follow one or more of several different mechanisms, including aggregation, protein unfolding, denaturation or redox reactivity. Aggregation based nonspecific inhibition is often tested utilizing chymotrypsin (although this system, would not be suitable for other serine proteases), which follows cleavage of a fluorescent substrate, by comparison of IC_{50} values with the normal assay data to identify suspicious similarities in activity. Other types of behavior exhibited by aggregation based inhibitors are: time-dependent inhibition, steep dose response curves, temperature or denaturant independence, dependence upon [E], strong dependence on ionic strength, or detergent and noncompetitive inhibition kinetics.

3. The stoichiometry of binding can be calculated from the theoretical binding capacity of SPR or OWG methods as follows:

$$\text{No. of binding sites} = \text{Binding Capacity(RU)} \times \text{MW}_{\text{ligand}}) / (\text{Activity} \times \text{MW}_{\text{analyte}} \times \text{RU}_{\text{immobilised ligand}})$$

4. Plasma protein binding in is most commonly quoted as %-free (or free fraction) where % free = compound concentration in the buffer chamber/compound concentration in chamber containing 100 % plasma. However, protein binding may also be quoted in terms of the first apparent binding constant K_{app}, which may be usefully compared to the binding constant for its target protein Although plasma is a mixture and compound may bind to multiple components, because binding to albumin tends to dominate and is typically in large excess over compound it is observed that protein binding often fits a simple 1:1 binding model reasonably well

$$K_{\text{app}} = \frac{[\text{PD}]}{[\text{D}][\text{P}]} \quad (1)$$

Substituting in equations for total plasma protein concentration $[\text{P}]_{\text{tot}} = [\text{P}] + [\text{PD}]$ (measured using a generic assay such as Bradford) and total drug concentration $[D]_{tot} = [D] + [PD]$ where [P] and [D] are unbound protein and drug, respectively, and [PD] is bound drug generates a quadratic equation:

$$[D]^2 + D([P]_{tot} - [D]_{tot} + 1/K_{app}) - [D]_{tot}/K_{app} = 0 \quad (2)$$

Which can readily be solved for free drug concentration [D] at any given protein or drug concentration where albumin remains in excess.

5. With all protein binding experiments it is important that the experimental conditions are carefully controlled. In particular, plasma when exposed to the atmosphere liberates CO_2 causing the pH to rise. Typically freshly defrosted plasma will exhibit a pH around 8 and this can continue to rise throughout a dialysis experiment if a CO_2 incubator is not used or the dialysis buffer has insufficient buffer capacity. For basic compounds this leads to significant over estimation of protein binding. To control for this it is important that all plasma is pH checked at the beginning of an experiment and adjusted to pH 7.4. Also for long dialysis experiments a sufficiently high concentration of buffer should be used to prevent pH drift.
6. Steep HPLC gradients are applied to short reversed phase columns (typically 30–50 × 2 mm internal diameter) packed with small particles (1.7–3 μm) to generate efficient separations with short analysis times (96). Often the separations are carried out in "staggered parallel mode" on dual column systems to minimize delays associated with column re-equilibration and injector load/wash cycles. Where circumstances allow, discrete samples may be pooled (also referred to as cassette analysis) prior to analysis to reduce analytical turnaround time thereby increasing assay capacity (97).

References

1. Kola I, Landis J (2004) Can the pharmaceutical industry reduce attrition rates? Nat Rev Drug Discov 8:711–715
2. Hopkins AL, Groom CR (2002) The druggable genome. Nat Rev Drug Discov 1:727–730
3. Edfeldt FBN, Folmer RHA, Breeze AL (2011) Fragment screening to predict druggability (ligandability) and lead discovery success. Drug Discov Today 16:284–287
4. Pérot S et al (2010) Druggable pockets and binding site centric chemical space: a paradigm shift in drug discovery. Drug Discov Today 15:656–667
5. Egner U, Hillig RC (2008) A structural biology view of target druggability. Expert Opin Drug Discov 3:391–401
6. Hajduk PJ, Huth JR, Tse C (2005) Predicting protein druggability. Drug Discov Today 10:1675–1682
7. Cheng AC et al (2007) Structure-based maximal affinity model predicts small-molecule druggability. Nat Biotechnol 25:71–75
8. Hajduk PJ, Huth JR, Fesik SW (2005) Druggability indices for protein targets derived from NMR-based screening data. J Med Chem 48:2518–2525
9. Fink T, Reymond JL (2007) Virtual exploration of the chemical universe up to 11 atoms of C, N, O, F: assembly of 26.4 million structures (110.9 million stereoisomers) and analysis for new ring systems, stereochemistry, physicochemical properties, compound classes, and drug discovery. J Chem Inf Model 47:342–353

10. Medina-Franco JL et al (2007) A similarity-based data-fusion approach to the visual characterization and comparison of compound databases. Chem Biol Drug Des 70:393–412
11. Vivat Hannah V et al (2010) Native MS: an 'ESI' way to support structure- and fragment-based drug discovery. Future Med Chem 2:35–50
12. Copeland R (2003) A mechanistic considerations in high-throughput screening. Anal Biochem 320:1–12
13. Campos-Olivas R (2010) NMR screening and hit validation in fragment-based drug discovery. Curr Top Med Chem 11:43–67
14. Pellecchia M et al (2008) Perspectives on NMR in drug discovery: a technique comes of age. Nat Rev Drug Discov 7:738–745
15. Billeter M, Wagner G, Wüthrich K (2008) Solution NMR structure determination of proteins revisited. J Biomol NMR 42:155–158
16. Seidler J et al (2003) Identification and prediction of promiscuous aggregating inhibitors among known drugs. J Med Chem 46:4477–4486
17. Dalvit C et al (2000) Identification of compounds with binding affinity to proteins via magnetisation transfer from bulk water. J Biomol NMR 18:65–68
18. Mayer M, Meyer B (1999) Characterisation of ligand binding by saturation transfer difference NMR spectroscopy. Angew Chem Int 38:1784–1788
19. Hajduk PJ, Olejniczak ET, Fesik SW (1997) One-dimensional relaxation- and diffusion-edited NMR methods for screening compounds that bind to macromolecules. J Am Chem Soc 119:12257–12261
20. Shuker SB et al (1996) Discovering high-affinity ligands for proteins: SAR by NMR. Science 274:1531–1534
21. Ludwig C, Guenther UL (2009) Ligand-based NMR methods for drug discovery. Front Biosci 14:4565–4574
22. Vanwetswinkel S et al (2005) TINS, target immobilised NMR screening: an efficient and sensitive method for ligand discovery. Chem Biol 12:207–216
23. Dalvit C et al (2005) Sensitivity improvement in ^{19}F NMR-based screening experiments: theoretical considerations and experimental applications. J Am Chem Soc 127:13380–13385
24. Jahnke W, Rüdisser S, Zurini M (2001) Spin label enhanced NMR screening. J Am Chem Soc 123:3149–3150
25. Moy FJ et al (2010) Novel synthesis and structural characterisation of a high-affinity paramagnetic kinase probe for the identification of non-ATP site binders by nuclear magnetic resonance. J Med Chem 53:1238–1249
26. Becattini B et al (2006) Structure-activity relationships by interligand NOE-based design and synthesis of antiapoptotic compounds targeting Bid. Proc Natl Acad Sci U S A 103:12602–12606
27. Sledz P et al (2010) Optimization of the interligand Overhauser effect for fragment linking: application to inhibitor discovery against Mycobacterium tuberculosis pantothenate synthetase. J Am Chem Soc 132:4544–4545
28. Danielson UH (2009) Fragment library screening and lead characterization using SPR biosensors. Curr Top Med Chem 9:1725–1735
29. Huber W, Mueller F (2006) Biomolecular interaction analysis in drug discovery using surface plasmon resonance technology. Curr Pharm Des 12:3999–4021
30. Mukundan H (2009) Waveguide-based biosensors for pathogen detection. Sensors 9:5783–5809
31. Fang Y et al (2006) Resonant waveguide grating biosensor for living cell sensing. Biophys J 91:1925–1940
32. Cunningham BT (2004) Label-free assays on the BIND system. J Biomol Screen 9:481–490
33. Dodgson K et al (2009) 100K well screen for a muscarinic receptor using the Epic label-free system—a reflection on the benefits of the label-free approach to screening seven-transmembrane receptors. J Recept Signal Transduct Res 29:163–172
34. Chan LL et al (2009) A method for identifying small molecule aggregators using photonic crystal biosensor microplates. Conf Proc IEEE Eng Med Biol Soc 2009:788–791
35. Heeres JT et al (2009) Identifying modulators of protein-protein interactions using photonic crystal biosensors. J Am Chem Soc 131:18202–18203
36. Chan LL et al (2008) A general method for discovering inhibitors of protein-DNA interactions using photonic crystal biosensors. ACS Chem Biol 3:437–448
37. Fenn JB et al (1989) Electrospray ionization for mass spectrometry of large biomolecules. Science 246:64–71
38. Bruylants G, Wouters J, Michaux C (2005) Differential scanning calorimetry in life science: thermodynamics, stability, molecular recognition and application in drug design. Curr Med Chem 12:2011–2020

39. Zhang R, Monsma F (2010) Fluorescence-based thermal shift assays. Curr Opin Drug Discov Devel 13:389–402
40. Holdgate GA et al (2010) Affinity-based, biophysical methods to detect and analyze ligand binding to recombinant proteins: matching high information content with high throughput. J Struct Biol 172:142–157
41. Muckenschnabel I et al (2004) SpeedScreen: label-free liquid chromatography-mass spectrometry-based high-throughput screening for the discovery of orphan protein ligands. Anal Biochem 324:241–249
42. Patriksson A, Marklund E, van der Spoel D (2007) Protein structures under electrospray conditions. Biochemistry 46:933–945
43. van den Heuvel RH, Heck AJ (2004) Native protein mass spectrometry: from intact oligomers to functional machineries. Curr Opin Chem Biol 8:519–526
44. Hofstadler SA, Sannes-Lowery KA (2007) Interrogation of noncovalent complexes by ESI-MS: a powerful platform for high throughput drug discovery. In: Wanner KT, Höfner G (eds) Mass spectrometry in medicinal chemistry, 1st edn. Wiley, Weinheim, pp 321–338
45. Coan KE, Shoichet BK (2008) Stoichiometry and physical chemistry of promiscuous aggregate-based inhibitors. J Am Chem Soc 130:9606–9612
46. Feng BY et al (2007) A high-throughput screen for aggregation-based inhibition in a large compound library. J Med Chem 50:2385–2390
47. Achyuthan KE, Whitten DG (2007) Design considerations for high throughput screening and in vitro diagnostic assays. Comb Chem High Throughput Screen 10:399–412
48. Renaud JP, Delsuc MA (2009) Biophysical technique for ligand screening and drug design. Curr Opin Pharmacol 9:622–628
49. Miller JR et al (2010) The use of biochemical and biophysical tools for triage of high-throughput sceening hits—a case study with *Escherichia coli* phosphopantetheine adenylyltransferase. Chem Biol Drug Des 75:444–454
50. Zhu Z, Cuozo J (2009) High-throughput affinity-based technologies for small-molecule drug discovery. J Biomol Screen 14:1157–1164
51. Holdgate G (2007) Thermodynamics of binding interactions in the rational drug design process. Expert Opin Drug Discov 2:1103–1114
52. Zehender H et al (2004) SpeedScreen: the “missing link” between genomics and lead discovery. J Biomol Screen 9:498–505
53. Giannetti AM, Koch BD, Browner MF (2008) Surface plasmon resonance based assay for the detection and characterization of promiscuous inhibitors. J Med Chem 51:574–580
54. Freyer MW, Lewis EA (2008) Isothermal titration calorimetry: experimental design, data analysis, and probing macromolecule/ligand binding and kinetic interactions. Methods Cell Biol 84:79–113
55. Velazquez-Campoy A et al (2004) Isothermal titration calorimetry. Curr Protoc Cell Biol Chapter 17, Unit 17.8
56. Breeze AL (2000) Isotope-filtered NMR methods for the study of biomolecular structure and interactions. Prog NMR Spectrosc 36:323–372
57. Tugarinov V, Kay LE (2005) Methyl groups as probes of structure and dynamics in NMR studies of high-molecular-weight proteins. Chembiochem 6:1567–1577
58. Li D, DeRose EF, London RE (1999) The inter-ligand Overhauser effect: a powerful new NMR approach for mapping structural relationships of macromolecular ligands. J Biomol NMR 15:71–76
59. Becattini B, Pellecchia M (2006) SAR by ILOEs: an NMR-based approach to reverse chemical genetics. Chem Eur J 12:2658–2662
60. Sanchez-Pedregal VM et al (2005) The INPHARMA method: protein-mediated interligand NOEs for pharmacophore mapping. Angew Chem Int Ed 44:4172–4175
61. Bartoschek S et al (2010) Drug design for G-protein-coupled receptors by a ligand-based NMR method. Angew Chem Int Ed 49:1426–1429
62. Ladbury JE, Klebe G, Freire E (2010) Adding calorimetric data to decision making in lead discovery: a hot tip. Nat Rev Drug Discov 9:23–27
63. Freire E (2009) A thermodynamic approach to the affinity optimization of drug candidates. Chem Biol Drug Des 74:468–472
64. Ruben AJ, Kiso Y, Freire E (2006) Overcoming roadblocks in lead optimization: a thermodynamic perspective. Chem Biol Drug Des 67:2–4
65. Freire E (2008) Do enthalpy and entropy distinguish first in class from best in class? Drug Discov Today 13:869–874
66. Ferenczy GG, Keserű GM (2010) Thermodynamics guided lead discovery and optimization. Drug Discov Today 15:919–932
67. Adams H et al (2001) Quantitative determination of intermolecular interactions with fluorinated aromatic rings. Chemistry 7:3494–3503

68. Swinney DC (2009) The role of binding kinetics in therapeutically useful drug action. Curr Opin Drug Discov Devel 12:31–39
69. Zhang R, Monsma F (2009) The importance of drug-target residence time. Curr Opin Drug Discov Devel 12:488–496
70. Tummino PJ, Copeland RA (2008) Residence time of receptor-ligand complexes and its effect on biological function. Biochemistry 47:5481–5492
71. Copeland RA, Pompliano DL, Meek TD (2006) Drug-target residence time and its implications for lead optimization. Nat Rev Drug Discov 5:730–739
72. Williams JW, Morrison JF, Duggleby RG (1979) Methotrexate, a high-affinity pseudosubstrate of dihydrofolate reductase. Biochemistry 18:2567–2573
73. Magotti P et al (2009) Structure-kinetic relationship analysis of the therapeutic complement inhibitor compstatin. J Mol Recognit 22:495–505
74. Lu H et al (2009) Slow-onset inhibition of the FabI enoyl reductase from Francisella tularensis: residence time and in vivo activity. ACS Chem Biol 4:221–231
75. Wan H, Rehngren M (2006) High-throughput screening of protein binding by equilibrium dialysis combined with liquid chromatography and mass spectroscopy. J Chromatogr A 1102:125–134
76. Lee KJ et al (2003) Modulation of nonspecific binding in ultrafiltration protein binding studies. Pharm Res 20:1015–1021
77. Rich RL et al (2001) High-resolution and high-throughput protocols for measuring drug/human serum albumin interactions using biacore. Anal Biochem 296:197–207
78. Valko K et al (2003) Fast gradient HPLC method to determine compounds binding to human serum albumin. Relationships with octanol/water and immobilized artificial membrane lipophilicity. J Pharm Sci 92:2236–2248
79. Eddershaw PJ, Beresford AP, Baylis MK (2000) ADME/PK as part of a rational approach to drug discovery. Drug Discov Today 5:409–414
80. Hop CE et al (2008) High throughput ADME screening: practical considerations, impact on the portfolio and enabler of in silico ADME models. Curr Drug Metab 9:847–853
81. Hsieh Y (2008) HPLC-MS/MS in drug metabolism and pharmacokinetic screening. Expert Opin Drug Metab Toxicol 4:93–101
82. Pang KS (2003) Modeling of intestinal drug absorption: roles of transporters and metabolic enzymes (for the Gillette review series). Drug Metab Dispos 31:1507–1519
83. Muenster U et al (2008) Characterization of substrates and inhibitors for the in vitro assessment of bcrp mediated drug-drug interactions. Pharm Res 25:2320–2326
84. Iwatsubo T et al (1997) Prediction of in vivo drug metabolism in the human liver from in vitro metabolism data. Pharmacol Ther 73:147–171
85. Prakash C, Shaffer CL, Nedderman A (2007) Analytical strategies for identifying drug metabolites. Mass Spectrom Rev 26:340–369
86. Riley RJ, McGinnity DF, Austin RP (2005) A unified model for predicting human hepatic, metabolic clearance from in vitro intrinsic clearance data in hepatocytes and microsomes. Drug Metab Dispos 33:1304–1311
87. Ito K et al (1998) Prediction of pharmacokinetic alterations caused by drug-drug interactions: metabolic interaction in the liver. Pharmacol Rev 50:387–412
88. Moody GC et al (1999) Fully automated analysis of activities catalysed by the major human liver cytochrome P450 (CYP) enzymes: assessment of human CYP inhibition potential. Xenobiotica 29:53–75
89. Trubetskoy OV, Gibson JR, Marks BD (2005) Highly miniaturized formats for in vitro drug metabolism assays using vivid fluorescent substrates and recombinant human cytochrome P450 enzymes. J Biomol Screen 10:56–66
90. Zientek M et al (2008) Development of an in vitro drug-drug interaction assay to simultaneously monitor five cytochrome P450 isoforms and performance assessment using drug library compounds. J Pharmacol Toxicol Methods 58:206–214
91. Wang RW et al (2000) Human cytochrome P-450 3A4: in vitro drug-drug interaction patterns are substrate-dependent. Drug Metab Dispos 28:360–366
92. Grimm SW et al (2009) The conduct of in vitro studies to address time-dependent inhibition of drug-metabolizing enzymes: a perspective of the pharmaceutical research and manufacturers of America. Drug Metab Dispos 37:1355–1370
93. Atkinson A, Kenny JR, Grime K (2005) Automated assessment of time-dependent inhibition of human cytochrome P450 enzymes using liquid chromatography-tandem mass spectrometry analysis. Drug Metab Dispos 33:1637–1647
94. Ma B, Prueksaritanont T, Lin JH (2000) Drug interactions with calcium channel blockers:

possible involvement of metabolite-intermediate complexation with CYP3A. Drug Metab Dispos 28:125–130

95. Grime KH et al (2009) Mechanism-based inhibition of cytochrome P450 enzymes: an evaluation of early decision making in vitro approaches and drug-drug interaction prediction methods. Eur J Pharm Sci 36:175–191

96. Tiller PR, Romanyshyn LA, Neue UD (2003) Fast LC/MS in the analysis of small molecules. Anal Bioanal Chem 375:788–802

97. Hsieh Y, Korfmacher A (2006) Increasing speed and throughput when using HPLC-MS/MS systems for drug metabolism and pharmacokinetic screening. Curr Drug Metab 7:479–489

Chapter 13

Biophysical Screening for the Discovery of Small-Molecule Ligands

Alessio Ciulli

Abstract

Discovering small-molecule chemical probes of protein function has great potential to elucidate biological pathways and to provide early-stage proof-of-concept for target validation. Discovery of such probes therefore underpins many of the chemical biology and drug discovery efforts in both academia and the pharmaceutical industry. The process generally begins with screening small molecules to identify *bona fide* "hits" that bind non-covalently to a target protein. This chapter is concerned with the application of biophysical and structural techniques to small-molecule ligand screening, and with the validation of hits from both structural (binding mode) and energetic (binding affinity) stand-points. The methods discussed include differential scanning fluorimetry (thermal shift), fluorescence polarization (FP), surface plasmon resonance, ligand-observed NMR spectroscopy, isothermal titration calorimetry, and protein X-ray crystallography. The principles of these techniques and the fundamental nature of the observables used to detect macromolecule-ligand binding are briefly outlined. The practicalities, advantages, and disadvantages of each technique are described, particularly in the context of detecting weak affinities, as relevant to fragment screening. Fluorescence-based methods, which offer an attractive combination of high throughput and low cost are discussed in detail. It is argued that applying a combination of different methods provides the most robust and effective way to identify high-quality starting points for follow-up medicinal chemistry and to build structure–activity relationships that better inform effective development of high-quality, cell-active chemical probes by structure-based drug design.

Key words Fragment screening, Biophysical techniques, Differential scanning fluorimetry, Thermal shift, Fluorescence polarization, Calorimetry, NMR spectroscopy, X-ray crystallography, Surface plasmon resonance

1 Introduction

The discovery and design of small molecules that modulate or probe biological systems motivates much of the present research in chemical biology and drug discovery. The spatial, temporal, and dose-dependent controls of biomolecular activity that are afforded by small molecules have advantages for systematic studies of complex biological processes in comparison to the more traditional

Mark A. Williams and Tina Daviter (eds.), *Protein-Ligand Interactions: Methods and Applications*, Methods in Molecular Biology, vol. 1008, DOI 10.1007/978-1-62703-398-5_13, © Springer Science+Business Media New York 2013

gene knockouts or RNAi approaches (1, 2). Potent, selective, and cell-penetrant small-molecule binders, often referred to as "chemical probes," provide powerful tools to aid elucidation of protein function inside the cell (3–5). In parallel, recent advances in our molecular understanding of many diseases have revealed many new potential targets for small-molecule intervention, significantly expanding the druggable genome (6, 7). The ability to rapidly and robustly discover lead compounds against this increasing range of targets would provide starting points for drug discovery that would significantly impact on our ability to develop the next generation of medicines for clinical application. The identification of biologically active small molecules is however expensive and highly demanding in terms of resources and know-how, inevitably requiring multidisciplinary approaches at the interface of chemistry and biology.

Since the late 1990s within the pharmaceutical industry, and in the past decade in academic and research institutions, significant investment and efforts have focused on high-throughput screening (HTS) of large library collections (>100,000 compounds). Typically, highly robotized complex bioassays are set up using the purified target protein and required labels or assay-components, or as target-based whole cell screens. Hits are identified that exhibit a statistically significant level of activity or inhibition at relatively low concentration (typically 10–100 μM) (8–11). Although these approaches have proven successful at identifying biologically active compounds, direct binding is rarely measured and the assays are consequently known to be significantly prone to artifacts that arise, e.g., from compound aggregation, interference of the compound with the assay or off-target effects (12–14). Significant hit triage efforts are therefore required to deconvolute the true mechanism that underpins the observed response in the assay. It is becoming increasingly apparent that hits identified from these high-throughput screens rarely behave as genuine, reversible small-molecule binders for a given protein target.

More recently, however, significant advances in analytical, biophysical and structural techniques for monitoring weak-to-moderate binding affinities of protein–ligand interactions have facilitated the development and success of fragment-based drug discovery. In fragment screening, compared to HTS, smaller libraries (usually ~1,000 and rarely >10,000) of compounds of relative small size (MW usually <300 Da) are screened at higher concentration (usually >0.5 mM) for direct, non-covalent binding to the target protein (15, 16). It is now widely accepted that *bona fide* binding hits, even if of low complexity and of weak affinities, represent high quality, attractive points for further medicinal chemistry optimization. More information on the concepts and applications of fragment-based drug discovery are available in several seminal papers and recent reviews (17–25).

One of the results of these recent developments is that fragment screening is now firmly established as an early-stage lead discovery approach, very often performed in parallel with HTS against the target of interest. A second corollary to this approach is that biophysical and structural methods, which were previously only used for quality controls or during the late stages of lead optimization, are now being increasingly used for screening and validation during the early stages of the discovery process. Albeit typically of lower throughput than bioassays used in HTS, biophysical and structural techniques are highly information-rich and thus very valuable early in the development process (15, 16, 26, 27). A number of advantages of biophysical and structural techniques over the complex or indirect bioassays used for HTS provide strong motivation for their increased use:

- They allow a direct measurement of binding, so are less prone to artifacts due to compound aggregation and interference with the assay;
- They are generally applicable to any protein target class, specifically they do not require an active enzyme or knowledge of the protein's function;
- They enable detection and characterization of low affinities, so are particularly amenable for screening fragment-libraries;
- Many biophysical techniques are available, each with different strengths and weaknesses, and it is valuable to apply multiple methods that monitor different "observables";
- Quantitative measurements of direct binding (K_d) or indirect dose–response effects (IC_{50}) provide reliable ways to develop structure–activity relationships early in the drug or probe development process.
- The structure of the protein–ligand complex or at least some details on the location of the binding site and the binding mode of the compound can be obtained. This information is often critical for subsequent optimization and development of the compounds.

A panel of biophysical techniques—fluorescence-based thermal denaturation/differential scanning fluorimetry (DSF) and fluorescence polarization (FP) assays, isothermal titration calorimetry (ITC), nuclear magnetic resonance spectroscopy (NMR), surface plasmon resonance (SPR), and protein X-ray crystallography (PX)—can be used to monitor and characterize protein–ligand interactions (Fig. 1). Here, I will briefly review the principles of their operation, the advantages and disadvantages of each technique, and the practicalities of their utilization in the context of screening, with a focus on the specific applications of these methods to fragment-based ligand discovery.

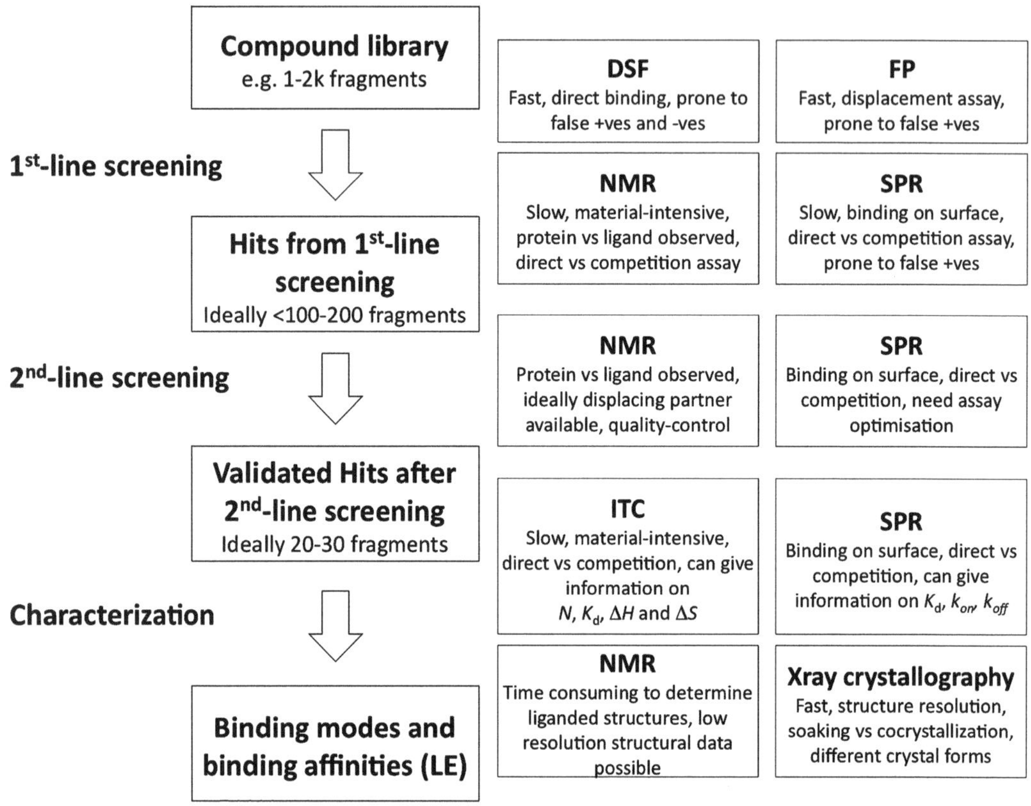

Fig. 1 Flowchart of a possible strategy for compound screening, validation, and characterization using biophysical techniques

1.1 Differential Scanning Fluorimetry (Thermal Shift) Assay

Proteins exist in thermodynamic equilibrium between multiple conformational states and the binding of a small-molecule to the protein will alter the populations of these states. In the simplest case we can consider a two-state system where there is only a folded (native) and an unfolded (denatured) state (Fig. 2). The population of the unfolded state of a protein increases as the temperature of the solution is increased. Usually, specific binding of a small-molecule to a structurally defined site of the native state of a protein will stabilize the native state more than any nonspecific interaction with the denatured state and hence increase the free energy difference between the two states, ΔG_{D-N} (Fig. 2). The effect of this will be to increase the population of the native state at all temperatures and will result in a shift of the melting temperature of the protein (T_m), the temperature at which there is 50 % denaturation, to a higher value. Consequently, by measuring T_m in the presence and absence of a potential ligand it is possible to detect any protein–ligand binding.

There are several long-established ways of measuring the T_m of a protein, e.g., through changes in its secondary structure content

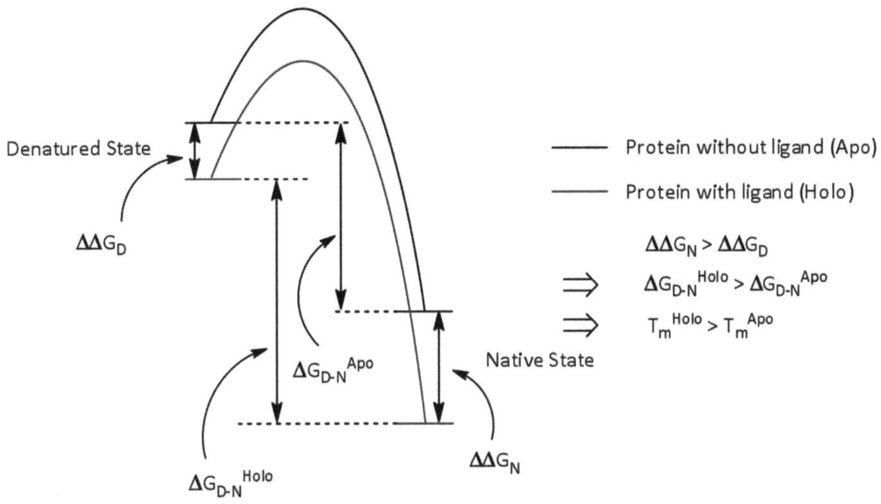

Fig. 2 Energy profile diagram showing protein stabilization by ligand binding. Stabilization of the native, folded state by specific binding of a ligand results in a greater free energy difference between native and denatured states (ΔG_{D-N}) and a higher proportion of folded protein at all temperatures

monitored by circular dichroism or infra-red spectroscopy, through monitoring the heat of the transition by differential scanning calorimetry, or through monitoring the temperature dependent changes in intrinsic fluorescence of a protein due to increased solvent exposure of tryptophan residues in the unfolded state. However, another approach, much better suited to HTS applications, is the so-called "thermal shift" or "thermofluor" DSF assay (28–31).

This, now popular, approach monitors the temperature dependence of the fluorescence signal of an environmentally sensitive fluorescent dye that binds preferentially to the denatured state of a protein. SYPRO Orange is a commercially available dye that is commonly used to measure the change between native and denatured states of a protein. The dye's fluorescence is quenched in aqueous solution. However, upon binding to a hydrophobic surface a fluorescent signal is emitted. Upon denaturation, the hydrophobic core of the protein becomes solvent exposed and hence the fluorescent dye now has a much larger hydrophobic surface area to bind to relative to a native state protein. Hence by monitoring the fluorescent signal, it is possible to determine the extent of denaturation. Examination of differences in the temperature dependent fluorescence profile of protein plus dye in the presence and absence of a potential ligand may reveal a change in T_m indicative of binding.

A plot of fluorescence signal against temperature should give a sigmoidal plot (Fig. 3a). The melting temperature is determined by the point of inflection of this curve. This can most easily be assessed by plotting the derivative of the fluorescent signal against temperature (Fig. 3b).

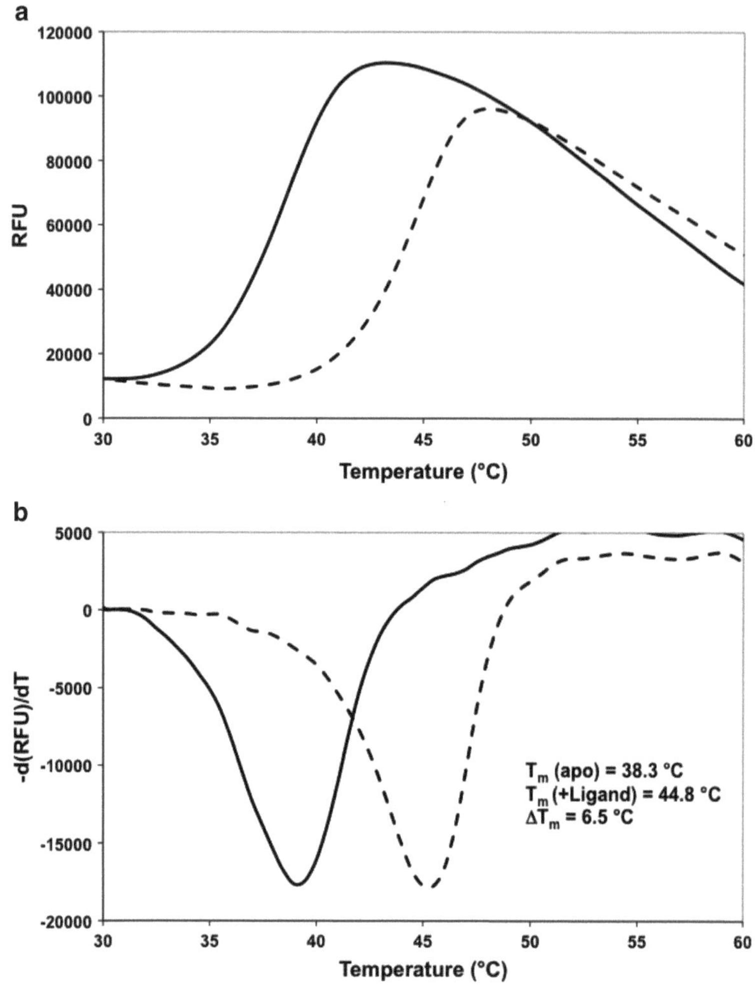

Fig. 3 Monotoring differential melting behavior of a protein and a protein–ligand complex by fluorescence. (**a**) Typical DSF plots of a protein in the absence (*solid line*) and presence (*dashed line*) of a binding ligand. (**b**) The derivative of the fluorescent signal is plotted against temperature. The minimum of this derivative plots allow convenient identification of the melting temperatures T_m, from which a thermal shift ΔT_m can be measured. Data are shown for the binding of a small molecule inhibitor against the enzyme pantothenate synthetase from *Mycobacterium tuberculosis* (50)

DSF is being extensively used in high-throughput assays for small-molecule hit identification as it can be easily implemented in microplate formats using a real-time thermal cycler instrument (see Subheadings 2.1 and 3.1 for details).

1.2 Fluorescence Polarization Assay

Measurements of FP and the related anisotropy reveal information on molecular mobility, which is dependent on size and shape. Specifically, the extent of polarization depends on the rotational correlation time of the fluorophore. Processes that significantly alter the rate of rotation of a fluorophore can be monitored through changes in polarization. Such process include the binding of a fluorescent ligand to a protein, in which case the ligand will rotate much more slowly, or changes in the shape or oligomeric state of an intrinsically fluorescent or fluorescently labeled protein induced by ligand binding. In FP, plane-polarized light is used to excite a fluorophore. Experimentally, the degree of polarization is determined from measurements of fluorescence intensities parallel (F_{II}) and perpendicular ($F\perp$) to a plane, and can be expressed in terms of FP (P) or anisotropy (r):

$$P = (F_{\mathrm{II}} - F\perp)/(F_{\mathrm{II}} + F\perp) \quad (1)$$

$$r = (F_{\mathrm{II}} - F\perp)/(F_{\mathrm{II}} + 2F\perp) \quad (2)$$

P and r are both relative quantities with little dependence on dye concentration or fluorescence intensity changes. Consequently, polarization-based readouts are less dye dependent and less susceptible to environmental interferences, such as pH changes, than assays based on fluorescence intensity measurements. The magnitude of both P and r increases as the rotation of the fluorophore slows (*see* **Note 1**).

Most frequently, an intrinsically fluorescent ligand is used or a fluorescent dye is attached to small, rapidly rotating molecules, e.g., peptides targeting a protein–protein interface (see Subheading 2.2 for details) (32). When these fluorescent probes are free in solution, rapidly rotating in the absence of their target protein, the initially photoselected orientational distribution becomes randomized prior to emission, resulting in low FP. Conversely, binding of the fluorescent probe to a large, slowly rotating protein maintains much of the high initial FP. FP therefore provides a direct readout of the extent of binding of a fluorescent probe to proteins and other biopolymers, providing a robust assay for screening small molecules that compete with the probe for the same binding site.

Since not all small molecules are fluorescent, FP assays for screening compound libraries tend to be carried out in a competitive inhibition mode by titrating different concentrations of small molecules against a sample containing standard concentrations of protein and a fluorescent-version of a known ligand to generate a dose–response curve, which can be used to determine an IC_{50} (and through back-calculation a K_d, see Subheading 3.2.2 for details) (33). The choice of these fixed concentrations of target protein and fluorescent ligand throughout the assay is very important and should be optimized taking into account several parameters (see Subheading 3.2.1 for details). This competition mode is particularly useful for screening as the small molecules to be tested do not themselves need to be fluorescent and the screen is selective for binding mechanisms which affect the binding of the known fluorescent ligand either by direct competition or allosterically.

1.3 Isothermal Titration Calorimetry

The strength of ITC lies in its ability to measure directly the heat associated with a chemical reaction in solution. Because measurable heat uptake or release accompanies almost all reactions, ITC is broadly applicable to characterization of protein–ligand interactions and relatively simple to carry out, as no fluorescent labels or modification of the protein or ligand for surface attachment are required. ITC is also the only method currently able to directly measure the enthalpy, ΔH, of a ligand binding to a protein (34).

An ITC experiment proceeds by injection of a solution containing one component of the reaction (usually the ligand) into a temperature controlled stirred-cell containing the other component. In the first few injections most of the ligand will bind to the protein, allowing measurement of the enthalpy, ΔH. As the experiment proceeds and the protein saturates with ligand the signal diminishes allowing the estimation of the affinity and stoichiometry. At the end of the titration full saturation is achieved and mainly the background heat of dilution is observed (Fig. 4a).

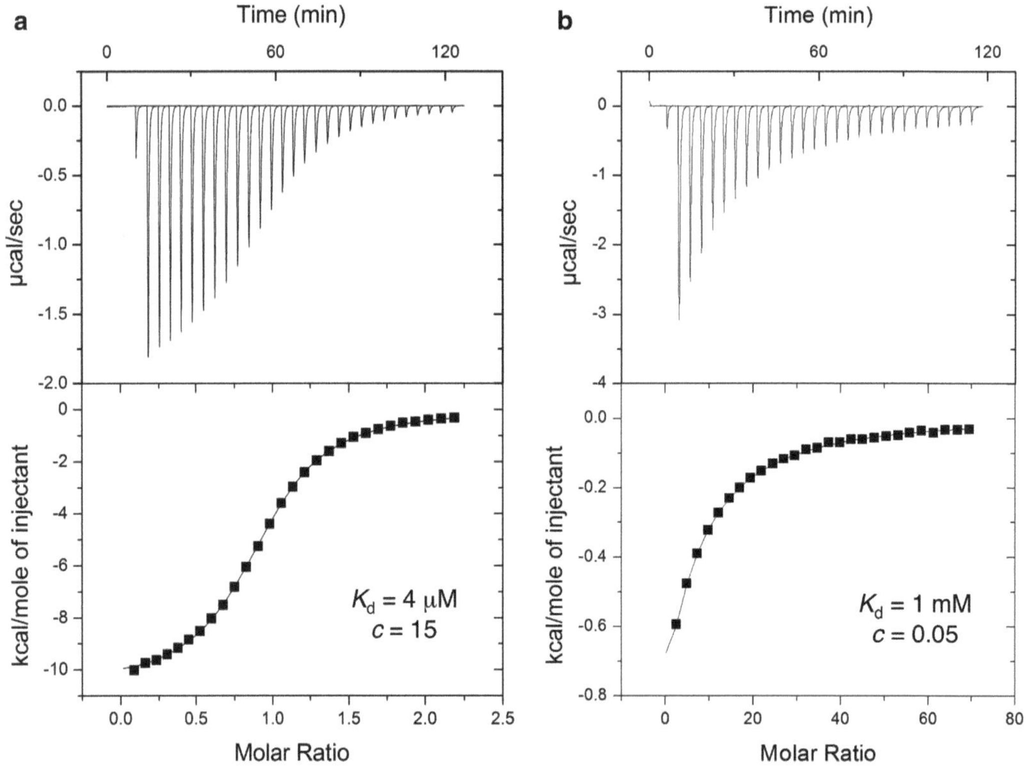

Fig. 4 Characterization of protein–ligand binding using ITC. Typical calorimetric titrations to study high affinity (**a**) and low affinity (**b**) interactions (i.e., high *c* and low *c* conditions, respectively). Low *c* curves are frequently observed with low affinity fragments. Data are shown for the binding of ATP (*left*) and of a small molecule fragment (*right*) against the enzyme pantothenate synthetase from *Mycobacterium tuberculosis* (50)

Analysis of the integrated heats from each of the injections can determine the association constant (K_a), the enthalpy of binding (ΔH) and the stoichiometry (n). The free energy change due to binding, ΔG, is directly related to K_a by the equation $\Delta G = -RT \ln(K_a)$, while the entropy of binding (ΔS) can be directly calculated using the thermodynamic equation $\Delta G = \Delta H - T\Delta S$.

A critical parameter that determines the shape of the binding isotherm is the so-called c-value,

$$c = nK_a P_T \quad (3)$$

where n is the stoichiometry and P_T the total protein concentration. There are two distinct "regimes" that we should consider separately: high c values ($c > 10$, see Fig. 4a) and low c values ($c < 10$), see Fig. 4b) (35). A comparison of how the different titration curves are predicted based on the different c values conditions is shown by the simulations in Fig. 5. The low affinity, low c, regime is much more common for fragment binding. The design and analysis of ITC experiments in both regimes are considered in detail in the Chapter 4.

1.4 Surface Plasmon Resonance

Surface plasmon resonance (SPR) is an optical technique based on the transfer of light (electromagnetic) energy to electrons in a thin layer of metal in contact with a solution. Gold is the preferred metal as it is compatible with a number of linking chemistries and will not oxidize over time.

In the standard SPR set up, a beam of polarized monochromatic light is shone through a prism at a thin-layer of gold coating one surface of the prism. The prism causes the light to be reflected at the gold-coated surface. However, light is not reflected precisely at the prism-gold junction, but it (or its electromagnetic field) penetrates some distance into and beyond the gold (in a phenomenon called evanescent wave formation). At a particular angle of incidence, absorption of some of the light by the electrons in the gold excites charged density waves, called "surface plasmons," which propagate along the metal surface. This absorption is maximum where transfer of momentum matches that of the plasmons. At this resonance condition, i.e., at a specific incident angle, the intensity of the reflected light is reduced sharply. The evanescent wave extends ~100–200 nm into the solution and decays exponentially away from it. Consequently, if the gold layer is sufficiently thin the resonance condition/angle depends not only on the metal, but also on the properties (refractive index) of the medium just above the gold surface. SPR is thus highly sensitive to changes in the environment close to the gold—aqueous solution interface, while processes in bulk solution have no influence on the angle of minimum reflectance. A change in the refraction index at the surface of the sensor (due for example to something binding near the surface) may hence be monitored as a shift in the resonance angle (36).

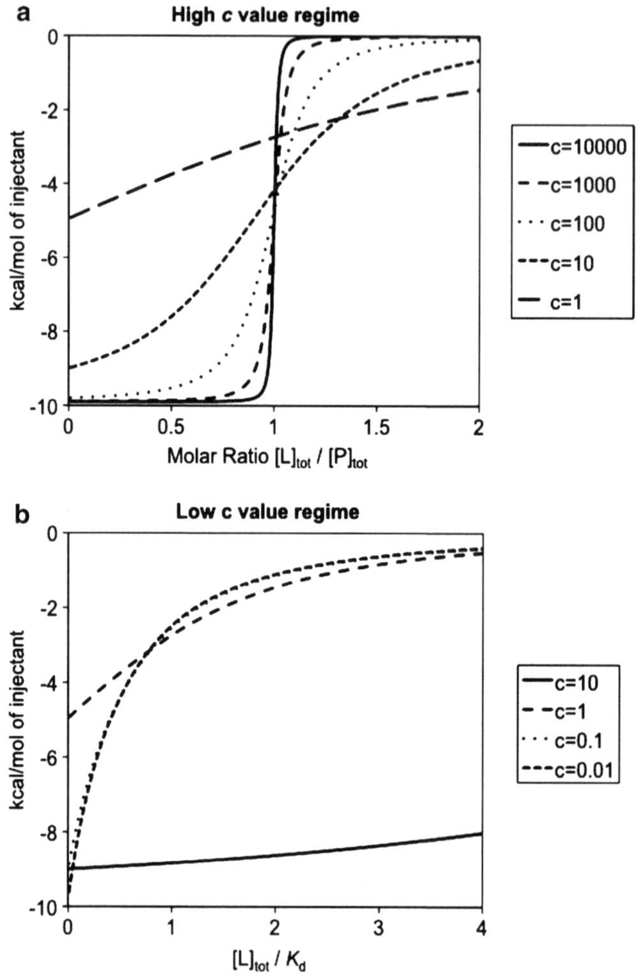

Fig. 5 Experimental design for ITC under (**a**) high and (**b**) low *c* value regimes, showing typical sigmoidal and hyperbolic curves, respectively. The resulting ΔH (kcal/mol) is plotted vs. the molar ratio of total ligand and protein concentrations (**a**) and the ratio between total ligand concentration and the dissociation constant (**b**). The simulated curves assume a ΔH of −10 kcal/mol except the last two curves in *panel b* ($c = 0.1$ and 0.01) in which case the curves are magnified by factors of 10 and 100, respectively, to emphasize their similar curved shapes (35)

In SPR, a "chip" is used which contains a glass surface that is coated by a thin layer of gold, required for the SPR response. A dextran matrix is covalently attached via a linker layer on the solution side of the gold film, to allow immobilization of receptor molecules, e.g., the protein on the surface (Fig. 6). When an analyte for example a ligand in solution binds to the protein the refractive index near the surface changes and an SPR shift is detected, which can be monitored in real time in a so-called "sensogram." Since the change in refractive index, i.e., the SPR signal, is proportional to

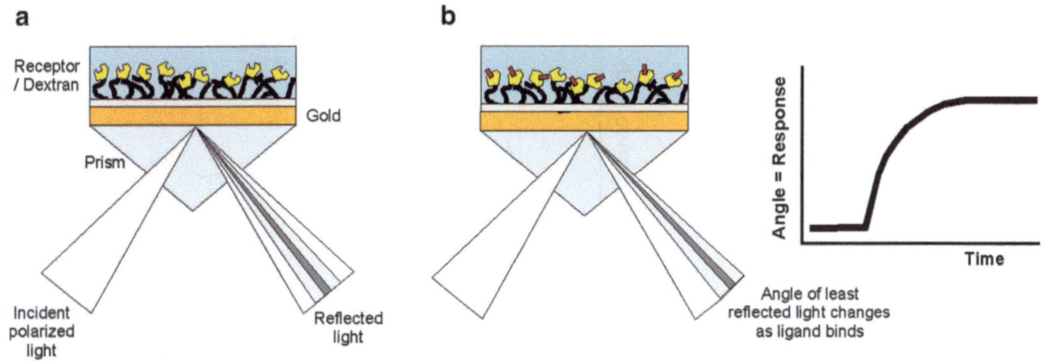

Fig. 6 Operation of a surface plasmon resonance biosensor. (**a**) The glass of the sensor chip is coated with a thin layer of gold, a dextran matrix is attached via a linker layer to the gold and receptors are cross-linked to the dextran. Light is reflected from the surface of a sensor chip and resonant absorption is seen at an angle dependent on the quantity of material bound at the surface. (**b**) When an interaction occurs, a shift in the resonant angle is observed that is proportional to the amount of material bound. A plot of resonance signal vs. time, the sensogram, can be monitored in real time

the mass bound at the surface, it is possible to measure the affinity and kinetic rate constant of the interaction.

Immobilization of the receptor molecule to the sensor surface is required and is of primary importance to the design of a successful assay. The coupling method must be efficient, must produce a highly stable association (to prevent signal drift) and must allow control of the amount of material immobilized (*see* **Note 2**).

Once the protein has been attached to the surface, the partner ligand can be flowed through the chip, and if a binding event occurs it can be directly monitored in a sensogram. A schematic representation of a typical sensogram trace is shown in Fig. 7 where signal increases until the protein binding sites are saturated, subsequently buffer without ligand is flowed over the chip and ligand is progressively removed. Both the on-rate and off-rate constants of the binding process can be determined and their ratio gives an accurate estimate of affinity. The ability to characterize slow off-rate ligands is particularly useful during the optimization of the pharmacokinetic properties of lead compounds and drugs.

Advances in instrument sensitivity and experimental design have allowed SPR to be established as a front line method for primary fragment screening; however, nonspecific effects due to the use of high concentrations of small molecules required to studying weak affinities of fragments can often be seen with this assay. Experimental details for small-molecule screening are given in the Chapter 6 including considerations particular to fragments.

1.5 Nuclear Magnetic Resonance Spectroscopy

NMR spectroscopy is a powerful technique to study protein–ligand and protein–protein interactions in a solution environment, and is being used extensively in the pharmaceutical industry for hit

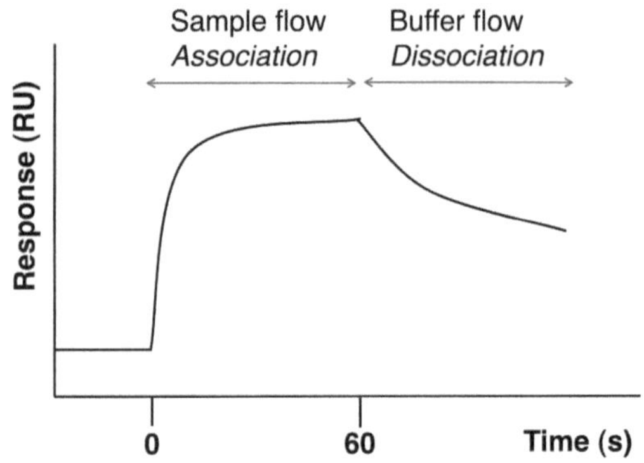

Fig. 7 Typical binding sensogram observed with an SPR biosensor. After baseline equilibration, at $t = 0$ s a solution containing one of the binding partners (e.g., a small molecule ligand) is flowed over the surface to which is attached the other binding partner. As the ligand binds to the protein, an increase in signal is observed due to the increase in material at the surface. Analysis of this part of the binding gives the observed rate constant (k_{obs}), from which the association rate constant of the interaction (k_{on}) is obtained, if the ligand concentration is known, using the equation $k_{obs} = k_{on}[L] + k_{off}$. The signal plateaus once equilibrium has been established, then (here at $t = 60$ s) buffer replaces the ligand solution and the protein–ligand complex starts dissociating. Analysis of this part of the binding curve gives the dissociation rate constant (k_{off}). The response level at equilibrium can yield the concentration of active ligand in the sample. The binding affinity can be calculated from the ratio of the rate constants ($K_d = 1/K_a = k_{off}/k_{on}$). A pulse of a regeneration solution (e.g., high salt, low pH, etc.) is then typically used to disrupt the non-covalent interaction and regenerate the surface. These types of curves are typically recorded over a range of ligand concentrations (affinity can also be determined from the ligand concentration dependence of the response), and often over a range of different temperatures, to allow for reliable determination of the kinetic and thermodynamic parameters

identification. Different experimental formats have been used which are based on observing either the NMR signals of the ligand or the protein. Improvements in instrumentation and advances in automation are facilitating rapid screening of increasingly large compound libraries (37).

Binding equilibria modulate both the frequency and width of NMR spectral lines in response to the rate of "chemical exchange" between the free and bound states of the ligand and receptor. Observation of these modulated spectral parameters forms the basis for all NMR screening experiments.

In a two-state equilibrium, ligand and protein molecules will exist in either a free (L, P) or complexed (PL) state. In the free state, both protein and ligand retain their intrinsic NMR parameters (e.g., chemical shifts, relaxation rates, translational and diffusion coefficients). In each other's presence, the mutual binding affinity of ligand and protein drives an exchange process that toggles both sets of molecules between the free and complexed states.

Under these conditions, the ligand transiently adopts NMR parameters characteristic of the typically much larger receptor. Alternatively, from the receptor's perspective, the ligand transiently perturbs the binding site microenvironment, and may alter the distribution of conformations sampled by the receptor molecules. In either case, the exchange modulates the NMR parameters of both molecules.

Since the ligand bound state is thermodynamically favored, the dissociation rate k_{off} is slower than association and, thus, k_{off} is the important limiting factor in defining the kinetics of the chemical exchange process. There are two distinct cases:

(a) Exchange is fast on the NMR time scale. Many cycles of protein–ligand formation and dissociation occur on the "NMR timescale," i.e., the reciprocal of the frequency differences of signals in the bound and free states. We refer to this as fast-exchange regime. This is the common scenario for moderate to weak affinity ligands, e.g., with a $k_{off} > 10^2\ s^{-1}$ (and typically a corresponding $K_d > 100\ \mu M$), such as fragments.

(b) The average lifetime of the protein–ligand complex is much longer than the NMR timescale, typically this may correspond to $k_{off} < 10\ s^{-1}$ and $K_d < 10\ \mu M$. This is the so-called slow-exchange regime.

Under the fast exchange regime, exchange-modulated NMR parameters can be described as simple sums. Therefore, a general NMR parameter Q becomes the simple fractional average of its value in the bound (f_{bound}) and free (f_{free}) populations:

$$Q_{avg} = f_{bound} \times Q_{bound} + f_{free} \times Q_{free} \quad (4)$$

Observed differences between Q_{avg} and Q_{free} provide a signature of binding and indicate a hit in a NMR screen based on that parameter (38).

NMR parameters that are employed as observables for ligand binding experiments in fragment screening include chemical shifts, relaxation times (*see* **Note 3**), and the nuclear overhauser effect (NOE).

1.5.1 Protein-Observed NMR Experiments

NMR spectroscopic techniques were amongst the first to be applied for fragment screening due to their flexibility and ability to detect weak interactions. Fesik and colleagues at Abbott Laboratories pioneered Structure–Activity Relationships (SAR) by NMR spectroscopy, in which perturbations of the two-dimensional ^1H-^{15}N HSQC (Heteronuclear Single Quantum Correlation) spectrum of a protein caused by fragment binding are used to obtain structure and affinity data (17). They demonstrated for the first time that this information could be used in a medium throughput format to detect hits and suggest ways to link them to form high-affinity leads. One of their first examples was the discovery of potent non-peptidic inhibitors of stromelysin, a zinc-dependent matrix metalloprotease and an important drug target (39).

This protein-based approach is readily implemented for screening libraries of compounds usually by adding several small molecules to the protein at a time, in order to improve throughput, and subsequently identifying hits by deconvoluting the positive mixtures. However, several drawbacks are associated with protein-based NMR methods.

(a) The size of the protein target should be <~40 kDa as larger proteins have weaker and overlapping NMR signals.

(b) A high concentration of protein is required (typically >0.1 mM).

(c) The protein needs to be labeled with isotopes, usually ^{15}N and often ^{13}C and/or ^{2}H.

(d) By measuring the changes in protein chemical shifts upon titrating a binding ligand it is possible to measure the K_d; however, this is only straightforward for binding of low-affinity ligands (fast-exchange and $K_d > [P]$).

(e) Some information on the ligand binding site can be obtained; however, this requires the peaks from the two-dimensional protein NMR spectra to be assigned (which can be a lengthy process).

1.5.2 Ligand-Observed NMR Spectroscopic Experiments

Most of the above limitations are in part addressed by NMR spectroscopic techniques that observe the signal of the ligand instead of the protein. Ligand-observed NMR experiments are generally faster, require less protein and enable direct identification of small molecule binders using simple one-dimensional spectra. These ligand-detected approaches render the molecular weight of the target protein irrelevant, making them of general applicability. Actually, in many ligand-observed NMR experiments the larger the protein target the better because complexation with larger proteins causes greater changes in the NMR parameters, and only very small proteins (MW < 10 kDa) may be problematic. Compounds can still be screened in mixtures, and, in my own experience, this works best with no more than three or four at a time in order to minimize their signal overlap (*see* **Note 4**) (40). Within mixtures, individual small molecules can be identified and binding characterized provided their chemical shifts in solution are known. NMR spectra of the ligand or the ligand-mixture are recorded in the presence of the protein and compared to control spectra recorded in the absence of the protein.

Most ligand-based NMR experiments that are employed in screening exploit the efficient transfer of the information on the ligand's bound state to the free ligand for detection in the fast-exchange regime. These experiments are typically carried out with

$L_T/P_T > 10$ (so the fraction of free ligand is always higher than that of bound ligand), and the binding compounds usually have $K_d > 10$ μM. The experimental conditions for fragment-based NMR screening are thus well suited to such fast exchange experiments.

Ligand-based NMR experiments have been described which take advantage of all these properties. Relaxation-edited NMR methods exploit the much faster signal relaxation (decay) of the bound ligand (41). Other techniques rely on the observation of change of sign of NOEs peaks for small organic molecules in the presence of the protein to detect binding. Two commonly used experiments are saturation transfer difference (STD) and Water-LOGSY. STD relies on transfer of magnetization (signal) directly from the protein to the bound ligand complex, e.g., by exciting the aliphatic methyl group region of the protein (Ala, Val, Leu, Ile) the NOE will result in transfer of magnetization to nearby protons of any bound ligand (42). In contrast, in WaterLOGSY the magnetization is transferred indirectly via water molecules at the binding site (43). Examples of the observations of ligand binding via these methods are shown in Fig. 8—in each case careful controls experiments are required to avoid false positive results, for example by repeating the experiments in the presence of a competitor ligand of known binding mode and affinity to provide some information on the specificity of the binding interaction. Further details can be found in the Chapter 14.

1.6 Protein X-Ray Crystallography

Historically, protein crystallography was a slow, resource-intensive and time-consuming technique that was used only for lead optimization. Recent years, however, have seen major advances in protein expression, methods for crystallization and structure determination, and it has become easier to access dedicated world-class facilities for X-ray data collection, e.g., at synchrotrons. These transformations have enabled X-ray crystallography to impact more broadly in the drug discovery process. The applications of X-ray crystallography as a screening tool for fragment-based drug discovery were pioneered in the late 1990s by Abbott in the USA and Astex Therapeutics in the UK (44, 45).

In order to apply protein X-ray crystallography in a screening context, the protein target must first be crystallized and its structure solved (*see* **Note 5**). Secondly, a well-established crystallization process and an ability to reliably obtain crystals of protein–ligand complexes is a crucial step in any screening efforts. Protein crystals are exposed to an X-ray source and a diffraction pattern is recorded. It is possible to use this information to reconstruct an electron density map of the molecule causing such a diffraction pattern, thereby allowing one to solve the crystal structure of the protein. If a small molecule is bound to the protein in the crystal, this can be rapidly identified by inspection of differences between the electron

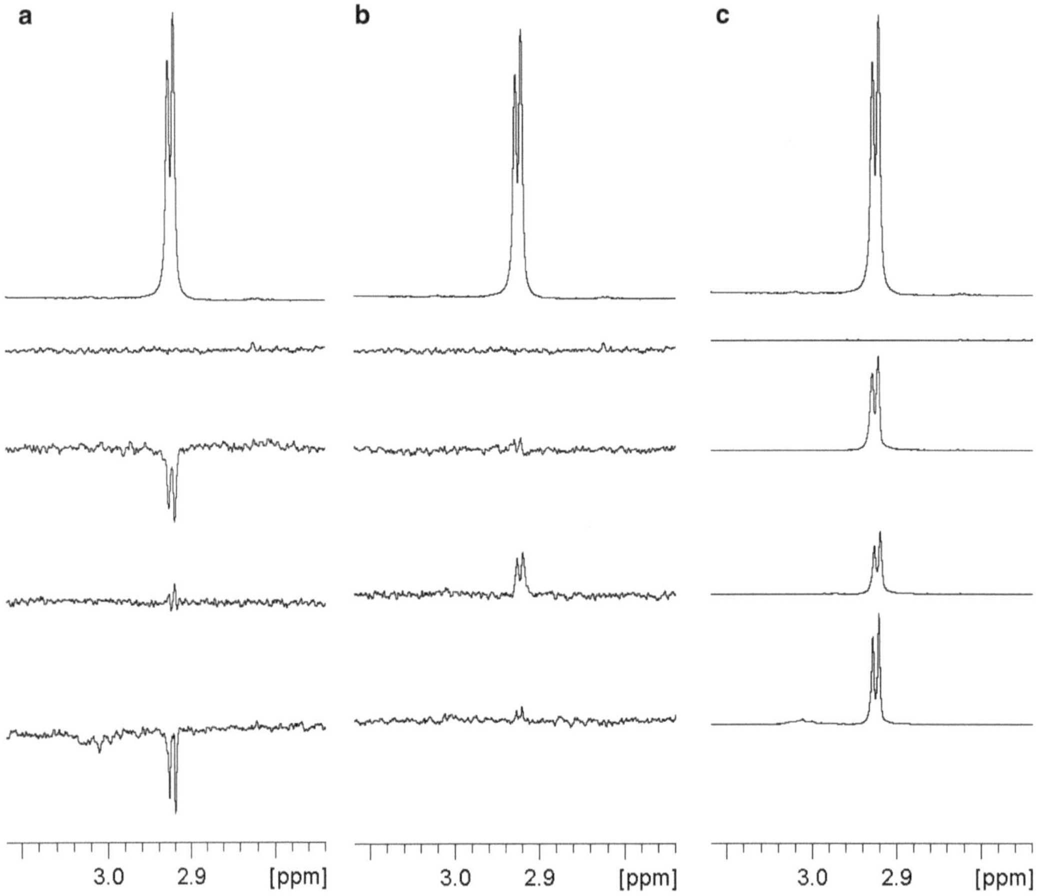

Fig. 8 Identification of ligand binding to a protein using 1D ^{1}H ligand-observed NMR spectroscopy. (**a**) WaterLOGSY, (**b**) STD, and (**c**) relaxation-edited binding and displacement experiments are shown. From top to bottom: normal ^{1}H spectrum of ligand in the absence of protein; a control spectrum of buffer alone; ligand in the absence of protein; ligand in the presence of protein; ligand in the presence of protein and a known high affinity binder (a "displacer"). The spectra show a doublet from a methyl group adjacent to an amide NH of a fragment ligand binding to the human bromodomain of BAZ2B protein (MW = 13.6 kDa, which is on the lower limit for ligand-based NMR techniques), displaced by a high-affinity peptide H3Kac14 (49). The restoration of a spectrum similar to ligand alone by addition of the displacer demonstrates that the ligand binds at the same site. A subsequent X-ray crystal structure of the protein-fragment complex demonstrated that the fragment bound at the Kac binding site of the bromodomain

density maps of the complex and the protein alone. Detailed examination of these differences and fitting of molecular models into the density enable the identification of the small molecule and its binding mode (Fig. 9).

In some cases, the crystal form of the unliganded protein is not suitable for soaking of small molecules, requiring laborious co-crystallization trials for each small molecule. General procedures for protein X-ray crystallography, ligand soaking and co-crystallization are covered in detail in Chapter 17 (*see* **Note 6**). Obtaining a

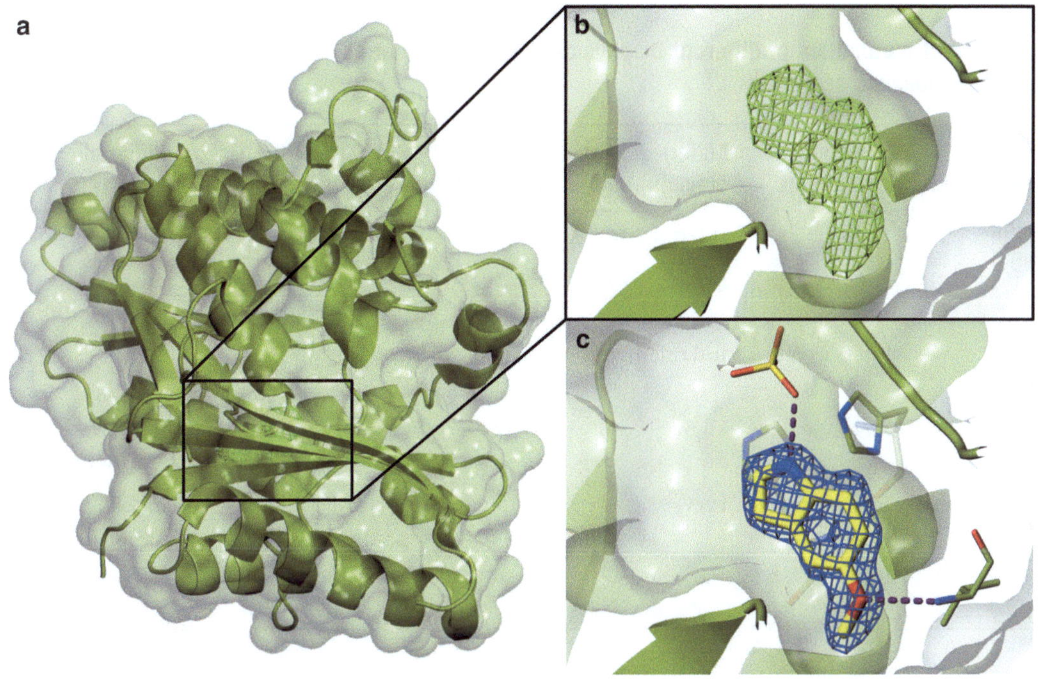

Fig. 9 Identification of a fragment bound to a protein by X-ray crystallography. (**a**) Crystal structure of a protein with a fragment bound. Protein backbone atoms are shown as green ribbons with a transparent superposed van der Waals surface. (**b**) Visualization of difference electron density (mesh) unexplained by the molecular model of the apo-protein highlights the location of the small molecule binding site. This initial difference $F_o - F_c$ electron density map (contoured to 3σ) allows rapid identification of the bound fragment. (**c**) Final $2F_o - F_c$ electron density map (mesh, and contoured to 1σ) upon further refinement including the bound ligand confirms the identity of the fragment and its binding mode. The figures show binding of 5-methoxyindole (a fragment) to the enzyme pantothenate synthetase from *Mycobacterium tuberculosis* (51)

crystal structure of a ligand bound to the target protein is important to inform rational design and careful optimization of the compounds affinities and physicochemical properties by medicinal chemistry based approaches. This is routinely conducted within structure-guided hit-to-lead and lead optimization programs for drug discovery (see Chapter 19).

1.7 Concluding Remarks

In summary, the biophysical techniques and approaches described herein have become widely implemented in small-molecule ligand discovery efforts in both academic and industrial research laboratories worldwide. Each biophysical technique described has diverse but highly complementary sets of capabilities (summarized in Table 1) and particular advantages and disadvantages with regard to compound screening, hit validation, and characterization (summarized in Table 2).

Based on the author's experience in participating to several small molecule discovery and drug design projects, it is crucial to

Table 1
Summary of parameters of different biophysical techniques that are important for small molecule screening

Technique	Screening throughput	Material consumption	Covalent immobilization	Detectable K_d range	Binding site information
DSF	High	Intermediate	None	Up to 5 mM	None
FP	High	Low	None	Down to K_d of probe	Limited to competition
NMR	Intermediate	Intermediate	None	Low nM—10 mM	Good
ITC	Low	High	None	Low nM—5 mM	Limited to competition
SPR	Intermediate	Low	Required	pM—2 mM	Limited to competition
PX	Low	Intermediate	None	Up to ligand solubility	Excellent

identify several *bona fide* binding ligands and to gain some structural information of their interactions with the target protein as early on in a program as possible. To this end, it is strongly argued that applying a combination of orthogonal methods provides the most robust and effective way to identify attractive starting points and to build structure–activity relationships early to better inform effective decision-making during the development of chemical probes by medicinal chemistry. One strategy that is proposed to achieve this is to apply an integrated screening cascade of biophysical techniques, starting from the more high-throughput methods, e.g., DSF and/or FP to screen across a large compound library, then enriching and validating the hit list, e.g., by NMR and/or SPR secondary screens, and ultimately using the more material intensive but more information-rich techniques, e.g., ITC and/or PX to characterize the compounds of interest, see Fig. 1 and (23). This strategy is broadly practicable in most research-intensive academic institutions, and can be feasible even within the constraints and often limited resources of academic laboratories.

To date, the implementation of such an integrated, multidisciplinary approach has been hampered, at least within academia, in part by the fact the individual laboratories have traditionally tended to develop strong, deep expertise in individual techniques, e.g., NMR vs. PX or ITC vs. SPR. In addition, many techniques were previously considered too expensive to justify their application in most projects. However, recent developments in instrumentation and automation, coupled to reduction to costs and increased sharing of technical expertise, e.g., across and between industry and

Table 2
Summary of relative advantages and disadvantages of different biophysical techniques with regard to small molecule screening, validation, and characterization

Technique	Advantages	Disadvantages
DSF	• High throughput • Applicable to most target proteins • Direct binding assay	• Prone to false positives and false negatives • Material intensive • Across-plate variability
FP	• High throughput • Applicable to most target proteins • Competition binding assay	• Prone to false positives and artifactual effects • Requires labelling of known ligand
NMR (ligand-observed)	• Intermediate throughput • Applicable to most target proteins (>10 kDa) • Provides quality control	• Prone to false positives due to compound aggregation or nonspecific effects
NMR (protein-observed)	• Intermediate throughput • Can identify binding site (need peak assignment) • Can measure K_d from ligand titrations	• Limited to small (<30 kDa) and soluble target proteins • Requires expensive isotope labelling of target proteins • Material intensive
ITC	• Direct/competition binding assays • Applicable to most target proteins • High information content (K_d, ΔH, ΔS and n)	• Low throughput • Material intensive • Requires large heat changes upon binding for reliable measurements
SPR	• Label-free detection and ease of automation • Applicable to most target proteins • Direct/competition binding assays • Low material consumption • High information content (K_d, k_{on}, k_{off} and n)	• Requires immobilization of one of the binding partner to a surface • Prone to artifacts due to compound aggregation or nonspecific effects • Requires time-consuming assay optimization
PX	• Intermediate throughput • Can rapidly identify binding site and ligand binding mode • Can directly identify ligand-induced conformational changes	• Limited to soluble target protein that can be crystallized • Requires expensive X-ray sources (in-house, access to synchrotrons) • Crystal packing may occlude binding site • High occupancy of the ligand binding site required

academia, are now making the entire platform of the biophysical techniques described here widely available to the research community. Furthermore, the range of techniques at the researcher's disposal is continually expanding with new methods such as Alpha Screen, Biolayer Interferometry, and Microscale Thermophoresis that are also emerging as suitable techniques for studying of protein–ligand interactions.

In the following sections the focus is on the practicalities of implementation of fluorescence-based techniques that are highly sensitive and relatively low cost making them suitable for wide application to ligand screening in academia and industry. These are usually the first to be applied to a new protein target, before moving on to more detailed characterization by ITC, SPR, NMR and crystallography as outlined above and considered in detail elsewhere in this volume.

2 Materials

2.1 Differential Scanning Fluorimetry

Thermal cyclers are available from a number of suppliers, including Roche Applied Sciences, e.g., LightCyler® 480 Real-Time PCR System; Stratagene, e.g., Mx3005P Real-Time QPCR Systems; Applied Biosystems®, e.g., real-time PCR instruments; and Bio-Rad, e.g., iCycler iQ5 PCR Thermal Cycler. Different instruments from different manufacterers may have different plate capabilities (e.g., 96-well vs. 384-well) as well as different ranges of temperature (e.g., a run can be started at 25 °C with a iCycler iQ5 but can only be started at 37 °C with a LightCycler 480). Screening assays are typically performed in 96-well thin-wall white PCR plates, but can also be implemented in a 384-well format.

Several environmentally sensitive fluorescent dyes are available for use in DSF; however, by far the most commonly used is Sypro Orange. The wavelengths for excitation and emission are 490 and 575 nm, respectively. Sypro Orange comes as a 5,000× concentrated solution in DMSO. Since this dye has the advantage of a high signal to noise ratio, a low final concentration in the range of 1–5× yields optimal data with most proteins. Several buffer types and concentrations should be screened in advance to identify optimal conditions for reproducible protein thermal melts. It may be best to avoid buffers such as Tris, which exhibit large temperature dependence of their pK_a. Protein is usually used at concentrations of 1–10 μM, and small molecule ligands should be used at concentrations around or above the (known or expected) dissociation constants of their protein–ligand complex. In case of high affinity ligands, i.e., $K_d < 10$ μM, a concentration of at least equal that of the protein should be used to ensure significant population of the bound state.

2.2 Fluorescence Polarization

2.2.1 Capabilities of Instruments Required and Suitable Models

FP assays can be carried out using any fluorescence microplate reader capable of orthogonal-plane optical fluorescence readouts, and with an excitation wavelength range typically in the region of 230–900 nm. Several instruments capable of fast readout for HTS are available from different manufacturers, including PHERAstar Microplate Reader models from BMG LABTECH, the Perkin Elmer Envision® Multilabel Reader, and the Tecan Pòlarion Fluorescence Microplate Reader.

Filters for excitation and emission are required with wavelengths that need to be specified depending on the dye of interest (see below). Screening assays are typically performed in 96-well or 384-well format, e.g., Corning black Nonbinding Surface Microplates, but can also be implemented in a 1,536-well format.

2.2.2 Fluorescent Dyes and Criteria for Selection

Fluorophores that can be used for competitive FP assays can be divided into intrinsic probes, meaning the ligand to be displaced in the assay is naturally fluorescent, and extrinsic probes, which need to be introduced in to the known ligand to aid its detection. The latter strategy is the most commonly used since it is generally applicable to most systems, but has the disadvantage that their introduction may perturb the system. This should be carefully checked in advance during the assay optimization (see Subheading 3.2.1 and Eq. 5). Fluorescent dyes typically used in FP assays include carboxyfluorescein (FAM) or fluorescein thioisocyanate (FITC) dyes. For screens of compounds that modulate protein–peptide interactions FAM or FITC can be conveniently attached at the N-terminus or at the C-terminus of the peptidic sequence of interest using solid-phase peptide synthesis and amide forming reactions. These dyes have excitation maxima around 490 nm, and emission maxima around 520–535 nm. Alternative dyes used include the whole range of Invitrogen's Alexa fluor dyes that span the visible spectrum.

Criteria for dye selection include consideration of the approximate absorption and emission maxima for the conjugates, which need to be compatible with the instrument being used and importantly with the available optical filters. An important parameter to be considered is the fluorophore brightness. This is expressed as the product of the extinction coefficient ε and the quantum yield Φ of the dye. The extinction coefficient ε (M^{-1} cm^{-1}) at a given wavelength, typically the maximum of absorbance, expresses the energy capture efficiency of the dye and is a constant value. The quantum yield expresses the emission efficiency of the dye, and is defined as the ratio between the number of photons emitted and the number of photons absorbed by the fluorophore. An ideal fluorophore would have a quantum yield close to one. The quantum yields of a given fluorophore can vary significantly with conditions since fluorescence is strongly influenced by local environment and, consequently, they are not usually reported as a constant. Other important parameters of the dyes to be considered include photostability, susceptibility to photobleaching, pH insensitivity, and water solubility.

FP assays tend to be pretty insensitive to the choice of buffers and pH, hence optimal conditions for the particular protein of interest should be readily used. The choice of protein concentration to be used is critical for dose response screening assays and is described in Subheading 3.2.1. The concentration of fluorescent ligand that is sufficient to give a good polarization signal can be as low as 1–10 nM, thereby reducing the amount of fluorescent labeled ligand required in the assay.

3 Methods

3.1 Differential Scanning Fluorimetry

Screening assays performed in 96-well plates require 50–100 μL per well, whereas only 10–20 μL solution per well would be required in a 384-well format. As the heating rates can be as fast as 2 degrees per minute, a single plate can be screened in less than 30 min. Once a plate has been prepared and before loading it in the thermal cycler, it is good practice to centrifuge the plate for about 1 min. It is mandatory to record several reference samples (i.e., multiple samples of protein in the absence of ligand and presence of equal amount of any co-solvent, e.g., DMSO), and recommended, if possible, to run negative control samples (protein in the presence of a ligand known not to bind to it) as well as positive control samples (protein in the presence of a known binder).

Ligands can be screened individually, e.g., one per well or present in more than one well to allow multiple reads; alternatively they can be initially pooled and screened as mixtures. If a significant thermal shift is detected with a hit pool, then the protein is subsequently screened again against all individual ligands in that pool to determine the specific binding ligand(s).

A number of software tools have been developed and are available to fit the raw data obtained from the protein thermal melt. Custom calculation software based on Microsoft Excel that can be used for data visualization and analysis are available from the Structural Genomics Consortium Oxford (ftp://ftp.sgc.ox.ac.uk/pub/biophysics), see also (46). Commercial software and instrument specific technical guidance is also available from the instrument manufacturers. The entire fluorescence raw data (RFU) from the thermal melt curve can be accurately fitted to by nonlinear regression using, e.g., KaleidaGraph (http://www.synergy.com), SigmaPlot (http://www.sigmaplot.com/), Prism (http://www.graphpad.com/), GraFit (http://www.erithacus.com/grafit), or Origin (http://www.originlab.com). This enables one to obtain the enthalpy of protein unfolding (ΔH_u), the heat capacity change upon protein unfolding (ΔC_{pu}), and T_m, according to the equation described by Lo et al. (30). Data fitting using this approach is complicated by the fact that most protein DSF curves do not exhibit the ideal sigmoidal shape with constant pre- and post-transitional fluorescence intensities. This is particularly the case in the high-temperature region of the curve, which often exhibits a progressive decrease in fluorescence due to protein aggregation following unfolding (see Fig. 3a). Given that the temperature midpoint for the unfolding transition (T_m) tends to be the only parameter that is monitored in ligand screening experiments, this is often more conveniently determined indirectly by identifying the inflection point of the RFU curve. For this purpose, most instrument softwares generate a negative first derivative plot of the melting curve raw fluorescence data ($-d\text{RFU}/dT$) against temperature, from which the T_m can be determined as the position of the minimum (see Fig. 3b).

3.2 Fluorescence Polarization

3.2.1 Determination of Optimal Protein Concentration for the Assay

To determine the optimum concentration of protein to use in subsequent screening experiments, a standard curve in which different concentrations of protein are titrated against a known concentration of fluorescently labeled peptide should be recorded. As the concentration of protein is increased, a large FP signal develops. This titration yields the dissociation constant K_d of the complex between the protein and the fluorescent peptide. The selected concentration of the fluorescent peptide must not be much greater in magnitude than $2 \times K_d$ to prevent the stoichiometric titration of the fluorescent ligand. The titration is then repeated in the presence of a fixed amount of a nonfluorescent form of the same peptide that would be expected to compete for binding. A plot similar to that shown in Fig. 10 can then be obtained in which there is a protein concentration dependent reduction in the observed FP signal due to displacement by the nonfluorescent peptide.

This second titration yields an "apparent" dissociation constant K_d^{app} for the fluorescent complex that should be larger than that measured in the absence of the unlabelled peptide. Using the following simple competition model Eq. 5 it is possible to calculate the K_d for the unlabelled peptide, allowing one to check that the introduction of the fluorophore has not dramatically affected the native interaction of the peptide:

$$K_d^U = \frac{K_d^L}{K_d^{app} - K_d^L} \times [U]_{tot} \tag{5}$$

where K_d^U (the dissociation constant of the unlabeled peptide) can be calculated using the known values $[U]_{tot}$ (the total concentration of the unlabeled peptide), K_d^L (the dissociation constant of the labeled peptide), and K_d^{app} (the apparent dissociation constant of the labeled peptide in the presence of a fixed concentration of unlabeled peptide).

At high protein concentrations, as shown in Fig. 10, addition of an inhibitor that binds to the same site as the fluorescent peptide does not alter the FP signal by much as there is sufficient protein that the fluorescent peptide is still mostly found in complexes. Thus, whereas high protein concentrations ensure a strong FP signal and hence a reliable assay, the percentage inhibition on addition of a relatively low affinity inhibitor is low. Conversely, at low protein concentrations, there is a large percentage change in the FP signal upon addition of a displacer but the absolute FP signals are much lower in intensity leading to a low signal–background ratio.

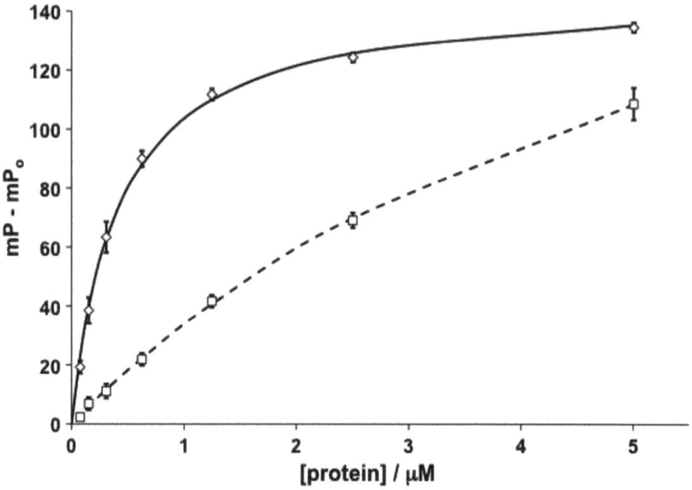

Fig. 10 Typical titration plots of protein into a constant concentration of fluorescent-labeled peptide (12 nM). The background polarization of the reference fluorophore in the absence of protein (mP_0) is subtracted from all data recorded in the absence (average data shown as diamonds, curve fit as *solid line*) and presence (average data shown as squares, curve fit as *dashed line*) of a fixed concentration (5 μM) of unlabelled peptide. Error bars represent +/− the standard deviation over three data point replicates. Data shown is for the binding of a FAM-labeled HIF-1α peptide (FAM-DEALA-Hyp-YIPD, $K_d = 400$ nM) to the ternary complex between von Hippel Lindau protein (pVHL) Elongin B and Elongin C, and competition with the respective unlabeled peptide (52)

A compromise concentration is needed in which a substantial absolute signal and large percentage change are possible.

In general, assay quality can be quantified using a *Z*-factor, also called Z' (47), which in this case is defined as:

$$Z' = 1 - \frac{3(\sigma_{\text{free}} + \sigma_{\text{bound}})}{mP_{\text{bound}} - mP_{\text{free}}} \tag{6}$$

where σ_{free}, σ_{bound}, mP_{free}, and mP_{bound} are the standard deviations and the means of the negative (free) and positive (bound) controls, respectively. A value greater than 0.7 is considered to yield a good assay format. Having both relatively small errors and a large signal strength maximizes the value of Z'. At least threefold repetition of the titrations above is necessary to provide error estimates for the values in this formula.

To determine an appropriate range of protein concentrations in which there is substantial absolute signal strength and good percentage inhibition, the parameter ($Z' \times$ % binding inhibition) should be obtained (Table 3). The largest values indicate the optimum protein concentration range. The data shown in Table 3, obtained from the plots in Fig. 10, suggest that protein concentrations between 300 nM and 1.2 μM would be near optimal in this case.

Table 3
Z', % Inhibition and $Z' \times$ % Inhibition as calculated from the data shown in Fig. 10

Protein conc. (nM)	Z'	% Inhibition	$Z' \times$ % Inhibition
5,000	0.95	19	18.27
2,500	0.95	44	42.02
1,250	0.95	63	58.06
625	0.88	76	66.37
313	0.70	82	57.44
156	0.56	82	46.33
78	0.54	88	47.35

$Z' \times$ % inhibition provides a good metric to identify the optimal range of protein concentration for use in screening

3.2.2 Determination of Dose–Response Curves

Using the optimal concentration of protein for this assay, determined as previously described, dose response behavior can be recorded for each inhibitor as required by measuring fluorescence of a fixed concentration of fluorescent ligand in the presence of fixed concentration of protein and varying concentrations of the compounds to be tested. The displacement of the fluorescent ligand will give a reduction in the FP signal and a dose response curve can be plotted, see Fig. 11, from which an IC_{50} can be measured. By definition an IC_{50} is the inhibitor concentration that displaces 50 % of the fluorescent ligand from the protein–ligand complex.

The K_d of the competitive inhibitor I can then be back calculated (33) from their IC_{50} by using

$$IC_{50} = \left(\frac{(2[P]_T - f_0[L]_T)(2 - f_0)}{2K_L f_0} - 1\right) \times K_I + \left([P]_T - \frac{K_L f_0}{2 - f_0} - \frac{f_0[L]_T}{2}\right) \quad (7)$$

In Eq. 7: $K_I = K_d$ of the competitive inhibitor I (to be calculated); $K_L = K_d$ of the fluorescent ligand L to be displaced (measured independently); $[P]_T$ = total protein concentration used in the assay; $[L]_T$ = total concentration of fluorescent ligand L used in the assay; f_0 = fractional saturation of fluorescent ligand, i.e., $[PL]/[L]_T$ ($[PL]$ is known based on the equilibrium constant K_L and the initial conditions of the assay).

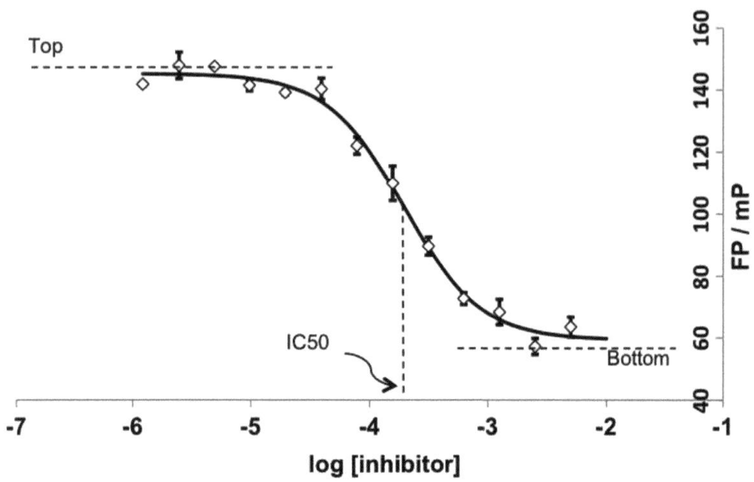

Fig. 11 A typical plot of fluorescence polarization (FP), in millipolarization (mP) unit, vs. log [inhibitor]. The IC_{50} is readily determined to be the midpoint of the sigmoidal dose response curve. Data in the y axis can also be plotted as (% Inhibition) calculated as 100 × (Top − mP)/(Top − Bottom), where Top and Bottom are the asymptotes of the sigmoidal curve. Data shown is for the dose–response inhibition of signal due to binding of a FAM-labeled HIF-1α peptide (FAM-DEALA-Hyp-YIPD) to the ternary complex between von Hippel Lindau protein (pVHL), Elongin B and Elongin C upon competition with a small molecule ligand (52)

Data on 8–10 compounds individually can be obtained in triplicate over a ~10^4–fold compound concentration range in under 10 min, e.g., in a 384-well plate by performing serial two- or threefold dilutions of an initial concentrated stock of the compounds to be tested. It is important to run appropriate controls during the assay, including: a sample of Fluorophore + Protein − Inhibitor (which will give the maximum signal); a sample of Fluorophore − Protein − Inhibitor (which will give the minimum signal); a sample of Fluorophore − Protein + Inhibitor (to exclude compound interference with the signal of the fluorescent dye).

FP has an upper limit of measuring IC_{50} that is dependent only on ligand solubility and co-solvent tolerance of the assay (*see* **Note 7**). For weak-affinity ligands, the assay starts to become unreliable nearing the IC_{50} upper limit due to the high concentrations of compound potentially interfering with the polarization signal (*see* **Note 8**). In contrast, the lower limit of measuring IC_{50} is dependent on the K_d of the fluorescent peptide (33). For potent inhibitors, IC_{50} values acquired in assays using a weak-binding (high K_d) fluorescent ligand are higher than IC_{50} values acquired in assays using a tight-binding (low K_d) fluorescent ligand. Consequently, the higher the affinity of the fluorescent ligand, the wider the range of inhibitor potency that can be resolved by the assay. This is particular important to establish reliable SAR during hit-to-lead and lead optimization campaigns. In our hands, the IC_{50} values obtained can fluctuate on a day-to-day basis and dependent on conditions, sometimes as much as tenfold. It is therefore very important to repeat titrations of a test inhibitor when comparing assays run in different days or using different stocks of binding partners. Even with the associated errors and limitations, FP is a useful technique as it enables a quick determination of IC_{50} with moderate accuracy and back calculation of K_d, if necessary.

4 Notes

1. Measured values of *P* in biophysical analysis are between 10 and 300 mP. Consequently, instruments capable of very precise measurements are required to monitor changes upon binding in some reactions. Fortunately, many instruments can now measure with a precision close to ±2 mP, which means that almost all reactions can be followed reliably.
2. Amine coupling is a common approach to immobilizing a protein on an SPR chip. Coupling can be easily achieved by reacting amine groups of the protein with carboxylate groups of the chip dextran matrix upon their activation using 1-ethyl-3-(3-dimethylaminopropyl)-carbodiimide (EDC) and *N*-hydroxysuccinimide (NHS). An alternative approach to amine coupling is to immobilize the protein via the free thiols of its cysteine residues. This is accomplished upon reacting the activated surface with 2-(2-pyridinyldithio)ethaneamine hydrochloride (PDEA), which contains an amine and a reactive disulfide for thiol-disulfide exchange, prior to protein immobilization.

   The progress of the activation and coupling process can be monitored through the SPR response itself, see Fig. 12 for an example.
3. Relaxation times: T_1 the longitudinal relaxation time. It depends upon the overall rate of tumbling of the protein. It can be modulated by local motions due to conformational flexibility. T_2 the transverse relaxation time. It is shorter than T_1 and is correlated with dynamic processes in the protein. T_2 determines the linewidth of the NMR signals and decreases with increasing size of the molecule. Generally, small molecules (<1,000 Da) exhibit slow relaxation, i.e., long longitudinal ($T_1 = 1 - 2$ s) and transverse ($T_2 = 1$ s) relaxation times and small and positive NOEs. In contrast, bound ligands share the NMR properties of the larger protein (>10,000 Da) that are distinctly different: fast relaxation, i.e., short relaxation times ($T_1 =$ 20–50 ms and $T_2 =$ 5–20 ms for a 30 kDa protein) and large and negative NOEs. These differences give rise to changes

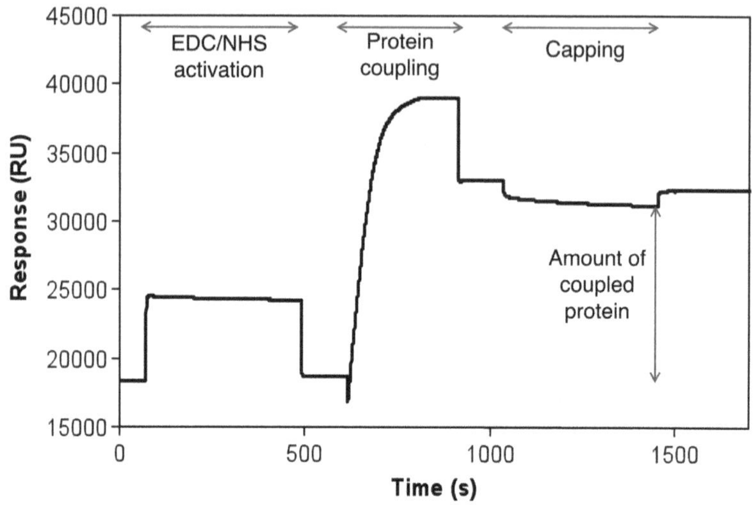

Fig. 12 Typical protein immobilization sensogram is shown as a plot of Response Unit (RU) against time. The following protein amine coupling procedure was used: 1 min equilibration followed by 7 min EDC/NHS activation; 2 min equilibration followed by 5 min protein injection; 2 min equilibration followed by 5 min ethanolamine capping; 5 min equilibration. A final coupling of 15,000 RU of a protein was achieved. One resonance unit (RU) represents the binding of approximately 1 pg protein/mm^2 and thus it is proportional to the mass of protein bound (36)

in the average NMR parameters of the ligands that can be monitored to assess binding to the protein target.

4. The exact concentrations of ligands and protein used will vary depending on the particular system being investigated, e.g., components solubility, on the number of ligands being tested per sample, and on the nature of the NMR experiment to be carried out. Typically, in small molecule fragment screens the ligands are tested at concentrations of 250 μM to 1 mM thereby allowing robust detection of fragment signals. To increase the throughput of the screening process, and to reduce the amount of protein and experimental time required, fragments can be mixed together in cocktails. Depending on the solubility of compounds, cocktails of up to 8–10 fragments can be used; however, in my own experience such large mixtures tend to be problematic due to extensive signal overlap from the ligands, which hampers identification of bound ligands, meaning subsequent deconvolution procedures would be required. In addition, such large mixtures tend to increase the possibility of different ligands interfering with each other in solution. For example it is known that organic salts tend to have lower solubilities in DMSO than the corresponding free acids and bases, and are thus more likely to lead to precipitation from

DMSO. Mixing acidic and basic fragments can lead to similar problems. These interferences can cause compound aggregation and ultimately affect the screen by increasing the number of false positives. For these reasons a maximum mixture size of 3–4 fragments is recommended.

5. Most known PX structures and their crystallization conditions are deposited in the Protein Data Bank (PDB) http://www.rcsb.org/ which is an open-access database (48). Structures and other information may also be available as coordinates in patent descriptions or from the primary literature.

6. Fragments or larger compounds are typically soaked as solutions in dimethylsulfoxide (DMSO), individually or as mixtures, into the target protein crystal by adding them in a solution containing the mother liquor used to crystallize the protein. The concentration of the molecular fragments is usually 20 mM or higher, even up to 200 mM, to allow for the weakness of the interaction (usually in the mM range). However, for stronger binding interactions (affinities in the micromolar range) lower concentrations can be used, albeit this should be in excess of the protein concentration (often in the low millimolar range) to allow saturation of the binding sites in the crystal. Protein crystals may not tolerate the high concentrations of compounds and DMSO (or other organic co-solvents) used and can be easily damaged during the soaking procedure. If a cryoprotectant solution is required prior to freezing crystals, the small molecules used for soaking may be included in such a solution (however, in the author's experience this has not been a strict requirement). If mixtures are used in the soaking protocol, the individual fragments should be unambiguously distinguishable from their electron density.

7. A maximum of 10 % v/v DMSO can be usually tolerated in an assay. However, the limit should be determined on a case by case basis, e.g., it has been observed that DMSO interferes with an assay by binding to the targeted binding site on the protein, thereby potentially competing with the ligand being screened (49). So care should be taken to exclude such DMSO interference, for example by performing a dose–response curve as a function of DMSO concentration.

8. All types of fluorescent-based assays, including FP assays, are prone to interference that can give rise to false positives, false negatives or other artifacts. For example, compound autofluorescence in a given well may be erroneously interpreted as displacement, giving rise to a false positive. This is because the additional fluorescent signal due to the compound is likely to be of low polarization and can exaggerate the apparent level of ligand displacement in that well. In contrast, compound

insolubility under the screening condition may result in excessive light scattering, which can increase the overall polarization measurement meaning such a compound may appear as an "activator." Compound absorbance at the excitation or emission wavelengths should instead have no effect on the observed polarization.

Acknowledgments

The author wishes to thank several members of the Ciulli and Abell research groups (Department of Chemistry) and the Blundell research group (Department of Biochemistry) for productive collaboration and invaluable discussions over the years. The Ciulli laboratory is funded primarily but not exclusively by the UK Biotechnology and Biological Sciences Research Council (BBSRC). The author thanks the BBSRC for the award of a David Phillips Fellowship (BB/G023123/1). The author did not receive any financial contribution or writing assistance for the production of this chapter and declares no competing financial interests.

References

1. Crews CM (2010) Targeting the undruggable proteome: the small molecules of my dreams. Chem Biol 17:551–555
2. Schreiber SL (2005) Small molecules: the missing link in the central dogma. Nat Chem Biol 1:64–66
3. Frye SV (2010) The art of the chemical probe. Nat Chem Biol 6:159–161
4. Edwards AM, Bountra C, Kerr DJ, Willson TM (2009) Open access chemical and clinical probes to support drug discovery. Nat Chem Biol 5:436–440
5. Cole PA (2008) Chemical probes for histone-modifying enzymes. Nat Chem Biol 4:590–597
6. Hopkins AL, Groom CR (2002) The druggable genome. Nat Rev Drug Discov 1:727–730
7. Russ AP, Lampel S (2005) The druggable genome: an update. Drug Discov Today 10:1607–1610
8. Broach JR, Thorner J (1996) High-throughput screening for drug discovery. Nature 384:14–16
9. Spencer RW (1999) High-throughput screening of historic collections: observations on file size, biological targets, and file diversity. Biotechnol Bioeng 61:61–67
10. Macarron R, Banks MN, Bojanic D et al (2011) Impact of high-throughput screening in biomedical research. Nat Rev Drug Discov 10:188–195
11. Bleicher KH, Böhm H-J, Müller K, Alanine AI (2003) Hit and lead generation: beyond high-throughput screening. Nat Rev Drug Discov 2:369–378
12. Shoichet BK (2006) Interpreting steep dose–response curves in early inhibitor discovery. J Med Chem 49:7274–7277
13. Babaoglu K, Simeonov A, Irwin JJ et al (2008) Comprehensive mechanistic analysis of hits from high-throughput and docking screens against beta-lactamase. J Med Chem 51:2502–2511
14. McGovern SL, Caselli E, Grigorieff N et al (2002) A common mechanism underlying promiscuous inhibitors from virtual and high-throughput screening. J Med Chem 45:1712–1722
15. Jhoti H, Cleasby A, Verdonk M, Williams G (2007) Fragment-based screening using X-ray crystallography and NMR spectroscopy. Curr Opin Chem Biol 11:485–493
16. Ciulli A, Blundell TL, Abell C (2008) Discovery and extrapolation of fragment structures towards drug design. In: Stroud RM, Finer-Moore J (eds) Computational and structural approaches to drug discovery: ligand–protein interactions. The Royal Society of Chemistry, Cambridge
17. Shuker SB, Hajduk PJ, Meadows RP, Fesik SW (1996) Discovering high-affinity ligands for

proteins: SAR by NMR. Science 274:1531–1534
18. Hann MM, Leach AR, Harper G (2001) Molecular complexity and its impact on the probability of finding leads for drug discovery. J Chem Inf Comput Sci 41:856–864
19. Blundell TL, Jhoti H, Abell C (2002) High-throughput crystallography for lead discovery in drug design. Nat Rev Drug Discov 1:45–54
20. Rees DC, Congreve M, Murray CW, Carr R (2004) Fragment-based lead discovery. Nat Rev Drug Discov 3:660–672
21. Hajduk PJ, Greer J (2007) A decade of fragment-based drug design: strategic advances and lessons learned. Nat Rev Drug Discov 6:211–219
22. Congreve M, Chessari G, Tisi D, Woodhead AJ (2008) Recent developments in fragment-based drug discovery. J Med Chem 51:3661–3680
23. Ciulli A, Abell C (2007) Fragment-based approaches to enzyme inhibition. Curr Opin Biotechnol 18:489–496
24. Murray CW, Rees DC (2009) The rise of fragment-based drug discovery. Nat Chem 1:187–192
25. Erlanson DA (2012) Introduction to fragment-based drug discovery. Top Curr Chem 317:1–32
26. Lundqvist T (2005) The devil is still in the details—driving early drug discovery forward with biophysical experimental methods. Curr Opin Drug Discov Devel 8:513–519
27. Ciulli A, Williams G, Smith AG, Blundell TL, Abell C (2006) Probing hot spots at protein--ligand binding sites: a fragment-based approach using biophysical methods. J Med Chem 49:4992–5000
28. Ericsson UB, Hallberg BM, Detitta GT, Dekker N, Nordlund P (2006) Thermofluor-based high-throughput stability optimization of proteins for structural studies. Anal Biochem 357:289–298
29. Cummings M, Farnum M, Nelen M (2006) Universal screening methods and applications of ThermoFluor®. J Biomol Screen 11:854–863
30. Lo M-C, Aulabaugh A, Jin G et al (2004) Evaluation of fluorescence-based thermal shift assays for hit identification in drug discovery. Anal Biochem 332:153–159
31. Kranz JK, Schalk-Hihi C (2011) Protein thermal shifts to identify low molecular weight fragments. Methods Enzymol 493:277–298
32. Reindl W, Strebhardt K, Berg T (2008) A high-throughput assay based on fluorescence polarization for inhibitors of the polo-box domain of polo-like kinase 1. Anal Biochem 383:205–209
33. Huang X (2003) Fluorescence polarization competition assay: the range of resolvable inhibitor potency is limited by the affinity of the fluorescent ligand. J Biomol Screen 8:34–38
34. Wiseman T, Williston S, Brandts JF, Lin LN (1989) Rapid measurement of binding constants and heats of binding using a new titration calorimeter. Anal Biochem 179:131–137
35. Turnbull WB, Daranas AH (2003) On the value of *c*: can low affinity systems be studied by isothermal titration calorimetry? J Am Chem Soc 125:14859–14866
36. Van der Merwe PA (2001) Surface plasmon resonance. In: Chowdhry B, Harding S (eds) Protein–ligand interactions: hydrodynamics and calorimetry. Oxford University Press, Oxford
37. Pellecchia M, Bertini I, Cowburn D et al (2008) Perspectives on NMR in drug discovery: a technique comes of age. Nat Rev Drug Discov 7(9):738–745
38. Lepre CA, Moore JM, Peng JW (2004) Theory and applications of NMR-based screening in pharmaceutical research. Chem Rev 104:3641–3676
39. Hajduk PJ, Sheppard G, Nettesheim D et al (1997) Discovery of potent nonpeptide inhibitors of stromelysin using SAR by NMR. J Am Chem Soc 119:5818–5827
40. Śledź P, Abell C, Ciulli A (2012) Ligand-observed NMR in fragment-based approaches. In: Bertini I, McGreevy K, Parigi G (eds) NMR of biomolecules: towards mechanistic systems biology. Wiley-VCH, Weinheim
41. Hajduk PJ, Olejniczak E, Fesik S (1997) One-dimensional relaxation- and diffusion-edited NMR methods for screening compounds that bind to macromolecules. J Am Chem Soc 119:12257–12261
42. Mayer M, Meyer B (1999) Characterization of ligand binding by saturation transfer difference NMR spectroscopy. Angew Chem Int Ed Engl 38:1784–1788
43. Dalvit C, Fogliatto G, Stewart A, Veronesi M, Stockman B (2001) WaterLOGSY as a method for primary NMR screening: practical aspects and range of applicability. J Biomol NMR 21:349–359
44. Nienaber VL, Richardson PL, Klighofer V et al (2000) Discovering novel ligands for macromolecules using X-ray crystallographic screening. Nat Biotechnol 18:1105–1108
45. Blundell TL, Abell C, Cleasby A et al (2002) High throughput X-ray crystallography for drug discovery. In: Flower DR (ed) Drug design: cutting edge approaches. The Royal Society of Chemistry, Cambridge

46. Niesen FH, Berglund H, Vedadi M (2007) The use of differential scanning fluorimetry to detect ligand interactions that promote protein stability. Nat Protocols 2:2212–2221
47. Zhang J, Chung T, Oldenburg K (1999) A simple statistical parameter for use in evaluation and validation of high throughput screening assays. J Biomol Screen 4:67–73
48. Berman HM, Westbrook J, Feng Z et al (2000) The Protein Data Bank. Nucleic Acids Res 28:235–242
49. Philpott M, Yang J, Tumber T et al (2011) Bromodomain-peptide displacement assays for interactome mapping and inhibitor discovery. Mol BioSyst 7:2899–2908
50. Ciulli A, Scott DE, Ando M et al (2008) Inhibition of Mycobacterium tuberculosis pantothenate synthetase by analogues of the reaction intermediate. ChemBioChem 9:2606–2611
51. Hung AW, Silvestre HL, Wen S et al (2009) Application of fragment growing and fragment linking to the discovery of inhibitors of Mycobacterium tuberculosis pantothenate synthetase. Angew Chem Int Ed Engl 48:8452–8456
52. Buckley DL, Van Molle I, Gareiss PC et al (2012) Targeting the von Hippel-Lindau E3 ubiquitin ligase using small molecules to disrupt the VHL/HIF-1α interaction. J Am Chem Soc 134:4465–4468

Chapter 14

Screening Protein–Small Molecule Interactions by NMR

Ben Davis

Abstract

Nuclear magnetic resonance (NMR) is well suited to probing the interactions between ligands and macromolecular receptors. It is a truly label-free technique, requiring only the presence of atoms (usually ^1H or ^{19}F) which give rise to observable resonances on either the ligand or the receptor. A number of parameters associated with these resonances can be used to distinguish rapidly tumbling compounds from ligands which bind to a macromolecular receptor. As such, NMR reports directly on the molecular components involved in the binding interaction whilst avoiding artifacts arising from the addition of an observable label. NMR is also unique amongst biophysical techniques in giving information on the chemical nature of almost all of the constituents present in the sample, thus allowing ready identification of sample, contaminants, degraded material and buffers. Solution phase NMR is also free of artifacts introduced by the presence of a solid support or matrix, although some interesting NMR techniques have been developed to identify ligand–receptor interactions in both solid and heterogeneous phase systems.

NMR can readily report on molecular interactions across a wide range of affinities and timescales. Although NMR is not an inherently sensitive technique, the development of cryogenic probeheads over the past decade has dramatically increased the range of applicability of the technique and reduced the stringent sample requirements that used to be regarded as an "Achilles' heel" of NMR. The last, but by no means the least, NMR has the ability to determine structural information at atomic resolution; this has proved to be particularly useful when applied to those protein–ligand systems which cannot be readily crystallized.

Key words Nuclear magnetic resonance, NMR, STD, Water-LOGSY, T_2, Relaxation-filtered, NOE

1 Introduction

The approaches used to screen for protein–ligand interactions by NMR can be split into two broad categories—those which observe the ligand component of the binding interaction, and those which observe the protein component. Ligand-observed experiments typically utilize the differential modulation of certain observables (such as relaxation rates or the transfer of magnetization) for compounds that form complexes with protein compared to compounds which do not bind. Protein-observed techniques usually rely on the observation of chemical shift perturbations (CSPs) in isotopically

Mark A. Williams and Tina Daviter (eds.), *Protein-Ligand Interactions: Methods and Applications*, Methods in Molecular Biology, vol. 1008, DOI 10.1007/978-1-62703-398-5_14,

labeled protein. Additionally, a number of NMR techniques are available which can be used to obtain structural data about the bound state of a ligand–receptor complex at various levels of resolution. An extensive discussion of the wide variety of NMR techniques used for the investigation of protein–ligand complexes is outside of the remit of this chapter, and in any case can be found in several recent reviews (1–5). Here, the focus is on the practical application of a selected set of widely used experiments for screening compound libraries to identify those compounds which bind to a target protein.

A common approach when screening potential ligands for interactions with receptors is to mix several compounds into a single sample and to identify ligands within the compounds present. This approach substantially reduces both time and resource required to screen a library for ligands, and is a valid methodology where care is taken to minimize spectral overlap and to ensure that no interaction occurs between compounds. The use and design of mixtures is discussed in more detail in Subheading 3.1, along with the necessary controls and caveats.

1.1 Ligand-Observed NMR Binding Experiments

Ligand-observed methods to detect binding by NMR rely on the differential behavior of small molecules and of macromolecules, and the transfer of macromolecule-like properties from the bound state to the free state of the ligand where they can be more readily observed. Typically, the ligand is in a large molar excess over the receptor. If the ligand rapidly transfers between bound and free states (i.e., is in the "fast exchange" regime), then during the course of the experiment many ligand molecules bind transiently to each receptor, and are each modulated by the slower tumbling of the macromolecule before dissociating back into solution. A consequence of this fast exchange is that there is a non-stoichiometric response from the bound population, and the signal is in effect amplified. This in turn results in high sensitivity and allows the detection of ligands at concentrations well below their K_D. The reliance of many ligand-observed NMR experiments on this amplification does impose the strict restriction that the ligands associate and dissociate rapidly on the NMR timescale. However, there is no upper limit on the size of the receptor (indeed, the technique has been successfully applied to systems as large as monoclonal antibodies, intact viruses and whole cells (6–9)), the protein does not need to be isotopically labeled and is used at relatively low concentrations. Given these many advantageous factors, the sensitivity and speed of the experiments, and because rapid association and dissociation is often readily achieved in practice, ligand-observed NMR experiments have become widely used to detect binding.

One of the major limitations of ligand-observed NMR binding experiments is that no information is obtained about the location of the ligand binding site, or indeed the specificity of the interaction, merely that there is an interaction between the macromolecule and the ligand. This can to some extent be circumvented by the use of a competition step (10), where a known ligand or protein is used to block a binding site and so prevent binding of the putative ligand to that site. If the ligand binds to the site of interest, then repeating the ligand-observed NMR experiment in the presence of a site-specific competitor should indicate a reduced or abolished bound population of the ligand. Care must obviously be taken to verify that the putative ligand is truly binding to the site of interest (ideally by determination of the crystal or NMR structure of the protein with the known ligand), but in general a competition step greatly improves the robustness and reliability of ligand-observed binding experiments.

Although ^{1}H is by far the most widely used nucleus in ligand-observed NMR binding experiments, because of the near ubiquitous presence of hydrogen in organic molecules and strong NMR signal of its nucleus, a number of robust and sensitive experiments rely on ^{19}F observation (see ref. (11) for an excellent review). The large chemical shift anisotropy component of ^{19}F relaxation makes T_2 relaxation-filtered, ^{19}F experiments particularly sensitive to binding to macromolecules (12). The combination of ^{19}F's wide chemical shift range and the fact that ^{19}F will be present only in selected molecules effectively removes the issues of spectral overlap and baseline distortion by buffers that can afflict ^{1}H based experiments. Further, a ^{19}F-labeled competitive binder or "spy probe" is used the FAXS experiment (fluorine chemical shift anisotropy and exchange for screening) (13, 14) to act as an indirect reporter for the binding of unlabeled ligands. However, the requirement to have ^{19}F present in the compound of interest or a spy probe obviously restricts the range of ligands to which this mode of detection is applicable.

Ligand-observed NMR binding experiments fall broadly into three categories:

1. Experiments where magnetization is transferred between the receptor and the ligand following selective excitation of one component. This transfer can be through direct excitation (e.g., the saturation transfer difference (STD) experiment (15)) or indirect (e.g., the water-LOGSY (Water-Ligand Observed via Gradient SpectroscopY) experiment (16) where the selective excitation is of the bulk solvent).
2. Experiments where the differential relaxation properties of magnetization in slowly tumbling macromolecules and in rapidly tumbling compounds are used to distinguish compounds which bind to macromolecules from those that do not (e.g., $T_{1\rho}$ or T_2 relaxation filtered 1D experiments (17, 18)). The differential effect can be enhanced by the use of relaxation agents such as spin probes (18) or paramagnetic metals (19, 20),

or by immobilizing the receptor onto a solid support in a mixed-phase system such as in the TINS (Target Immobilized NMR Screening) method (21).

3. Experiments where ligands are distinguished by the modulation of other physical properties due to their interaction with a macromolecular receptor, for example by using diffusion filters (17) or magnetization transfer following the selective purging of a population based on diffusion behavior (22, 23). This group of experiments is, however, not widely used, and is not discussed further.

The most commonly used ligand observed NMR experiments used to detect binding of a ligand to a receptor are the STD, water-LOGSY and relaxation filtered experiments; these are discussed in greater detail below.

1.1.1 STD

The STD experiment (24, 25) is widely used to detect interactions between a receptor and any ligand. A subset of the receptor's protons, typically the region of the NMR spectrum corresponding to methyl groups buried in the hydrophobic core of the protein (−0.5 to 0.3 ppm), are saturated by a long train of selective pulses. During this saturation period, the magnetization is transferred, by dipole–dipole interactions between neighboring ^{1}H nuclei, throughout the protein and also to any protons in ligands bound to the protein. In this "on-resonance" spectrum, resonances from any ligands are therefore fully or partially saturated, dependent on the kinetics of both binding and relaxation, while the resonances of non-binding compounds will remain unsaturated.

The experiment is then repeated with the selective pulses set to a frequency which does not correspond to any chemical shift present in the sample, for example −40 ppm. This "off-resonance" experiment produces a reference spectrum in which the signals of neither receptor nor ligand nor non-binding molecules are saturated.

If the on-resonance (saturated) experiment is then subtracted from the off-resonance (unsaturated) experiment, the resulting difference spectrum contains signal from saturated ligands and receptor, but with no signal from non-binding molecules (Fig. 1a, b). This subtraction can be done either post-acquisition, having acquired the separate on- and off-resonance spectra (e.g., as implemented in the Bruker stddiff pulse sequences) or "on the fly" during acquisition, in which case only the resultant difference experiment will be recorded (e.g., the Varian dpfgse_satxfer pulse sequence).

1.1.2 Water-LOGSY

The Water-Ligand Observed via Gradient SpectroscopY (water-LOGSY) experiment (16, 26) uses selective excitation of the bulk solvent (typically H_2O), followed by a magnetization transfer step to all solutes. The difference between the sign of magnetization transfer from solvent to small-molecules compared to the transfer

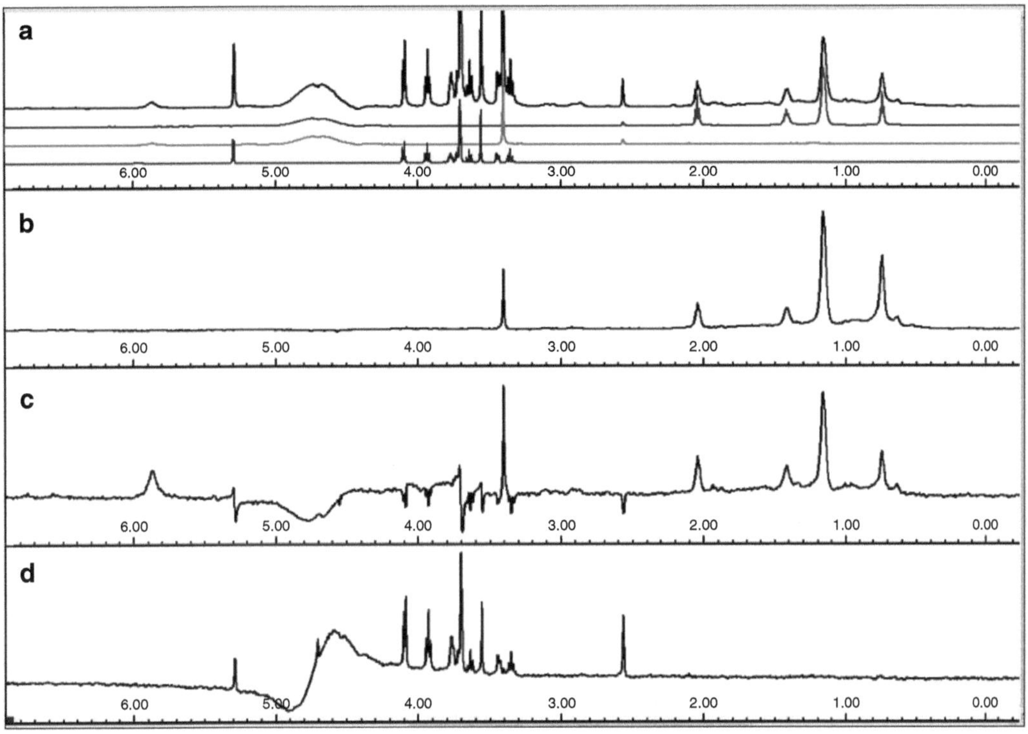

Fig. 1 1D ^{1}H and ligand-observed NMR binding experiments using a sample of avidin with three small molecule compounds (10 μM avidin, 500 μM octanoic acid, 500 μM 2-imidazolidinone, 500 μM sucrose in 10 mM potassium phosphate pH 7.5). (**a**) The excitation sculpted 1D spectrum of the sample (*top*), along with reference spectra acquired under identical conditions for each of the small molecules; octanoic acid (*second down*), 2-imidazolidinone (*third*) and sucrose (*fourth*). (**b**) The STD spectrum of the avidin:small molecule sample, showing a clear signal from the 2 ligands (octanoic acid and 2-imidazolidinone) and no signal from the non-binding sucrose. A saturation time of 2.2 s was used. (**c**) The water-LOGSY spectrum of the sample, showing positive signal from the ligands and negative signal from the sucrose and residual protonated DMSO resonance (2.59 ppm). A positive signal is also observed at 5.85 ppm but this is attributable to a labile proton of the 2-imidazolidinone, which is in chemical exchange with the bulk solvent, and is also present in this experiment in the absence of protein. A mixing time of 2 s was used. (**d**) The CPMG-filtered 1D spectrum of the sample, showing signal from the sucrose and DMSO but no signal from the ligands. A mixing time of 800 ms was used. The spectral features around 4.8 ppm observed in most spectra are a result of the residual signal from water

to macromolecules via the nuclear Overhauser effect (NOE) (27) enables ligands to be readily identified. Slowly tumbling macromolecules and bound ligands give rise to a large positive signal, while rapidly tumbling molecules free in solution give rise to small negative signal. The presence of any bound water molecules within the active sites of receptors also contributes to the large positive signal observed for ligands.

The water-LOGSY experiment is rapid, and is also sensitive compared to the STD. However, a number of caveats must be considered when analyzing the data obtained.

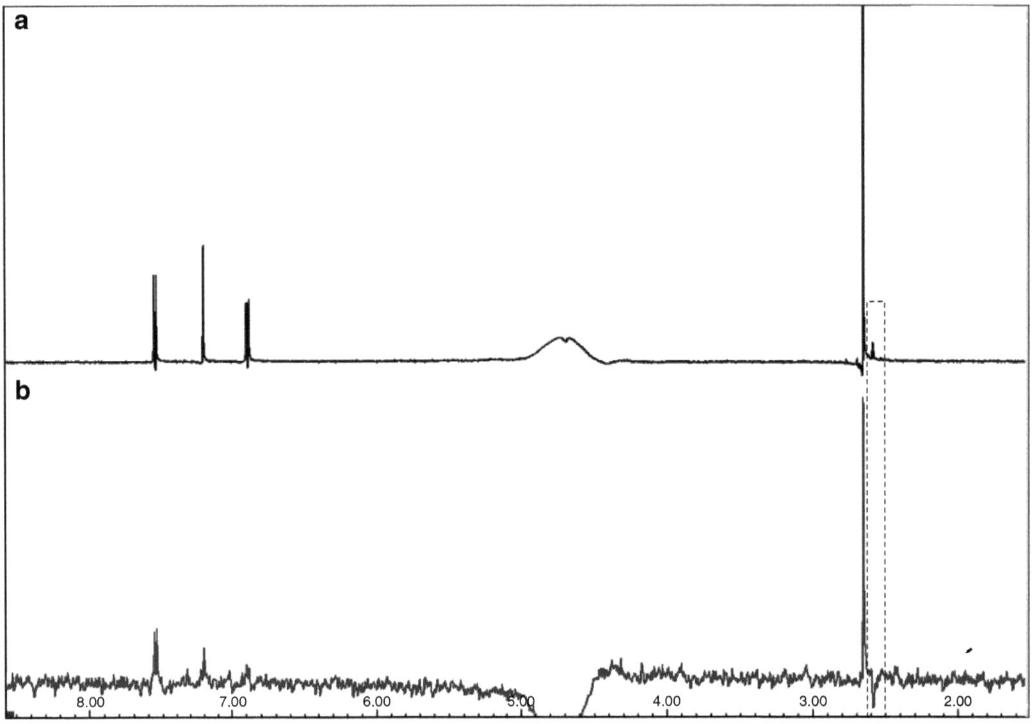

Fig. 2 1D ^{1}H (**a**) and water-LOGSY (**b**) spectra of a compound (500 μM compound, 90 % H_2O, 10 % D_2O, 50 mM potassium phosphate pH 7.5, 25 °C). The compound alone has a weak positive water-LOGSY spectrum similar to that expected for a ligand bound to a macromolecule. In contrast the residual protonated DMSO resonance (2.59 ppm) is negative (highlighted in *boxed area*) as expected. This spectrum implies the presence of aggregated particulates in the solution comprising the compound itself or of impurities with which it interacts. The compound has no exchangeable resonances, no macromolecule is present in the sample, and the sample contained no visible precipitate

Any labile protons (e.g., NH or OH groups) may also give rise to positive signals, through chemical exchange of protons between ligand and water, regardless of whether they arise from ligands or non-binding compounds (Fig. 1c). However, since these positive signals are restricted to a subset of the protons within a compound they are usually readily identified.

More problematically, some compounds give rise to completely positive water-LOGSY spectra even in the absence of any macromolecular receptor (Fig. 2). This may be due to interactions between a population of solution-phase compound and microscopic precipitates or aggregates formed by the compound in the buffer used. In order to avoid this artifact being mistaken for binding, it is good practice to acquire water-LOGSY spectra on the compound in the relevant buffer in the absence of any macromolecule, and to change conditions or compound if positive water-LOGSY spectra are observed.

Secondly, the resultant water-LOGSY spectrum contains signal from both the bound and free populations of a ligand. Thus, if only a small population is bound (high K_D, or high molar excess of ligand), the contribution from the free population will dominate the observed spectrum and the peaks will be negative even though the ligand is binding to the receptor. It is therefore simpler to interpret results in experiments which perturb the relative populations of the free and bound state, for example by adding a known potent competitor molecule, and to use any observed increase in a free population (more negative signal) as an indication that a bound population has been reduced.

1.1.3 Relaxation Filtered Experiments

Experiments based around the differential relaxation rates of rapidly and slowly tumbling molecules have been widely used for many years to distinguish ligands from non-binding compounds (17, 18). Both $T_{1\rho}$ and T_2 relaxation experiments are common, restricting the available magnetization relaxation pathways in order to selectively attenuate or remove signal from macromolecules and ligands and thus filter the observed spectrum to show only non-binding compounds. However, there is the potential for widely varying relaxation rates for individual resonances within a molecule, dependent on local interactions between nuclei and electrons, dynamics and environment. Consequently, these relaxation experiments are also most reliably used in conjunction with a competition step, where an increase in the compound's signal on addition of a potent competitor implies an increase in the unbound population and therefore validates the compound as a ligand (Fig. 3).

As discussed above, the use of a known potent competitor ligand dramatically increases the reliability of ligand observed NMR binding experiments, in addition to giving an indication as to the location of the ligand binding site. A potential competitor molecule should be characterized thoroughly in terms of its purity and stability as discussed in Subheading 3.1, and, ideally, its binding site and affinity confirmed by one or more biophysical techniques. Where a known and characterized competitor is available for the site of interest, the ligand-observed experiments should be acquired on a sample containing the protein and ligand, and then repeated on a sample containing the protein, ligand and competitor. This latter step is not possible in all situations, for example where no known ligand exists for a site, and in such a case it is possible obtain similar validation by screening the ligand for binding to a mutant protein in which the suspected binding site has been removed or modified through site directed mutagenesis.

1.2 Protein-Observed NMR Binding Experiments

Experiments which detect binding through the observation of signals or perturbed signals on the protein typically require higher concentrations of macromolecule, since this is the component which is observed. The exact concentration required will obviously depend on the experimental setup and the nature of the receptor, but is typically in the range of 25–500 μM. Additionally, except in

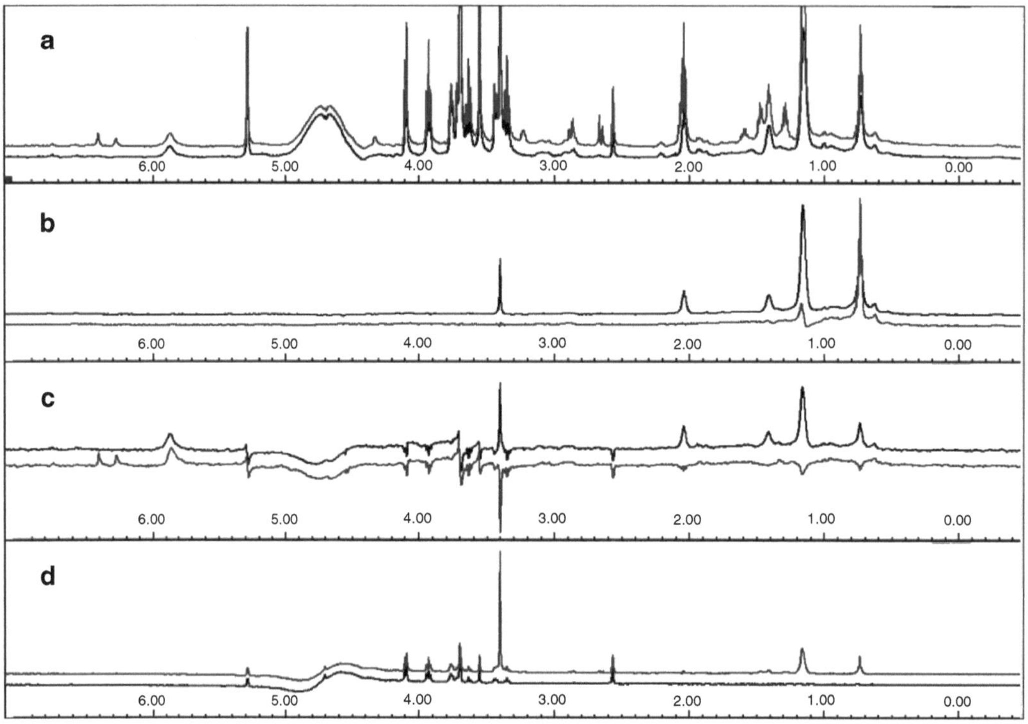

Fig. 3 1D ^{1}H and ligand-observed NMR binding experiments of 10 μM avidin, 500 μM octanoic acid, 500 μM 2-imidazolidinone, 500 μM sucrose in 10 mM potassium phosphate pH 7.5, before and after the addition of 100 μM biotin. All experimental conditions are identical to those used in Fig. 1. (**a**) The excitation sculpted 1D spectrum of the sample before (*lower*) and after (*upper*) the addition of biotin. Additional resonances from the free biotin are clearly visible in the second spectrum. (**b**) The STD spectrum of the sample, showing clear signal from ligands before the addition of biotin (*upper*). After addition of the biotin (*lower*), no signal is visible from the 2-imidazolidinone at center of spectrum. Some signal is visible from octanoic acid, but (unlike in Fig. 1b) this saturation has not propagated across the entire molecule and is attributable to direct irradiation of the methyl group in the octanoic acid (at 0.75 ppm) whose chemical shift is close to that of the protein methyl groups that are the principal target of the saturating irradiation. This could be interpreted as indicating binding were the direct irradiation of the methyl group not considered. No signal is observed from the biotin which has very high affinity and is not in fast exchange between free and bound states. (**c**) The water-LOGSY spectrum of the sample (*upper*), showing positive signal from the 2-imidazolidinone and octanoic acid, and negative signal from the sucrose and residual protonated DMSO resonance (2.59 ppm). After addition of the biotin (*lower*) negative peaks are observed for these compounds, indicating that they are no longer binding. Positive signal is also observed from the labile protons of the biotin (6.3 ppm) as a consequence of chemical exchange, while no signal is observed from other biotin protons. (**d**) The CPMG-filtered 1D spectrum of the sample, showing signal from the sucrose and DMSO (*lower*), and also the appearance of signal from the displaced octanoic acid and 2-imidazolidinone (*upper*) after addition of biotin

some rather unusual cases, the receptor is required to be isotopically labeled in order to clearly resolve which signals are perturbed by the ligand. These factors constitute a rather strict requirement for high levels of protein expression in a system which allows ready isotopic labeling. Because resonances become increasingly broad and overlapped with increasing size of protein, a further restriction on the applicability of protein-observed techniques is an upper size limit for the target. Typically, this would be in the range of 20–30 kDa if ^{15}N-^{1}H HSQC spectra are used, 30–60 kDa for ^{15}N-^{1}H TROSY spectra, and ~100 kDa or more if ^{13}C-^{1}H HSQC experiments are used in conjunction with the selective labeling of certain aliphatic side chains (28).

Protein-observed techniques are not limited by the requirement for weak binding (fast on/off rates) which is a fundamental restriction of ligand-observed methodologies. However, the restrictions discussed above limit the application of protein-observed techniques, although where they can be used these methods yield data on both the affinity of the ligand (through analysis of the magnitude of CSPs with titration of the ligand concentration) and the location of the ligand binding site (through mapping of the CSPs onto the protein structure) (29, 30). This identification of the ligand binding site through CSP mapping is a key feature of protein-observed techniques, and one which can be reliably used to identify both active- and allosteric-site binding ligands (31).

In order to map a ligand binding site, a number of approaches are available. If a sequence specific assignment and three-dimensional structure have been determined, the binding site can simply be identified by mapping the location of the assigned shifted residues on the structure. By using the differential CSP patterns induced by structurally related ligands (e.g., AMPPNP, ADP, AMP, adenosine, and adenine to probe an ATP binding protein), an indication of the location and orientation of the ligand within the binding site can be determined. If no structure is available, the binding site can be inferred using the similarities of observed shifts with those produced by ligands with a known binding site. If no previously known ligands exist, and a sequence specific assignment is not possible, then simply mapping the ligand-induced CSPs can prove useful in order to confirm a consistent binding site for the putative ligands, even if the location and nature of the binding site is not known. However, some care must be taken with CSP mapping, since the observed perturbations can extend well beyond the ligand binding site, particularly when propagated through elements of secondary structure.

An alternative approach to CSP mapping is the transferred cross-saturation method (28), originally designed to probe the large interface formed by a protein–protein interaction. Selective saturation is achieved by uniformly ^{15}N,^{2}H labeling the receptor, exchanging protons back onto the protein's labile amide groups by dissolving in H_2O and then saturating the aliphatic proton region of the ligand. Amide (HN) resonances in the binding site interface can then be identified directly through loss of signal due to nearby aliphatic protons of the bound ligand. This effect is short-ranged and thus more localized to the binding site than simple chemical shift changes. Practical application of transferred cross-saturation requires a robust protein expression system in order to achieve the high levels of deuteration required for the technique to work.

1.3 Structural Characterization of Protein–Ligand Complexes

High resolution 3-dimensional structural data are key to fully characterizing a protein–ligand complex. Structures are typically determined using X-ray crystallography. However, there are situations where it is not possible to obtain a high resolution crystal structure of the ligand–receptor complex, and in these instances it can be useful to obtain NMR data in order to assist structural characterization of the complex. While a de novo structure determination by NMR gives the most robust and complete structural characterization in the absence of an X-ray structure, this is often not practicable given experimental and/or other resource limitations. Over the past decade, a number of alternative NMR techniques have been developed which can yield valuable structural information about a protein–ligand complex quickly and with fewer experimental constraints than a de novo structure determination (28). Although these techniques are outside the remit of this chapter, a brief overview of some of the more widely used methods is given below since several of the experimental approaches used in screening can also provide structural characterization beyond the mapping of a ligand binding site on a receptor by CSPs. Whereas CSP based mapping is discussed under Subheading 3.3, it is worth noting here that docking of a ligand onto a receptor using only CSP data is possible (e.g., using the HADDOCK software package (32)), and has been successfully used to model the structures of ligand–receptor complexes (33, 34).

Broadly, and analogously to the NMR experiments used to detect binding, the NMR techniques used to characterize the structure of protein–ligand complexes can be classified into ligand-observed and protein-observed experiments. The limitations and practicalities of each type of technique are also analogous to those discussed for detecting binding. Ligand-observed experiments require fast exchange on the NMR timescale between the free and bound states, but have no limitations on the upper size limit of the receptor (indeed, they often work better for larger receptors). Protein-observed experiments can be used for ligands which are tight binding, and so are in slow exchange between free and bound states, but typically require isotopic labeling and are restricted to smaller proteins (generally 40 kDa or below).

1.3.1 Ligand-Observed Experiments

NMR parameters which persist from the bound to the free state under favorable circumstances, such as the NOE, can be used to obtain structural data about the small population of ligand in a bound state whilst observing the ligand in the free state.

The transferred NOE (trNOE) experiment is widely and routinely used to characterize the bound state of ligands (35, 36). Because of the long correlation time (τ_c) of the receptor compared to the free compound, the bound state of the ligand gives rise to large, positive NOEs, whilst the free state gives rise to weak negative NOEs. The observation and assignment of positive NOEs from a ligand can thus be used to impose constraints of the conformation of the ligand adopted in the binding site of the receptor (37).

Related to the trNOE experiment are two techniques based around the transfer of magnetization between ligands in the bound state, either directly between ligands bound simultaneously in proximal binding sites (the ILOE, inter-ligand NOE, experiment (38, 39)) or in a protein-mediated manner through ligands binding sequentially to the same site (the INPHARMA, inter-ligand NOESY for pharmacophore mapping, experiment (40, 41)). Both of these experiments rely on the transferred NOE effect to give structural data which can be used to model the ligands either into a known binding site, or relative to the other ligand in an unknown binding site. Observation of ILOE or INPHARMA signal is somewhat dependent on the k_{on} and k_{off} rates for the complex, but where signal is observed it can provide valuable structural data about the binding site (42–44) (Fig. 4).

The STD experiment discussed under Subheading 1.1 can be quantitated to determine a map of the proximity of regions of the ligand to the receptor (25, 42, 43, 45). Atoms close to the ligand–receptor interface experience a greater degree of saturation than those remote from the interface, resulting in a gradation of the observed STD effect across the ligand. Because of T1 relaxation during the saturation period, the longitudinal relaxation rates of the ligand resonances must be taken into account; this dependence of the STD amplitude on proximity to the protein–ligand interface can then be used to infer the binding mode or to distinguish between otherwise viable docking poses (46).

1.3.2 Protein-Observed Experiments

Observation of intermolecular contacts through the protein component of a ligand–protein complex yields valuable structural data which can be difficult to obtain through other methods. If the protein is relatively small (typically less than 40 kDa) and can be economically ^{13}C and ^{15}N labeled, then a $^{13}C^{15}N$-filtered, ^{12}C-detected 3D HMQC-NOESY experiment, which will detect NOEs between the protein and ligand resonances, can give structural data on the interface between the protein and the ligand at an atomic resolution. In conjunction with a known assignment and structural model of the protein alone this data can then be used to guide the docking and modeling of the protein–ligand complex (47). The unambiguous determination of both the ligand and protein components of an intermolecular NOE allows rapid and robust discrimination of possible binding modes.

2 Materials

NMR spectrometers are available from a number of suppliers, primarily Varian, now a part of Agilent Technologies (http://www.agilent.com), Bruker (http://www.bruker.com/) and Jeol

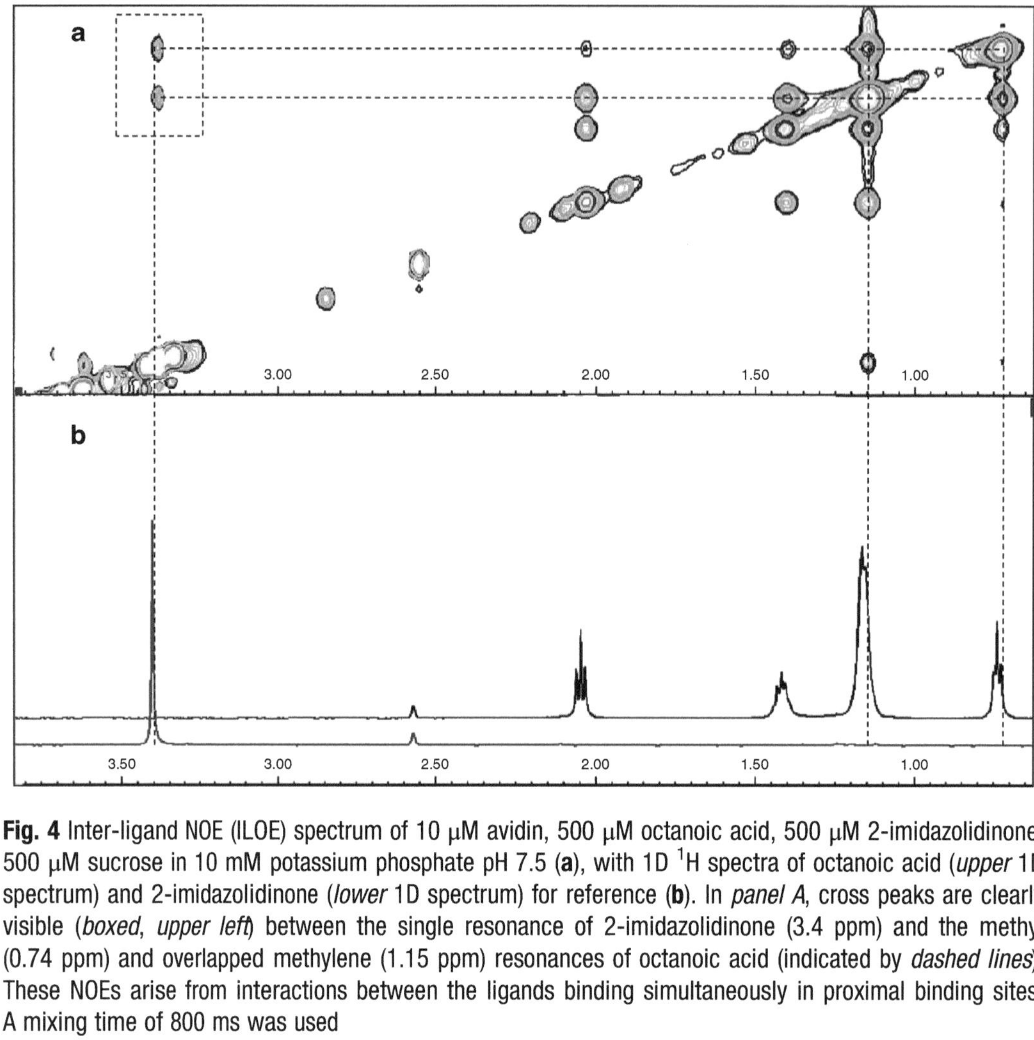

Fig. 4 Inter-ligand NOE (ILOE) spectrum of 10 μM avidin, 500 μM octanoic acid, 500 μM 2-imidazolidinone, 500 μM sucrose in 10 mM potassium phosphate pH 7.5 (**a**), with 1D ^{1}H spectra of octanoic acid (*upper* 1D spectrum) and 2-imidazolidinone (*lower* 1D spectrum) for reference (**b**). In *panel A*, cross peaks are clearly visible (*boxed*, *upper left*) between the single resonance of 2-imidazolidinone (3.4 ppm) and the methyl (0.74 ppm) and overlapped methylene (1.15 ppm) resonances of octanoic acid (indicated by *dashed lines*). These NOEs arise from interactions between the ligands binding simultaneously in proximal binding sites. A mixing time of 800 ms was used

(http://www.jeol.com/). Since Bruker and Varian are the most common instruments used for biomolecular NMR, further specific discussion is limited to these two manufacturers.

2.1 Pulse Sequences

Both Varian and Bruker have pulse sequences for most, if not all, of the NMR experiments discussed below. In some instances these are supplied with the spectrometer, whilst others may have to be requested from the relevant vendor's applications support group. Modifications to the pulse sequence (such as changing the water suppression method) are readily carried out by the user or the vendor's application's engineers.

Since most data will be acquired in H_2O based solutions, robust solvent suppression is vital and will be assumed to be used throughout this discussion. A number of water suppression methods have been developed; most of these are suitable for use in NMR ligand binding experiments, although we have found that excitation sculpting (48) is particularly useful for ligand-observed experiments acquired under automation.

2.2 Automation

If medium- to high-throughput screening of ligands by NMR is to be undertaken, automated sample changing, acquisition and processing is required. Both Varian and Bruker spectrometers can be fitted with proprietary automated sample changers, some of which are able to refrigerate the samples and/or prepare samples in an automated manner (e.g., Bruker's SampleJet (http://www.bruker.com/products/mr/nmr/automation/automation.html). Modern NMR probes are capable of performing automated tuning and matching; shimming, pulse calibration and spectral processing can also be automated. As discussed in the pulse sequence section (Subheading 2.1), the vendor's application support group will be able to assist with the implementation of these techniques.

2.3 Standard Samples

Numerous standard samples exist that can be used to implement and test NMR experiments designed to detect ligand binding to a macromolecular receptor. The choice of test sample depends largely on the type of ligand binding experiments it is intended to be used for. Several options are discussed below. Avidin ubiquitin and calbindin D9k samples are stable at room temperature for at least a week.

- Ligand-observed NMR sample: 10 μM avidin, 500 μM octanoic acid, 500 μM 2-imidazolidinone, 500 μM sucrose, 10 mM potassium phosphate pH 7.5, 10 % D_2O (*see* **Note 1**).
- Protein-observed NMR sample A: 100 μM isotopically labeled ubiquitin, 10 mM potassium phosphate pH 6.0, 10 % D_2O (*see* **Note 2**).
- Protein-observed NMR sample B: 100 μM isotopically labeled calbindin-D9K, 10 mM potassium phosphate pH 6.0, 10 % D_2O (*see* **Note 3**).

2.4 Analysis Software

In principle, any NMR software which allows comparison of several 1D and/or 2D spectra can be used to analyze ligand binding experiments. The software associated with the NMR spectrometer (VNMR for Varian and TOPSPIN for Bruker spectrometers) can readily be used for simple inspection and comparison of spectra. Other NMR analysis software packages, such as Analysis (CCPN) or MNova (Mestrelab Research), are available and are also capable of being used to analyze ligand binding experiments.

For large-scale analysis, the software package AMIX (Bruker) allows for the direct comparison of many 1D and 2D spectra, along with a searchable, user-maintained database of compound spectra. This is particularly useful when screening large numbers of compounds against several macromolecular targets. AMIX also has capabilities for semi-automated analysis of sets of ligand-observed binding experiments where competition is used to distinguish specific binding interactions and of ^{15}N-^{1}H HSQC titrations. ACD NMR software (ACD/1D NMR Manager) also has spectral database functionality, but its features are more focused on use as an analytical NMR tool rather than as a screening application.

3 Methods

The acquisition of NMR binding experiments to detect ligand interactions is similar to the acquisition of any other NMR data in terms of shimming, pulse calibration, solvent suppression and so on. These general features of NMR experiments have been extensively discussed elsewhere (49–52), and only specific aspects of experimental procedure relating to the detection of ligand–protein interactions will be covered below.

3.1 Sample Validation

One of the key features of NMR is the ability to characterize all soluble components of a solution which contain observable nuclei. Thus, by acquiring simple 1D ^{1}H spectra, it is possible to confirm that the target protein is in solution and is folded correctly, that any compounds and buffers are present in solution at the correct concentration and that their spectra are consistent with their structure and that there are no contaminants which may affect the experimental observations (*see* **Note 4**). Crucially, such sample validation and verification substantially increases the robustness of ligand identification by NMR.

Prior to attempting to determine whether a compound is interacting with a macromolecular receptor, there are several important controls which must be considered. While this is particularly true for ligand-observed NMR experiments, since small changes in the NMR spectra can readily be misinterpreted as an indication of binding, protein-observed spectra can also suffer from artifacts associated with the addition of compounds to the solution. The following controls should be carried out when preparing any screening experiment.

1. Acquire regular 1D ^{1}H spectra of all compounds individually in the absence of protein as a reference, ideally in the same buffer as will be used in the binding experiment. The 1D ^{1}H spectrum of each compound should be confirmed as being consistent with its chemical structure (*see* **Note 5**).
2. Ensure that the compound is soluble at the concentration used by comparison to an internal standard such as 1 mM TSP (*see* **Note 6**).

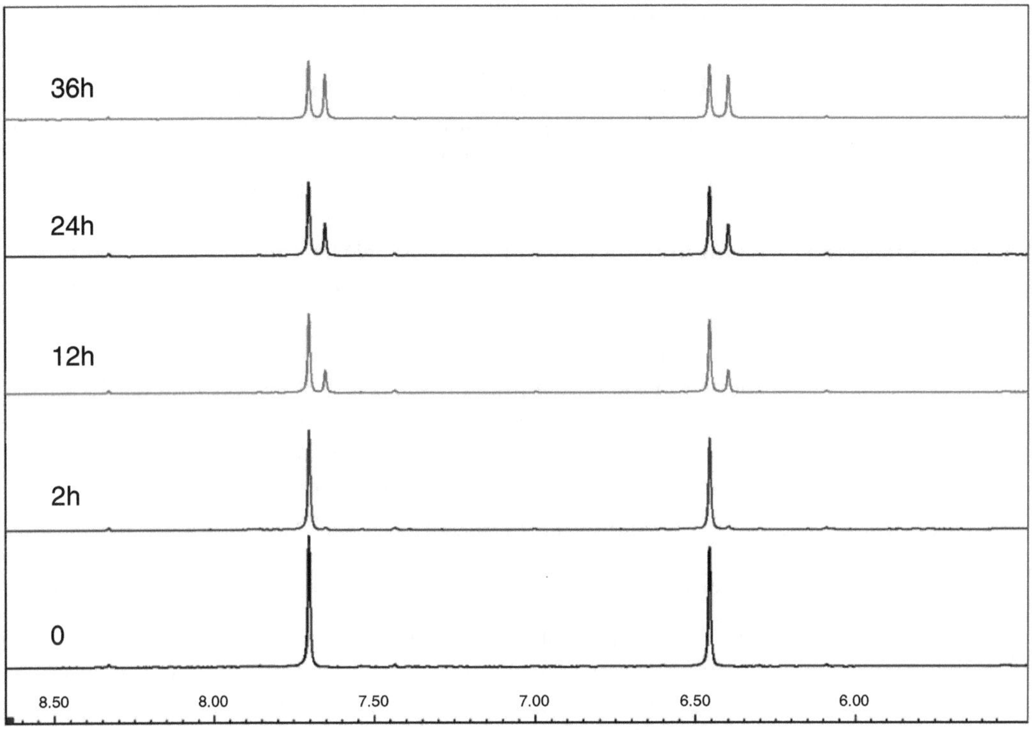

Fig. 5 1D ^{1}H spectra of a compound in aqueous solution (500 μM compound, 90 % H_2O, 10 % D_2O, 50 mM potassium phosphate pH 7.5, 25 °C) acquired over time. The compound can be seen to degrade rapidly on standing in aqueous solution. Any binding experiments performed on this compound would ideally be completed within 4 h of the sample being prepared from a DMSO stock or some allowance must be made in interpretation of screening results

3. Ensure that the compound spectrum does not change with time. The compound must be stable and soluble under the experimental conditions for at least the expected duration of the experiment, and ideally for substantially longer (Fig. 5) (*see* **Note 7**).
4. Acquire 1D ^{1}H spectra of the protein in the relevant buffer but without the addition of compounds. Acquire these spectra periodically in order to ensure that the protein spectrum does not change over the intended timescale of the experiment (*see* **Note 8**).
5. If screening for ligands using a mixture of compounds in the same sample, ensure that spectral dispersion in the 1D ^{1}H NMR spectrum is sufficient to allow unambiguous identification of all ligands present (*see* **Note 9**). It is also worthwhile confirming, through comparison with the reference spectra for individual compounds, that all compounds expected to be in solution are present at the correct concentration (to identify errors which can occur during sample preparation), and that no

interactions occur between any of the compounds present in the mixture.

6. Acquire regular 1D ^{1}H spectra before acquiring any other more complex NMR experiment. Repeat acquisition of 1D ^{1}H spectrum periodically and, particularly, after addition of any competitor compound (if used) (*see* **Note 10**).

3.2 Ligand-Observed NMR Binding Experiments

Many different ligand-observed NMR binding experiments have been proposed in the literature. Because the majority of these experiments are rapid, and have similar sample requirements, it is common practice to acquire two or more different binding experiments on the same sample while it is in the spectrometer. Comparison and combination of the data from these various experiments is then used to identify ligands with a greater degree of certainty, and reduced risk of false positives and negatives, than would be the case if only a single experiment were to be used.

In order to acquire data from ligand-observed NMR binding experiments, the following general protocol should be applied.

1. Prepare a sample of compound(s) and protein in a suitable buffer (*see* **Note 11**) at the required concentrations and stoichiometry (*see* **Note 12**). Typical conditions might be 10 μM protein, 500 μM to 1 mM compound, 50 mM potassium phosphate pH 7.5, although this will depend greatly on the requirements of the protein. Data is generally acquired at 25 °C, although acquiring water-LOGSY data at 5 °C has been reported to give an improved signal–noise ratio at the expense of substantially increased sample equilibration time (53).
2. Acquire a 1D ^{1}H NMR spectrum with water suppression, followed by one or more ligand-observed NMR binding experiments (Fig. 1). A positive result from several NMR binding experiments is generally a more robust indication of a true ligand than a single experiment, reducing the risk of false positive or negative data.
3. Optimize experimental parameters in step 2 in order to maximize signal in the NMR experiment(s) (*see* **Note 13**).
4. Add known potent competitor ligand, if available, and repeat the experiments acquired in step 2 above (Fig. 3). Putative ligands which are competitive with this known ligand will show reduced or abolished binding (*see* **Note 14**). Addition of a competition step allows the identification of ligands which bind specifically to the site of the known ligand or which are displaced from another specific site through allosteric effects.

3.3 Protein-Observed NMR Binding Experiments

Protein-observed NMR binding experiments typically rely on the identification of ligand-induced CSPs of resonances in heteronuclear correlation spectra. Reference spectra are acquired for the protein in the relevant buffer, and these spectra compared with spectra acquired on the protein in the presence of one or more putative ligands.

Since most protein-observed NMR binding experiments rely on indirect CSPs observed through heteronuclear correlation spectra rather than the direct modulation of ligand physical properties as is the case in ligand-observed NMR binding experiments, it is important to rule out factors other than binding which could influence the observed chemical shifts. The two most common of these factors are small pH changes caused by the addition of ligands and the binding of DMSO (DMSO is often used as a solvent for the ligand stock solution and frequently binds in the active sites of proteins) (*see* **Note 15**).

A general method for identifying ligands using protein-observed NMR binding experiments would be:

1. Prepare an NMR sample containing protein at sufficient concentration to observe the required experiment in a timescale which is appropriate for the number of samples required, given constraints of protein and ligand solubility and availability. Typical sample conditions would be 50–100 μM protein, with ligand concentrations being titrated from zero to 1 mM; if ^{15}N-^{1}H HSQC spectra are acquired on a 600 MHz spectrometer with a cryoprobe, a typical acquisition time for such a sample would be in the region of 15–60 min. The buffer used is not as crucial as in ligand-observed experiments, since buffer peaks will not be observed, but the buffer should be optimized for maximum protein stability.
2. Acquire 1D ^{1}H and heteronuclear spectra (*see* **Note 16**) on the freshly prepared protein-only sample. From here onwards, 2D ^{15}N-^{1}H HSQC spectra will be used as an example, but a large number of different experiments are available in the literature which could be used to detect ligand binding.
3. Repeat these experiments over time until initial signs of degradation (typically the appearance of small but sharp peaks around 8 ppm) or precipitation (often indicated by a general broadening or loss of intensity of protein resonances) are observed (Fig. 6). This will give an indication of the timescale available for ligand binding experiments. If the protein is not soluble or stable in the initial buffer, optimization of the conditions is required.
4. In parallel with steps 2 and 3, make up fresh samples of isotopically labeled protein and titrate pH and, if the ligand stock solution is dissolved in DMSO, titrate DMSO. Acquire 1D ^{1}H and 2D ^{15}N-^{1}H HSQC spectra at each titration point. Identify resonances affected by small changes in pH and/or [DMSO], and also the extent to which pH and [DMSO] can be changed before the protein degrades or precipitates (Fig. 7).

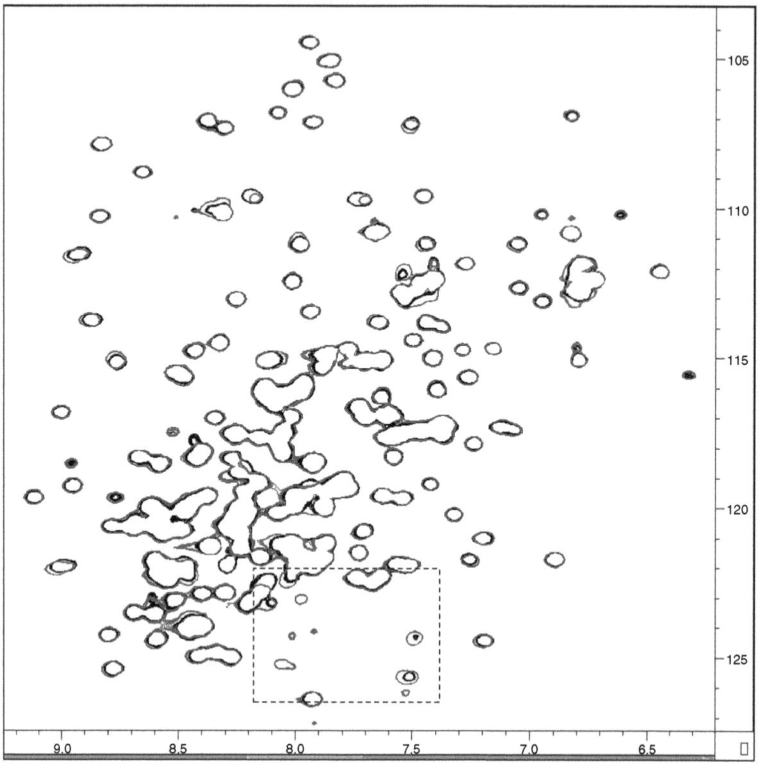

Fig. 6 ^{15}N-^{1}H HSQC spectra of a protein when freshly prepared (*black*) and after 48 h at room temperature (*grey*). Additional peaks are clearly visible in the second spectrum between 7.45 and 8.05 ppm (*boxed region*), indicative of the presence of additional random coil peptide chain. Subsequent, SDS-PAGE analysis showed that the protein was partially degraded

5. Given the constraints of protein solubility and stability, allowable pH and DMSO limits and available spectrometer time, add compound(s) to a fresh protein sample and acquire 1D ^{1}H and 2D ^{15}N-^{1}H HSQC spectra (*see* **Note 17**). These compounds should be previously characterized in terms of chemical identity, solubility and stability as discussed in Subheading 3.1. Titrate the ligand concentration (either in the same sample or in different samples, dependent on experimental requirements) and observe patterns of CSPs. These CSPs can be used to infer the site of binding and/or to determine the K_D of the ligand (54). If medium- or high-throughput screening of ligands is being carried out, addition of a single concentration of compound(s) can be performed rather than a titration. However, this is often be more prone to erroneous interpretation of artifactual shifts such as solvent binding or pH changes, and will not give any data about exchange regime or ligand K_D. For these reasons, it is recommended to perform titrations where possible.

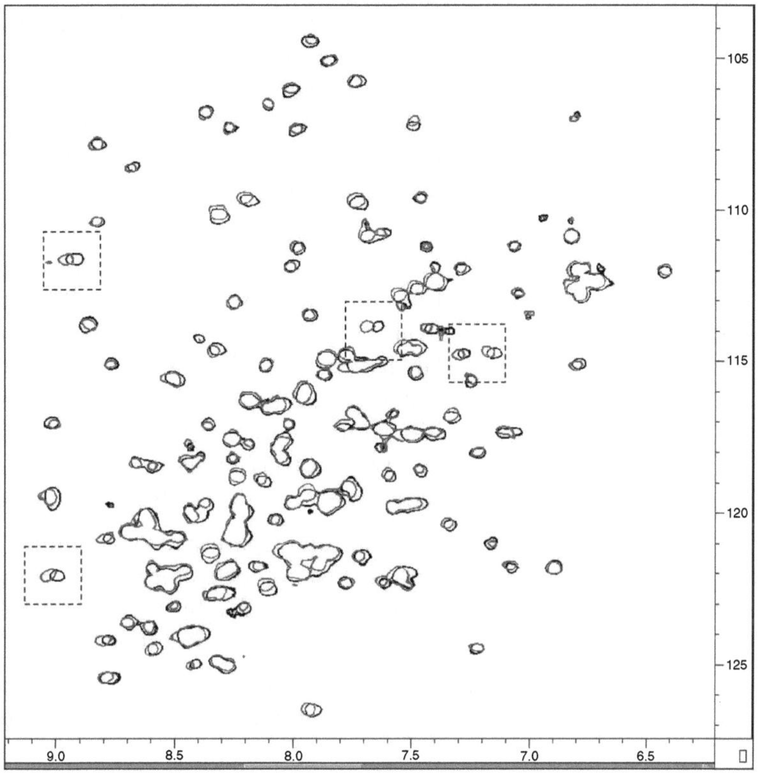

Fig. 7 ^{15}N-^{1}H HSQC spectra of a protein showing the effect of increasing concentrations of DMSO. For clarity, only 0 % DMSO (*black*) and 4 % (*grey*) are shown; additional spectra were acquired in the presence of 0.5 %, 1 %, and 2 % DMSO. Spectra acquired at concentrations of DMSO below 1 % are unchanged. At higher concentrations of DMSO, significant shifts are seen in a number of resonances (*boxed*), indicating an interaction with DMSO. The pH was unchanged during the titration. Consequently, all subsequent experiments were performed at [DMSO] of 1 % or below

4 Notes

1. The Ligand-Observed Sample is used to show clear binding signal for both ligands (Fig. 1). The addition of 50 μM biotin to the sample demonstrates the effect of adding a potent competitor which fully occupies the ligand binding site (Fig. 3). This sample can also be used to test the ILOE experiment, with ILOE peaks visible between the methylene resonance of 2-imidazolidinone and methylene and methyl resonances of octanoic acid (Fig. 4).
2. Since most protein-observed experiments observe only the protein, rather than interactions between the protein and a ligand,

they can be implemented using any correctly isotopically labeled protein sample. ^{15}N or $^{13}C^{15}N$ labeled ubiquitin is a common standard sample for testing heteronuclear experiments, but does not bind a small molecular ligand. Isotopically labeled ubiquitin is available from VLI research (http://www.vli-research.com/). If the effect on the protein of binding a ligand is required to be seen, Protein Sample B should be used.

3. Protein Sample B can be used to test ^{15}N-^{1}H HSQC or ^{13}C-^{1}H HSQC experiments used to detect ligand binding by titrating Ca^{2+} into the NMR sample. Calbindin-D9K can be purchased from ProtEra (http://www.proterasrl.com/) as ^{15}N or $^{13}C^{15}N$ labeled material dependent on the type of experiment to be tested. Since the ligand in this case has no protons, this sample is not suitable for testing the $^{13}C^{15}N$-filtered, ^{12}C-detected 3D HMQC-NOESY experiment, or any other experiment which requires observation of both ligand and protein.
4. Since in any experimental technique errors can and do arise, and unexpected behavior is often shown by the receptor and/or the compounds, this aspect of NMR can be used to identify and rectify errors before they become an issue.
5. This step is key. It has been known for vendors to supply the wrong compound, or for mistakes to occur during labeling processes. If the compound spectrum is ambiguous, or is inconsistent with the proposed structure, acquire data in organic solvents to confirm the structure of the compound.
6. Alternatively, the compound can be titrated into solution and spectra acquired at each concentration. The point at which the peak integrals deviate from the expected value can be taken as the limiting solubility of the compound under this buffer condition.
7. Many compounds are not stable in aqueous solution over protracted periods of time. If unexpected changes in the compound spectrum are observed, it may also be worth reexamining the stock solution of the compound.
8. Depending on the size of the target protein and the dispersion in the 1D ^{1}H spectrum, relaxation or diffusion filtered 1D experiments may be required to suppress ligand signals and allow observation of the protein spectrum in order to confirm that the protein is folded and present in solution. However, a simple 1D ^{1}H spectrum with solvent suppression is often sufficient, although a reasonably high number of scans (such as 128 or 256) may be required if the protein is at a low concentration.
9. If the compounds present in the mixture are chemically diverse, and the buffer spectrum is minimal, 5–15 compounds can typically be mixed in a cocktail without substantial spectral overlap. However, if the compounds are chemically similar or

substantial buffer resonances are present in the spectrum, careful selection of compounds based on chemical shift dispersion may be required. In this case, the reference sample spectra acquired during the validation of the individual compounds should be used to design compound mixtures with acceptable spectral overlap.

10. Acquisition of regular 1D ^{1}H spectra allows monitoring of the sample over time, and ensures that any changes which occur, such as protein or ligand precipitation or change in quality of the shimming, are recognized rapidly. If using a competitor ligand it confirms that the correct competitor ligand has been added at the correct concentration.

11. Since most ligand-observed experiments are ^{1}H detected, resonances from any buffer components will be observed; this is particularly problematic for buffers such as tris (tris(hydroxymethyl)aminomethane) or glycerol. When present at high concentration, these resonances can dominate the acquired spectra, causing baseline distortion and rendering regions of the spectrum unusable. This can be a serious issue for ligands which contain resonances only in the 3–5 ppm range, which is where the majority of buffer resonances occur. Phosphate buffers can be used to circumvent this issue, but in cases where the protein is not stable in phosphate buffer (or directly binds phosphate) then careful selection of buffers to reduce signal overlap or the use of deuterated buffers should be considered. Obviously, ^{19}F detected experiments do not suffer from distortion by buffer signals, and this can be a major argument for their use in certain circumstances.

12. Typically, ligand-observed binding experiments are acquired on samples which contain a substantial excess of ligand over receptor. This increases the signal amplification discussed above, and reduces the degree of spectral interference from resonances of the receptor. The optimal relative stoichiometry of ligand and receptor depends on the experiment(s) to be acquired. For experiments such as the STD, where the observed signal is a function of the bound population, the ligand can be in a high molar excess over the protein—values for [L]/[P] of 100–200 are not unusual. However, experiments such as the water-LOGSY contain contributions from both the free and bound populations of a ligand. As the molar ratio increases, the negative signal contribution from the unbound population will increase and dominate the recorded spectrum (26). Ligand–protein molar ratios ([L]/[P]) of 10–50 are therefore more typically used, dependent on the K_D and solubility of the ligand. Where both types of experiment are to be acquired, a molar ratio of [L]/[P] of 50–100 is a

Table 1
Typical experimental parameters for ligand-observe NMR screening experiments

Experiment	Parameter	Typical values	Notes
STD	Mixing time	1–2.5 s	
	On-resonance saturation frequency	−1.0 to 0.7 ppm	Depends on receptor spectrum and size
	Off-resonance saturation frequency	−40 ppm	Must be in region of spectrum free of resonances
Water-LOGSY	Mixing time	1–2.5 s	
Relaxation filter	Relaxation delay	200–800 ms	Varying delay will select different affinity ranges

good initial value (e.g., 10 μM protein and 500 μM ligand), and can modified subsequently if found to be sub-optimal.

13. Most, if not all, ligand-observed NMR binding experiments have one or more experimental parameters which require optimization and which will affect the quality of the observed spectrum. The more important of these are given in Table 1, along with some typical values.
14. The Gaddum equation

$$f = \frac{[A]}{[A] + K_A\left(1 + \frac{[B]}{K_B}\right)}$$

describes the behavior of two ligands with a common binding site, where f is the fractional occupancy of the receptor to ligand A, and K_A and K_B the equilibrium dissociation constants of A and B, respectively. Ideally a 90 % reduction in the bound population should be obtained. In practice, it is usually best to use the most potent competitor available. Regardless of the potency of the competitor, a minimum concentration of two to three times the receptor concentration should be used to allow for experimental errors.
15. In order to rule out these effects, both pH and DMSO (if used) should be titrated against the protein and the relevant spectra acquired at each point. The pH and [DMSO] should then, if possible, be kept within a range where no spectral perturbation is observed or a spectrum of appropriate pH and [DMSO] be used as a reference in assessing evidence of binding. Signals from protonated buffers in 1D ^{1}H spectra are often pH dependent and can, when present, be used to identify the precise pH of each sample.

16. The exact choice of the experiment used depends greatly on the nature of the target protein, but ^{15}N-^{1}H HSQC or ^{13}C-^{1}H HSQC spectra are the most commonly used techniques for proteins below 30–40 kDa (37). For larger proteins, ^{15}N-^{1}H TROSY spectra should be acquired (55). The selective ^{13}C labeling of methyl groups in valine, isoleucine, and leucine can be used to give good quality ^{13}C-^{1}H HMQC spectra of proteins of 100 kDa or more (56, 57). In certain cases the simple 1D ^{1}H spectra may contain strongly shifted and well resolved methyl resonances which are perturbed on ligand binding, although in general proton spectra of proteins contain too much spectral overlap to be of use in identifying ligand binding.
17. If the ^{15}N-^{1}H HSQC is of substantial duration (e.g., >2 h), acquire a 1D ^{1}H spectrum after the correlation spectrum in order to confirm that the sample conditions have not changed during the course of the experiment.

References

1. Nietlispach D et al (2004) Structure determination of protein complexes by NMR. Meth Mol Biol 278:255–288
2. Powers R (2009) Advances in nuclear magnetic resonance for drug discovery. Expert Opin Drug Discov 4:1077–1098
3. Ludwig C, Guenther UL (2009) Ligand based NMR methods for drug discovery. Front Biosci 14:4565–4574
4. Middleton DA (2006) NMR methods for characterising ligand–receptor and drug–membrane interactions in pharmaceutical research. Annu Rep NMR Spectrosc 60:39–75
5. Meyer B, Peters T (2003) NMR spectroscopy techniques for screening and identifying ligand binding to protein receptors. Angew Chem Int Ed Engl 42:864–890
6. Benie AJ, Moser R, Bauml E, Blaas D, Peters T (2003) Virus–ligand interactions: identification and characterization of ligand binding by NMR spectroscopy. J Am Chem Soc 125:14–15
7. Gharbi-Benarous J et al (2004) Epitope mapping of the phosphorylation motif of the HIV-1 protein Vpu bound to the selective monoclonal antibody using TRNOESY and STD NMR spectroscopy. Biochemistry 43:14555–14565
8. Claasen B, Axmann M, Meinecke R, Meyer B (2005) Direct observation of ligand binding to membrane proteins in living cells by a saturation transfer double difference (STDD) NMR spectroscopy method shows a significantly higher affinity of integrin alpha(IIb)beta3 in native platelets than in liposomes. J Am Chem Soc 127:916–919
9. Meinecke R, Meyer B (2001) Determination of the binding specificity of an integral membrane protein by saturation transfer difference NMR: RGD peptide ligands binding to integrin alphaIIbbeta3. J Med Chem 44:3059–3065
10. Hubbard RE, Davis B, Chen I, Drysdale MJ (2007) The SeeDs approach: integrating fragments into drug discovery. Curr Top Med Chem 7:1568–1581
11. Dalvit C (2007) Ligand- and substrate-based ^{19}F NMR screening: principles and applications to drug discovery. Prog Nucl Magn Reson Spectrosc 51:243–271
12. Dalvit C et al (2005) Sensitivity improvement in 19F NMR-based screening experiments: theoretical considerations and experimental applications. J Am Chem Soc 127:13380–13385
13. Dalvit C, Flocco M, Veronesi M, Stockman BJ (2002) Fluorine-NMR competition binding experiments for high-throughput screening of large compound mixtures. Comb Chem High Throughput Screen 5:605–611
14. Dalvit C, Fagerness PE, Hadden DT, Sarver RW, Stockman BJ (2003) Fluorine-NMR experiments for high-throughput screening: theoretical aspects, practical considerations, and range of applicability. J Am Chem Soc 125:7696–7703
15. Mayer M, Meyer B (1999) Characterization of ligand binding by saturation transfer difference

NMR spectroscopy. Angew Chem Int Ed Engl 38:1784–1788
16. Dalvit C et al (2000) Identification of compounds with binding affinity to proteins via magnetization transfer from bulk water. J Biomol NMR 18:65–68
17. Hajduk PJ, Olejniczak ET, Fesik SW (1997) One-dimensional relaxation- and diffusion-edited NMR methods for screening compounds that bind to macromolecules. J Am Chem Soc 119:12257–12261
18. Rossi C, Donati A, Sansoni MR (1992) Nuclear magnetic resonance as a tool for the identification of specific DNA–ligand interaction. Chem Phys Lett 189:278–280
19. Bertini I, Fragai M, Lee YM, Luchinat C, Terni B (2004) Paramagnetic metal ions in ligand screening: the Co(II) matrix metalloproteinase 12. Angew Chem Int Ed Engl 43:2254–2256
20. Pellecchia M, Sem DS, Wuthrich K (2002) NMR in drug discovery. Nat Rev Drug Discov 1:211–219
21. Vanwetswinkel S et al (2005) TINS, target immobilized NMR screening: an efficient and sensitive method for ligand discovery. Chem Biol 12:207–216
22. Chen A, Shapiro MJ (1998) NOE pumping: a novel NMR technique for identification of compounds with binding affinity to macromolecules. J Am Chem Soc 120:10258–10259
23. Chen A, Shapiro MJ (2000) NOE pumping as a high-throughput method to determine compounds with binding affinity to macromolecules by NMR. J Am Chem Soc 122:414–415
24. Jayalakshmi V, Krishna NR (2002) Complete relaxation and conformational exchange matrix (CORCEMA) analysis of intermolecular saturation transfer effects in reversibly forming ligand–receptor complexes. J Magn Reson 155:106–118
25. Meyer B et al (2004) Saturation transfer difference NMR spectroscopy for identifying ligand epitopes and binding specificities. Ernst Schering Res Found Workshop 44:149–167
26. Dalvit C, Fogliatto G, Stewart A, Veronesi M, Stockman B (2001) WaterLOGSY as a method for primary NMR screening: practical aspects and range of applicability. J Biomol NMR 21:349–359
27. Neuhaus D (2011) Nuclear overhauser effect. Encyclopedia Magn Reson. doi:10.1002/9780470034590.emrstm0350.pub2
28. Pellecchia M et al (2002) NMR-based structural characterization of large protein–ligand interactions. J Biomol NMR 22:165–173
29. Moore CD et al (2009) Structural and biophysical characterization of XIAP BIR3 G306E mutant: insights in protein dynamics and application for fragment-based drug design. Chem Biol Drug Des 74:212–223
30. Veldkamp CT, Ziarek JJ, Peterson FC, Volkman BF, Chen Y (2010) Targeting SDF-1/CXCL12 with a ligand that prevents activation of CXCR4 through structure-based drug design. J Am Chem Soc 132:7242–7243
31. Balogh E, Wu D, Zhou G, Gochin M (2009) NMR second site screening for structure determination of ligands bound in the hydrophobic pocket of HIV-1 gp41. J Am Chem Soc 131:2821–2823
32. de Vries SJ, van Dijk M, Bonvin AM (2010) The HADDOCK web server for data-driven biomolecular docking. Nat Protoc 5:883–897
33. Medek A, Hajduk PJ, Mack J, Fesik SW (2000) The use of differential chemical shifts for determining the binding site location and orientation of protein-bound ligands. J Am Chem Soc 122:1241–1242
34. Wang YS et al (2010) Application of fragment-based NMR screening, X-ray crystallography, structure-based design, and focused chemical library design to identify novel microM leads for the development of nM BACE-1 (beta-site APP cleaving enzyme 1) inhibitors. J Med Chem 53:942–950
35. Zhou CC, Swaney SM, Shinabarger DL, Stockman BJ (2002) ^{1}H nuclear magnetic resonance study of oxazolidinone binding to bacterial ribosomes. Antimicrob Agents Chemother 46:625–629
36. Pristovsek P, Simcic S, Wraber B, Urleb U (2005) Structure of a synthetic fragment of the lipopolysaccharide (LPS) binding protein when bound to LPS and design of a peptidic LPS inhibitor. J Med Chem 48:7911–7914
37. Catoire LJ et al (2010) Structure of a GPCR ligand in its receptor-bound state: leukotriene B4 adopts a highly constrained conformation when associated to human BLT2. J Am Chem Soc 132:9049–9057
38. Becattini B et al (2004) Targeting apoptosis via chemical design inhibition of bid-induced cell death by small organic molecules. Chem Biol 11:1107–1117
39. Becattini B, Pellecchia M (2006) SAR by ILOEs: an NMR-based approach to reverse chemical genetics. Chem Eur J 12:2658–2662
40. Orts J, Griesinger C, Carlomagno T (2009) The INPHARMA technique for pharmacophore mapping: a theoretical guide to the method. J Magn Reson 200:64–73

41. Sanchez-Pedregal VM et al (2005) The INPHARMA method: protein-mediated interligand NOEs for pharmacophore mapping. Angew Chem Int Ed Engl 44:4172–4175
42. Bartoschek S et al (2010) Drug design for G-protein-coupled receptors by a ligand-based NMR method. Angew Chem Int Ed Engl 49:1426–1429
43. Becattini B et al (2006) Structure-activity relationships by interligand NOE-based design and synthesis of antiapoptotic compounds targeting Bid. Proc Natl Acad Sci U S A 103:12602–12606
44. Sledz P et al (2010) Optimization of the interligand overhauser effect for fragment linking: application to inhibitor discovery against Mycobacterium tuberculosis pantothenate synthetase. J Am Chem Soc 132:4544–4545
45. Mayer M, Meyer B (2001) Group epitope mapping by saturation transfer difference NMR to identify segments of a ligand in direct contact with a protein receptor. J Am Chem Soc 123:6108–6117
46. Kemper S et al (2010) Group epitope mapping considering relaxation of the ligand (GEM-CRL): including longitudinal relaxation rates in the analysis of saturation transfer difference (STD) experiments. J Magn Reson 203:1–10
47. Constantine KL, Davis ME, Metzler WJ, Mueller L, Claus BL (2006) Protein–ligand NOE matching: a high-throughput method for binding pose evaluation that does not require protein NMR resonance assignments. J Am Chem Soc 128:7252–7263
48. Hwang TL, Shaka AJ (1995) Water suppression using excitation sculpting with gradients. J Magn Reson A 112:275–279
49. Berger S, Braun S (2004) 200 and more NMR experiments: a practical course, 3rd edn. Wiley, Weinheim
50. Downing AK (2004) Protein NMR techniques, 2nd edn. Humana, New Jersey
51. Cavanagh J, Fairbrother W, Palmer AG, Skelton NJ (1995) Protein NMR spectroscopy: principles and practice. Academic Press, San Diego
52. Wuthrich K (1976) NMR in biological research: peptides and proteins. North-Holland, Amsterdam
53. Murray CW et al (2010) Fragment-based drug discovery applied to Hsp90. Discovery of two lead series with high ligand efficiency. J Med Chem 53:5942–5955
54. Fielding L (2007) NMR methods for the determination of protein–ligand dissociation constants. Prog Nucl Magn Reson Spectrosc 51:219–242
55. Kelly MJ et al (2001) The NMR structure of the 47-kDa dimeric enzyme 3,4-dihydroxy-2-butanone-4-phosphate synthase and ligand binding studies reveal the location of the active site. Proc Natl Acad Sci U S A 98:13025–13030
56. Tugarinov V, Hwang PM, Ollerenshaw JE, Kay LE (2003) Cross-correlated relaxation enhanced 1H-13C NMR spectroscopy of methyl groups in very high molecular weight proteins and protein complexes. J Am Chem Soc 125:10420–10428
57. Tugarinov V, Kay LE (2005) Methyl groups as probes of structure and dynamics in NMR studies of high-molecular-weight proteins. Chembiochem 6:1567–1577

Part V

Molecules in Native Environments

Chapter 15

Model Membrane Systems

Heiko Keller, Remigiusz Worch, and Petra Schwille

Abstract

The context of the membrane is crucial for the interaction of many membrane proteins with their ligands. However, many detailed studies cannot be carried out in living cells. Therefore, studying these interactions requires model membrane systems that are compatible with the used analytical method. A big variety of these methods is available, each of which has its advantages and disadvantages. This chapter gives an overview over the existing techniques, a basic introduction into work with lipids, and detailed protocols for selected methods.

Key words Model membranes, Artificial lipid membranes, Liposomes, Small unilamellar vesicles, Giant unilamellar vesicles, Supported lipid bilayers, Nanodiscs, Plasma membrane sheets, Receptor–ligand interaction

1 Introduction

Model membrane systems can be classified according to their shape and origin. Depending on the shape, model membranes are compatible with a wide variety of methods, ranging from biochemical and biophysical solution assays to microscopy. So far, many applications have been focused on studying the membranes themselves, rather than membrane-associated processes like protein–protein or protein–ligand interactions. However, many model membranes with embedded proteins should be amenable to these studies, too, i.e., be suitable samples for many of the techniques discussed in this book. In this chapter, we discuss how some practical limitations that often restrict the use of model membranes can be overcome. After introducing various membrane structures, an overview is given over alternative ways of their production. Finally, several established protocols, which are suitable for beginners in the field, are described in detail.

Mark A. Williams and Tina Daviter (eds.), *Protein-Ligand Interactions: Methods and Applications*, Methods in Molecular Biology, vol. 1008, DOI 10.1007/978-1-62703-398-5_15, © Springer Science+Business Media New York 2013

1.1 Model Systems and Compatible Analytical Methods

1.1.1 Membrane Structures

Small unilamellar vesicles (SUVs) and large unilamellar vesicles (LUVs) have diameters in the range of 20–100 nm or 50–200 nm, respectively. They are commonly used for assays in bulk suspensions. They are relatively robust and can be handled like most other aqueous solutions or suspensions. Furthermore, the small size allows for a high membrane surface area per volume. Many biochemical or biophysical methods can be used for assay read-out. One simple example is a binding assay, in which vesicles and unbound proteins are separated by centrifugation and the protein samples are further analyzed and quantified (1). As another example, vesicle fusion mediated by protein–protein or protein–ligand interactions can be analyzed in a fluorometer- or microscopy-based assay. Here, two kinds of vesicles are labeled with different suitable fluorophores to result in fluorescence resonance energy transfer after vesicle fusion (2–4). For fluorescence spectrometric analysis, it has to be taken into account that vesicles are a source of light scattering (*see* Chapter 2).

Despite their suggestive name, LUVs are not necessarily bigger than SUVs. The main difference is that the size distribution of LUVs is almost monodisperse and well controlled. Therefore, LUVs are mostly used if the biological process to be studied is dependent on the membrane curvature (5). Protein affinity for membranes is most likely to be influenced by membrane curvature, if big groups like amphipathic helices are inserted into the membrane.

The analysis by light microscopy requires giant unilamellar vesicles (GUVs) with vesicle diameters well above the diffraction limit. Alternatively, flat supported membranes like supported lipid bilayers (SLBs) can be used, which can extend over several millimeters. The support makes SLBs more robust than GUVs, and therefore accessible to methods like atomic force microscopy (AFM). The trade-off is that the support may change membrane properties like protein or lipid mobilities, and restrict potentially interesting deformations like vesicle budding.

Flat suspended membranes, also termed black lipid membranes, cover a small aperture in a thin plastic sheet, which separates two reservoirs of buffer. They are assembled from individual monolayers. Suspended membranes are mainly used for applications in electrophysiology, because both sides are accessible. This advantage is gained on the expense of stability, as minute mechanical disturbances can cause a pressure difference between the buffer reservoirs resulting in membrane rupture. These specialized methods are well described, for example, in refs. (6–8). Relatively novel model membrane systems are nanodiscs. These are flat bilayer fragments stabilized by a scaffold protein coat, governing both stability and a narrow size distribution around a diameter of ~10 nm (9, 10). They are more stable than conventional liposomes, which in combination with monodispersity makes them applicable to a range of biochemical and biophysical experiments with integral and peripheral membrane proteins (11–13). The bilayers of nanodiscs are self-assembled directly from detergent-solubilized lipid and protein components via slow removal of the detergent.

1.1.2 Lateral Heterogeneity

In the plasma membrane of living cells, membrane components can partition into nanometer-scale domains termed lipid rafts. Several reasons for this lateral heterogeneity are discussed such as liquid-ordered (L_o) and liquid-disordered (L_d) phase coexistence, critical fluctuations, or colloidal phenomena (14–16). Macroscopically separated phases can be observed in several membrane model systems, for example SLBs, GUVs, and giant plasma membrane vesicles (GPMVs). These coexisting phases are used as models for lipid rafts and non-raft membranes, respectively. Therefore, fluorescence microscopy imaging and related techniques on these model systems are valuable tools for studying lipid and protein properties.

The increasing application of fluorescence-based methods is based on the availability of various lipophilic fluorescent dyes that can easily be incorporated into the membrane (*see* **Note 1**).

GUVs are especially useful for microscopic studies because of their size, membrane curvature, and presence of a free-standing bilayer. Up to now, a range of different membranes has been investigated, including simple binary mixtures, as well as protein and lipid extracts from native cell membranes. Depending on the composition and temperature, a great variety of domain patterns has been observed, which can be explained as a consequence of the competition between the line tension of the lipid domains and the long-range electrostatic interaction between different domains (17, 18).

The scale of domains observed by light microscopy techniques is restricted to the diffraction limit. Therefore, structures significantly smaller than the micrometer range cannot be detected, unless super-resolution techniques or AFM are used. In the case of AFM, a monolayer or bilayer has to be prepared on a solid support, usually on mica crystals. Different domains can be detected due to their difference in height, providing additional information about the lateral structure organization of the membrane. Thanks to the improvement in instrumentation, it is possible to mount an AFM scanning unit on the top of an inverted fluorescent microscope and effectively combine these two methods, reducing the artifacts resulting from the use of a single technique (19, 20).

1.2 Origin of Model Membranes

1.2.1 Synthetic Model Membranes

Model membranes can be produced either in a synthetic "bottom-up" approach from purified or synthetic lipids and proteins, or they can be derived "top-down" from cells (*see* Fig. 1). Synthetic model membranes are produced "bottom-up" and hence allow for the full control of protein and lipid composition. This requires the purification or synthesis of all components and a subsequent

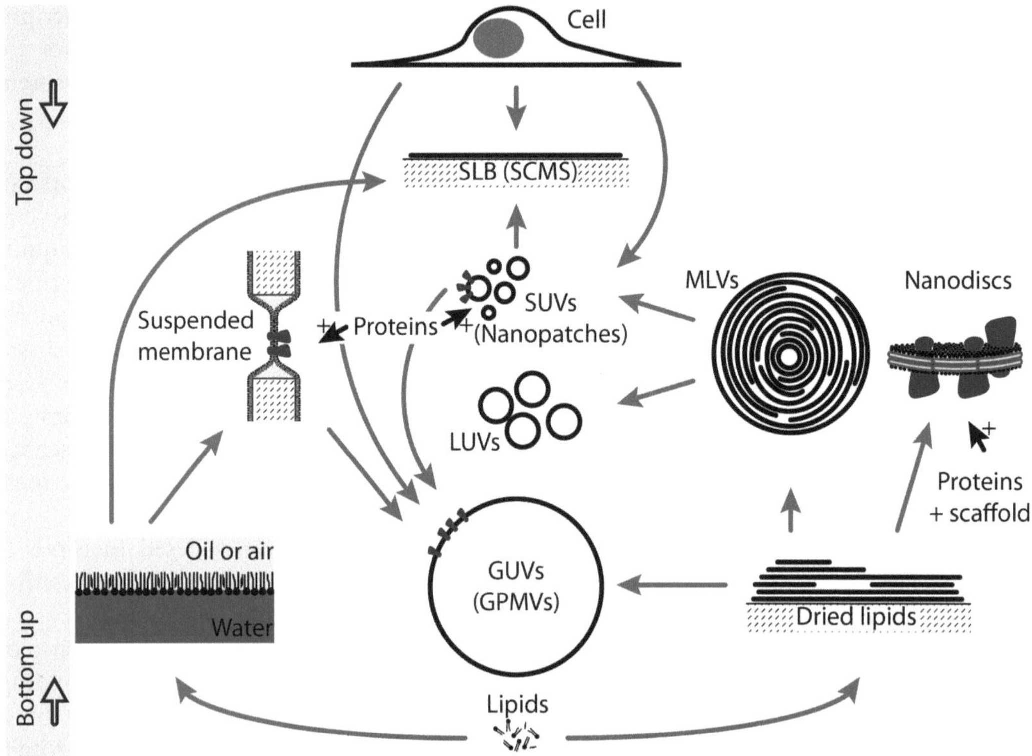

Fig. 1 Model membrane systems. Cell-derived model systems (in *brackets*) are named differently than corresponding synthetic systems. See Table 1 for methods and abbreviations. Items are not drawn to scale

reconstitution into functional membranes. Membrane formation is achieved via self-organization of lipids. When suitable lipids are dried and rehydrated, they form multilamellar assemblies (i.e., stacks of bilayers). These are then transformed into SUVs, LUVs, GUVs, or SLBs by mechanical deformation (*see* Table 1 for abbreviations). All these model systems ideally consist of only one bilayer (i.e., they are unilamellar) to avoid artifacts in subsequent assays. However, a certain fraction of bilamellar or multilamellar vesicles (MLVs) usually exists. SUVs have a diameter of typically less than 100 nm and are the smallest stable vesicles with the highest membrane curvature. They can be produced by uncontrolled mechanical disruption via sonication. LUVs are produced in a more controlled two-step process (21). First, the stacks of bilayers are resuspended as big MLVs by vortexing and freeze–thaw cycles. Second, these MLVs are passed through filters with the desired pore size. This process, termed extrusion, yields narrow size distributions of vesicles up to 200 nm in diameter. Above this size, this method yields wider size distributions and an increasing fraction of MLVs. GUVs with a size of a few μm to 100 μm are produced most efficiently by electroformation. Here, a low frequency alternating electric field is used to gently agitate the rehydrated stack of bilayers. This causes

Table 1
An overview of commonly used model membrane systems

Origin	Synthetic		
Structure	**From bilayers**	**From monolayers**	**Cell-derived**
Vesicles, nm range *for binding assays, biochemistry*	Small unilamellar vesicles (SUVs) by sonication, large unilamellar vesicles (LUVs) by extrusion		Nanopatches
Vesicles, μm range *for microscopy*	Giant unilamellar vesicles (GUVs) by electroformation	Giant unilamellar vesicles (GUVs) by microfluidic jetting, double emulsion	Giant plasma membrane vesicles (GMPVs)
Flat supported membranes *for microscopy, AFM*	Supported lipid bilayers (SLBs) by vesicle fusion	Supported lipid bilayers (SLBs) by Langmuir–Blodgett	Supported cell-membrane sheets (SCMS) by coverslip rip-off or sonication
	Tethered lipid bilayers by combination of techniques		
Flat suspended membranes *for electrophysiology*		Black lipid membranes (also: suspended lipid membranes)	Patch clamp rip-off
Nanodiscs *for biophysics, biochemistry*	Nanodiscs: flat, scaffolded bilayer patches		

budding of vesicles from the surface of the stack. Subsequent vesicle fusion results in giant vesicles. SLBs can be produced from SUVs by vesicle rupture and fusion on a suitable flat substrate. Potentially unfavorable direct interactions of the lower monolayer with the support can be reduced by using various polymer cushions as an intermediate layer (22). These model membranes are termed tethered lipid bilayers.

Although numerous phenomena taking place in membranes can be studied in pure lipid systems, the influence of proteins cannot be neglected in a real biological membrane. Taking into account the lateral concentration of proteins (around 30,000 per μm^2), the biological membrane should be regarded as a lipid-protein composite, rather than a dilute solution of proteins in a lipid solvent (23). For example, proteins can significantly alter the phase-separation behavior of membranes, which has been shown by microscopy and by measuring interaction energies (24–26). All described standard protocols for synthetic model membranes work well for systems containing only lipids or lipids and peripheral membrane proteins, which are added to the vesicles after their formation. In contrast, the purification and reconstitution of trans-membrane proteins mostly require techniques using detergents

(27–29). Proteins are purified in presence of detergents and afterwards mixed with lipids and detergents or detergent-destabilized SUVs. Subsequently, detergents are slowly removed, so that lipids and transmembrane proteins co-assemble into small vesicles (30). These SUVs can then be transformed into GUVs by electroformation or into SLBs by vesicle fusion. However, the reconstitution of each transmembrane protein requires extensive optimization of detergents and procedures.

A second approach to form membranes from dissolved lipids uses the self-assembly of lipids into monolayers at phase boundaries. At the interface of water and air or nonpolar solvents, the hydrophilic head group of lipids will be immersed into the aqueous phase while the hydrocarbon chains will extend into the non-polar phase or air. This leads to the formation of a lipid monolayer. According to the Langmuir–Blodgett method, first a lipid monolayer is assembled on the surface of a water bath. This is then transferred to a substrate like a coverslip by retraction from the water bath. The second monolayer is added by subsequent re-immersion (31). Similarly, lipid bilayers can be assembled over apertures. These are termed suspended bilayers or black lipid membranes. Alternatively, monolayers form at the interface of water droplets and a bulk oil phase. A lipid bilayer forms when two of these droplets touch each other or when a droplet is transferred into a second bulk water phase. The incorporation of transmembrane proteins into these model membranes is only possible if these proteins spontaneously insert into the already formed lipid bilayers.

1.2.2 Cell-Derived Model Membranes

Cell-derived model membranes are increasingly used as alternatives to synthetic systems. Their main advantage is that endogenous and exogenously expressed transmembrane proteins are already embedded in the original cellular membranes and do not have to be reconstituted. Most cell-derived model membranes represent parts of dead or dying cells devoid of cytoskeleton. Therefore, they are simpler than living cellular membranes because of the lack of dynamic processes and sub-μm structures like protrusions or ruffles. The compositional complexity of cellular membranes and derived model systems is similar, but the control over membrane components is limited. Therefore, cell-derived model membranes represent an interesting intermediate between living cells and synthetic model membranes.

Model membranes are obtained from several membrane compartments of living cells by various mechanical and chemical treatments. A straightforward approach to isolate plasma membrane-derived model systems is to attach a coverslip or a patch-clamp pipette to the cell surface and rip off a membrane patch. Alternatively, chemical treatments can induce the blebbing of cells. These blebs, termed GPMVs, can then be detached as vesicles with a diameter of 5–20 μm. Other membrane compartments can be isolated by subcellular fractionation protocols often followed by purification procedures (32–35).

1.3 Application to Studying Receptor–Ligand Interactions

The quantitative characterization of receptor–ligand interactions in terms of affinity and efficacy is of a great pharmacological interest, especially in the systems in which the functionality of receptors is preserved in their natural membrane environment. In this section, we introduce fluorescence correlation spectroscopy (FCS) and its application to receptor–ligand studies.

1.3.1 Fluorescence Correlation Spectroscopy on Membranes

FCS is based on registering the fluctuations of fluorescence intensity coming from labeled molecules diffusing through a small, open volume, usually realized by focused laser light in a confocal setup. Besides the diffusion coefficient and the average number of particles, a wide range of dynamic parameters can be exploited (directed transport velocities, kinetic rates of photophysical processes, lifetime of the triplet state, etc.). These parameters are obtained by numerical nonlinear curve fitting, assuming a certain model according to the diffusion type (3D, 2D) and/or aforementioned processes. Due to limited space in this chapter, we refer the reader to general articles on FCS (36–38), focusing here on the topics directly related to the membrane fluidity and protein–ligand interactions.

The mobility of fluorescent lipophilic probes in model membranes of GUVs allowed conclusions about sterol–lipid interactions, which are supposed to result in the formation of a L_o phase, characterized by a rather high degree of lipid mobility in spite of high structural order. Indeed, in mixtures of phospholipids with low phase transition temperatures (T_m), the presence of cholesterol causes a smooth transition from L_d to L_o phase. In contrast, for membranes containing sphingomyelin, DPPC, and DSPC, cholesterol increases mobility, exerting a fluidizing effect on the membrane (39). The diffusion coefficients (D) of lipids in the membrane are usually in the order of few $\mu m^2/s$, depending on the fluidity of the membrane. Similar values were reported for proteins reconstituted in GUVs, displaying a weak dependence on the hydrodynamic radius (R) of the protein R ($D \sim \ln(1/R)$), in accordance with the Saffman–Delbrück model (40). For comparison, proteins in live cell plasma membranes usually show slower diffusion by at least an order of magnitude. Recently several novel modifications of the FCS technique were applied for the measurements on membranes to overcome technical difficulties related with the use of the "standard" single-focus FCS approach (reviewed in ref. (37)).

1.3.2 Binding of a Fluorescently Labeled Ligand to Lipids or Membrane Proteins

The simplest and most straightforward way of performing ligand binding studies by FCS is to probe changes in diffusion when soluble molecules (D of the order of 100 $\mu m^2/s$ for nucleotides to tens of $\mu m^2/s$ for proteins) get significantly slowed down by binding to a membrane protein. Measuring the autocorrelation curve for different ligand concentrations gives monotonically increasing fractions of slower diffusion, related to the bound state. Taking the values from nonlinear curve fitting, a binding isotherm can be obtained, facilitating the determination of K_d for a given receptor–ligand pair. Such an approach was mainly used for ligand–receptor interaction studies in living cells for different classes of membrane proteins like tyrosine kinase receptors, ion channels, and different seven transmembrane spanning receptors (for review *see* ref. (41)). Nevertheless, the strategy described above can in principle be applied to any other model membrane system, allowing for protein–lipid and receptor–ligand interactions.

Despite the feasibility of a single color approach exclusively monitoring the labeled ligand, a dual-color variant of FCS (fluorescence cross-correlation spectroscopy, FCCS) may be beneficial, as by detection of co-diffusing molecule complexes, it is possible to distinguish from nonspecific membrane binding (36). However, this strategy also requires a fluorescently labeled receptor with a spectrally distinct emission from the ligand fluorescence. As an introduction to the FCCS technique, we refer the reader to the following articles (36, 42).

Binding studies can also be performed using systems in which the slow diffusion in a bound state is caused by a liposome (SUVs, LUVs with reconstituted receptors or nanopatches) or a nanodisc moving as complete single entity. A rough estimation using the Stokes–Einstein relationship shows that a 100 nm liposome in aqueous buffer will have a diffusion coefficient of ~2 $\mu m^2/s$, which is similar to the diffusion of a lipid dye in bilayers but easily distinguishable from freely diffusing molecules. Such an approach can be exemplified by FCCS studies on functionalized biotin–streptavidin complexes in nanopatches (43), as well as binding of both agonists and antagonists to the human μ-opioid receptor (44). A single color approach was applied for ligand binding analysis by estrogen receptor beta attached to functionalized 40 nm neutravidin coated beads (nanospheres) (45). A similar approach using nanodiscs was applied to islet amyloid polypeptide (IAPP) interaction studies with the membrane (13). In principle, nanodiscs with reconstituted proteins may be used for binding assays, like in the case of membrane-embedded core translocon SecYEG channel and its cytosolic partner SecA (12).

Ligand–receptor interactions regarded as the first step of signal transduction, may initiate different scenarios for the receptor dimerization. Some proteins follow the classical ligand dependent oligomerization and cross-activation scheme, whereas others oligomerize prior to ligand binding at the plasma membrane. Recently, a novel experimental assay for transmembrane helix-driven receptor dimerization using scanning FCCS in GPMVs was proposed (46). The influence on receptor dimerization induced by a bound ligand or the presence of other factors like designed peptides potentially disrupting the complex may be assessed by this method (47).

An especially interesting line of experiments is a combination of model membranes with cellular studies, like in the case of the immunological synapse reconstituted on a patterned SLBs (48). Such an approach allowed revealing a causal relation between the radial positions of T-cell receptors (TCR) and signaling activity. In particular, a prolonged signaling from TCR microclusters that had been mechanically trapped in the peripheral regions of the synapse was observed.

Another example of combining synthetic model membranes and cellular material is the recruitment of clathrin coat machineries to SUVs and GUVs (49, 50). A "bait" peptide originating from a receptor protein was chemically linked to a lipid and all other components were recruited from cytosolic extracts. This allowed for mass spectrometric analysis of the components as well as for microscopic analysis of the distribution of components between vesicle bud and neck.

In summary, many important features of biological membranes can be reconstructed in model membranes. In combination with constantly developed and improved analytical techniques, they can be used as platforms for all kinds of interaction studies. They can yield insights into membrane organization, which are not accessible in living cells.

2 Materials

2.1 Lipids

The choice of lipids is crucial for the properties of the model membrane, but unfortunately there is no standard mixture to use. Often lipids are viewed as some kind of two-dimensional solvent for the proteins of interest. In contrast to aqueous solutions, which mainly consist of water, there are thousands of lipids that make up a membrane without any specific one being the major component (51). Moreover, the lipid compositions of tissues, cells, subcellular compartments, and even of the two leaflets of one bilayer membrane can differ to a great extent from each other. Therefore, one strategy is to use total lipid extracts from isolated cells or subcellular fractions of interest. The strategy mostly used is to emulate the lipid composition of a certain membrane by a very simple mixture of few lipid components. Here, the physical properties of the lipids are the main criteria (52). Head groups of phospholipids define the charge of the membrane and are often specifically bound by proteins (*see* Fig. 2). Lipids are saturated if they do not contain C–C double bonds in their carbohydrate chain (e.g., POPC). The degree of saturation and the length of the carbohydrate chains determine the transition temperature of the lipid from the rigid gel phase to a fluid phase, which represents a biologically relevant state.

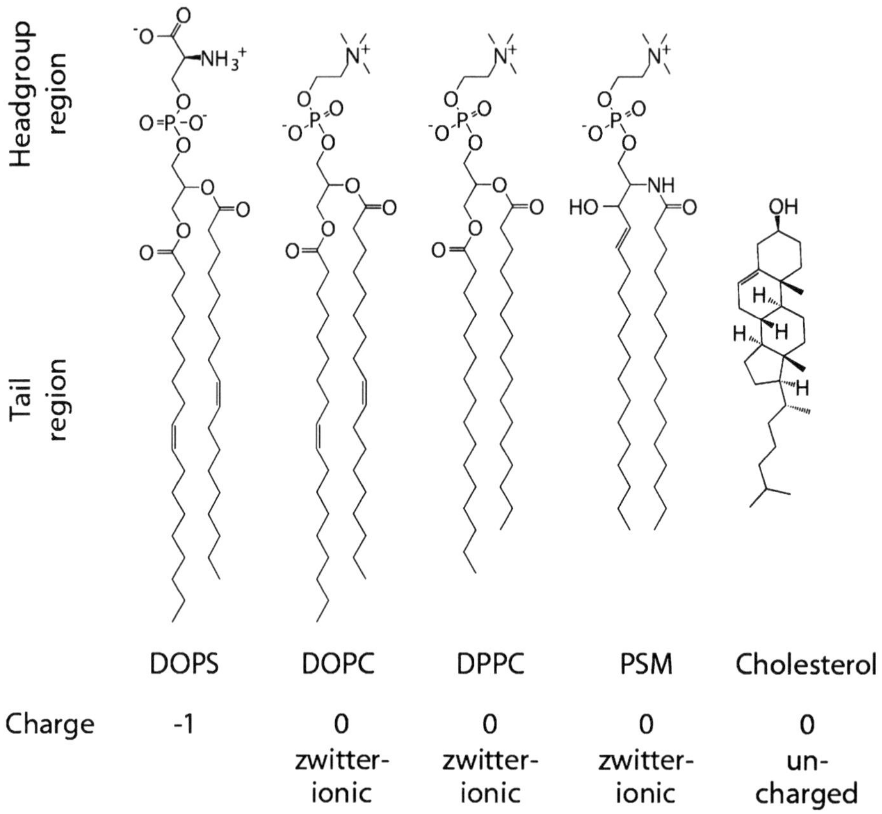

Fig. 2 Structures of some commonly used lipids. Dioleoyl phosphatidylserine [DOPS, PS(18:1/18:1)], dioleoyl phosphatidylcholine [DOPC, PC(18:1/18:1)], dipalmitoyl phosphatidylcholine [DPPC, PC(16:0/16:0)], palmitoyl sphingomyelin [PSM, SM(d18:1/16:0)]. Indicated charges apply at physiological pH (7.0–7.5)

Moreover, membranes with only three components can show the coexistence of several lipid phases (53). These phase-separated membranes are often used as models for lipid rafts. Another essential group besides phospholipids is sterols. Cholesterol influences the rigidity, stability, and phase-separation behavior of membranes. Its content increases throughout the secretory pathway from endoplasmic reticulum to the plasma membrane, where it reaches a molar fraction of 40 % of all lipids (54).

In the last few years, lipidome analysis by mass spectrometry has provided a much more detailed insight into the lipid composition of various membranes (34, 54–57). This provides the basis for a more targeted and well-founded selection of lipids for in vitro experiments.

Most lipids are available in high quality from Avanti Polar Lipids (Alabaster, AL, USA) as crude tissue extracts, fractionated tissue extracts containing only one lipid class with a mixture of different hydrocarbon chains or completely synthesized molecules. Several special lipids are provided by smaller manufacturers, e.g., phosphoinositides from Echelon Biosciences (Salt Lake City,

UT, USA). A big selection of common fluorescent lipid analogs is commercially available (*see* **Note 1**). Many lipids are very sensitive to oxidation, which can change their properties drastically (*see* **Note 2**).

2.2 Organic Solvents

Most lipids are insoluble in water and therefore require the use of organic solvents (*see* **Note 3**). Commonly used solvent mixtures are pure chloroform for most neutral lipids (e.g., most phosphatidyl cholines, cholesterol), a 2:1 (v/v) mixture of chloroform and methanol for charged or less hydrophobic lipids (e.g., most phosphatidyl serines), or chloroform–methanol–water 20:9:1 (v/v) for highly charged lipids or short acyl chain analogs (e.g., phosphatidyl inositol bisphosphate, dimyristoyl phosphatidyl serine).

2.3 Aqueous Solutions

Most of the described methods to produce synthetic model membranes do not require any specific buffer. Instead, the aqueous solution can be chosen according to the desired experiment. It is recommended to add a pH buffer to avoid undefined protonation states or hydrolysis of lipid head groups. Furthermore, salt should be added in physiological concentrations. Divalent cations can cause the clustering of negatively charged lipids and should therefore be avoided if such lipids are used (58). In the case of GUV electroformation, a solution of low electric conductivity has to be used and the osmolarity has to be considered if solutes should later be added to the GUV suspension. Osmolarity differences of more than 10 mOsm can cause substantial leaking or bursting of GUVs depending on the membrane composition (*see* **Note 4**).

Following specific buffers are used:

Phosphate-buffered saline (PBS): 137 mM NaCl, 2.7 mM KCl, 4.3 mM Na_2HPO_4, 1.4 mM KH_2PO_4, pH 7.4; PBS with 0.9 mM Ca^{2+}, and 0.5 mM Mg^{2+}; Giant Plasma Membrane Vesicle buffer (GPMV buffer): 150 mM NaCl, 2 mM $CaCl_2$, and 10 mM HEPES-NaOH, pH 7.4.

2.4 Devices and Other Materials

General lipid work requires a vacuum pump and a desiccator without desiccant to remove organic solvents, preferably a rotary vane oil vacuum pump providing a good vacuum. Bottled nitrogen or argon gas including pressure regulators are needed for storing lipids in a protective atmosphere. Further devices depend on the specific protocol. Bath sonicators or tip sonicators are used to form SUVs and, in case of the latter, to disrupt cells. Extrusion of LUVs can be achieved with syringe extruders (e.g., Mini-Extruder by Avanti Polar Lipids) or pressure-driven extruders (e.g., LIPEX extruder by Northern Lipids). The main advantage of the pressure-driven extruder is the better control of extrusion pressure and therefore vesicle size distribution. GUV electroformation requires an AC voltage generator and electroformation chambers, which are not commercially available (*see* **Note 5**). Glass, which has been activated, for example in an oxygen plasma cleaner, or freshly cleaved mica is used as a support for SLBs and plasma membrane sheets. Cell fractionation for lipid purification or direct recovery of vesicles is performed with ultracentrifuges.

3 Methods

This chapter describes some established standard protocols, which are rather easy to learn for beginners, and discusses their limitations. For newly developed or extended methods, which overcome some of the discussed limitations but require some technical experience, the reader is referred to primary articles.

3.1 SUVs by Sonication

Most of the described protocols are based on the self-assembly of lipids into stacks of bilayers. For this, a lipid mixture of choice is prepared in organic solvents. Then, the lipid solution is dried on the bottom of a glass vial by evaporation of the solvents under a nitrogen stream. This leads to the formation of stacks of lipid bilayers. Traces of solvents are removed by exsiccation in a vacuum for 1 h. Subsequently, a buffer of choice is added for rehydration (*see* Subheading 2). For many applications, 1 mg of lipids in 1 ml of buffer is a good starting point. The rehydration should be performed at a temperature above the melting temperatures of all lipid components or phase transition temperature of the mixture. 65 °C is recommended in case these temperatures are not known. The glass vial is vigorously vortexed every few minutes until all lipids are detached from the glass surface. This results in a turbid suspension of MLVs. If necessary, these MLVs can be transferred into plastic tubes, but part of the lipid material present in bigger sticky aggregates may be lost. MLVs are then converted into SUVs by application of ultrasound at the same temperature until the suspension appears clear. This usually requires 15–60 min in a sonicator bath or 10–60 pulses using a tip sonicator. If a tip sonicator is used, small pieces of metal, which originate from the sonicator tip, have to be removed by centrifugation for 10 min at 10,000 × *g*. Remaining bigger vesicles or lipid aggregates can be removed by centrifugation at 100,000 × *g* for 1 h if desired. SUVs should be used within one day because they aggregate over time and lipids are gradually oxidized and hydrolyzed. If the size distribution of vesicles is not important, they can be stored at 4 °C for a few days or at −20 to −80 °C for a few weeks. Thawed aliquots have to be briefly sonicated again.

3.2 LUVs by Extrusion

This protocol starts with resuspended MLVs, which are obtained by vortexing as described in the SUV protocol. The concentration should be 1–50 mg/ml depending on the extrusion device and intended experiments. The MLVs are subjected to several freeze–thaw cycles to increase their hydration and loosen the bilayer packing. MLVs are filled into cryotubes and frozen in liquid nitrogen. Then they are rapidly thawed in a water bath with a temperature above the lipids' phase transition temperatures. This is repeated five more times. The resulting vesicle suspension is afterwards converted into LUVs by repeated extrusion through a polycarbonate membrane with a pore size between 30 and 200 nm. An uneven number of repetitions are performed to collect LUVs from the formerly empty reservoir without remnant debris (typically 31). The size distribution can be determined most reliably with dynamic light scattering. It should be a narrow peak around the membrane pore size. Extruders are operated according to the manufacturer's instructions. Briefly, MLV suspensions are repeatedly passed through the polycarbonate membrane at a temperature above the lipids' phase transition temperatures. Thereby, the opaque MLV suspension turns into a clear LUV suspension, although some turbidity may remain at high lipid concentrations. Storage of LUVs is not recommended as their size distribution will gradually change.

An extension of this protocol allows for single vesicle analysis instead of bulk studies (59). For this, LUVs are immobilized on a surface in low densities via biotinylated lipids. Individual fluorescence intensities can be measured by microscopy.

3.3 GUVs by Electroformation

The most effective and reproducible way to produce GUVs is electroformation (60). The effects of protein–protein and protein–lipid interaction in domain assemblies and minimizing artifacts in GUV preparation and imaging were described in two recent reviews (61, 62). Electroformation requires a special device with two electrodes in distance of usually 1–2 mm in a compartment filled with aqueous buffer. Additionally, an AC voltage generator is required. The most common electroformation devices use either platinum wires or indium tin oxide (ITO)-coated glass slides as electrodes (*see* **Note 5**).

In the first step, electrodes are heated above the lipids' phase transition temperatures. A lipid mixture in organic solvents is evenly spread on the electrodes and dried for 5 min. 1 μl or 2 μl of a 10 mg/ml solution are recommended for platinum wires or ITO coverslips, respectively, per electrode. Vacuum is applied for at least 1 h at room temperature to remove residual solvent. In the second step, the electroformation chamber is assembled and filled with an aqueous solution of low ionic strength. The chamber is heated above the lipids' phase transition temperatures and an alternating current of 10 Hz and ca. 0.75 V/mm is applied for 1–3 h. Optionally, the frequency can be reduced to 2 Hz for additional 10 min to detach the GUVs from the electrode surface. Finally, GUVs can be harvested after cooling to room temperature. Single-use plastic aspirators or cut pipette tips will impose least shear stress on GUVs during handling.

Advanced protocols for experienced users: A modified electroformation protocol tolerates the presence of physiological salt concentrations during GUV formation (63). It was demonstrated to work with several membrane compositions, but it remains to be seen if it works as robustly as the basic protocol. A microfluidic device was used to produce very big GUVs by shooting a well-controlled jet of buffer through a suspended membrane (64). This method, which is analogous to blowing soap bubbles, allows for the controlled incorporation of various components. However, it requires a self-built electronically controlled microfluidic device.

3.4 SLBs by Vesicle Fusion

One common method to produce SLBs is to collapse SUVs on a glass coverslip. This takes place by itself for positively charged and partially also for neutral SUVs. These vesicles adhere to the negatively charged coverslip, rupture, and fuse to form a continuous membrane. Millimolar concentrations of Ca^{2+} or other divalent cations are necessary to overcome electrostatic repulsion when negatively charged lipids are used (65). Ca^{2+} has additional effects on negatively charged lipids. These lipids are partially sequestered during SLB formation on the membrane leaflet facing the support (66). Furthermore, Ca^{2+} clusters highly negatively charged lipids such as $PI(4, 5)P_2$ at low micromolar concentrations (58). Alternatively, high ionic strength can induce the rupture of SUVs by screening repulsive charges of vesicles and support (67). The quality of SLBs can be increased by using ultra-flat freshly cleaved mica instead of glass as a support. This leads to a more even membrane and avoids spots of stronger contact to the support and reduced lipid and protein mobility.

Freshly cleaved mica is glued on a glass coverslip within an open chamber (*see* **Note 6**). The chamber is heated to $T > T_m$ and filled with SUV suspension. An SLB with a diameter of 10 mm requires a SUV suspension containing approximately 40 μg of lipids at a concentration of for example 4 mg/ml (more than 0.1 mg/ml). 3 mM Ca^{2+} is added to induce vesicle fusion and SLB formation. After 15 min, the chamber is rinsed thoroughly with pre-warmed buffer to remove Ca^{2+} and unfused vesicles. It is important that the chamber is always partially filled to cover the SLB and that the membrane is not touched.

An advanced modification of SLBs is polymer-tethered bilayers (22, 68). In this case, membranes are not in direct contact with the support but are fixed to a polymer cushion. This improves the unrestricted diffusion of membrane components.

3.5 Nanopatches

The benefits of this technique are that the membrane proteins remain in their native lipid environment, and production of vesicles requires only few, easy to perform steps. Prepared material can be used for further studies, including protein purification or lipid analysis. In general, cell membranes are lysed by sonication in PBS and collected by pelleting in ultracentrifugation, after removal of large organelles (35). All the steps should be performed at 4 °C with cooled PBS.

The protocol starts by harvesting the cells cultured on Petri dishes or in culture flasks. Depending on the adhesive properties of the cell line, either mechanical scraping or trypsination can be used. Cells are washed in PBS three times by centrifugation (1,000 × *g* for 2 min). The last suspension is sonicated, resulting in a clearer solution (*see* **Note 7**). At this step plasma membranes are disrupted into small soluble fragments, whereas the large organelles remain intact and can be removed by pelleting in one centrifugation step (10,000 × *g* for 40 min). Remaining supernatant is ultracentrifuged (100,000 × *g* for 40 min) to isolate membrane lysate. The remaining pellet is resuspended in a small volume of PBS and passed through needles of increasing gauge (e.g., 20, 25, 27; at least five times each) to disperse aggregates of membrane patches. As the last optional step, bath sonication (10–15 min) can be applied. Like SUVs, nanopatches can be stored at 4 °C for a few days or at −20 to −80 °C for a few weeks.

3.6 Giant Plasma Membrane Vesicle

The method to produce GPMVs was first described by Scott and modified by Baumgart (69, 70). It is based on incubating cells with chemicals that block sulfhydryl groups in a Ca^{2+}-containing buffer. The mechanism is not understood completely. The treatment causes a Ca^{2+} influx, the degradation of phosphatidylinositol-4,5-bisphosphate ($PI(4,5)P_2$), and the exposure of phosphatidylserine on the outside of the cells (71). The loss of $PI(4,5)P_2$ probably destabilizes the cytoskeleton and leads to the formation of very big blebs, which can be detached as GPMVs.

Cells are seeded to reach a density of 30 % after one day. They are then washed two times with GPMV buffer (*see* Subheading 2). GPMVs are induced by incubation in blebbing solution, which consists of 2 mM *N*-ethylmaleimide in GPMV buffer for 2 h at 37 °C. The more common protocol using 2 mM DTT and 25 mM formaldehyde instead of *N*-ethylmaleimide should be avoided although it may yield more vesicles depending on the cell line used. DTT causes the loss of palmitoyl membrane anchors, an important post-translational modification of many membrane-associated proteins (72). GPMVs can optionally be detached by gentle swirling of the flask and decanted. A further purification is not necessary in most cases. Lipid phase separation can be induced by cooling GPMVs to approximately 10 °C.

An alternative to GPMVs are plasma membrane spheres (PMSs) (73). An extended incubation time in a certain buffer leads to swelling of whole A431 cells into spheres. Phase separation can be induced at 37 °C by cross-linking the glycosphingolipid GM1, which is present at very high levels in this cell line. This system is not as well characterized as GPMVs, but it could be shown that transmembrane proteins are correctly sorted into membrane domains according to their raft affinity.

3.7 Plasma Membrane Sheets by Rip-Off

Plasma membrane sheets are planar cell-derived supported membranes prepared by direct detachment using a poly-L-lysine coated coverslip (74). The resulting bilayers conserve fluidity and asymmetry of membrane leaflets allowing the studies of membrane components. The benefit of this system is the reduction of cellular autofluorescence, lack of cytoskeleton and, what is unique, the access to the intracellular leaflet, opening the rare possibilities of interaction studies involving the cytoplasmatic receptor tails, which is difficult to achieve in a noncellular system.

Glass coverslips are washed once with ethanol and once with water and dried in a stream of air or nitrogen. Further the coverslips are treated with oxygen plasma-cleaner to activate their surface. The covering volume of poly-L-lysine (0.1 mg/ml) is disposed onto the coverslip surface and is incubated for an hour. Rinsing with PBS is applied just before use. The cells, expressing membrane proteins of interest, are grown to high confluency (>80 %) and washed with PBS with Ca^{2+} and Mg^{2+} to induce their osmotic swelling. The prepared coverslips are placed on the top of the cells, applying gentle pressure for 3 min (*see* **Note 8**). The coverslips are removed, ripping-off the upper membrane, followed by rinsing with PBS. The coverslips can be mounted in microscopic chambers of choice, depending on the purpose of experiment. Patches should be immersed in PBS or any kind of aqueous buffer to prevent drying. Membrane sheets can be easily labeled with any kind of lipophilic tracer according to the manufacturer's instructions (*see* **Note 1**).

3.8 Nanodiscs

Nanodisc formation is achieved by self-assembly of detergent-solubilized lipids and membrane scaffold proteins (MSP) upon detergent removal. The key component of MSP is a fragment of the human apolipoprotein-I (9). Currently the Sligar lab has assembled a library of MSP constructs of varying lengths, that can be used do modulate nanodisc size (10).

The protocol starts with rehydration of lipid film obtained from an appropriate amount of lipids (e.g., 200 μl of 200 mg/ml POPC) as described for liposome formation. A mixture of lipids and MSP proteins is solubilized with detergent at approx. 1:65:100 MSP–lipid–detergent molar ratio. After 30–90 min of incubation (optionally with gentle shaking) at 4 °C, the self-assembly is initiated by detergent removal. This step can be realized either by the hydrophobic adsorption of polystyrene resins (Bio-Beads) or by dialysis against the buffer without detergent. In the case of Bio-Beads, they are washed with a working buffer directly before use and an appropriate amount (usually Bio-Bead–detergent ratio of 10 (w/w) 27) is added to the mixture, followed by 4–12 h incubation with gentle shaking or stirring. Bio-Beads are separated by sedimentation and removed. The size and monodispersity of nanodiscs is verified using size-exclusion chromatography on a Superdex 200 10/300 column (GE Healthcare). Fractions are pooled, optionally concentrated and stored at −80 °C.

4 Notes

1. Membrane dyes: Fluorescent membrane dyes typically show strong partitioning into the liquid disordered phase (L_d) (75). However, it has to be kept in mind that the partitioning of fluorescent probes does not depend on the membrane phase state, but on the local chemical environment of the lipid domains and therefore a probe characterization before imaging experiments is important (17). Furthermore, it has to be taken into consideration that lipid modifications often change the behavior of the labeled lipid, e.g., that its partitioning between lipid phases does not necessarily reflect the partitioning of the native lipid.

   A common choice for membrane fluorescent labeling is the family of long-chain dialkylcarbocyanines, like DiO, DiI, DiD, and DiA (Invitrogen, Carlsbad, CA, USA). These are fluorescent lipid analogs with a broad range of excitation and emission wavelengths. Another possibility is to use lipids chemically labeled with fluorescent molecules like Rhodamine and Bodipy (Invitrogen, Carlsbad, CA, USA, Avanti Polar Lipids, Alabaster, AL, USA). These lipids can be labeled either on the head group or on the alkyl chains.
2. Lipid stability: Especially unsaturated lipids oxidize readily, which causes a considerable change in physical properties such as the melting temperature. Therefore, care has to be taken to prevent oxidation. Lipids should always be stored at −20 to −80 °C under inert gas such as nitrogen or argon. Furthermore, only chloroform should be used that is stabilized with ethanol (76). Without stabilization or after the approximate shelf-life of 2 years, reactive and toxic phosgene can form. Furthermore, photooxidation can occur during microscopy or oxidation on electrodes during electroformation of GUVs (77). Here, conditions have to be optimized for sensitive experiments. This can include the use of degassed buffers, an enzymatic oxygen scavenger system, or the minimization of light exposure.
3. Handling of organic solvents: Organic solvents should never be stored in plastic tubes but instead in glass vials with a teflon-lined cap. Most accurate pipetting of volatile solvents can be achieved with positive displacement pipettes with glass capillary

tips or Hamilton syringes. However, normal pipettes may be used with limited precision and increased corrosion risk. To prevent dripping, the atmosphere inside the pipette has to be saturated with organic solvent vapor by pipetting the solvent up and down a few times.

Up to a few milliliters of organic solvent can easily be evaporated under a stream of nitrogen. This allows for the preparation of mixtures in volumes that are easy to handle and subsequent reconstitution at the desired concentration. Often traces of organic solvent have to be removed completely, e.g., before rehydration of lipid samples in aqueous buffers. This requires desiccation for 1–4 h in a good vacuum. To be on the safe side, a vacuum of <10 mbar is recommended, such as provided by a rotary vane oil vacuum pump. Many organic solvents are toxic or harmful and should always be handled under a fume hood. Only very small amounts may be exceptionally worked with in a well-ventilated area.

4. Buffers for standard electroformation: The best choice is deionized and degassed water. Up to approximately 20 mM of ions, like a dilute pH buffer, can be added without compromising the electroformation efficiency substantially. Neutral solutes like sucrose or glucose can be used to adjust osmolarity and density of the solution. It is for example convenient to form GUVs in a 200 mM sucrose solution and add the GUV suspension to a 200 mM glucose solution in an imaging chamber. Then the GUVs will sediment to the coverslip within a few minutes and can be observed under a microscope.

5. GUV electroformation chambers: Platinum wire chambers have been reported to have a lower risk of lipid oxidation while ITO coverslip chambers allow for direct observation of GUVs within the production chamber. The latter reduces the risk of damaging GUVs during transfer, especially for diameters above 20 μm.

   Chambers are not commercially available but can be assembled according to the scheme below (*see* Fig. 3). For indium tin oxide (ITO) chambers, glass slides (e.g., 26 × 75 mm) with a 100 nm ITO coating (commercially available) are assembled using screw clamps with the ITO sides facing each other and a 1 mm thick Teflon O-ring as a spacer. Lipids are dried previously on the area within the O-ring. The buffer reservoir is filled completely and without air bubbles through a cut in the O-ring. During electroformation, the ends of the O-ring are closed to avoid evaporation or leaking. For observation with high numerical aperture objectives, ITO-coated coverslips mounted in a flow chamber are used instead of glass slides (78). Pt wire chambers consist of Pt electrodes with a diameter of 0.5 mm in a distance of ca. 2 mm in any kind of closed chamber.

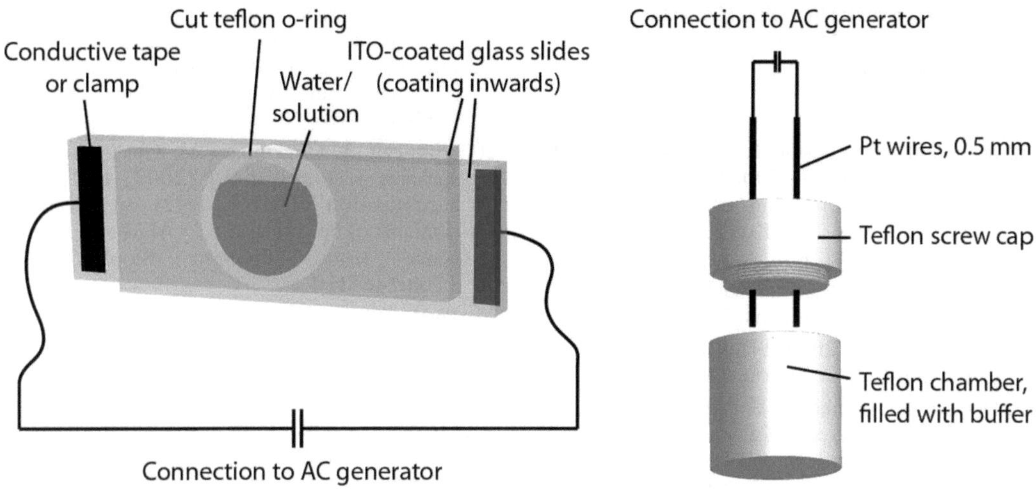

Fig. 3 Simple GUV chambers. Indium tin oxide (ITO)-coated conductive glass slides or platinum wires, respectively, are used as electrodes, which are connected to an alternating current generator

Round Teflon screw top chambers are convenient to use and can be fitted into standard heating blocks.

6. Chambers for SLBs: Any chamber that is accessible for rinsing can be used for preparing SLBs. Convenient examples are 8-well Labtek chambers or similar self-build designs, which allow for smaller volumes. For this end, a mica sheet is glued on a coverslip and a plastic ring (e.g., the upper part of a cut small centrifuge tube) is glued on top. UV-curable optical adhesive is recommended as glue for microscopy applications (e.g., Norland Optical Adhesive 63).
7. The time and power for sonication can be judged by observing the solution under the light microscope. The cell nuclei should be still visible and small membrane pieces should float freely in the solution.
8. The pressure on the coverslip can be applied by a 1 l bottle of water placed on a plastic cap spacer. When removing the glass, sliding movements should be avoided to prevent sheer forces removing the patches from the glass.

References

1. Saksena S, Wahlman J, Teis D, Johnson AE, Emr SD (2009) Functional reconstitution of ESCRT-III assembly and disassembly. Cell 136:97–109
2. Struck DK, Hoekstra D, Pagano RE (1981) Use of resonance energy transfer to monitor membrane fusion. Biochemistry 20:4093–4099
3. Yoon TY, Okumus B, Zhang F, Shin YK, Ha T (2006) Multiple intermediates in SNARE-induced membrane fusion. Proc Natl Acad Sci USA 103:19731–19736
4. Gong Y, Luo Y, Bong D (2006) Membrane activation:ΓÇë selective vesicle fusion via small molecule recognition. J Am Chem Soc 128:14430–14431

5. Peter BJ, Kent HM, Mills IG, Vallis Y, Butler PJ, Evans PR, McMahon HT (2004) BAR domains as sensors of membrane curvature: the amphiphysin BAR structure. Science 303:495–499
6. Ingolfson H, Kapoor R, Collingwood SA, Sparre Andersen O (2008) Single molecule methods for monitoring changes in bilayer elastic properties. J Vis Exp 21:e1032
7. Kapoor R, Kim H, Ingolfson H, Andersen OS (2008) Preparation of artificial bilayers for electrophysiology experiments. J Vis Exp 20:e1033
8. Winterhalter M (2000) Black lipid membranes. Curr Opin Colloid Interface Sci 5:250–255
9. Denisov IG, Grinkova YV, Lazarides AA, Sligar SG (2004) Directed self-assembly of monodisperse phospholipid bilayer Nanodiscs with controlled size. J Am Chem Soc 126:3477–3487
10. Nath A, Trexler AJ, Koo P, Miranker AD, Atkins WM, Rhoades E (2010) Chapter 6—single-molecule fluorescence spectroscopy using phospholipid bilayer nanodiscs. Single Mol Tools: Fluorescence Based Approaches Part A 472:89–117
11. Nath A, Atkins WM, Sligar SG (2007) Applications of phospholipid bilayer nanodiscs in the study of membranes and membrane proteins. Biochemistry 46:2059–2069
12. Alami M, Dalal K, Lelj-Garolla B, Sligar SG, Duong F (2007) Nanodiscs unravel the interaction between the SecYEG channel and its cytosolic partner SecA. EMBO J 26:1995–2004
13. Knight JD, Hebda JA, Miranker AD (2006) Conserved and cooperative assembly of membrane-bound alpha-helical states of islet amyloid polypeptide. Biochemistry 45:9496–9508
14. Simons K, Gerl MJ (2010) Revitalizing membrane rafts: new tools and insights. Nat Rev Mol Cell Biol 11:688–699
15. Destainville N, Dumas F, Salome L (2008) What do diffusion measurements tell us about membrane compartmentalisation? Emergence of the role of interprotein interactions. J Chem Biol 1:37–48
16. Veatch SL, Soubias O, Keller SL, Gawrisch K (2007) Critical fluctuations in domain-forming lipid mixtures. Proc Natl Acad Sci USA 104:17650–17655
17. Bagatolli LA (2006) To see or not to see: lateral organization of biological membranes and fluorescence microscopy. Biochim Biophys Acta 1758:1541–1556
18. Mouritsen OG (2005) Life—as a matter of fat: the emerging science of lipidomics. Springer, Berlin
19. Chiantia S, Ries J, Kahya N, Schwille P (2006) Combined AFM and two-focus SFCS study of raft-exhibiting model membranes. Chemphyschem: Eur J Chem Phys Physical Chem 7:2409–2418
20. Choucair A, Chakrapani M, Chakravarthy B, Katsaras J, Johnston LJ (2007) Preferential accumulation of Abeta(1-42) on gel phase domains of lipid bilayers: an AFM and fluorescence study. Biochim Biophys Acta 1768:146–154
21. Patty PJ, Frisken BJ (2003) The pressure-dependence of the size of extruded vesicles. Biophys J 85:996–1004
22. Knoll W, Bender K, Forch R, Frank C, Gotz H, Heibel C, Jenkins T, Jonas U, Kibrom A, Kugler R, Naumann C, Naumann R, Reisinger A, Ruhe J, Schiller S, Sinner EK (2010) Polymer-tethered bimolecular lipid membranes. In: Peter Meier W, Knoll W (eds) Advances in polymer science, polymer membranes/biomembranes. Springer, Berlin
23. Jacobson K, Mouritsen OG, Anderson RG (2007) Lipid rafts: at a crossroad between cell biology and physics. Nat Cell Biol 9:7–14
24. Almeida P, Pokorny A, Hinderliter A (2005) Thermodynamics of membrane domains. Biochim Biophys Acta Biomembr 1720:1–13
25. Hammond AT, Heberle FA, Baumgart T, Holowka D, Baird B, Feigenson GW (2005) Crosslinking a lipid raft component triggers liquid ordered-liquid disordered phase separation in model plasma membranes. Proc Natl Acad Sci USA 102:6320–6325
26. Johnson SA, Stinson BM, Go MS, Carmona LM, Reminick JI, Fang X, Baumgart T (2010) Temperature-dependent phase behavior and protein partitioning in giant plasma membrane vesicles. Biochim Biophys Acta 1798:1427–1435
27. Rigaud JL, Levy D (2003) Reconstitution of membrane proteins into liposomes. Methods Enzymol 372:65–86
28. Kahya N, Pécheur EI, de Boeij WP, Wiersma DA, Hoekstra D (2001) Reconstitution of membrane proteins into giant unilamellar vesicles via peptide-induced fusion. Biophys J 81:1464–1474
29. Girard P, Pecreaux J, Lenoir G, Falson P, Rigaud JL, Bassereau P (2004) A new method for the reconstitution of membrane proteins into giant unilamellar vesicles. Biophys J 87:419–429
30. Ollivon M, Lesieur S, Grabielle-Madelmont C, Paternostre M (2000) Vesicle reconstitution from lipid-detergent mixed micelles. Biochim Biophys Acta Biomembr 1508:34–50

31. Blodgett KB (1935) Films built by depositing successive monomolecular layers on a solid surface. J Am Chem Soc 57:1007–1022
32. Hartl P, Olson E, Dang T, Forbes DJ (1994) Nuclear assembly with lambda DNA in fractionated Xenopus egg extracts: an unexpected role for glycogen in formation of a higher order chromatin intermediate. J Cell Biol 124:235–248
33. Klemm RW, Ejsing CS, Surma MA, Kaiser HJ, Gerl MJ, Sampaio JL, de Robillard Q, Ferguson C, Proszynski TJ, Shevchenko A, Simons K (2009) Segregation of sphingolipids and sterols during formation of secretory vesicles at the trans-Golgi network. J Cell Biol 185:601–612
34. Brugger B, Sandhoff R, Wegehingel S, Gorgas K, Malsam J, Helms JB, Lehmann WD, Nickel W, Wieland FT (2000) Evidence for segregation of sphingomyelin and cholesterol during formation of COPI-coated vesicles. J Cell Biol 151:507–518
35. Swift JL, Burger MC, Cramb DT (2009) A quantum dot-labeled ligand-receptor binding assay for G protein-coupled receptors contained in minimally purified membrane nanopatches. Meth Mol Biol 552:329–341
36. Bacia K, Kim SA, Schwille P (2006) Fluorescence cross-correlation spectroscopy in living cells. Nat Meth 3:83–89
37. Chiantia S, Ries J, Schwille P (2009) Fluorescence correlation spectroscopy in membrane structure elucidation. Biochim Biophys Acta Biomembr 1788:225–233
38. Petrov EP, Schwille P (2008) State of the art and novel trends in fluorescence correlation spectroscopy. In: Resch-Genger U (ed) Springer series on fluorescence volume 6 part B, standardization and quality assurance in fluorescence measurements II. Springer, Berlin
39. Kahya N, Scherfeld D, Bacia K, Schwille P (2004) Lipid domain formation and dynamics in giant unilamellar vesicles explored by fluorescence correlation spectroscopy. J Struct Biol 147:77–89
40. Ramadurai S, Holt A, Krasnikov V, van den Bogaart G, Killian JA, Poolman B (2009) Lateral diffusion of membrane proteins. J Am Chem Soc 131:12650–12656
41. Briddon SJ, Hill SJ (2007) Pharmacology under the microscope: the use of fluorescence correlation spectroscopy to determine the properties of ligand-receptor complexes. Trends Pharmacol Sci 28:637–645
42. Weidemann T, Wachsmuth M, Tewes M, Rippe K, Langowski J (2002) Analysis of ligand binding by two-colour fluorescence cross-correlation spectroscopy. Single Mol 3:49–61
43. Swift JL, Heuff R, Cramb DT (2006) A two-photon excitation fluorescence cross-correlation assay for a model ligand-receptor binding system using quantum dots. Biophys J 90:1396–1410
44. Swift JL, Burger MC, Massotte D, Dahms TE, Cramb DT (2007) Two-photon excitation fluorescence cross-correlation assay for ligand-receptor binding: cell membrane nanopatches containing the human micro-opioid receptor. Anal Chem 79:6783–6791
45. Allen N, Thompson N (2006) Ligand binding by estrogen receptor beta attached to nanospheres measured by fluorescence correlation spectroscopy. Cytometry 69A:524–532
46. Worch R, Bokel C, Hofinger S, Schwille P, Weidemann T (2010) Focus on composition and interaction potential of single-pass transmembrane domains. Proteomics 10:4196–4208
47. Yin H, Slusky JS, Berger BW, Walters RS, Vilaire G, Litvinov RI, Lear JD, Caputo GA, Bennett JS, DeGrado WF (2007) Computational design of peptides that target transmembrane helices. Science 315:1817–1822
48. Mossman KD, Campi G, Groves JT, Dustin ML (2005) Altered TCR signaling from geometrically repatterned immunological synapses. Science 310:1191–1193
49. Anitei M, Stange C, Parshina I, Baust T, Schenck A, Raposo G, Kirchhausen T, Hoflack B (2010) Protein complexes containing CYFIP/Sra/PIR121 coordinate Arf1 and Rac1 signalling during clathrin-AP-1-coated carrier biogenesis at the TGN. Nat Cell Biol 12:330–340
50. Baust T, Czupalla C, Krause E, Bourel-Bonnet L, Hoflack B (2006) Proteomic analysis of adaptor protein 1A coats selectively assembled on liposomes. Proc Natl Acad Sci USA 103:3159–3164
51. Yetukuri L, Ekroos K, Vidal-Puig A, Oresic M (2008) Informatics and computational strategies for the study of lipids. Mol Biosyst 4:121–127
52. Marsh D (1990) CRC handbook of lipid bilayers. CRC press, Boca Raton
53. Feigenson GW (2006) Phase behavior of lipid mixtures. Nat Chem Biol 2:560–563
54. Kalvodova L, Sampaio JL, Cordo S, Ejsing CS, Shevchenko A, Simons K (2009) The lipidomes of VSV, SFV and the host plasma membrane analyzed by quantitative shotgun mass spectrometry. J Virol 83:7996–8003
55. Shevchenko A, Simons K (2010) Lipidomics: coming to grips with lipid diversity. Nat Rev Mol Cell Biol 11:593–598

56. Ejsing CS, Sampaio JL, Surendranath V, Duchoslav E, Ekroos K, Klemm RW, Simons K, Shevchenko A (2009) Global analysis of the yeast lipidome by quantitative shotgun mass spectrometry. Proc Natl Acad Sci USA 106:2136–2141
57. Chan R, Uchil PD, Jin J, Shui GH, Ott DE, Mothes W, Wenk MR (2008) Retroviruses human immunodeficiency virus and murine leukemia virus are enriched in phosphoinositides. J Virol 82:11228–11238
58. Carvalho K, Ramos L, Roy C, Picart C (2008) Giant unilamellar vesicles containing phosphatidylinositol(4,5) bisphosphate: characterization and functionality. Biophys J 95:4348–4360
59. Bhatia VK, Madsen KL, Bolinger PY, Kunding A, Hedegard P, Gether U, Stamou D (2009) Amphipathic motifs in BAR domains are essential for membrane curvature sensing. EMBO J 28:3303–3314
60. Angelova MI, Dimitrov DS (1986) Liposome electroformation. Faraday Discuss 81:303–311
61. Kahya N (2010) Protein–protein and protein-–lipid interactions in domain-assembly: lessons from giant unilamellar vesicles. Biochim Biophys Acta 1798:1392–1398
62. Morales-Penningston NF, Wu J, Farkas ER, Goh SL, Konyakhina TM, Zheng JY, Webb WW, Feigenson GW (2010) GUV preparation and imaging: minimizing artifacts. Biochim Biophys Acta 1798:1324–1332
63. Pott T, Bouvrais H, Meleard P (2008) Giant unilamellar vesicle formation under physiologically relevant conditions. Chem Phys Lipids 154:115–119
64. Stachowiak JC, Richmond DL, Li TH, Liu AP, Parekh SH, Fletcher DA (2008) Unilamellar vesicle formation and encapsulation by microfluidic jetting. Proc Natl Acad Sci USA 105:4697–4702
65. Nollert P, Kiefer H, Jähnig F (1995) Lipid vesicle adsorption versus formation of planar bilayers on solid surfaces. Biophys J 69:1447–1455
66. Richter RP, Maury N, Brisson AR (2005) On the effect of the solid support on the interleaflet distribution of lipids in supported lipid bilayers. Langmuir 21:299–304
67. Cremer PS, Boxer SG (1999) Formation and spreading of lipid bilayers on planar glass supports. J Phys Chem B 103:2554–2559
68. Tanaka M, Sackmann E (2005) Polymer-supported membranes as models of the cell surface. Nature 437:656–663
69. Baumgart T, Hammond AT, Sengupta P, Hess ST, Holowka DA, Baird BA, Webb WW (2007) Large-scale fluid/fluid phase separation of proteins and lipids in giant plasma membrane vesicles. Proc Natl Acad Sci USA 104:3165–3170
70. Scott RE, Maercklein PB (1979) Plasma-membrane vesiculation in 3T3-cells and Sv3T3 cells 2. Factors affecting the process of vesiculation. J Cell Sci 35:245–252
71. Keller H, Lorizate M, Schwille P (2009) PI (4,5)P-2 degradation promotes the formation of cytoskeleton-free model membrane systems. Chemphyschem 10:2805–2812
72. Levental I, Lingwood D, Grzybek M, Coskun U, Simons K (2010) Palmitoylation regulates raft affinity for the majority of integral raft proteins. Proc Natl Acad Sci USA 107:22050–22054
73. Lingwood D, Ries J, Schwille P, Simons K (2008) Plasma membranes are poised for activation of raft phase coalescence at physiological temperature. Proc Natl Acad Sci USA 105:10005–10010
74. Perez JB, Martinez K, Segura JM, Vogel H (2006) Supported cell-membrane sheets for functional fluorescence imaging of membrane proteins. Adv Funct Mater 16:306–312
75. Baumgart T, Hunt G, Farkas ER, Webb WW, Feigenson GW (2007) Fluorescence probe partitioning between Lo/Ld phases in lipid membranes. Biochim Biophys Acta 1768:2182–2194
76. Fuhrmann A, Gerzabek MH, Watzinger A (2009) Effects of different chloroform stabilizers on the extraction efficiencies of phospholipid fatty acids from soils. Soil Biol Biochem 41:428–430
77. Ayuyan AG, Cohen FS (2006) Lipid peroxides promote large rafts: effects of excitation of probes in fluorescence microscopy and electrochemical reactions during vesicle formation. Biophys J 91:2172–2183
78. Kahya N, Scherfeld D, Bacia K, Poolman B, Schwille P (2003) Probing lipid mobility of raft-exhibiting model membranes by fluorescence correlation spectroscopy. J Biol Chem 278:28109–28115

Chapter 16

Quantitative Fluorescence Co-localization to Study Protein–Receptor Complexes

Shanica N. Pompey, Peter Michaely, and Katherine Luby-Phelps

Abstract

Fluorescence microscopy can be used to assess quantitatively the interaction between a ligand and its receptor, between two macromolecules, or between a macromolecule and a particular intracellular compartment by co-localization analysis. In general, this analysis involves tagging potential interacting partners with distinct fluorophores—by direct labeling of a small ligand, by expression of fluorescent cDNA constructs, by immunofluorescence labeling, or by some combination of these methods. Pairwise comparison of the fluorescence intensity of the two fluorophores at each pixel in a two channel digital image of the sample reveals regions where both are present. With appropriate protocols, the image data can be interpreted to indicate where the potential interacting partners are co-localized. Keeping in mind the limited resolution of the light microscope, co-localization is often used to support the claim that two molecules are interacting.

All quantitative methods for evaluating co-localization begin with identifying the pixels where the intensities of both color channels are above background. Typically this involves two sequential image segmentation steps: the first to exclude pixels where neither channel is above background, and the second to set overlap thresholds that exclude pixels where only one color channel is present. Following segmentation, various quantitative measures can be computed to describe the remaining subset of pixels where the two color channels overlap. These metrics range from simple calculation of the fraction of pixels where overlap occurs to more sophisticated image correlation metrics. Additional constraints may be employed to distinguish true co-localization from random overlap. Finally, an image map showing only the co-localized pixels may be displayed as an additional image channel in order to visualize the spatial distribution of co-localized pixels. Several commercial and open source software solutions provide this type of co-localization analysis, making image segmentation and calculation of metrics relatively straightforward. As an example, we provide a protocol for the time-dependent co-localization of fluorescently tagged lipoproteins with LDL receptor (LDLR) and with the early endosome marker EEA1.

Key words Fluorescence, Co-localization, Confocal, Deconvolution, Immunofluorescence, Digital imaging, LDL, VLDL, LDL receptor

Mark A. Williams and Tina Daviter (eds.), *Protein-Ligand Interactions: Methods and Applications*, Methods in Molecular Biology, vol. 1008, DOI 10.1007/978-1-62703-398-5_16, © Springer Science+Business Media New York 2013

1 Introduction

1.1 Fluorescence Microscopy

Over the past two decades, technical advances have led to a revival of light microscopy as an important tool for biomedical research. Development of sensitive, high resolution cameras with a linear light response, and the availability of stable fluorescent tags with emission maxima spanning the visible spectrum from near UV to far red, make it possible to visualize specific molecules in a microscope specimen. Improved optical methods such as confocal, multiphoton and deconvolution microscopy that decrease the contribution of signal from above and below the focal plane permit the location of fluorescently tagged molecules with submicrometer accuracy in three dimensions. Simultaneous visualization of two molecules whose fluorescent tags have distinct emission spectra allows the investigation of molecular associations in situ, commonly referred to as co-localization.

1.2 Image Analysis for Fluorescence Co-localization

The simplest and most commonly employed method of determining co-localization is to display the image of one fluorophore in the red channel of an RGB image, display the image of the other in the green channel, and look for yellow pixels in the merged image. While this approach yields a qualitative assessment of co-localization, it often provides a poor estimation of the degree of co-localization. Because the appearance of yellow depends on the relative intensities of red and green, co-localization may be underestimated when there is wide disparity in the intensities of the two fluorophores. Such qualitative assessments can also overestimate co-localization, particularly in regions with dense populations of discrete structures that cannot be spatially resolved in the fluorescence microscope. In these regions yellow color may give a false impression of co-localization where none exists. For this reason, apparent co-localization in the Golgi region of tissue culture cells should be treated with suspicion. Fortunately, because the fluorescence intensity is a function of the number of fluorophores present, a more objective evaluation of co-localization is possible.

1.2.1 Common Quantitative Measures of Co-localization

For convenience, this discussion refers to the images of the two potentially co-localized species as the red and green channel, even though the original fluorescence tags may be any two colors, and in any case only monochrome images should be used for co-localization images. Finally, although we use the two-dimensional term, pixel, rather than the three-dimensional term, voxel, to refer to the elements of the intensity array that comprises a digital image, the discussion applies equally to both two- and three-dimensional datasets.

Cytofluorogram

Similar to how cells with different fluorescent tags are displayed in flow cytometry, the intensity of the red and green signals at each

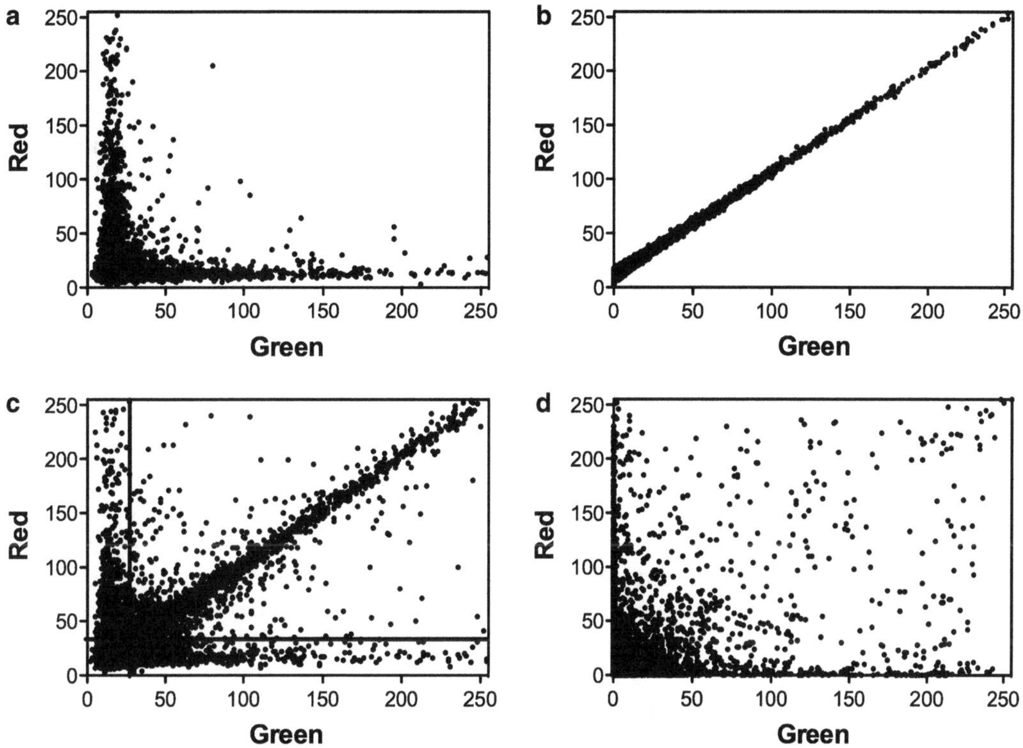

Fig. 1 Simulated cytofluorograms or scatterplots showing different cases of overlap between two color channels. (**a**) Only random overlap between red and *green signal*. (**b**) Strong overlap between the *red* and *green channels*. (**c**) Mixture of overlapping with nonoverlapping signal. (**d**) Actual data from an image with partial overlap, variable stoichiometry, and random noise. Scatterplots were generated in ImageJ using the JaCoP plug-in and the data were replotted in Prism (GraphPad)

pixel in an image may be plotted as a two-dimensional histogram, referred to as a cytofluorogram or scatter plot. In the absence of co-localization, the scatter plot will be bi-lobed as shown (Fig. 1a). Pixels with only green will be spread out along the green intensity axis and pixels with only red will be spread out along the red intensity axis. The center of the plot is empty because no pixels have both red and green. In the ideal case of stoichiometric co-localization, the scatterplot will be a straight line (Fig. 1b) because all pixels have both red and green at a fixed intensity ratio. The slope of this line gives the ratio of intensities in the two color channels, which may reflect the stoichiometry of the molecular association between the two species. Multiple discrete stoichiometries are sometimes apparent as distinct lobes of differing slope in the scatterplot. A tri-lobed scatterplot (Fig. 1c) indicates a mixture of co-localization with no co-localization. By choosing threshold values to exclude pixels where only one of the two channels is present, (indicated by the solid black lines in Fig. 1c) the pixels where the two color channels overlap can be examined selectively to extract quantitative information regarding the degree of overlap and the spatial pattern of overlap. In practice, the actual scatter plot often resembles a cloud of pixels with no clear delineation between co-localized and non-co-localized pixels (Fig. 1d). This may arise from either co-localization with no fixed stoichiometry or the effect of noise in the image or both. Objective methods of obtaining the overlap threshold under these circumstances are discussed later.

Fraction of Overlapping Pixels

Once thresholds have been set to exclude pixels lacking significant overlap, the extent of overlap can be computed for each channel. This is usually expressed as the fraction of pixels in each channel that overlap with signal from the other channel, that is, the number of pixels above the overlap threshold normalized by the total number of pixels in that channel. For example, a value of 0.5 for the red channel indicates 50 % of the pixels in the red channel also have green signal above the overlap threshold set for the green channel. The fractional overlap for the two channels will not necessarily be the same. If the binding partner in the red channel is widely distributed, while the binding partner in the green channel is only present when red is present, the % overlap for the red channel will be low, while that for the green channel will be very high.

Thresholded Manders' Coefficients (M1 and M2)

The variations of pixel intensities across a fluorescence microscope image directly indicate the relative number concentration of the fluorescent tag. With appropriate calibration standards, the fluorescence intensity at any given pixel can be interpreted in terms of the absolute number concentration of the tagged molecule, although this is seldom done. An intensity-based co-localization metric known as the Manders' coefficient incorporates the relationship between intensity and concentration and thus provides more meaningful biological information than the fraction of overlapping pixels for characterizing the co-incidence of the red and green channels. The thresholded Manders' Coefficient for each channel (designated M1 and M2) is effectively the sum of the fluorescence intensities of the overlapping pixels normalized by the sum of the intensities of all the pixels (1). As for the pixel overlap coefficient, the values for the two channels will not necessarily be the same.

Making the assumption that the fluorescence intensities at each pixel are proportional to the concentration of each species, the Manders' coefficients can be interpreted as the fraction of molecules of one species that overlap with the other. These values may differ dramatically from the fraction of overlapping pixels—a relatively small fraction of very bright pixels may result in a relatively high value for Manders' coefficient.

Pearson's Coefficient

The fraction of overlapping pixels is often referred to as % co-localization but it does not distinguish between random overlap and true co-localization. A metric commonly applied to identify truly co-localized pixels is the Pearson Coefficient (PC). This is a standard statistical measure widely used in image correlation analysis and introduced to fluorescence microscopy by Manders et al. (2). The PC tests whether there is a linear correlation between the intensity of the red signal and the intensity of the green signal at every pixel. The PC may be computed over all pixels in the image or in user specified regions of interest. It ranges from −1 to 1: a PC of 0 indicates random overlap, and a PC of 1 indicates perfect co-localization with a constant stoichiometry of red and green. A PC of −1 suggests a mutually exclusive localization of the two signals. It should be noted that in cases where the intensities of the two channels are not related by a constant ratio, a low positive value of the PC does not necessarily indicate poor co-localization, it may merely indicate that the intensities of the two color channels are not strongly correlated.

Interpretation of PC

The PC is often used to compare the co-localization of two different pairs of binding partners: for example, the binding of a ligand with a wild type protein vs. a mutant lacking a functional binding domain. Although in principle a PC $>$ 0 indicates co-localization, significant noise in the images will make it difficult to determine at what value the PC becomes significantly different from 0. A method developed by Costes (3) provides an objective evaluation of PC significance. In this approach, the pixels in one channel are scrambled and the PC is calculated. If 95 out of 100 randomized images give a PC less than the PC calculated for the original image, the PC for the original image is considered significant of co-localization with 95 % confidence. An alternative use of the data from randomized images is to plot the distribution of PCs for all the randomized images and obtain the standard deviation from a Gaussian fit to the data. A PC for the original image several fold higher than the standard deviation for the random images is likely to indicate significant co-localization. For high quality data, a PC as low as 0.1 may easily be as much as 50-fold higher than the standard deviation of the distribution of PCs for randomized images.

Co-localization Threshold Selection

A method proposed by Costes et al. (3) uses the PC to determine automatic and relatively objective intensity thresholds for excluding pixels with no co-localization. This is especially useful in cases like Fig. 1d where the scatterplot does not clearly indicate where the thresholds should be drawn. Beginning with the highest intensity values for one channel, the Costes method progressively lowers the threshold values for each channel until the PC for pixels below threshold in both channels reaches zero. At this point, the pixels below threshold in either channel are considered not co-localized and are excluded from the calculation of PC. The PC for the remaining, co-localized pixels can be dramatically better than for the dataset overall in cases where true co-localization occurs in only a few regions of the image. One caveat of the Costes method is that because PC assumes a linear relationship between red and green intensities, the threshold values for the red and green channels are not set independently. An additional issue is that dim pixels with true co-localization may be excluded by this method.

Other Measures of Co-localization

Additional measures of co-localization based on pixel intensity correlations have been proposed and are available in some software packages. Two common examples are van Steensel's Cross Correlation Function (4) and Li's Intensity Correlation Analysis (5). A more detailed discussion of these and other intensity based co-localization methods is available in two excellent reviews (6, 7).

Object-Based Methods

An entirely different approach to co-localization analysis is to build isosurface objects in each color channel and then calculate distance maps between the objects in one channel and the objects in the other. The results can be used to find the nearest green neighbor for each red object or to find all red objects within a certain distance of green objects. This approach is very useful for co-localization when the color channels overlap only partially or not at all—for example, when the two fluorescent tags are on opposite sides of the limiting membrane of the same vesicle. Strategies for interpreting the results of this kind of analysis are discussed in (6).

1.3 Limitations of Fluorescence Co-localization Analysis

None of the co-localization methods described above is completely unambiguous, especially in the presence of noise and/or when the stoichiometry of red to green is variable. Automated determination of thresholds may fail to find appropriate thresholds of co-localization or the thresholds may be set too low, resulting in overestimation of the degree of co-localization. The limitations of co-localization analysis have been rigorously investigated and are discussed in (1–7). Development of more reliable co-localization metrics remains an active field of research, e.g., (8, 9).

Even under ideal conditions, the limited spatial resolution of fluorescence microscopy dictates that fluorescence co-localization should never be regarded as definitive evidence of an interaction between two species. Co-localization data should always be supported by genetic or biochemical evidence of interaction, such as yeast two-hybrid, co-immunoprecipitation or cross-linking studies. For standard fluorescence microscopy, the diffraction theory of image formation according to Abbé and the Nyquist sampling theorem dictate that resolution is limited by the wavelength of the light and the calibrated pixel size in the image. The latter depends on the numerical aperture of the objective lens (NA), the image magnification, and the physical dimensions of pixels on the camera chip. For visible light, the highest NA lenses, and cameras with the smallest physical pixels, the resolution is limited to around 0.2 μm in *xy* and approximately 0.4 μm along the *z* axis. Thus, each voxel has a volume $\geq 0.016\ \mu m^3$. To put this in perspective, each voxel is large enough to contain 100 synaptic vesicles. Other microscopic methods with better resolution such as electron microscopy and Förster Resonance Energy Transfer (FRET) allow more confident determination of actual interactions between two species. Emerging super-resolution technologies, such as STED, STORM, PALM, or SIM (10), may also lead to improved resolution of co-localization.

1.4 Software for Co-localization

Most available software packages for analysis of fluorescence microscope images include some form of co-localization analysis. Four commercial packages with very sophisticated co-localization capability are Huygens (SVI, Inc), Imaris (Bitplane), Autoquant X (Media Cybernetics), and Volocity (Perkin Elmer). In addition, co-localization plug-ins are available for the open source java application ImageJ (http://rsbweb.nih.gov/ij/) and its various distributions, including Fiji (http://pacific.mpi-cbg.de/wiki/index.php/Fiji). Two plug-ins that implement all the co-localization methods described above are JaCoP (ImageJ) and Coloc_2 (Fiji). Different implementations of the same co-localization algorithms by these different packages can yield somewhat different values of PC and other metrics (Table 1). In general, this seems to reflect different criteria used for excluding background pixels and for setting intensity thresholds. The documentation for all the co-localization analysis packages listed above is available online (12) and offers a very useful resource for understanding the differences in how various features are implemented.

1.5 The LDLR-EEA1 Interaction as an Example

Defects in uptake of serum lipoproteins such as LDL and VLDL result in hyperlipidemia disorders implicated in heart disease. Although it is well known that uptake and metabolism of these lipoproteins depends on the LDL receptor (LDLR), the detailed regulation of uptake at the cellular level remains an area of active investigation. Time-dependent co-localization analysis can be used to map out the progress through the uptake pathway under different experimental conditions. Here we present results showing co-localization of the LDL with LDLR at the plasma membrane of the cell before uptake begins and the time course of association with the early endosomal compartment as the LDL is internalized by endocytosis.

2 Materials

1. Delipidated Medium: Dulbecco's MEM or Eagle's MEM, supplemented with 10 % (v/v) fetal lipoprotein poor bovine serum, 20 mM HEPES pH 7.5, penicillin G (100 units/ml) and streptomycin (100 μg/ml).
2. Cells lipoprotein starved for 24 hr in delipidated Dulbecco's MEM on 12 mm round #1.5 (0.17 mm) coverslips in 24-well microtiter plates.

Table 1
Comparison of co-localization metrics analyzed with different software implementations of co-localization analysis. The overall PC for the raw images, the PC calculated after the images were intensity thresholded, and the thresholded Manders'coefficients (M1 and M2) are shown for the same pair of images analyzed with Coloc 2 or Colocalization Analysis in Fiji, JaCoP in ImageJ and Imaris Coloc. The threshold values for each channel and the method used to choose the ROI are indicated for each result. The differences illustrate the dependence of the results on the algorithm used

Method	PC overall	PC thresholded	Thresh Manders'	Thresh	ROI
Fiji1.45k > analyze > colocalization > coloc 2 (32-bit images)	0.62	0.47	Ch1: 0.965 Ch2: 0.955	Ch1: 288 Ch2: 367	None but zero–zero pixels are explicitly excluded in the source code
Fiji1.45k > analyze > colocalization > colocalization threshold (16-bit images)	0.656	0.513	Ch1: 0.94 Ch2: 0.94	Ch1: 389 Ch2: 853	Exclude zero–zero pixels checked
JaCoP plug-in (16-bit images)	0.621	0.491	Ch1: 0.999 Ch2: 0.964	Ch1: 2 Ch2: 856	Masked both channels with threshold = 257 to exclude area outside cells
Imaris > coloc (32-bit images)	0.62	0.397	Ch1: 0.72 Ch2: 0.76	Ch1: 504 Ch2: 756	Masked with red channel, threshold = 110 (Costes automatic thresholding failed below 110)

3. Fluorescently labeled LDL and/or VLDL (11).
4. Ice-cold delipidated Eagle's MEM
5. Ice-cold PBS.
6. 3 % paraformaldehye in PBS, chilled to 4 °C.
7. Digitonin (Sigma Cat# D141).
8. 5 % Normal Goat Serum + 1 % BSA in phosphate buffered saline (PBS) (for blocking).
9. 5 % Normal Goat Serum + 0.1 % BSA in PBS (for antibody dilution).
10. Primary/secondary antibodies (*see* **Note 1**): Rabbit polyclonal antibody to bovine LDLR was a gift of Joachim Herz; mouse monoclonal antibody to EEA1 was from BD Biosciences, (Cat# 610457); goat anti-rabbit Alexa 647 was from Invitrogen (Cat# A21236).

11. Fluorescent nuclear stain: DAPI (Invitrogen, Cat# D3571).
12. Aquamount (Polysciences Inc., Cat# 18606).
13. Digital fluorescence imaging system (widefield or confocal).
14. ImageJ with JaCoP or Fiji with Coloc_2 or other image analysis software.

3 Methods

3.1 Overview of Procedure

In general the protocol can be divided into three major steps: sample preparation, image acquisition and image analysis. Sample preparation includes both the pulse chase labeling of live cells with fluorescent lipoprotein and immunofluorescence labeling of other markers in cells after uptake is stopped by fixation. Meaningful co-localization analysis requires careful attention to critical details at each step as outlined in the Notes.

3.2 Pulse Chase Uptake of Lipoprotein

1. Remove cells from 37 °C incubator and replace delipidated Dulbecco's MEM with ice-cold delipidated Eagle's MEM (*see* **Note 2**).
2. Chill cells for 10 min on ice to block endocytosis.
3. Dilute lipoprotein to 10 μg/ml in ice-cold delipidated Eagle's MEM.
4. Add lipoprotein solution to cells and place in 4 °C cold room, 1.5 h on ice to allow lipoprotein binding to receptors.
5. Rinse cells with ice-cold delipidated Eagle's MEM, 1× quickly.
6. Add warm delipidated Dulbecco's MEM and immediately place in 37 °C incubator for desired chase time. The selection of time points will be determined by which stage of uptake is of interest.

3.3 Immunofluorescence Staining

1. At each time point, remove a dish of cells from the incubator, and quickly rinse 3× with ice-cold PBS to stop uptake.
2. Fix cells with 3 % PFA in ice-cold PBS, 20 min on ice.
3. Rinse with ice-cold PBS, 2× quickly.
4. Permeablize cells with digitonin (10 μg/ml in PBS), 2 min on ice.
5. Rinse with ice-cold PBS, 3× quickly.
6. Block cells with 5 % Normal Goat Serum + 1 % BSA in PBS, 1 hr at room temperature (500 μl per well).
7. Add primary antibody diluted 1:100 in 5 % Normal Goat Serum + 0.1 % BSA in PBS and incubate 1 h at room temperature (150–200 μl of antibody solution, dropwise per well) (*see* **Note 3**).

8. Rinse with 0.1 % BSA in PBS, 3× quickly or move coverslips to fresh wells with 0.1 % BSA in PBS.
9. Incubate with secondary antibody diluted 1:400 in 5 % Normal Goat Serum + 0.1 % BSA in PBS, 1 h at room temperature (200 μl of antibody solution, dropwise per well) (*see* **Note 3**).
10. Rinse with 0.1 % BSA in PBS, 1× quickly.
11. Wash with 0.1 % BSA in PBS, 3× 5 min each.
12. Rinse with PBS, 1× quickly.
13. Incubate with DAPI in PBS (3 μl/ml) 10 min at room temperature (500 μl solution per well).
14. Rinse with PBS, 3× quickly.
15. Add 10 μl Aquamount to a glass slide. With fine tweezers, remove coverslip from well and invert on drop to mount (*see* **Note 4**).
16. Verify the specificity of your labeling (*see* **Note 1**).
17. Minimize background fluorescence on the coverslip (*see* **Note 3**).

3.4 Image Acquisition

1. Optimize imaging parameters (*see* **Notes 5–9**). Once imaging parameters have been optimized, these conditions should be used for all samples that will be compared: for example, comparison of control with experiment, wildtype with mutant, or monitoring changes in co-localization over time (*see* **Note 5**).
2. Acquire wide-field *z* stacks or confocal slices (*see* **Note 10**).
3. Register the two color channels in *x*, *y*, and *z* if necessary (*see* **Note 11**).
4. Deconvolve wide-field *z* stacks (*see* **Note 12**) (Fig. 2).

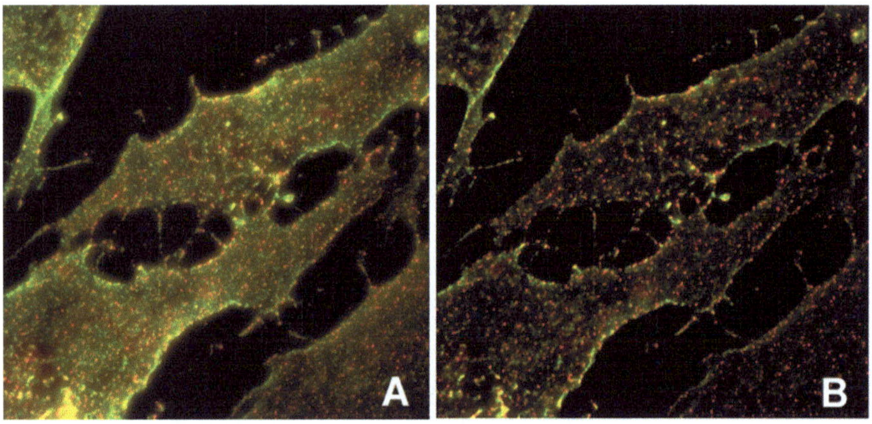

Fig. 2 Maximum intensity projections of a 15-slice wide-field epifluorescence *z* stack showing LDL (*red*) and LDLR (*green*) before (**a**) and after (**b**) deconvolution. Image stacks were deconvolved using the blind deconvolution algorithm with the default settings in Autoquant X (Media Cybernetics)

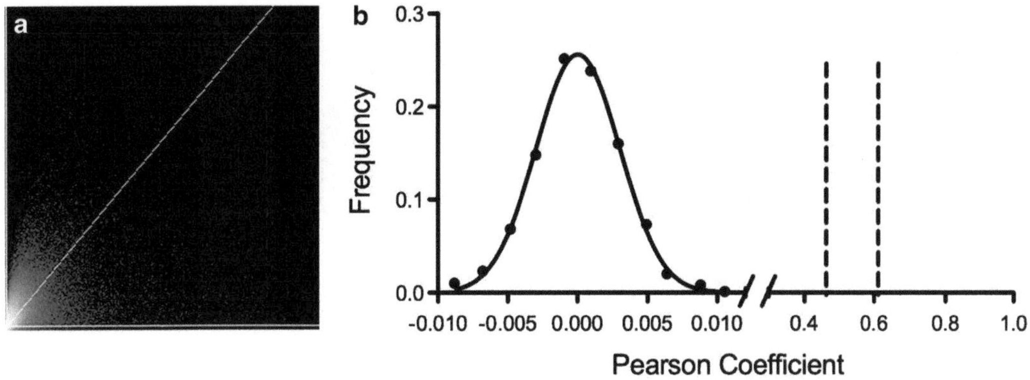

Fig. 3 (**a**) Cytofluorogram of *z* slice number 7 from the deconvolved image stack in Fig. 2. Horizontal and vertical white lines show the overlap thresholds determined using Fiji 1.45 k > Analyze > Colocalization > Colocalization Threshold to find the Costes automatic thresholds. Zero–zero pixels were excluded from the analysis. Diagonal white line has a slope of 1.63 indicating the best fitting stoichiometry between LDL and LDLR intensity. (**b**) PC significance test. The frequency distribution of PCs for 1,000 randomizations of the deconvolved *z* slice was obtained using JaCoP plug-in for ImageJ. Prism software (GraphPad) was used to fit the data to a Gaussian function (*solid line*). The fitted curve had a mean of zero and a standard deviation of 0.003 ($R^2 = 0.9965$). *Dashed lines* represent the PCs determined for the *z* slice using Fiji1.45k > Analyze > Colocalization > Coloc 2 before (PC = 0.62) and after (PC = 0.47) applying Costes' thresholds. The PC after thresholding is 150-fold greater than the standard deviation of the distribution of PCs calculated for the randomized images, indicating highly significant co-localization

3.5 Image Analysis

1. Choose an intensity threshold that excludes zero–zero pixels where no true signal is present in either channel (*see* **Note 13**).
2. Apply Costes method of automated overlap threshold selection (Fig. 3a).
3. Calculate the PC for the pixels above the overlap threshold (*see* Table 1) (*see* **Note 14**).
4. Calculate the Manders coefficients for the pixels above the overlap threshold in each color channel (*see* Table 1).
5. Apply Costes randomization to check the significance of the PC (Fig. 3b).
6. Build co-localization channel (Fig. 4).

Data obtained from different images is compared on the basis of the PC (*see* **Note 5**), e.g., a change in PC over time indicates a change in co-localization over time, which may be interpreted judiciously to indicate association or disassociation of ligand and receptor. For example, Fig. 5 indicates co-localization of LDL with the early endosome marker EEA1 peaks 10 min after enodcytosis is started by warming the cells to 37 °C, consistent with the well established pathway for uptake of LDL by endocytosis.

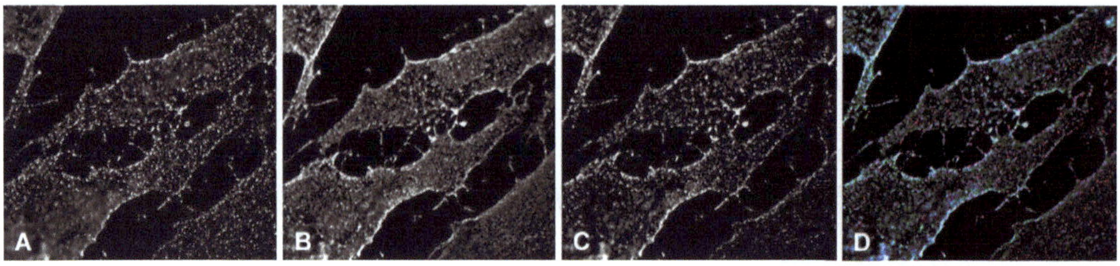

Fig. 4 Co-localized voxels in the deconvolved *z* slice used for Fig. 3 and Table 1 visualized as a separate channel. The co-localization channel was generated using Imaris Coloc. (**a**) *Red channel* (**b**) *Green channel* (**c**) Co-localization channel (**d**) Merge

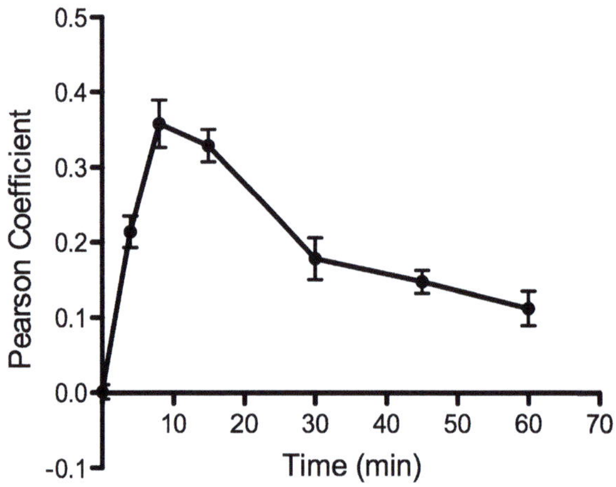

Fig. 5 Timecourse of co-localization of LDL with the early endosome marker, EEA1. Ten *z* stacks for each timepoint of a pulse-chase lipoprotein uptake experiment (except the zero timepoint, for which $n = 5$) were analyzed using Imaris Coloc (Bitplane). Mean PC ± SEM is plotted vs. time. Camera gain, exposure times, binning, objective magnification, *z* stack spacing and number of slices, deconvolution settings and co-localization analysis parameters were identical for all datasets

4 Notes

1. For immunofluorescence localization, the specificity of the antibodies/ligands should be demonstrated on samples lacking the antigen, for example by gene silencing or using tissue/cells from a knockout animal. Similar controls should be performed for fluorescently tagged ligands. For expression tags, it is best to demonstrate that the tagged proteins are functional and that their localization reflects the localization of the endogenous protein.
2. For optimal pulse chase experiments it is essential that the cells be kept on ice at 4 °C at all times before starting the chase, and also from the time when the chase is stopped until the fixation is completed. If necessary, the results can be improved by adding salt to the ice.

3. Spinning both primary and secondary antibodies for 5 min at 4 °C in a microfuge at high speed before applying to the cells is recommended to prevent deposition of fluorescent aggregates on the sample.
4. To minimize air bubbles in the sample, use only 10 μl of mounting medium and lever the coverslip onto the drop of mounting medium slowly from the side opposite the tweezers, allowing air bubbles to escape from the side where the tweezers grip the coverslip.
5. Single cell image analysis is subject to high variability due to cellular heterogeneity. As a result, the value of PC may vary widely from cell to cell. However, with high quality data and sufficient sample size, quantitative comparison of co-localization under different experimental conditions is possible as long as the image acquisition and processing parameters are identical for all datasets (Fig. 5).
6. Minimize imaging noise. Keep the gain of the detector as low as possible. Select excitation intensities and exposure times to get the widest possible range of signal intensities without saturating the detector. For best results, it is not advisable to use an offset on the bottom end of the intensity range.
7. Avoid detector saturation (intensity clipping). Quantification of fluorescence intensities assumes there is a linear relationship between the number of photons reaching the detector and the electronic output of the detector. However, the light response of all detectors has a maximum. Light intensities above this maximum will result in no further increase in electronic signal from the detector. This is referred to as saturation or intensity clipping. Extreme levels of saturation can also lead to spillover of signal from one pixel to another, effectively increasing the size of objects in the image.
8. Verify absence of crosstalk between channels. Image a red-only and a green-only sample in both channels using the imaging conditions you have established. Signal from the red-only sample should not appear in the green channel and vice versa. If necessary, the imaging conditions for each channel can be adjusted to minimize crosstalk.
9. Use a monochrome camera with 12-bit or 14-bit depth. Use of a color camera can introduce channel bleedthrough depending on the particular technology it uses to acquire RGB images. In addition, the RGB images acquired by color cameras are usually only 8-bits in each color channel, limiting the range of pixel intensities to 256. Research grade monochrome CCD cameras and photomultiplier tubes are inherently grayscale devices, and most support 12- or 14-bit images, increasing the range of possible intensities to 4,096 or 16, 384 gray levels, respectively.

The image channels can be digitally pseudocolored red or green post-acquisition if desired.

10. Be sure to save files in an uncompressed or lossless compressed file format. Usually, the best choice is a 16-bit TIFF file. Be sure to choose an option that saves the raw intensity values. The raw intensities will not be preserved when saving a 12- or 14-bit image as an 8-bit file, saving the images in JPEG format, or saving a screen shot of what is displayed on the monitor.
11. Some imaging systems may introduce a pixel shift between color channels that must be corrected to ensure high quality results of co-localization analysis. In particular, objective lenses that are not corrected for chromatic aberration across the entire spectrum of available fluorophores can produce very large shifts in the z direction, especially for UV or far-red emitting fluorophores. Because such shifts generally are systematic, once pixel shift has been characterized for a set of imaging conditions, subsequent images can be aligned using image analysis software. *Z*-stacks of beads that are fluorescent in multiple color channels, such as Tetraspeck fluorescent microspheres (Invitrogen, Cat # T14792), can be used to determine the degree of shift in *x*, *y*, and *z*.
12. Co-localization analysis for a pair of wide-field epifluorescence images may give a false impression of co-localization because each pixel contains signal from above and below the plane of focus. In that case, true co-localization cannot be distinguished from coincidental overlap of objects located in different focal planes. This limitation becomes increasingly important as the thickness of the sample increases. For this reason, co-localization should only be computed for pairs of confocal slices, confocal *z* stacks or wide-field epifluorescence *z* stacks that have first been deconvolved to restore out of focus photons to their slice of origin. Even confocal *z* stacks may benefit from deconvolution. Total Internal Reflection Fluorescence (TIRF) microscopy images have better depth discrimination than any of these methods and are also suitable for co-localization analysis.
13. Exclusion of background pixels may be accomplished by intensity thresholding, by drawing a region of interest on one of the images or by masking the image pair with a third color channel. For example, DAPI staining is often used as a mask to restrict the analysis to pixels within the nuclear region. In Fig. 4, the red channel was used to mask the dataset.
14. Table 1 illustrates how different implementations of the same co-localization algorithms may give somewhat different results. One source of differences is that Colocalization Threshold and JaCoP both require that the original 32-bit images be converted

to 16-bit images, which changed the intensity values in the images, affecting the results of the calculations. The available documentation for each method (12) does not reveal why the thresholded Manders' coefficients generated by Imaris Coloc are much lower than the others in Table 1. Some algorithms report non-zero values for the PC of the pixels below threshold, suggesting they have reached a local minimum that does not necessarily produce the optimal thresholds. For example, according to its source code, Coloc 2 reports the best thresholds after 30 iterations, rather than the threshold where PC becomes zero. In contrast, both Imaris Coloc and JaCoP (which takes a lot longer to arrive at a solution) actually require PC ≤ 0 for the pixels below threshold. In the case of Imaris Coloc, failure to reach this value within the maximum number of iterations results in a failure to find automatic thresholds. While the implementation of co-localization in open source software is very useful and more affordable than a commercial package like Imaris, the user should be aware that code for ImageJ and Fiji is under continuous development and updates may include changes and bugs in the way algorithms are implemented.

References

1. Manders EM, Stap J, Brakenhoff GJ, van Driel R, Aten JA (1992) Dynamics of three-dimensional replication patterns during the S-phase, analysed by double labelling of DNA and confocal microscopy. J Cell Sci 103(Pt 3):857–862
2. Manders EMM, Verbeek FJ, Aten JA (1993) Measurement of co-localization of objects in dual-colour confocal images. J Microsc 169:375–382
3. Costes SV, Daelemans D, Cho EH, Dobbin Z, Pavlakis G, Lockett S (2004) Automatic and quantitative measurement of protein–protein colocalization in live cells. Biophys J 86:3993–4003
4. van Steensel B, van Binnendijk EP, Hornsby CD, van der Voort HT, Krozowski ZS, de Kloet ER, van Driel R (1996) Partial colocalization of glucocorticoid and mineralocorticoid receptors in discrete compartments in nuclei of rat hippocampus neurons. J Cell Sci 109(Pt 4):787–792
5. Li Q, Lau A, Morris TJ, Guo L, Fordyce CB, Stanley EF (2004) A syntaxin 1, Galpha(o), and N-type calcium channel complex at a presynaptic nerve terminal: analysis by quantitative immunocolocalization. J Neurosci 24:4070–4081
6. Bolte S, Cordelieres FP (2006) A guided tour into subcellular colocalization analysis in light microscopy. J Microsc 224:213–232
7. Comeau JW, Costantino S, Wiseman PW (2006) A guide to accurate fluorescence microscopy colocalization measurements. Biophys J 91:4611–4622
8. Adler J, Pagakis SN, Parmryd I (2008) Replicate-based noise corrected correlation for accurate measurements of colocalization. J Microsc 230:121–133
9. RamÍRez O, GarcÍA A, Rojas R, Couve A, HÄRtel S (2010) Confined displacement algorithm determines true and random colocalization in fluorescence microscopy. J Microsc 239:173–183.
10. Toomre D, Bewersdorf J (2010) A new wave of cellular imaging. Annu Rev Cell Dev Biol 26:285–314.
11. Zhao Z, Michaely P (2009) The role of calcium in lipoprotein release by the low-density lipoprotein receptor. Biochemistry 48:7313–7324
12. Image documentation for JaCoP can be found here: http://imagejdocu.tudor.lu/lib/exe/fetch.php?media=plugin:analysis:jacop_2.0:just_another_colocalization_plugin:jacop_ij conf2008.pdf Documentation and a link to the source code for Coloc 2, Colocalization Threshold and Colocalization Test can be found here: http://fiji.sc/wiki/Colocalization_Analysis

Part VI

Structural and Computational Methods

Chapter 17

Studying Protein–Ligand Interactions Using X-Ray Crystallography

Andrew P. Turnbull and Paul Emsley

Abstract

X-ray crystallography is a powerful technique for studying protein–ligand interactions. Advances in techniques have meant that it is now possible to routinely determine the structures of ligand complexes in the majority of cases where crystallization conditions and protein structures are already known. Ligand soaking or cocrystallization, together with the potential use of molecular replacement, provides data for determining the structures of a protein in complex with ligands. Furthermore, advances in protein structure model building facilitate automatic ligand fitting to residual electron density in the protein–ligand complex.

Key words X-ray crystallography, Crystallization, Molecular replacement, Ligand fitting, COOT

1 Introduction

Over the past decade, technological advancements in high-throughput structure determination in the field of X-ray crystallography have lead to a rapid increase in the number of three-dimensional protein structures available in the protein database (PDB; http://www.pdb.org; (1)). In excess of 80,000 protein, protein–protein and protein–ligand X-ray structures have been deposited in the PDB, covering a broad range of protein fold space with most classes of proteins having representative structures. In many cases, structures deposited in the PDB can be used as initial "search models" to determine the structures of unknown proteins using the molecular replacement technique. Furthermore, these structures can be used to facilitate the study of protein–ligand interactions via soaking and cocrystallization experiments to glean insights into structure–function relationships. In addition, these techniques are used within the pharmaceutical industry to support structure aided drug design and fragment screening programs.

Mark A. Williams and Tina Daviter (eds.), *Protein-Ligand Interactions: Methods and Applications*, Methods in Molecular Biology, vol. 1008, DOI 10.1007/978-1-62703-398-5_17, © Springer Science+Business Media New York 2013

This chapter focuses on the techniques available to obtain protein–ligand crystals, determine their structures using molecular replacement, and fit ligands into their associated electron density map using model building techniques.

2 Materials

1. Purified, crystallization-grade protein.
2. Crystal screening reagents including coarse screens (*see* **Note 1**).
3. 24-Well Linbro trays and siliconized glass cover slides or 96-well SBS crystallization plates (e.g., the MRC crystallization plate) and access to a crystallization robot for automated 96-well plate preparation.
4. Temperature controlled room or incubator to maintain the crystal growth temperature.
5. Cold light source stereo microscope for inspecting the results from crystallization experiments.
6. Cryoloops, tools, and protectants for handling, manipulating, and cryoprotecting crystals prior to flash freezing in liquid nitrogen.
7. Seed bead kit (Hampton Research) for the generation of crystal seed stock for microseeding experiments.
8. An "in-house" X-ray source or access to a macromolecular beamline at a synchrotron source for screening crystals and subsequent data acquisition.
9. Crystallographic programs including the CCP4 suite (2) and COOT (3, 4) (*see* **Note 2** for a list of useful crystallographic programs and resources).

3 Methods

3.1 Setting up Crystallization Experiments

The most popular method for setting up crystallization experiments is vapor diffusion using hanging or sitting drops. 24-well Linbro plates are commonly used in the laboratory for basic hanging drop crystallization setups. Typically, 0.5 or 1 μl of protein is mixed with an equal amount of reservoir solution on a siliconized glass coverslip, which is suspended over the reservoir solution. The concentration difference between the drop and reservoir leads to vapor diffusion from the drop until the solution concentration matches that of the reservoir. Consequently, the concentration of the protein and precipitant in the drop increase. If the protein becomes supersaturated, crystals may form.

Once established, crystallization protocols are generally reproducible. However, there are instances when once-successful crystallizations fail to repeat. Several factors including minor differences in the protein composition and purity, and failure to reproduce the exact crystallization conditions can lead to non-reproducibility. In order to grow good quality crystals, the experimental conditions must be optimized by screening a broad range of conditions in which each crystallization parameter is varied. Even in cases where the crystallization conditions are already known, it is advisable to set up a customized fine screen in which the parameters are varied around the pre-established conditions. Parameters affecting crystallization include the protein and precipitant concentrations, temperature and pH. Many laboratories now have crystallization robots such as the Mosquito® (TTP Labtech) and the Phoenix™ RE (Rigaku Corporation), which can accurately and reproducibly dispense very small volumes (nanoliters in size) into 96-well plates for automated screening and optimization of crystallization conditions.

3.2 Obtaining Protein–Ligand Crystals

Soaking and cocrystallization are the two most widely used approaches to obtain cocrystal structures of protein–ligand complexes. These methodologies require the ligand to bind to the protein either before or after crystallization. In soaking, the ligand is incubated with preformed crystals of either the ligand-free protein (the apoenzyme form) or cocrystals of a previously bound ligand that can be displaced with a new one. In the cocrystallization method, the protein is incubated with excess ligand to form a complex prior to setting up the crystallization experiments. Compared to soaking, cocrystallization is more likely to lead to full occupancy of the ligand binding sites within the crystal. Soaking, however, is applicable in cases where cocrystallization has failed and a supply of soakable crystals is available.

3.2.1 Soaking Ligands into Existing Crystals

Soaking represents the simplest technique for obtaining protein-ligand crystals (*see* Fig. 1). Protein crystals typically contain between 30 and 70 % solvent, forming channels running through the crystal lattice that allow small molecules to diffuse into the crystal and potentially bind. Several factors must be taken into consideration to ensure that the ligand is successfully incorporated into the crystal. The soak time and ligand concentrations need to be optimized since many proteins are sensitive to organic solvents, such as dimethyl sulfoxide (DMSO), used to solubilize ligands that are insoluble in water. Typically, a final concentration of ≤ 5 % DMSO is used in soaking experiments, although some crystals are capable of tolerating much higher solvent concentrations which may improve ligand solubility. Furthermore, protein crystals need to be equilibrated in stabilizing solution prior to soaking to prevent them from dissolving. It is usually sufficient to use the reservoir solution to stabilize

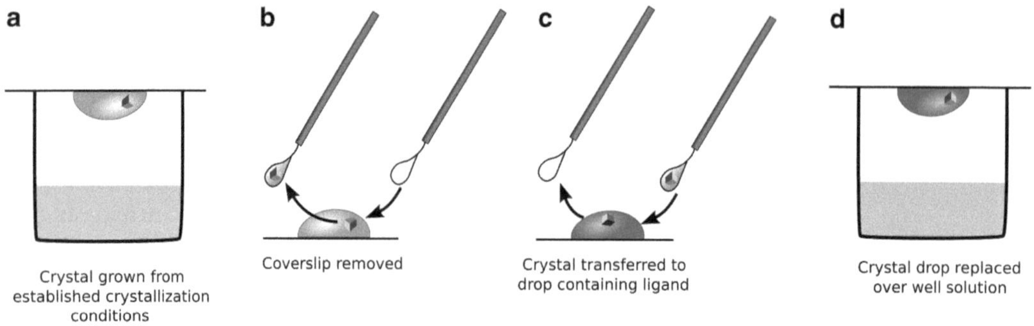

Fig. 1 Schematic illustration outlining the major steps involved in soaking ligands into existing crystals. In this example, an apoenzyme crystal has been grown in a 24-well Linbro plate using established crystallization conditions (**a**). The coverslip is removed and inverted, and the crystal is picked up from the drop using a cryoloop (**b**). The crystal is subsequently transferred to a drop containing the ligand (**c**).Usually a 50–100 mM stock of ligand in 100 % DMSO is diluted to a final concentration of ~2 mM using the reservoir solution from which the crystal was grown to stabilize the crystal and prevent it dissolving. Finally, the coverslip is placed over the reservoir solution (**d**) and the crystal is allowed to incubate with the ligand before being harvested for data collection

the crystal because the precipitant concentration in the drop is approximately the same as that in the reservoir. If ligand binding induces a conformational shift in the protein that is not compatible with crystal lattice packing interactions, crystal cracking can occur and may result in the crystal no longer being suitable for data collection. Such crystals can be gently cross-linked with glutaraldehyde (5) prior to soaking to increase resistance to any lattice disorder induced by ligand binding (*see* **Note 3**). In some cases, short soak times of minutes are necessary to achieve good occupancy for a ligand soaked into a crystal, whereas, in other cases, hours or even days of soaking are required to obtain full occupancy of the ligand binding site.

3.2.2 Cocrystallization

It is not always possible to soak ligands into a preexisting crystal system because the crystal lattice may have protein–protein packing interactions restricting access to the binding site. Additionally, the ligand may have low solubility in aqueous solution or the protein may be prone to aggregation. In such cases, cocrystallization can be utilized as an alternative means of obtaining protein–ligand complex crystals. In cocrystallization experiments, the ligand is added to the protein prior to crystallization. The protein sample must be homogeneous (based on gel filtration and mass spectrometry analysis), highly pure (>95 % purity based on SDS-PAGE), and stable (based on activity measurements) prior to setting up the cocrystallization experiments. The UV absorbance at 280 nm and the molar extinction coefficient calculated from the protein sequence can be used to determine the protein concentration (*see* Chapter 2).

In order to estimate the amount of ligand to use for cocrystallization, the dissociation constant, K_d, can be considered ((6); *see* **Note 4** and Chapter 1). The fractional saturation of the protein by the ligand is influenced by the ratio of the concentration of free ligand to the K_d. This ratio is affected by the affinity and concentration of the ligand and its solubility. To ensure approximately 90 % fractional saturation of the ligand binding site, the free ligand concentration must be in excess of the free protein concentration and, at equilibrium, the free ligand concentration should not deplete to less than $10 \times K_d$. As a rule of thumb, in cases where the K_d is not known, up to tenfold molar excess of ligand should be added to the protein solution to ensure sufficient saturation of the ligand binding site.

A concentrated stock solution of ligand (usually at 50 or 100 mM concentration) should be prepared to prevent dilution of the protein upon addition of the ligand. If the ligand is insoluble in aqueous solution, organic solvents such as DMSO, ethanol and low molecular weight polyethylene glycols (PEG 200, PEG 300, or PEG 400) can be used to solubilize the compound (*see* **Note 5**). Following incubation, centrifuging the sample at 9,600×*g* for 5 min at 4 °C will remove any precipitate prior to setting up the cocrystallization experiments. The crystallization experiments can either be set up based on fine screening around previously reported conditions (e.g., sampling precipitant concentration and pH in finer steps) or in coarse screens to cover a broad range of conditions (*see* **Note 1**).

3.2.3 Ligand Exchange in Crystals

There are instances when it may be relatively straight forward to grow crystals that contain a natural ligand or an inhibitor but difficult to obtain crystals with a new ligand. In such cases, it may be possible to soak a new ligand into existing cocrystals so that the initial ligand is displaced—this technique is known as ligand exchange (*see* Fig. 1). In order for this technique to be successful, the ligand binding site must be accessible to the new ligand and the crystal lattice interactions must be compatible with any conformational changes occurring on ligand binding. Additionally, the binding constant of the new ligand has to be taken into consideration since it is often difficult to substitute a new lower affinity ligand into a crystal with a much higher affinity ligand already bound.

3.2.4 Seeding

Sometimes it is not possible to obtain coliganded structures with soaking and cocrystallization experiments fail to produce crystals. In such cases, the seeding technique can be used (7). Seeding decouples nucleation from crystal growth and is normally carried out using previously obtained crystals. Homogeneous seeding techniques include microseeding, streak seeding, and macroseeding (*see* Fig. 2). In all these techniques, crystal seeds are added to a previously equilibrated, undersaturated drop that contains protein and ligand. Microseeding involves crushing a seed crystal in stabilizing solution, spinning down at 1–2 × *g* and making up serial dilutions (usually 10–1,000-fold) to inoculate equilibrated drops.

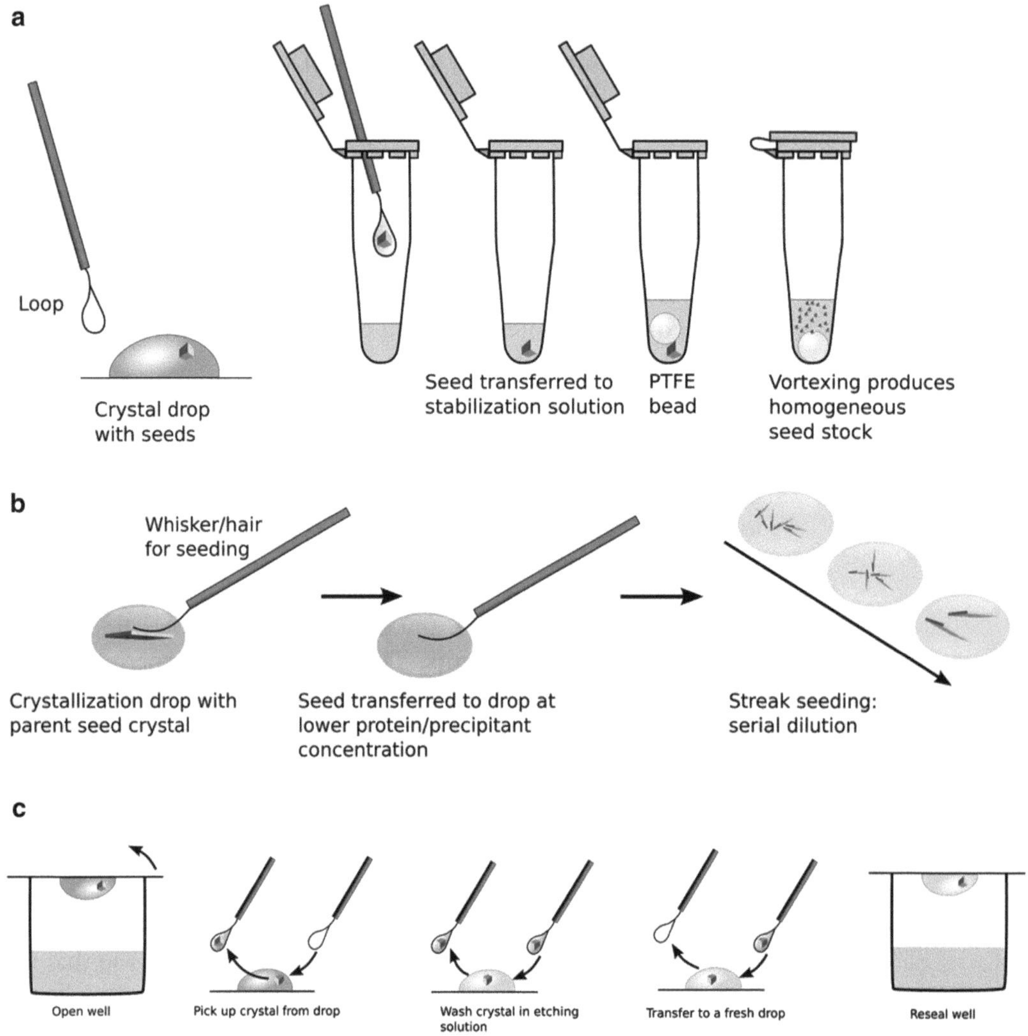

Fig. 2 Schematic representations of alternative seeding protocols. (**a**) Microseeding: a preformed crystal is harvested using a mounted cryoloop and transferred into a 1.5 ml microcentrifuge tube containing 50 μl stabilizing solution (usually reservoir solution) and a PTFE seed bead (Hampton Research). The tube is vortexed for 90 s and a further 450 μl stabilizing solution is subsequently added to produce a homogeneous seed stock. A number of serial dilutions of the seed stock can be made (for example 1:10, 1:100, and 1:1,000) and finally a small quantity of the microseed solution is added to the protein crystallization drop. The number of serial dilutions is variable and needs to be optimized to reduce the number of seeds in microseeding to produce the best quality crystals. (**b**) Streak seeding: a clean cat whisker/hair is used to gently touch a parent seed crystal and dislodge seeds. The seeds remain attached to the whisker and are transferred to a new protein-precipitant drop by running the cat whisker through the drop in one smooth motion to draw a streak line. The loaded cat whisker can be used to seed more than one drop in rapid succession in order to transfer progressively fewer seeds to the subsequent drops. The drops are resealed and subsequently checked for signs of crystals forming along the streak seeding track at a later date. Many crystals take a few to several days to grow, even from streak seeding. If the drops remain empty, the precipitant concentration used is too low (seeds dissolved) or the seed stock is too dilute. If there are too many nuclei, the seed stock is too concentrated. (**c**) Macroseeding: the fundamental difference between micro- and macroseeding is that, in macroseeding, you can see the seed. To be successful with macroseeding, it is necessary to choose a high quality crystal that is free from defects. The crystal is picked up in a cryoloop and washed several times in an etching solution (i.e., undersaturated condition to slightly melt the crystal) to clean the growing surface. Removing the impurity layer on the surface of the crystal allows growth to resume by transferring the seed to a pre-equilibrated crystallization drop

Alternatively, stock for microseeding can be conveniently generated using Hampton Research's Seed Bead kit (*see* Fig. 2a). For streak seeding, a crystallization drop is made up at lower protein or precipitant concentration, a fiber (cat whisker or hair) is touched onto the seed crystal and is drawn through the equilibrated drop (*see* Fig. 2b). Crystals should grow along the streak and additionally self nucleating crystals may appear in parts of the drop further away from the streakline. For macroseeding, the crystal is moved into an etching solution prior to placing it into a fresh drop (*see* Fig. 2c). If no crystal seeds are available, heterogeneous agents such as silica and horse hair can be incorporated in the crystallization experiment to induce growth of protein crystals (8).

3.3 Crystal Looping, Freezing, Mounting and Data Collection

Once crystallization experiments have been set up, each well in the plate needs to be inspected under a cold light source stereo microscope to check for the appearance of crystals over a time course from days to several weeks. When crystals are observed, it is advisable to seek expert help from a member of the laboratory or department with a crystallographic background to aid crystal mounting, freezing and data collection. Suitable crystals can be stabilized and cryoprotected with a suitable agent (examples include glycerol and ethylene glycol), mounted in a cryoloop and flash frozen in liquid nitrogen (*see* **Note 6**). The frozen crystal can be screened either in-house or at a synchrotron source to evaluate its diffraction quality and collect data. The data set can subsequently be integrated (for example, using MOSFLM (9)) and scaled using the CCP4 suite of programs and its graphical user interface (2, 10) to generate an MTZ file containing the experimental reflection data. Considerable effort has gone into making the programs easy to use for novices and the CCP4 suite can be downloaded for free for academic users from the CCP4 Web site (http://www.ccp4.ac.uk/).

3.4 Structure Determination by Molecular Replacement

Molecular replacement represents one of the most powerful phasing techniques for X-ray crystallography and is fundamentally a method of solving unknown crystal structures when a suitable structure of a related protein is available that can be used as a search model.

3.4.1 Phasing and Molecular Replacement

The construction of the electron density map in a crystal unit cell is performed by consideration of the "structure factors" that represent complex numbers. These complex numbers can be described in terms of an amplitude and phase [*see* Fig. 3; refer to ref. 11 and the interactive "Crystallography 101" Web tutorial (*see* **Note 2**) for a detailed description of crystallographic theory]. The measured data constitute the intensities, which are proportional to the square of the structure factor amplitudes and do not directly contain phase information. There are a number of techniques that can be used to obtain the necessary phase information, one of which is "molecular replacement."

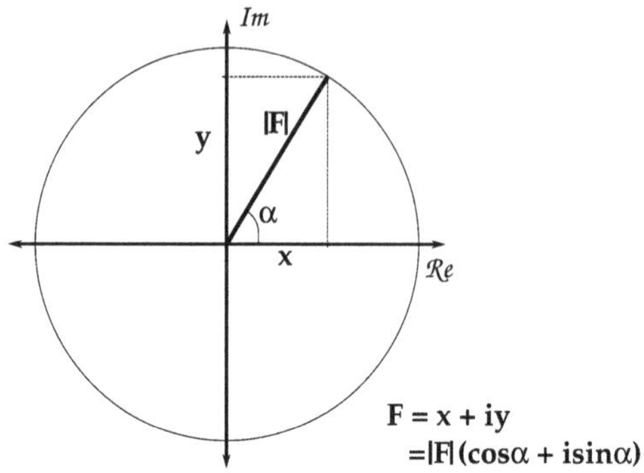

Fig. 3 A structure factor can be defined in polar coordinates by its Argand diagram plot. An Argand diagram represents complex numbers as points in the complex plane using the *x*-axis as the real axis and *y*-axis as the imaginary axis. A structure factor can be defined by its amplitude (|F|) and phase angle (α) from the positive real axis

3.4.2 What Is Molecular Replacement?

Molecular replacement is the positioning of a molecule similar to the target protein in the unit cell. Both the position and orientation need to be determined, corresponding to a six-dimensional search (12). The placement of the similar molecule in the unit cell enables initial estimates of the phases to be calculated (in general, a phase probability distribution).

In cases where ligand complexes are being studied of a protein whose structure has previously been determined and which is available in the PDB, the search model will be represented by the protein in the apoenzyme form and/or in complex with a ligand. In cases where no structural data are available, it is often possible to find an approximate model of the protein under investigation in the PDB. The coordinates for structures deposited in the PDB can be freely downloaded from the database.

3.4.3 Finding a Search Model

There is a high correlation between the level of sequence identity and the degree of structural similarity between related proteins: those with at least 30 % sequence identity represent suitable candidates for use in molecular replacement. The search model can be generated manually, based on an alignment of the primary sequence of the target protein to homologous structures in the PDB using NCBI-BLASTp (http://blast.ncbi.nlm.nih.gov/Blast.cgi?PAGE=Proteins) and the COBALT multiple protein alignment tool (http://www.ncbi.nlm.nih.gov/tools/cobalt/), followed by removal of any segments of low sequence identity by deleting the corresponding residues in the PDB file using a text editor. Alternatively, programs such as CCP4s CHAINSAW (13) can take the input sequence alignment and modify the model by pruning off non-conserved residues. Also the sculptor module of PHASER (16) may be used to generate polished models ready for molecular replacement. Additionally, automated homology based modeling servers such as Swiss-model (http://www.swissmodel.expasy.org/), 3D-JIGSAW (http://bmm.cancerresearchuk.org/3djigsaw/), and I-TASSER (http://zhanglab.ccmb.med.umich.edu/I-TASSER/) can be used to generate three-dimensional models from the primary amino acid sequence of the target protein. These Web-based servers only require the primary sequence of the target protein as input in order to generate an appropriate homology model.

Automated methods for solving protein structures given the target sequence and experimental structure factors are also under development and include BALBES ((14); http://www.ccp4.ac.uk/BALBESSERV) and MrBUMP ((15); http://www.ccp4.ac.uk/MrBUMP/). These programs do not require user intervention when running complicated combinations of jobs such as searching for homologous structures, molecular replacement and refinement.

3.4.4 Programs for Molecular Replacement

The mathematics of the molecular replacement fitting problem means that the six-dimensional search can be broken down into two sequential three-dimensional searches, i.e., a rotation and a translation search which, on their own, are considerably more computationally tractable. Recent advances in molecular replacement include the implementation of maximum likelihood-based algorithms in the program PHASER, leading to dramatic improvements in the success rate of this technique (16).

3.4.5 Molecular Replacement Using PHASER

Although PHASER is available as a stand-alone program, it can be conveniently run from the CCP4i user interface by selecting the "Molecular Replacement" module and opening the "Phaser MR" task window (*see* Fig. 4). In most cases, it is capable of solving structures using the "automated search" mode. The MTZ file containing the processed diffraction data must be input and the structure factor amplitudes (F) and their standard deviations (SIGF) selected. Additionally, PHASER must be given the homology models to use for molecular replacement and their sequence identities to the target protein as a percentage. The models can be selected in the "define ensembles (models)" tab. Ensembles of aligned models can be used to improve the signal by selecting "Add superimposed PDB file to the ensemble." The sequence file for the target in FASTA format, its molecular weight or number of residues and the expected number of molecules in the asymmetric unit should be input in the "Define composition of the asymmetric unit" tab. Finally, in the "Search parameters" tab, the ensemble for performing the molecular replacement can be selected as well as the number of copies to search for in the asymmetric unit. When solving a known protein complexed with a ligand (assuming that the space group and cell parameters are the same), the expected asymmetric unit contents will be the same as the number of monomers in the PDB file. After the job is complete, PHASER outputs a

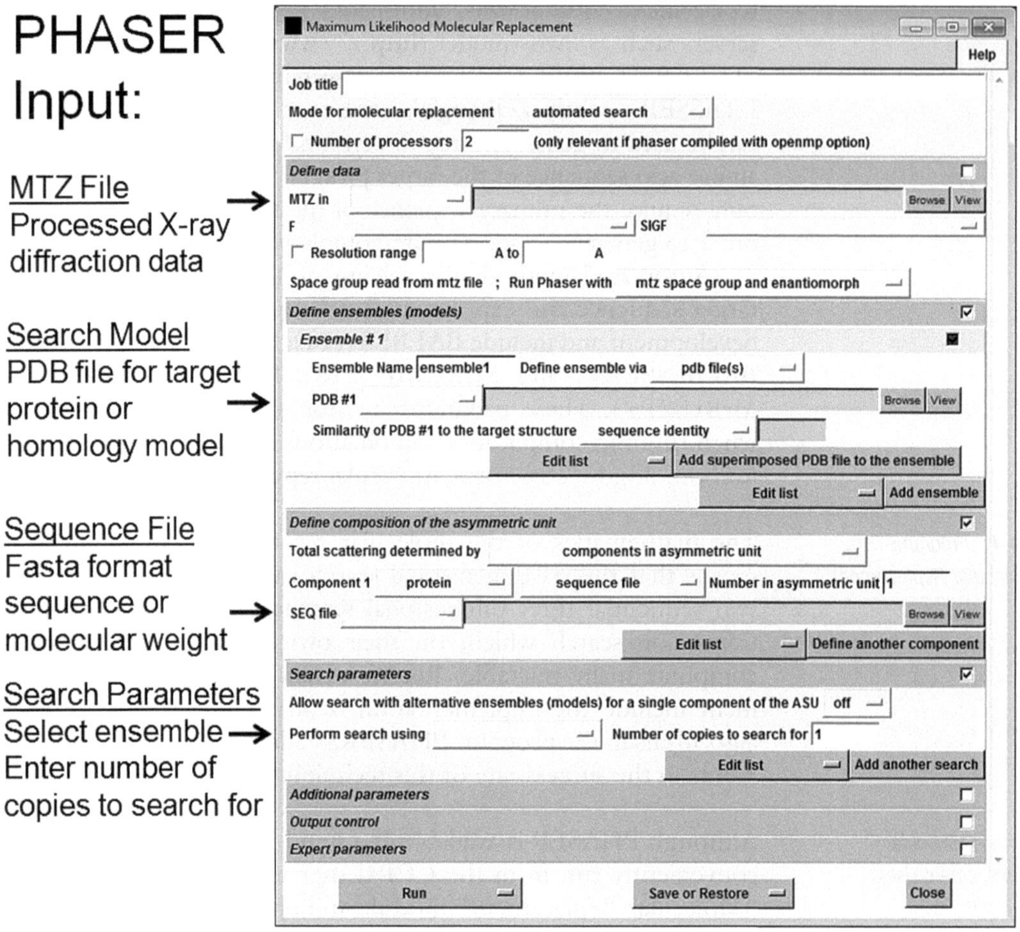

Fig. 4 Running PHASER using the CCP4i graphical user interface. The starting point is an MTZ file containing the protein structure factors and an appropriate search model. The "Ensemble" keyword can be used to specify search models. The packing criterion in the "Additional parameters" tab determines the maximum number of allowable Cα atom clashes in the packing function for solutions (default = 10) and can be increased for low homology models. A solution is feasible when the *Z*-score is higher than seven and distinct from all other solutions and the LLG is positive. When the PHASER run is complete, PDB and MTZ files for the solutions are output, which can be inspected directly in COOT

.sol file containing the molecular replacement solution. The quality of the solution is reflected by the values for the log likelihood gain (LLG) and the Z-score. LLG provides an indication of how much better the solution is compared to a random solution, whereas the Z-score indicates how many standard deviations the solution is above the mean. A definite solution should have a positive LLG and a Z-score greater than seven. Another test for a correct solution is to see whether the molecule packs correctly in the unit cell, in other words to check that no part of the molecule occupies the same space as another molecule or a symmetry-related copy of itself. The packing function throws out any solutions that do overlap. PHASER outputs a PDB file containing the correctly positioned model and an MTZ file containing the original data plus calculated structure factors from the model and columns of map coefficients.

These PDB and phased MTZ files can be opened and inspected directly in COOT to show the models and maps. The maps generated from a molecular replacement solution are practically unbiased and, as such, are useful for visual inspection. What one looks for is a lack of large negative peaks in the difference map (which, if present, may indicate that the protein has undergone a conformational change upon ligand binding) and the appearance of positive density for features not added to the search model (e.g., large side chains, the ligand, or even missing loops in chemically sensible configurations).

3.5 Ligand Fitting

3.5.1 Defining Your Ligand

The definition of the correct geometry for a novel ligand is often the most manual and time-consuming operation in the study of protein–ligand complexes using X-ray crystallography. There are a number of tools that can help with the correct assignment of the geometry but there is no substitute for a detailed understanding of the chemistry of the ligand. In many cases, the description of a ligand is stored as a SMILES string (*S*implified *M*olecular *I*nput *L*ine *E*ntry *S*ystem; Daylight Chemical Information Systems, Inc., CA). SMILES is a linear notation for chemical structures that can be interpreted by the modeling system as the basis of two-dimensional and ultimately three-dimensional molecular construction. Simple examples of SMILES strings include acetic acid "CC(=O)O" and benzene "c1ccccc1." There are a number of programs that can generate the description of a ligand. Currently, one of the most widely used in industry is CORINA (http://www.molecular-networks.com/products/corina; Molecular Networks GmbH, Germany). In addition, PRODRG (a ligand dictionary generating program) is incorporated into the CCP4 suite of programs and is also accessible via the Web server http://davapc1.bioch.dundee.ac.uk/prodrg/ (17). The construction of the two-dimensional description for PRODRG is simplified using a Web-based JME molecular editor front end (courtesy of Peter Ertl, Novartis). PRODRG outputs a low energy conformer PDB file and the REFMAC5 cif dictionary defining the geometrical restraints. Such restraints include bond lengths, bond angles, planes, and chiral centers. Alternatively, the programs JLIGAND or SKETCHER and LIBCHECK, part of the CCP4 suite (2), can be used to generate the coordinates and dictionary files.

3.5.2 Automated Ligand Fitting Using COOT

The program COOT is freely available from http://biop.ox.ac.uk/coot/ and can be used for manual and automated model building, model completion and validation. COOT displays maps and models and allows model manipulations such as idealization, real space refinement, manual rotation/translation, rigid-body fitting, ligand searching, adding water molecules, residue mutations and side-chain rotamer fitting.

The ligand fitting algorithm in COOT provides the ability to either search for a single ligand type or a cocktail of different ligand types and to fit them into unknown features in the electron density map. COOT tries to fit every ligand type to every potential ligand feature in the electron density map, scores each potential ligand type to each feature and reports the best fitting ligand in each site. It also takes into account any residual electron density. For example, if a phosphate ion is fitted into the electron density for an ATP molecule, the fit will be good for the five atoms of the phosphate ion but the score will be down weighted because there will be residual density not accounted for by these atoms.

COOT also builds the ligands into the crystal space which takes into account the crystal symmetry wrapping of the asymmetric unit. This sophisticated understanding of crystal symmetry enhances the site identification and ensures that only one real space representation of map features is built. The CLIPPER tools (18) can be used to rapidly query any point in space by mapping the grid point back into the asymmetric unit. The maps appear to be infinite and the electron density value can be queried anywhere in space and, given those grid coordinates, CLIPPER will quickly map back into the asymmetric unit where it keeps its electron density description. Using this method, the blobs of residual electron density corresponding to potential ligand binding sites can be identified and ranked according to the amount of electron density they encapsulate.

Software that generates structure factor coefficients from atomic models usually generates coefficients for two types of map. One is the direct or atomic map (the best estimate of the true electron density across the unit cell) and the other is a difference map, which represents the differences between the data and the model. Where the difference map is positive, the data suggests adding extra features (atoms) to the model. Where the difference map is negative, the suggestion is that the model has an erroneously positioned atom or atoms near or in the negative peak and that such atoms should be moved or removed from the model. Therefore, when one is hoping that the experimental data represents a protein-ligand complex, the idea is to search the peaks in the difference map hoping to see a positive peak in the ligand binding site. The electron density levels in a difference map generally have a near-Gaussian histogram distribution and so a reasonable level to look for peaks, as a rule of thumb, is above 3.5 r.m.s.d. (where r.m.s.d. is the standard deviation of all of the electron density points in the unit cell).

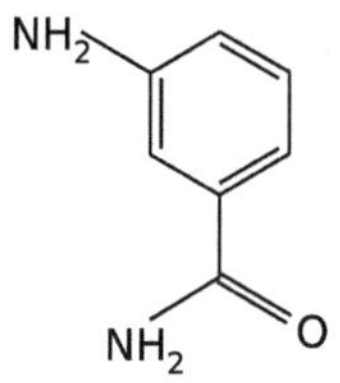

Fig. 5 Chemical structure of 3-aminobenzamide

Extra electron density can be identified by loading the PDB and MTZ files for the protein–ligand complex, running the COOT menu options "Validate → Unmodelled blobs...", selecting the weighted difference density map mFo-DFcalc with amplitudes DELFWT and phases PHDELWT, and finding significant blobs above a 3.5 r.m.s.d. level. After visual inspection of these blobs, the coordinates of the ligand can be automatically fitted using the menu option "Calculate → Other Modelling Tools → Find Ligands..". The user is subsequently presented with a list of acceptable ligands which can be examined manually to identify those best fitting the difference electron density. The automated ligand fitting algorithm in COOT doesn't always fit the ligand into the electron density correctly. More manual tools in COOT to help out in such cases include "nudge ligand," "fit ligand," and "jiggle-fit" which translate or rotate the ligand around its individual components. Additionally, the "Real Space Refine Zone" tool can be used to improve the fit to the electron density whilst maintaining consistency with the known geometry. Real space refinement is the use of the map in addition to geometry terms to improve the positions of the atoms.

If the ligand type already has an existing example in the PDB, its three letter code can be used to retrieve the ligand using the "File→ Get Monomer..." option in COOT. If it is not known whether the monomer already exists or not, the monomer library can be searched using the molecule name and the "File→ Search Monomer Library..." option.

3.6 How to Fit a Ligand Using COOT: A Worked Example

3.6.1 Defining the Search Ligand and Its Geometry

To illustrate ligand fitting, the following example will be presented in which 3-aminobenzamide, (*see* Fig. 5; SMILES string definition "NC(=O)c1cccc(N)c1") is fitted into the electron density using data for the catalytic domain of poly(ADP-ribose)polymerase 2 (PDB code = 3KCZ; (19)). This emulates a case where the ligand type did not already exist in the PDB dictionary. Cocrystals of the protein–ligand complex were grown by the sitting drop method of vapor diffusion at 4 °C in drops comprising 100 nl protein solution (30 mg/ml, corresponding to approximately 700 μM) plus 10 mM 3-aminobenzamide (i.e., 14-fold molar excess of ligand), and 200 nl of reservoir solution (22 % PEG3350 in 0.1 M Tris–HCl, pH 9.0).

Prior to data collection, these crystals were stabilized in reservoir solution supplemented with 25 % glycerol as cryoprotectant and flash frozen in liquid nitrogen.

First, a three-dimensional model of the ligand has to be made and its geometric restraints generated. Two methods of doing this will be described. First, if the SMILES string is to hand or easy to determine then one can use CCP4's LIBCHECK via the COOT interface (File → SMILES) which will result in a new molecule for the ligand being displayed and the restraints generated and read.

Alternatively, if the SMILES string is not known, then PRODRG with its front-end JME editor can be used to draw a standard chemical pseudo-3D (using wedge bond representation) description of the ligand. PRODRG provides a Web page that produces output in a number of different formats. Alternatively, PRODRG can be run within the CCP4 interface. Generally speaking, it is recommended to use a PDB file with all hydrogens and the REFMAC5 dictionary file (which is in standard mmCIF format). The PDB and dictionary files output from PRODRG can be read into COOT. PRODRG users should note however, that PRODRG will produce a model that is fully hydrogenated (with the exception that carboxylates are not protonated), so that the resulting model represents the molecule at extremely low pH. Therefore users may need to delete one of several hydrogens, bonded to amine nitrogens for example, to create a more physiological representation.

3.6.2 Fitting the Ligand

When an MTZ file is auto-opened in COOT, generally two maps are displayed: a map that represents where the atoms are (2mFo-DFc) and a difference map (mFo-DFcalc, representing the differences between the model and the data). It is preferable to use the former map for ligand fitting since this map is usually less noisy than the difference map. As a rule of thumb, the first guess map level should be about 1.0 sigma (and about 3.5 if the difference map is to be used) to identify potential ligand binding sites. This level is used to find clusters of grid points that represent a blob into which the ligand will be fitted. In a noisy map, the density for the ligand may be broken and so a lower sigma cut-off level should be used. If the ligand chemical structure contains a heavy atom, the ligand site can be easily identified by looking for high peaks in the difference map. In this case, and in cases where one has prior knowledge about the probable binding site, the ligand fitting algorithm can be told to search at just one particular site rather than searching the entire map for potential ligand binding sites.

In order to fit the 3-aminobenzamide ligand in COOT, use the menu option "Calculate → Other Modelling Tools → Find Ligands…". On pressing "Find 'Em!" a dialog box opens with a ranking list of the ligand hits in the density map. Clicking on the buttons for these hits navigates the view to the appropriate region in the map. In this example, there are four ligand hits. The first two are the 3-aminobenzamide binding sites in the two unique monomers in the asymmetric unit corresponding to chains "A" and "B." For the two remaining hits, the density is not well modeled by the ligand based on simple visual inspection—the shape of the density does not match the shape of the ligand and not all of the atoms of the 3-aminobenzamide are covered by the density at the 1 r.m.s.d. level. COOT has a default rule that fitted ligands have at least 75 % of their atoms in positive density—this is deliberately generous so that one gets false positives rather than false negatives. These two remaining blobs are in fact glycerol molecules derived from the cryoprotecting solution (*see* **Note 7**).

3.6.3 Tidying Up the Structure

The top two hits for the 3-aminobenzamide should be merged into the parent molecule and the other two hits should be replaced with glycerol and merged (option "Calculate → Merge Molecules…"). When the ligands and glycerol have been merged, then for each of them, the fit to the map can be improved using the real space refinement tool. Select "Regularize" from the "Model/Fit/Refine" menu and click on two atoms to define the zone that will be regularized. At the end of the regularization, the intermediate atoms' positions are displayed in white and a dialog box opens in which the regularization can be accepted or rejected.

3.7 Analyzing the Structure

3.7.1 Structure Validation

To a large extent, the ligand can be treated as just another "residue" in the macromolecule for validation. The geometry of a ligand can be evaluated to determine how well it corresponds to the restraints. COOT provides "Geometry Analysis" validation which represents the ligand as a whole (and can be visually compared with the other residues in the macromolecule). For a more detailed description of the geometric distortions compared to the restraints, REFMAC5 can be used with the keyword options "LIST ALL" with "NCYCLES 0" (so that REFMAC5 merely analyses the geometry and does not do any refinement). The geometry of the ligand compared with its restraints should be at least of average quality compared to the other residues being restrained. It is often the case that the density for the ligand is poorer than that for average residues.

3.7.2 Evaluating the Ligand Interactions

It is becoming increasingly routine to analyze and refine ligand structures using hydrogens. To some extent, this is as a result of the software facilitating their use and partly due to the growing realization that better structures can be achieved by using hydrogens. The most widely used program to analyze inter-residue interactions when hydrogens are part of the model is PROBE (20, 21), part of the Molprobity suite and the various programs used to

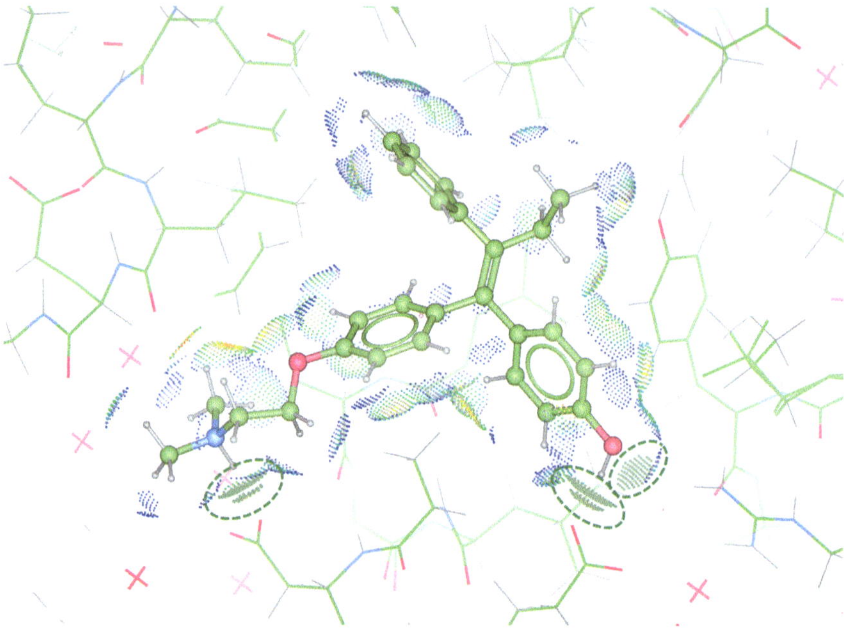

Fig. 6 Crystal structure of estrogen related receptor-gamma ligand binding domain complexed with 4-hydroxy-tamoxifen (PDB code = 2GPU) highlighting the atomic van der Waals (vdW) surface (represented as *spheres of small dots*). Figure prepared using the "Validate> Probe clashes" option in COOT. This validation tool provides a visual representation of the all-atom contacts by representing with colored dots and short lines the interacting surfaces of the molecule (in this case, we select only the interacting surfaces between the protein and the ligand). This representation was generated using REDUCE to add hydrogens to the ligand in the context of the binding site and PROBE to generated the dots of the interacting surfaces. The different interaction types and distances have different color designations. Here we outline, with a *broken-line ellipse*, the hydrogen bond interactions between the ligand and the protein

visualize its output (for example: KING, O and COOT). If the model does not include hydrogens and PROBE is to be used for analysis, one would generally use REDUCE (22) to add them. So that the Molprobity tools know the connectivity of the hydrogen atoms on the ligand, a dictionary can be provided to REDUCE using the "-DB" keyword (if one has read the dictionary for the ligand into COOT, it will automatically handle the dictionary for Molprobity tools). Thus the interactions of the ligand (including hydrogens) and the binding site can be analyzed in detail (*see* Fig. 6).

Another tool in Coot that is useful for interpreting protein–ligand interactions is "Highlight Interesting Site" (Extensions → Representations → Highlight Interesting Site (here)...) - this represents the ligand in purple ball and stick and the residues of the ligand environment in atom colors ball and stick. Additionally, the environment distances are displayed so that one can see the charged and uncharged interactions of the ligand and the protein (*see* Fig. 7). Furthermore, to show the electrostatic surface of the

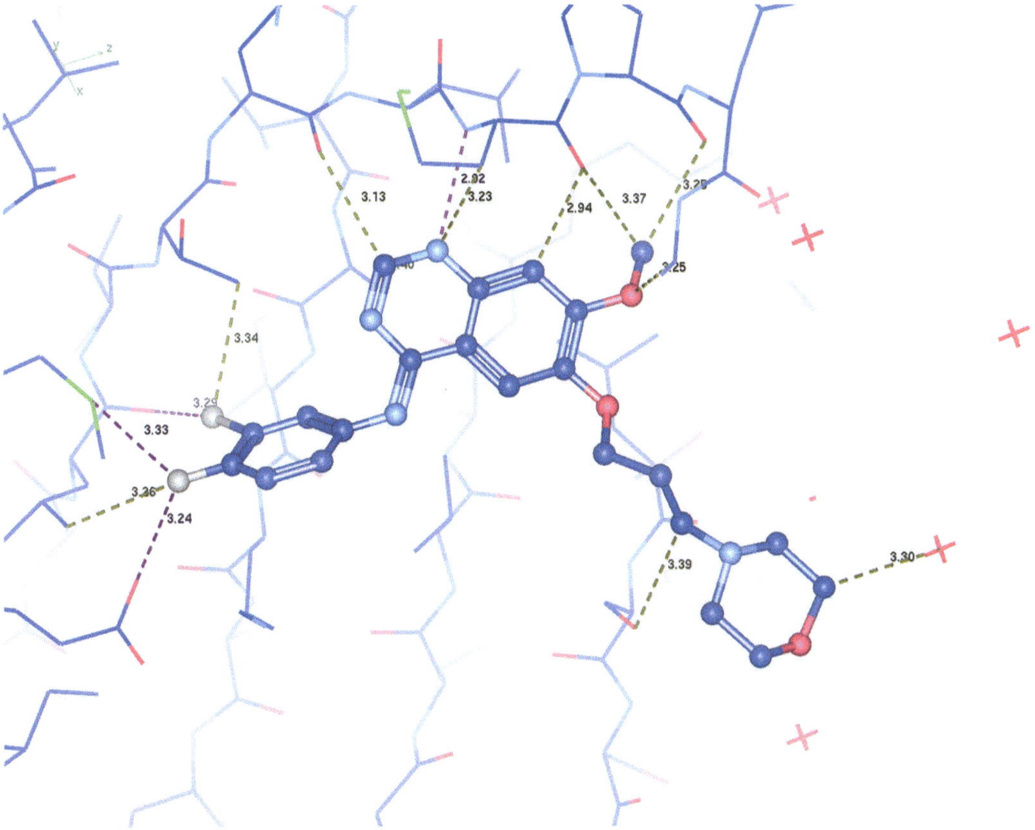

Fig. 7 Representation of EGFR kinase in complex with gefitinib (PDB code = 2ITO). Figure prepared using the "Extensions> Representations> Highlight Interesting Site (here)..." option in COOT. Protein–ligand hydrogen bonds are represented by *dashed lines* and their associated bond distances. Bumps between non-hydrogen bonding groups are also shown

binding site around the ligand select "Extensions → Representations → Clipped Surface Here (around this residue)..." (*see* Fig. 8).

4 Notes

1. There are several coarse screens commercially available from vendors including Hampton Research, Jena Biosciences GmbH, and Molecular Dimensions Ltd. A good starting point is to use a combination of the PACT and JCSG+ screens, each comprising 96 conditions, which provide a useful minimal screening strategy for small academic laboratories (23).
2. Crystallographic programs and resources:

   3D-JIGSAW. A Web-based server that builds three-dimensional models for proteins based on homologues of known structure (http://bmm.cancerresearchuk.org/3djigsaw/).

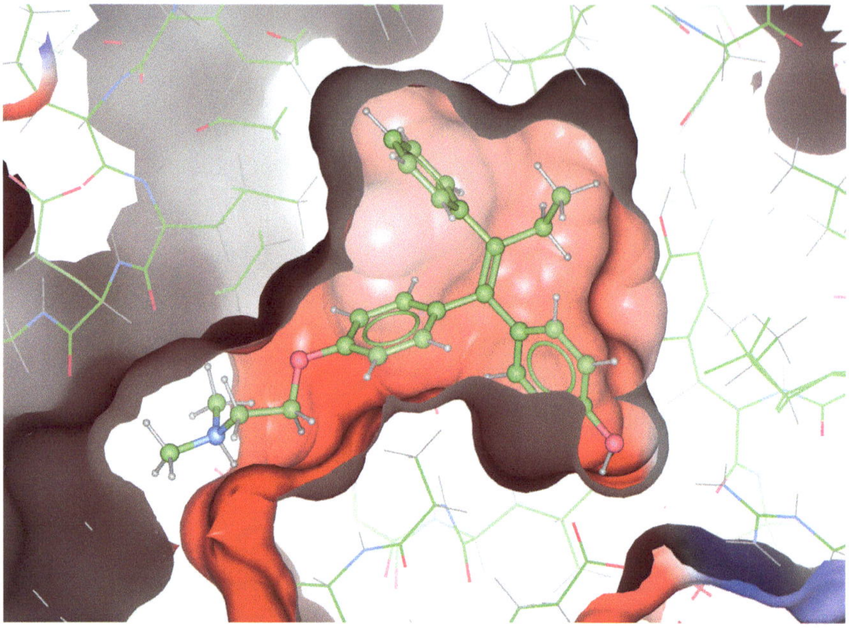

Fig. 8 Clipped surface representation of the 4-hydroxy-tamoxifen binding site in the estrogen related receptor-gamma ligand binding domain (PDB code = 2GPU). Figure prepared using the "Extensions> Representations> Clipped Surface Here (around this residue)..." option in COOT. The electrostatic potential is mapped to the protein surface and represented in color gradation from *blue* to *red* (this information is lost in this black and white image). Such a representation is useful to analyze and optimize charged interactions between the ligand and the protein

BALBES. An automated molecular replacement pipeline for solving protein structures using X-ray crystallographic data (http://www.ccp4.ac.uk/BALBESSERV).

CCP4 SUITE. A collection of programs covering most of the computations required for macromolecular crystallography, which can be downloaded from http://www.ccp4.ac.uk/.

CHAINSAW. A molecular replacement utility which takes an alignment between target and model sequences and modifies the model PDB file by pruning off non-conserved residues. CHAINSAW is integrated in the CCP4 suite of programs for X-ray crystallography (http://www.ccp4.ac.uk/html/chainsaw.html).

COBALT. A multiple protein sequence alignment tool (http://www.ncbi.nlm.nih.gov/tools/cobalt/).

COOT. A molecular graphics application for model building, model completion and validation, particularly using X-ray data. The Linux and Windows executables are freely downloadable from http://biop.ox.ac.uk/coot/.

CORINA. A program widely used in the pharmaceutical sector to convert the 2D structures of drug-like molecules into 3D structures (http://www.molecular-networks.com/products/corina).

CRYSTALLOGRAPHY 101. An Interactive Web Tutorial by Bernhard Rupp covering protein crystallography theory and techniques (http://www.ruppweb.org/Xray/101index.html).

I-TASSER. A Web-based server for predicting a protein's structure and function from its primary amino acid sequence (http://zhanglab.ccmb.med.umich.edu/I-TASSER/).

JLIGAND. An interface to create the description of ligands and covalent links that can be used in refinement. The program can be downloaded from http://www.ysbl.york.ac.uk/mxstat/JLigand/.

MOLPROBITY. A Web application for evaluating the accuracy of a macromolecular structure (http://molprobity.biochem.duke.edu/).

MOSFLM. A program for integrating crystal diffraction data. The program is free to both academics and industry (http://www.mrc-lmb.cam.ac.uk/harry/frames/).

MrBUMP. An automated system for molecular replacement (http://www.ccp4.ac.uk/MrBUMP/).

BLAST. A Web application for querying a sequence against the Protein DataBase to identify homologous structures for use as search models in molecular replacement (http://blast.ncbi.nlm.nih.gov/).

PHASER. A program for phasing macromolecular crystal structures with maximum likelihood methods, available through the Phenix and CCP4 software suites (http://www.phaser.cimr.cam.ac.uk/index.php/Phaser_Crystallographic_Software).

PRODRG. A ligand dictionary generating program. PRODRG is accessible via the Web server http://davapc1.bioch.dundee.ac.uk/prodrg/ and is also available as a standalone version.

PROTEIN DATABASE (*PDB*). Archive of experimentally determined structures (http://www.rcsb.org/pdb/home/home.do).

SWISS-MODEL. A fully automated protein structure homology-modeling server (http://www.swissmodel.expasy.org/).

3. To cross-link crystals obtained from a hanging drop plate, place a micro-bridge containing 2–5 μl of 25 % glutaraldehyde into the well and reseal the drop. Allow the glutaraldehyde to diffuse

into the sample drop containing the crystal for 30 min–6 h. The cross-linking time is primarily influenced by the number of lysine residues on the surface of the protein and temperature. It is important to avoid buffers containing ammonium sulfate or Tris (tris(hydroxymethyl)aminomethane) since amines interfere with cross-linking.

4. K_d is a measure of the affinity between the ligand and protein and is defined as $K_d = [P][L]/[PL]$ where [P], [L], and [PL] represent molar concentrations of protein, ligand, and the protein–ligand complex respectively. The K_d represents the concentration of ligand at which the ligand binding site is half occupied, i.e., [PL] = [P]. The fractional saturation of the protein is influenced by the ratio of the concentration of free ligand ([L]) to the K_d.
5. In cases where a ligand is extremely insoluble, it can be added to dilute protein at 4–10 times molar ratio (corresponding to the ligands maximum solubility level) and incubated at 4 °C overnight to ensure complete binding of the ligand to the protein. Subsequently, the protein–ligand sample can be coconcentrated before setting up the crystallization experiments. This will increase the likelihood of the ligand binding to free protein and also helps avoid DMSO shock.
6. Most data are collected at cryo-temperatures because mounting crystals in loops is easier than mounting in capillaries and the cryo-temperatures reduce radiation damage.
7. When glycerol is used as a cryoprotectant to inhibit ice formation on flash freezing the crystal, it is often found in the resulting electron density map despite only exposing the crystal to the cryoprotecting solution for a number of seconds. Occasionally, other components of the crystallization solution (including PEGs, salts, and small molecules) serendipitously cocrystallize with the macromolecule. Furthermore, ligands can be quite often carried by the protein through protein purification and can be seen in the resultant electron density map, despite not having been added in the crystallization solution [e.g., $NADP^+$ carried over in the structure of *Neisseria meningitidis* pyrroline-5-carboxylate reductase (PDB code = 2AG8) and 2′-deoxy-GTP carried over in the crystal structure of *Helicobacter pylori* protein HP0184 (PDB code = 2ATZ)].

References

1. Berman HM et al (2002) The protein data bank. Acta Crystallogr D Biol Crystallogr D58:899–907
2. Collaborative Computational Project, Number 4 (1994) The CCP4 suite: programs for protein crystallography. Acta Crystallogr D Biol Crystallogr D50:760–763
3. Emsley P, Cowtan K (2004) Coot: model-building tools for molecular graphics. Acta Crystallogr D Biol Crystallogr D60:2126–2132

4. Emsley P et al (2010) Features and development of *Coot*. Acta Crystallogr D Biol Crystallogr D66:486–501
5. Lusty CJ (1999) A gentle vapor-diffusion technique for cross-linking of protein crystals for cryocrystallography. J Appl Crystallogr 32:106–112
6. Danley DE (2006) Crystallization to obtain protein–ligand complexes for structure-aided drug design. Acta Crystallogr D Biol Crystallogr D62:569–575
7. Bergfors T (2007) Succeeding with seeding: some practical advice. In: Read R, Sussman J (eds) Evolving methods for macromolecular crystallography. Springer, Dordrecht, pp 1–10
8. D'Arcy A, Mac Sweeney A, Haber A (2003) Using natural seeding material to generate nucleation in protein crystallization experiments. Acta Crystallogr D Biol Crystallogr D59:1343–1346
9. Leslie AGW (1992) Recent changes to the MOSFLM package for processing film and image plate data. Jnt CCP4/ESF-EACMB Newslett Protein Crystallogr 26.
10. Potterton E et al (2003) A graphical user interface to the CCP4 program suite. Acta Crystallogr D Biol Crystallogr D59:1131–1137
11. Rupp B (2009) Biomolecular crystallography: principles, practice, and application to structural biology. Garland Science, New York. ISBN 978-0-8153-4081-2
12. Evans P, McCoy A (2008) An introduction to molecular replacement. Acta Crystallogr D Biol Crystallogr D64:1–10
13. Stein N (2008) CHAINSAW: a program for mutating pdb files used as templates in molecular replacement. J Appl Crystallogr 41:641–643
14. Long F et al (2008) BALBES: a molecular replacement pipeline. Acta Crystallogr D Biol Crystallogr D64:125–132
15. Keegan RM, Winn MD (2007) Automated search-model discovery and preparation for structure solution by molecular replacement. Acta Crystallogr D Biol Crystallogr D63:447–457
16. McCoy AJ et al (2007) Phaser crystallographic software. J Appl Crystallogr 40:658–674
17. Schuettelkopf AW, van Aalten DMF (2004) PRODRG: a tool for high-throughput crystallography of protein-ligand complexes. Acta Crystallogr D Biol Crystallogr D60:1355–1363
18. Cowtan K (2003) The clipper C++ libraries for x-ray crystallography. IUCr Computing Commission Newsletter 2:4–9
19. Karlberg T et al (2010) Crystal structure of the catalytic domain of human PARP2 in complex with PARP inhibitor ABT-888. Biochem 49:1056–1058
20. Chen VB et al (2010) MolProbity: all-atom structure validation for macromolecular crystallography. Acta Crystallogr D Biol Crystallogr D66:12–21
21. Word JM et al (1999) Visualizing and quantifying molecular goodness-of-fit: small-probe contact dots with explicit hydrogen atoms. J Mol Biol 285:1711–1733
22. Word JM et al (1999) Asparagine and glutamine: using hydrogen atom contacts in the choice of side-chain amide orientation. J Mol Biol 285:1735–1747
23. Newman J et al (2005) Towards rationalization of crystallization screening for small- to medium-sized academic laboratories: the PACT/JCSG+ strategy. Acta Crystallogr D Biol Crystallogr D61:1426–1431

Chapter 18

Molecular Fields in Ligand Discovery

Paul J. Gane and A.W. Edith Chan

Abstract

The discovery of novel biologically active small molecules is now a technologically and economically viable proposition for academic and small biotechnology laboratories wishing to build on their biological research into target proteins. Such small molecules may be useful reagents for further biological research or may form the basis for early-stage drug discovery. The availability of specialized virtual screening software to filter large molecular libraries into manageable numbers of compounds for biological assays is the driving force for finding novel ligands. The main focus of this chapter is the basis and use of molecular field methods to assess the interactions that may be made by small molecules. Molecular field based measures of capability and similarity of interaction may be used to discover novel ligands and expand ligand series for potential use as future therapies.

Key words Intermolecular interactions, Drug discovery, Molecular fields, Virtual screening, Bioisosteres, Scaffold hopping

1 Introduction

> I would like our field to be effective, one that contributes as much as possible to the most important industry on earth—the discovery of these amazing small molecules with their potential for dramatic effects on health and well-being. Anthony Nicholls (1).

The search for new "amazing molecules" is the science and art of drug discovery. Why art? Because at each stage within the drug development pipeline decisions are made by individuals with a personal bias based upon years of experience; this subjectivity, call it art, is not easy to quantify even less, codify. However, the greater our knowledge of science the better these decisions become and in modern drug discovery a good understanding of the molecular interactions which occur between a drug and its target molecule can prove crucial for success.

Mark A. Williams and Tina Daviter (eds.), *Protein-Ligand Interactions: Methods and Applications*, Methods in Molecular Biology, vol. 1008, DOI 10.1007/978-1-62703-398-5_18,

In most cases these intermolecular interactions are non-covalent and weak, much weaker than the bonds between their constituent atoms. It is this very weakness which provides a low energy barrier for fast dynamic reactions to occur allowing life as we know it to exist. Our understanding of these interactions is still incomplete and our algorithms are based on approximations and assumptions but despite this, computational chemistry or molecular modelling in general has become the leading tool in drug discovery.

There are many excellent textbooks and articles (2–5) describing the vast range of computational methods available including chapters within this volume; in an attempt to avoid repetition, only a brief outline of intermolecular interactions will be given followed by a more detailed description of one particular approach which strives to visualize intermolecular interactions from a ligand molecule's point of view.

1.1 Intermolecular Interactions

Let us step back for a moment and try to recall exactly what those intermolecular interactions are. The potential energy curve, often called a Lennard-Jones potential (Fig. 1), occurs in many introductory texts, and shows the effect of a long range attractive force bringing the molecules together until they reach a minimum energy at a separation r_m (the force being zero at this point), and that as the molecules are pushed closer together a repulsive force rapidly rises to oppose the attraction (N.B. that the energy is still negative until the collision diameter (σ) is reached). The total intermolecular energy is a combination of the attraction and repulsion curves shown as dot-dash lines (Fig. 1). Note that here the molecules themselves are assumed to be spherical so there are no orientation parameters to consider but in real, flexible molecules this simple energy curve becomes a complex potential energy surface.

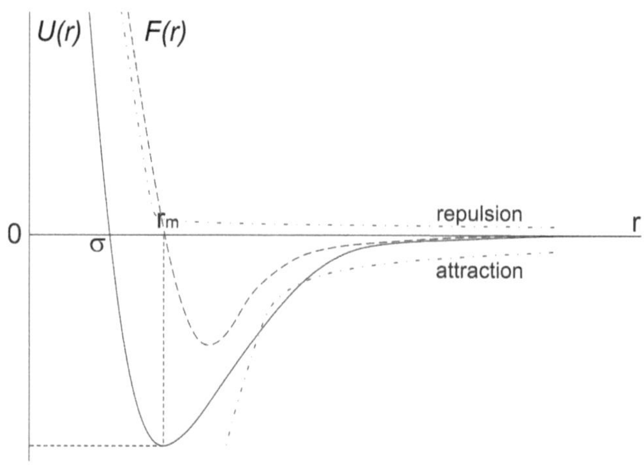

Fig. 1 The Lennard-Jones potential. Potential energy, *U*(*r*), and force, *F*(*r*), versus inter-atomic/molecular distance (*r*) for neutral particles

The explanation of the multiple interactions which contribute to the complete intermolecular potential energy surface was not possible until the development of quantum mechanics. The attractive forces are a result of both classical electrostatics and quantum effects, whereas the very short-range repulsive force is largely due to the Pauli exclusion principle which disallows full electron shells from overlapping. The attractive forces can be divided into a number of components:

- Charge–charge (Coulombic) interactions, if both molecules carry a net charge.
- Dipole–dipole (or multipole) interactions (Keesom forces), if the molecules are polarized, i.e., they have a time-averaged asymmetric distribution of charge.
- A charge or dipole (or multipole) in one molecule can distort the electron cloud of a neighboring neutral molecule causing it to become temporarily polarized, the two molecules are then attracted to interact with the each other (Debye force). This is known as induction.
- Even if the molecules do not carry a charge and are not polarized there is a still a substantial, but short-range, interaction between all molecules, this is a result of quantum fluctuations in the electron orbitals which can induce correlated fluctuations in neighboring molecules causing them attract via so called dispersion energy (London forces).

Coulombic interactions, i.e., those between charges (monopoles), weaken only slowly over distance (r) as a function of $1/r$, so they are very long range. Dipole–dipole is reduced more rapidly by a factor of $1/r^3$. So polar molecules first interact by virtue of the electrostatic forces which rapidly get stronger as the molecules approach each other, as the distance reduces to around 6 Å the short-range dispersion forces ($1/r^6$) begin to add significantly to the overall binding energy but if they get closer still within 2 Å or so the large repulsion ($1/r^{12}$) counteracts the attraction leading to the equilibrium distance or van der Waals radius, identified as r_m on the Lennard-Jones curve. Another term often used to describe the collective effect of the non-Coulomb terms is the "van der Waals interactions," which represents an aggregate of Keesom, Debye, and London forces.

The size of the contributions of each component to the total force or potential depends upon the chemical nature of the interacting molecules. For example, water molecules interact mainly by hydrogen bonds which can be considered as largely electrostatic/dipole–dipole interactions; in fact each water molecule usually participates in 3–4 hydrogen bonds, and this dominates their intermolecular interactions and explains many physicochemical properties of water. On the other hand, dispersion energy is the major contributor to HCl interactions, even though it is also a polar molecule (6). The reason is twofold. First, chlorine is more polarizable than oxygen and as polarizability increases dispersion becomes stronger. Second, the larger atomic volume of chlorine reduces the ability of its lone pairs to form strong hydrogen bonds.

In addition to electrostatic, induction, dispersion and repulsion terms some non-covalent interactions (e.g., hydrogen bonds or halogen bonds) may have a significant quantum mechanical contribution due to partial sharing of electrons between atoms (7). However, such quantum mechanical effects are not usually explicitly modelled, but their effect is approximately included via some modification of the parameters describing the interactions of the groups involved.

We consider some of the more specific interactions which exist between a small molecule (drug) and its protein target in the next section. The reader is directed to the outstanding review by Bissantz (8) which describes these interactions from the perspective of a medicinal chemist. Chan et al. (9) have analyzed the Protein Data Bank (PDB) (10) to identify the most likely interacting small molecular fragments for a given amino acid, which is both instructive and provides a basis for assessing the relative importance of these interactions.

1.2 Hydrogen Bonds, Salt Bridges, and Weak Hydrogen Bonds

A "classical" hydrogen bond (H-bond) D–H···:A is formed between the donor D, an electronegative atom, which induces a partially positively charge on its covalently bonded H, and acceptor A, an electronegative atom with a lone pair, giving $D^{\delta-}$–$H^{\delta+}$···: $A^{\delta-}$, an electrostatically attractive situation. The H-bond donor and acceptor in biological macromolecules is in most cases N or O, but S–H and even C–H are seen to act as donors and make low energy or weak hydrogen bonds (11).

Accounting for H-bonds in protein–ligand interactions is not as simple as one might think, because they may encompass a wide variety of chemical groups interacting via a hydrogen atom. In fact, H-bonds form a continuum of interactions from those that are barely distinguishable from dispersion alone, through to polar and electrostatic charge interactions and on to almost complete formation of a covalent bond. A number of classification schemes have emerged (12). Perhaps the simplest is to divide them into three classes according to the strength of the bond: weak (<4 kcal/mol), moderate (4–15 kcal/mol), and strong (15–45 kcal/mol) according to the nature of donor and acceptor groups (13). (N.B. strong H-bonds are very rare in protein–ligand interactions). However, Gilli et al. (12) emphasize that for any given D–H···:A system the energy may vary substantially for a given environment. For example, in water the dimer HO–H···OH_2 is typically less than 5 kcal/mol, but in acidic or basic environments, we have protonated [H_2O···H···OH_2]+ or deprotonated [H–O···H···O–H]– forms, with energies reaching 32 kcal/mol. They conclude that

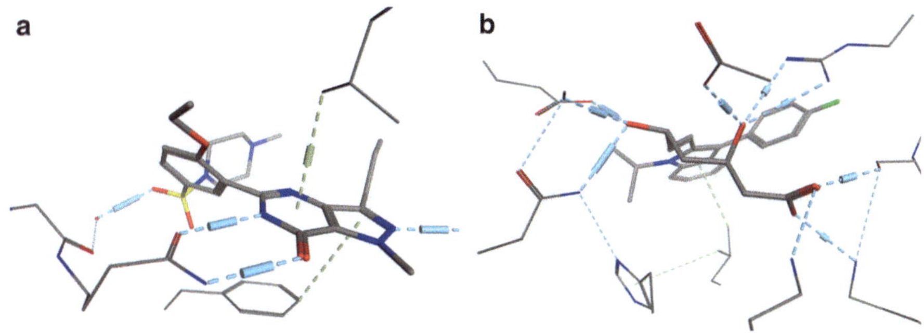

Fig. 2 Hydrogen bonding in protein–ligand interactions. (**a**) Weak and moderate strength H-bonds between sildenafil (Viagra™) and its protein target phosphodiesterase 5A (PDB entry 1TBF). In addition to H-bonds to polar amino-acid side chains, weak bonds are identified to the aromatic heterocycle of sildenafil. (**b**) Many moderate and strong H-bonds involving charged protein groups (lysine, arginine) mediate the interactions between the cholesterol lowering drug fluvastatin and its target HMG-CoA reductase (PDB entry 1HWI)

although the electronegativity of the donor and acceptor atoms guides H-bond strength, it is controlled in many situations by their acid–base dissociation constants.

Changes in hydrogen bonding number and strength are almost ubiquitous in the formation of protein–ligand complexes. It would seem, therefore, that H-bonds are likely to be important to ligand binding; however, the energetic contribution they give to affinity may be much less than is often assumed because H-bonds made with a ligand often replace those previously made with water (14). H-bonds are, however, extremely important for directionality and recognition in protein–ligand interactions (15). Also, the dipolar character of H-bonds may mean that H-bond formation preferentially reduces off-rates of ligands. H-bonding ligands displaced 1 or 2 Å by thermal fluctuations may be restrained by the $1/r^3$ potential and come back into contact rather than escaping into the bulk solvent (Fig. 2).

1.3 Halogen Bonds

The synthetic chemist has long added halogens to compounds with the intent of improving ADME properties and they were also thought to provide hydrophobic bulk. For some time it has been observed that direct interactions (X-bonds) exist between organic halides (X-bond donors) and protein electronegative atoms such as O, S, or N (X-bond acceptors) forming, for example, a C–X···O bond. At first sight this is counterintuitive given the high electronegativity of the halogens, however, for Cl, Br, and I the electrons in the C–X bond are asymmetrically distributed such that the electrons surround the axis of the bond, but are withdrawn from its terminus, producing a partial positive charge known as a sigma hole at the halogen atom. This effect becomes larger with increasing size of the halogen. The partial positive end that can interact

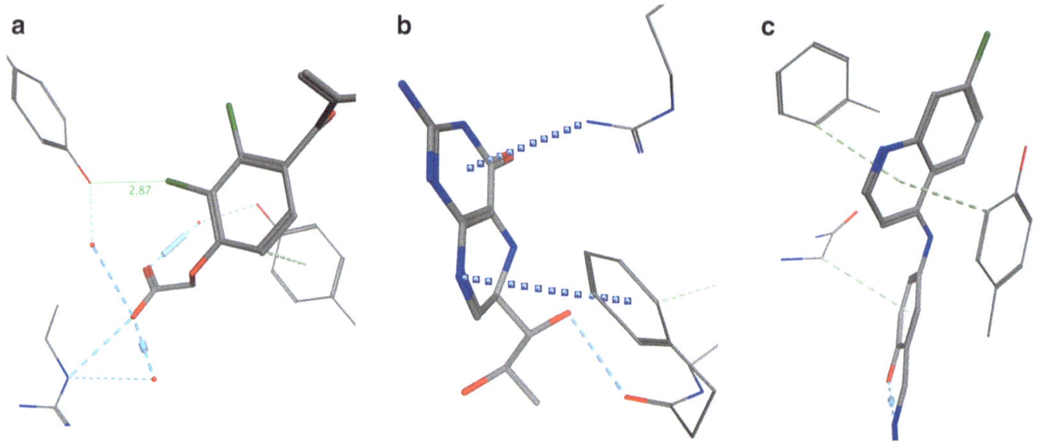

Fig. 3 Other electrostatic interactions in protein–ligand interactions. (**a**) A halogen bond between ethacrynic acid and a tyrosine of human glutathione transferase P1-1 identified by the short distance (2.87 Å) between the tyrosine OH and chlorine substituent of the phenolic ring (PDB entry 3N9J). (**b**) Cation–π interactions (*small spheres*) between biopterin and nitric oxide synthase (PDB entry 1DMI). The positive guanidinium-group of arginine and a positive nitrogen of the pteridine ring are attracted to delocalized π electrons in aromatic rings (**c**) Aromatic–aromatic interactions create offset stacking in the amodiaquine and histamine methyltransferase complex (PDB entry 2AOU)

with a lone pair or aromatic π-cloud (showing parallels with H-bonds) results in close approach of halide and electronegative atoms (*see* Fig. 3a). Fluorine's small size and extremely high electronegativity prevents it from forming halogen bonds. Parisini et al. (16) highlight examples of X-ray crystal structures of small halogenated molecules including drugs bound to proteins with these directional halogen bonds. Calculating the electrostatic potential surrounding the halogen containing compounds requires quantum mechanical methods as shown by Riley and Hobza (17), who also describe further examples from the PDB.

1.4 Aromatic Interactions

Aromatic systems, such as benzene, are defined by delocalized π electrons above and below the ring of bonded atoms, sandwiching partial positively charged σ bonds around the circumference of the ring. This simple model is sufficient to explain the interactions between aromatic rings (as a consequence of π–π interactions) and of other functional groups interacting with aromatic π rings (18).

A benzene dimer is observed crystallographically and is largely in two forms, slipped-parallel and T-shaped. Direct overlay of one ring onto another is unstable, due to the electrostatic repulsion of the π electrons in the two molecules. The major component of attraction appears to be dispersion forces with electrostatics controlling the molecules' orientation. These dispersion and electrostatic terms probably control other aromatic interactions. A good review is given by Tsuzuki (19) who describes not only π–π interactions, but also OH–π, NH–π, CH–π, and cation–π interactions, which involve electrostatic attraction to the π electrons and are also of interest in protein–ligand and drug design applications (20). Functional substituents to the aromatic group can have electron-withdrawing or electron-donating abilities, which modulate the π interactions. An interesting, if unusual, example is hexafluorobenzene which has an inverted electron distribution, due to the highly electronegative fluorine atoms, will parallel stack with benzene with a high energy of interaction of 8 kcal/mol. Aromatic heterocycles, which are found in the majority of current small molecule drugs, have distinct electronic properties that may require special consideration (19).

With the exception of the cation–π interaction, aromatic interactions are energetically only one half to one quarter the value of a weak H-bond, but they are nonetheless important in orientating and directing of ligands within a binding site (*see* Fig. 3b, c) (21).

1.5 Hydrophobic Interactions and the Energetics of Binding

Water, the "lubricant of life" (22) is not merely a passive solvent in which physicochemical interactions occur, but substantially contributes to the balance of forces/energies which compete during ligand binding. Water molecules which surround the ligand and occupy the binding site are displaced to allow direct association between the protein and the ligand. Changes in the interactions of these displaced water molecules, consequently contribute to binding energetics. Water molecules which interact with solutes are considered to be more ordered than bulk water. Hence, when they are displaced they are thought to provide a large positive entropic component which favors binding. In addition, the nonpolar surfaces of ligand and protein interact via dispersion forces which although individually weak are additive and sum over substantial surface areas providing a large enthalpic component to binding. The ligand, when bound, is in a fixed conformation along with the interacting residues of the protein—this loss of mobility lowers the entropy and directly opposes binding. The more tightly bound the ligand, the greater the enthalpy of binding and the more restricted its motion and the higher the entropic penalty. This trade-off between the two components of the free energy of binding is known as entropy–enthalpy compensation.

Detailed features of the interactions in a particular protein–ligand complex are significant in determining the net contribution of water to binding. Other factors to consider are the size of the ligand and its shape in relation to the geometry of the binding cavity and the surface it presents to interfacial water (water molecules trapped between ligand and protein) (23). Myslinski et al. (24) conclude that displacing bound water does not necessarily result in favorable entropy to binding if particular bound water molecules are less ordered than is generally thought.

It is also important to remember that in solution both water and ions dramatically reduce any long-range Coulombic interactions, and that conversely the environment found within certain binding sites may enhance the electrostatics, altering pK_as and chemical reactivity.

1.6 Conclusion

Although, we have more or less sophisticated ideas of the types of interactions that are significant in the association of proteins and small molecules, the prediction of the free energy of binding of small-molecule ligands to a protein is still an unsolved problem, primarily because of the non-additivity and multiplicity of the interactions involved (25). Particular difficulties are found with calculating solvation–desolvation of both ligand and protein and the role of counterions; flexibility and conformational changes; long-range electrostatics; changes in ionization/pK_a of ligand functional groups and binding site residues and the kinetics associated with some H-bonds (26). Given these difficulties and limitations, it is remarkable how successful many of the computational tools considered in the following sections are in aiding ligand and drug discovery.

2 In Silico Methods for Ligand and Early-Stage Drug Discovery

The route from first identifying a protein target, to a useful biological reagent may be long, and that to a marketed drug is certainly long and tortuous with many feedback cycles. Since in both ligand and drug discovery the initial goal is to identify a small molecule with high affinity and specificity for a protein binding-site many of the early stages of the process (Fig. 4) are similar.

2.1 Finding Hits with Virtual Screening

High throughput screening (HTS), the testing of tens of thousands to millions of compounds for activity against a chosen target, is the most common strategy in the pharmaceutical industry for the initial identification of small molecules that are active against a protein target. However, HTS is immensely expensive in terms of time, labor, and resources and is not possible in an academic environment or small companies, and not justifiable outside of drug development programs. To make matters worse the success rate of HTS is often poor, so why make the haystack larger when looking for a needle?

A more "rational" approach is to assay a small subset of compounds which have been preselected to have a much higher likelihood of binding to the chosen target. This is most often achieved with computational filtering or "virtual screening" of a large database of available chemical compounds. There are many virtual screening methods and many choices of program for each given method.

Docking, fragment building and de novo ligand generation are routine methods for using structure to screening large compound libraries or to build novel ligands into binding sites (27).

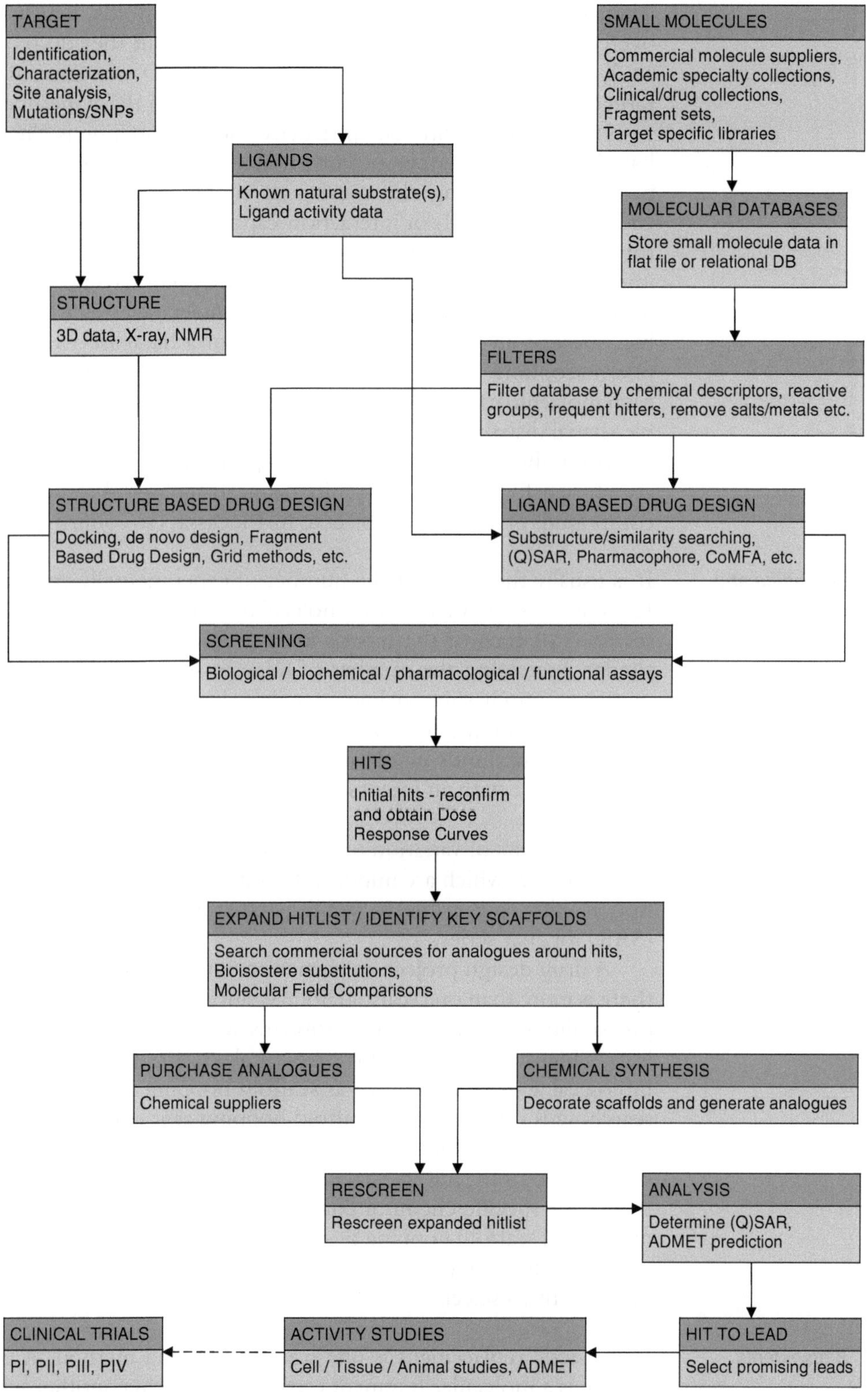

Fig. 4 Flow diagram of the drug discovery process. A similar process applies to the discovery of any potent ligand but note that many feedback cycles occur throughout the process and the later stages greatly simplified

Their great advantage is in discovering new scaffolds, which may overcome possible patent and synthetic issues, but they do require an X-ray, NMR, or homology model of the 3D structure of the target protein (*see* Chapter 19 for more detail on these methods).

In contrast, traditional "pharmacophore" searching is "ligand-based" where a number of known active compounds are superposed by optimizing the overlap of simple descriptors of their chemical groups and/or interaction-capable sites. Regions of structure or functionality which are in common are regarded to be important for activity and a pharmacophore point having particular properties is placed there. Because many molecules are rather flexible, successfully matching molecules from a database to a set of pharmacophore points requires that sufficient conformations of each molecule have been generated to adequately sample the possible spatial distributions of functionality.

A number of the top scoring compounds from a virtual screen are chosen for a preliminary biochemical/biological assay. Active compounds or "hits" are subject to further assays to confirm activity.

2.2 Expanding and Diversifying Hits

It is usually the case that the number of experimentally confirmed HTS hits is small or the core molecular unit (better known as the scaffold) of each of the hits or known binders is chemically very similar. Increasing the number of hits can sometimes simply be a matter of purchasing analogues which share the scaffold and re-assaying. An online tool, ZINC (28) has revolutionized the purchase of compounds by collating over 100 different vendors on a single website with an integrated similarity and substructure search applet.

The results of variation on a scaffold can point to fragments of the molecules which are important for activity and those which are not, making it possible to build a structure–activity relationship (SAR) for that series.

A drug design project, in particular, is more likely to succeed if there is more than one scaffold library, since diversity gives a greater probability of later avoiding problems with toxicity and metabolism. Identifying different structural scaffolds with similar biological activities is known as scaffold hopping. Pharmacophore searching is often used for scaffold hopping as it compares generic properties such as location of H-bond donors/acceptors, presence of aromatic rings, full/partial charges, hydrophobicity, etc., rather than their specific chemical constituents. Li et al. (29) propose a measure of scaffold hopping by quantifying the differences between scaffolds which may be of great benefit in deciding on the diversity of scaffolds to select.

2.3 Bioisosteres

Bioisosteric replacement is one scaffold hopping method. A bioisostere is a molecular fragment capable of replacing another whilst retaining similar biological function of the molecule. Bioisosteres may also be known as nonclassical isosteres to differentiate them from the stricter definition of an isostere which requires a chemical and a physical match such as volume and shape. There are many known "standard" bioisosteres for all common functional groups as well as ring equivalents, peptide bond substitutions, isomer and bond inversions; for an excellent and thorough guide see Wermuth (30) (pp 290–342), also summary tables of bioisosteres are given in David Young's book (5) (pp 60–64). It is important to recognize that some bioisosteres will work well with some protein targets but not others. Molecular field software has proven useful in discovering new bioisosteres and is exemplified in the program FieldStere (*see* Subheading 3.3).

2.4 Molecular Fields in Virtual Screening

The complementary matching of electrostatic, steric, and hydrophobic properties of a ligand to its protein binding site is the basis of drug action. Drugs with similar structures usually have similar biological functions; this is the basis of drug design. It is well known that some compounds with different chemical structures can have the same biological effect; this is the basis of bioisosteric replacement; and is possible despite chemical differences because molecules' electrostatic and hydrophobic properties may in fact be similar. Comparing different compounds by the spatial distribution of their properties, i.e., their "molecular fields" rather than by structure alone, has resulted in the discovery of compounds having different core structures (scaffolds) but comparable biological activity (31).

The challenge is to calculate the molecular field properties quickly and to score them quantitatively so that they accurately distinguish differences and similarities between one molecule and another. However, fields vary continuously with conformation and across the molecular surface making comparisons between molecules extremely difficult. Fields can be sampled at defined points on a 3-dimensional grid and compare values at grid points. However, the grid has to be at a high enough resolution to capture rapid changes in the field (this is very apparent in protein electrostatics calculations (32)) and consequently large amounts of grid data are generated and computational comparison is expensive.

A solution to the comparison problem is to look only at the extrema (minima and maxima) of these fields where intermolecular interactions are most likely to occur (33). The IsoStar database catalogues the likely interaction of chemical functional groups between small molecules and between ligands and proteins (34). The calculation and comparison of extrema is embodied in the Cresset (35) collection of programs based on the extended electron density (XED) force field by Vinter (36). The majority of molecular mechanics force fields mimic the unequal distribution of electrons in a molecule by applying atom-centered partial charges. XED constructs a negative set of electron pseudo-orbitals around the

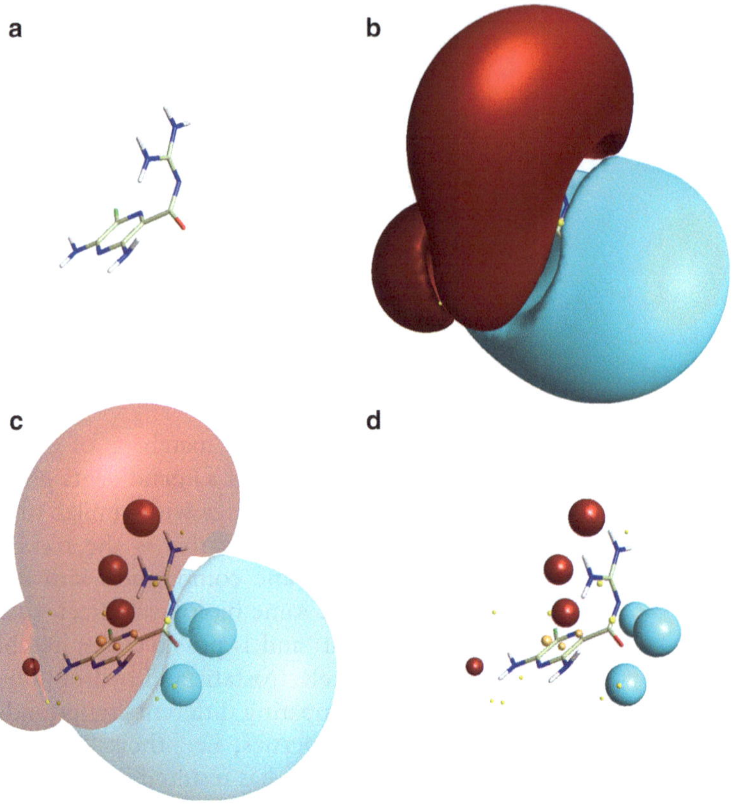

Fig. 5 The diuretic amiloride represented (**a**) as a stick model, (**b**) electrostatic contours (*dark*/*red* positive, *light*/*cyan* negative), (**c**) reducing the fields to Field Points, (**d**) final display of Field Points including *smaller spheres* representing van der Waals attractive and hydrophobic interaction sites. Sphere size is in proportion to the magnitude of the extrema

positively charged nuclei to give neutral atoms and then distributes partial charges appropriately at the orbital locations to mimic the electron distribution. This gives a much more accurate description of the molecular electrostatic potential, particularly of π-clouds and lone pairs.

The Cresset definition of extrema is called Field Points, and the number calculated per molecule is approximately equal to the heavy atom count. This is a vast reduction in data points compared to using the full electrostatic field grid, making this a computationally tractable approach to in silico HTS. The paper by Cheeseright et al. (37) details the molecular mechanics potentials used in calculating the steric (van der Waals), Coulombic, and hydrophobic fields via molecular "probes" first demonstrated by the ground-breaking work of Goodford (38) in visualizing protein–ligand interactions (Fig. 5).

Virtual screening requires that once the Field Points have been calculated for a given molecule they need to be compared and scored to each member of the chemical library which is to be screened. The set of Field Points are compared between two molecules by sampling conformations to maximize the overlay of the fields and find the optimal alignment. This is in contrast to aligning molecules by structure; here it is the fields that are compared, and therefore, it should find matches between molecules of very different structural classes but having a similar number and distribution of Field Points. Field Point techniques are thus perfectly suited for scaffold hopping.

Cresset has developed a commercial package of programs with provision for free teaching licenses and substantial academic discounts. This approach by software vendors, others of note are OpenEye (39), MOE (40), CCDC (41), etc., allows even modestly funded academic labs or institutions to establish ligand and early stage drug discovery programs (42).

3 Methods

This section will focus on the Cresset suite of software aimed at discovering small molecules by virtual screening, scaffold hopping and identifying new bioisosteres. It is designed to be an introduction for chemists and modellers unfamiliar with the software, not an exhaustive description of every parameter, which are detailed in each of the Cresset Software Manuals.

Three of the Field programs from Cresset (FieldTemplater, FieldAlign, and FieldStere) are standalone and can be run on most platforms. FieldScreen, however, requires a Linux cluster and extra software tools, it will not be discussed further here, but Cresset do provide a number of unique options for running FieldScreen which as the name suggests, is a high throughput screening tool. A separate visualization program, FieldView is freely available for calculating and viewing fields and Field Points.

FieldTemplater is an advanced pharmacophore matching tool coupled with conformer generation. If the bioactive conformation of a ligand (or ligands) is known it can be used as a 3D template to compare with other molecules. FieldTemplater assumes that all ligands bind in a similar fashion to the template. In many projects there is no known conformation to work from, but aligning the Fields of a few active compounds reveals Field Points in common and suggests those regions which are responsible for activity. Thus, FieldTemplater performs conformational searching to optimize the overlap of Field Points and predict the bioactive conformations. Any matches can be confirmed by visualizing their Field Points. These conformations can then be used in the FieldAlign program.

FieldAlign will align a small library of compounds (up to 500) against a single conformation of a known 3D structure or against a prealigned set from FieldTemplater, and rank them according to their Field Point similarities. Because of the computational expense of 3D field alignment, for large compound libraries of many thousands to millions of molecules other screening methods need to be used first such as docking or traditional pharmacophore filtering. These preselected hits can be then rescreened with FieldAlign to obtain an alternative and potentially more accurate ranking using molecular field similarity.

FieldStere is a bioisostere search routine, which goes beyond the traditional functional replacements substituting selected portions of your molecule with its own database of Field fragments and comparing the total fields of the new molecule and the parent, with the aim of retaining the integrity of the original activity. FieldStere is especially useful in proposing diverse substitutions which are not apparent unless the Field Points are superposed.

3.1 FieldTemplater

Input: A file of known ligands in mol, mol2, or sdf format (*see* **Note 1**).

- Start FieldTemplater (a Wizard window should appear, use this for the first trial).
- You are asked to load one or more molecules via the Import bar.
- Once these have been added, click Next. Constraints can be added to each molecule to enforce overlay of particular functional groups in the situation where you know those groups interact with the same region of the binding site.
- Now you can select the "quality" level of the run. This refers to the extent of shortcuts taken in the calculation protocols, with higher quality settings corresponding to greater sampling of conformations and more precise field alignments. Each of the settings (quick, normal, accurate) are appropriate to different scenarios, e.g., quick may be adequate for a fragment or to get an initial filter of a complex system prior to running the more computationally demanding and time consuming "accurate" setting.
- The results appear as sets of conformations ranked according to field similarity, shape similarity or combined similarity. Field similarity is based purely on the electrostatic match of two molecules, whereas shape similarity is based on their Gaussian volume overlap. Both similarity measures have their merits, but as the contribution to binding of electrostatics and shape will be dependent of the chemotype of the ligand and the nature of the target often the combined score is most useful.
- You can save the project and then visually inspect the superposition, manually making different templates visible or not. It is common to manually select templates that have high similarity scores or match some other data on known binding mode or simply through chemical intuition. Finally "visible" templates can be selected for export to an sdf file. If "visible" is not selected it will save all generated conformers (Fig. 6).

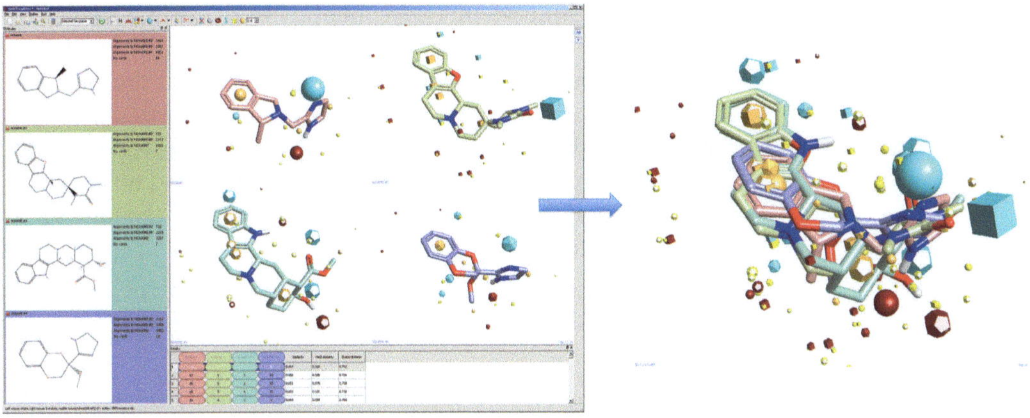

Fig. 6 FieldTemplate creation with Cresset software. The interaction FieldPoints for four α2 adrenergic antagonists (*left*) can be used to superimpose them (*right*) despite their belonging to two distinct chemical classes

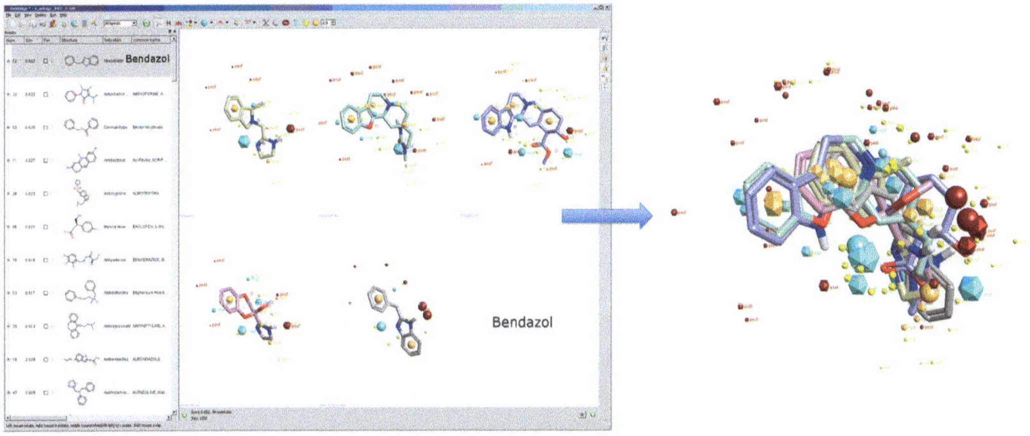

Fig. 7 Field Align. Searching a database of existing drugs with the template created from the four molecules in Fig. 18.6 identifies bendazol (an antihelmintic) as the top hit (*bottom center*). Bendazol is superposed with the four reference molecules (*bottom right*)

3.2 FieldAlign

- Input a single or small set of reference structures preferably in its/their bioactive conformation(s) or a computed conformation from FieldTemplater as sdf, mol, mol2, or SMILES format (*see* **Note 2**).
- Next add a database in sdf (*see* **Note 3** on database preparation). For the example in Fig. 7 a filtered version of the John Hopkins Clinical Collection (43) of known drugs was used.
- Protonation and tautomerization are issues of which the user must be aware. FieldAlign has built in routines for automatic

protonation, but where experimental information is available it is better to specify the protonation state manually using the software's molecular editor window. Tautomers are not dealt with automatically, but can also be manually specified in the molecular editor—this is best done by including any possible/likely tautomers as separate molecular entities. The screening outcome may be very sensitive to tautomers as they invariably produce different field patterns.

- FieldAlign ranks the members of the input database according to their Field similarity score with the reference molecule(s) or template.
- Clicking the + symbol expands different conformational hits for that molecule.
- Highlighting any molecule from the database given in the left-hand window immediately places it into the main viewing window aligned with the reference(s). This enables straightforward visual inspection.

FieldAlign is not a "high throughput" screening program, and will only process 500 molecules per run. It is best to use FieldAlign on pre-filtered sets of molecules, e.g., as a post-docking/pharmacophore screen. FieldAlign will provide matches with different chemical functionality from that of the reference(s) some of these new scaffolds will potentially have similar biological activity.

3.3 FieldStere

FieldStere is a fragment replacement program which compares the Fields of its internal list of fragments with a selected part of the input molecule of interest. If a field match is found, it replaces the selected part with the new fragment and field scores and compares the entire field of the new molecule with the entire field of the original input molecule.

- Input a reference molecule (again in sdf, mol, mol2, or SMILES format).
- You can allow FieldStere to choose the protonation state.
- Drag the cursor around the section of the molecule you wish to replace.
- Now select the database(s) to include in your search, there are five to choose from very commonly found fragments to very rare ones. The more databases that are included the longer the processing takes. There are a number of scoring and filtering parameters that can be adjusted by the user before running FieldStere, e.g., which atoms to or not to use to reconnect fragments, the size of fragments to consider, their logP, number of rotatable bonds, whether fragments should be incorporated into rings, etc.

The results are presented in multiple windows (Fig. 8), the main graphics display the original reference molecule overlaid or

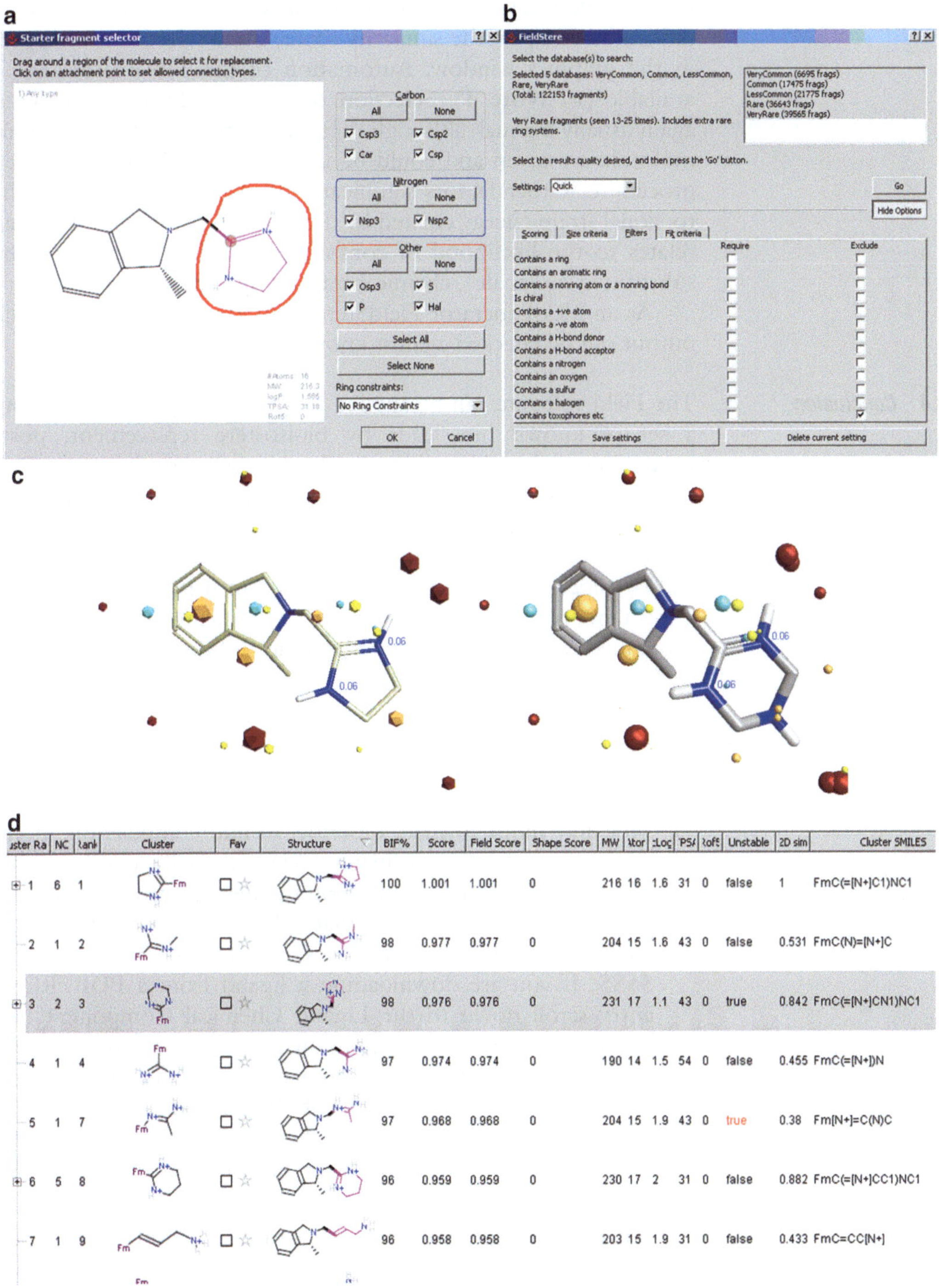

Fig. 8 FieldStere—bioisostere replacement based on fragment selection. (**a**) Drag cursor around chosen fragment, an imidazoline which will be replaced by its steric and electrostatic equivalents (as determined by FieldPoints) drawn from the built-in chemical group database. Bioisostere selection is highly customizable. (**b**) Settings can be adjusted in a series of menus for scoring, size, filters, etc. Filters are useful for selecting particular chemotypes and are comparable to pharmacophore constraints. (**c**) The reference molecule (*left*) with a modified version in which the fragment is replaced by a selected bioisostere (*right*). (**d**) Thumbnails of the all matching fragments are displayed in a sorted list along with physicochemical data

side by side with its new fragment. The candidate fragments are ranked based upon the similarity score, when selected they appear in the graphics window. Automation of the above processes is available (*see* **Note 4**). Selection of these modified molecules is again usually manual, although their similarity scores to the original molecule are given and could be used as a filter in an automated procedure. Other factors significant in selection may include the torsional strain energy created by the fragment substitution (which relates to the likelihood of conformation being adopted by the substituted molecule), chemical tractability and novelty.

As an idea generator FieldStere is an excellent tool and often output can be checked against known SAR for validation.

3.4 Conclusion

The Field software can be applied to scaffold hopping, expanding a set of known hits/leads by bioisostere replacement, post-processing of virtual screening results, bioactive conformer generation, and many other applications which require new ideas to suggest new chemistry or to overcome existing patents.

Of course, given the many software products available, some will be superior for specialist functions, but it is the focus on the importance of molecular field extrema which allows these particular programs to rapidly generate unique solutions to molecular similarity.

4 Notes

1. Mol and mol2 formats describe atomic composition, bond connectivity, and coordinates (44), originally developed by Molecular Design Ltd and Tripos, they are now an industry standard. Structure data format (sdf) files are simply concatenated mol or mol2 format entries separated by a line containing $$$$. If you are downloading a ligand from a PDB/RCSB entry scroll down to the Ligand Chemical Component box and click the Download pulldown and save as an sdf. Note this will include *x*, *y*, *z*, coordinates as well as SMILES and InChI string representations of the molecular architecture. Alternatively you can extract a ligand from a pdb format by using a text editor or from modelling software and modify the file format.
2. Where a receptor/protein structure is known "excluded volumes" can also be used as a steric boundary during FieldAlign or FieldStere runs. When available such boundaries can be used to prevent selection of molecules or fragments which would otherwise clash with protein atoms.
3. Molecular databases must be prepared beforehand using filters to limit molecular weight. This is particularly important if your

database of structures came from a general chemicals catalogue or a set of therapeutics which cover large peptides and even proteins. It is best to set an upper and lower MW filter as this will also remove salts, metals and other counterions which are not usually of interest. Chemically reactive compounds and those which are known to interfere with assays are usually removed as are a group molecules known as "frequent hitters," which are known to bind to many protein families and may diminish selectivity, "privileged scaffolds," commonly occurring core structures may be omitted or retained depending on synthetic or patent strategies (45). Which compounds are included or excluded may vary according to established procedure, personal chemical bias, or the protein target under consideration.

4. Non-graphical command-line tools are also available for FieldStere, FieldAlign, and the XedTools (*see* **Note 5**). These command-line tools provide added control, increased speed, and automation of repeated applications of procedures to different ligands. Field and Xed programs are also integrated into data analysis pipeline tools such as Knime (http://www.knime.org) and PipelinePilot (Accelrys, Inc.).
5. XedTools provide forcefield based manipulation of molecules. Xedmin uses the XED force field to energy minimize ligand conformations with or without the protein present. XedeX generates diverse families of distinct, energy minimized conformations of small molecules for use in subsequent Field template construction or alignment.

Acknowledgments

OpenEye for free access to their full suite of programs, MOE and GOLD for their generous academic pricing, Cresset for their free teaching licenses and support. Andy Vinter and Martin Slater for their excellent advice.

References

1. Nicholls A (2011) What do we know? Simple statistical techniques that help. In: Bajorath J (ed) Chemoinformatics and computational chemical biology. Springer Science, New York
2. Merz KM et al (2010) Drug design structure- and ligand-based approaches. CUP, Cambridge
3. Leach AR (2001) Molecular modelling: principles and applications, 2nd edn. Longman, Edinburgh
4. Gasteiger J, Engel T (2004) Chemoinformatics: a textbook. Wiley, Wienheim
5. Young DC (2009) Computational drug design: a guide for computational and medicinal chemists. Wiley, New Jersey
6. Rigby M et al (1986) The forces between molecules. OUP, Oxford
7. Stone AJ (2008) Intermolecular potentials. Science 321:787–789

8. Bissantz C, Kuhn B, Stahl M (2010) A medicinal chemist's guide to molecular interactions. J Med Chem 53:5061–5084
9. Chan AW, Laskowski RA, Selwood DL (2010) Chemical fragments that hydrogen bond to Asp, Glu, Arg and His sidechains in protein binding sites. J Med Chem 53:3086–3094
10. Protein DataBank (PDB) at RCSB: http://www.rcsb.org/
11. Williams MA, Ladbury JE (2003) Hydrogen bonds in protein–ligand complexes. In: Böhm HJ, Schneider G (eds) Protein–ligand interactions from molecular recognition to drug design. Wiley, Weinheim
12. Gilli G, Gilli P (2009) The nature of the hydrogen bond: outline of a comprehensive hydrogen bond theory. OUP, Oxford
13. Labowski SJ (2006) Hydrogen bonding: new insights. Springer, Netherlands
14. Foloppe N (2005) Structure-based design of novel Chk1 inhibitors: Insights into hydrogen bonding and protein–ligand affinity. J Med Chem 48:4332–4345
15. Kubinyi H (2008) The changing landscape in drug discovery. In: Stroud RM, Finer-Moore J (eds) Computational and structural approaches to drug discovery. RSC, Cambridge
16. Parisini E et al (2011) Halogen bonding in halocarbon–protein complexes: a structural survey. Chem Soc Rev 40:2267–2278
17. Riley KE, Hobza P (2011) Strength and character of halogen bonds in protein–ligand complexes. Cryst Growth Des 11:4272–4278
18. Hunter CA, Sanders JKM (1990) The nature of π–π Interactions. J Am Chem Soc 112:5525–5534
19. Tsuzuki S (2005) Interactions with aromatic rings. Struc Bond 115:149–193
20. Gooding M (2011) Exploring the interaction between siRNA and the SMoC biomolecule transporters: implications for small molecule-mediated delivery of siRNA. Chem Biol Drug Des 79:9–21
21. Lanzarotti E (2011) Aromatic–aromatic interactions in proteins: beyond the dimer. J Chem Inf Model 51:1623–1633
22. Barron LD, Hecht L, Wilson G (1997) The lubricant of life: a proposal that solvent water promotes extremely fast conformational fluctuations in mobile heteropolypeptide structure. Biochemistry 36:13143–13147
23. Setny P, Baron R, McCammon JA (2010) How can hydrophobic association be enthalpy driven? J Chem Theory Comput 6:2866–2871
24. Myslinski JM et al (2011) Protein–ligand interactions: thermodynamic effects associated with increasing nonpolar surface area. J Am Chem Soc 133:18518–18521
25. Hassan SA et al (2005) Computer simulation of protein–ligand interactions: challenges and applications. In: Nienhaus GU (ed) Protein–ligand Interactions. Humana, New Jersey
26. Schmidtke P et al (2011) Shielded hydrogen bonds as structural determinants of binding kinetics: application in drug design. J Am Chem Soc 133:18903–18910
27. Cavasotto CN, Phatak SS (2011) Docking methods for structure-based library design. In: Zhou JZ (ed) Chemical library design. Springer, New York
28. Irwin JJ, Shoichet BK (2005) ZINC – a free database of commercially available compounds for virtual screening. J Chem Inf Model 45:177–182
29. Li R et al (2011) Development of a method to consistently quantify the structural distance between scaffolds and to assess scaffold hopping potential. J Chem Inf Model 57:2507–2514
30. Wermuth CG (2008) The practice of medicinal chemistry. Academic, London
31. Low CMR (2005) Scaffold hopping with molecular field points: identification of a cholecystokinin-2 (CCK2) receptor pharmacophore and its use in the design of a prototypical series of pyrrole- and imidazole-based CCK2 antagonists. J Med Chem 48:6790–6802
32. Warwicker J (1994) Improved continuum electrostatic modelling in proteins, with comparison to experiment. J Mol Biol 236:887–903
33. Apaya RP (1995) The matching of electrostatic extrema: a useful method in drug design? A study of phosphodiesterase III inhibitors. J Comput Aided Mol Des 9:33–43
34. Bruno IJ et al (1997) Isostar: a library of information about non-bonded interactions. J Comput Aided Mol Des 11:525–537, http://www.ccdc.cam.ac.uk/products/csd_system/isostar/
35. Cresset software http://www.cresset-group.com
36. Vinter JG (1994) Extended electron distributions applied to molecular mechanics of intermolecular interactions. J Comput Aided Mol Des 8:653–668
37. Cheeseright T et al (2006) Molecular field extrema as descriptors of biological activity: definition and validation. J Chem Inf Model 46:665–676
38. Goodford PJ (1985) A computational procedure for determining energetically favorable binding sites on biologically important macromolecules. J Med Chem 28:849–857

39. OpenEye http://www.eyesopen.com/
40. MOE (Molecular Operating Environment) http://www.chemcomp.com/software.htm
41. GOLD http://www.ccdc.cam.ac.uk/products/life_sciences/gold/
42. Gane PJ, Chan E, Selwood D. ChemiBank http://www.ucl.ac.uk/chemibank
43. John Hopkins clinical collection http://htc.wustl.edu/library/JHCCL.html
44. Mol2 format http://tripos.com/tripos_resources/fileroot/pdfs/mol2_format.pdf
45. Rishton GM (2003) Non-leadlikeness and leadlikeness in biochemical screening. DDT 8:86–96

Chapter 19

Structure-Based Virtual Screening for Novel Ligands

William R. Pitt, Mark D. Calmiano, Boris Kroeplien, Richard D. Taylor, James P. Turner, and Michael A. King

Abstract

Computational searches for novel ligands for a given protein binding site have become ubiquitous in the pharmaceutical industry, and are potentially equally useful in helping identify small-molecule tools for biology. Here we describe the steps needed to carry out a high-throughput docking (HTD) or three-dimensional (3D) pharmacophore virtual screen starting with a model of the target protein's structure. The advice given is, in most cases, software independent but some tips are provided which apply only to certain popular programs. Useful work can be carried out using free programs on a modest workstation. Of course, any resultant "hits" remain in the virtual world until they are experimentally tested.

Key words Virtual screening, Docking, Pharmacophores, Molecular modelling, Computational chemistry

1 Introduction

The term "virtual screening" covers any computational selection process whereby small molecules are chosen for experimental testing. The main advantages of virtual over experimental high-throughput screening are that much larger numbers of compounds can be evaluated, and physical samples are only required for the relatively small numbers of compounds in the final selection. Here we describe the virtual screening techniques that can be applied when a crystal structure of a target protein is available. The two main approaches used are high-throughput docking (HTD) (Fig. 1) and structure-based pharmacophore searching. Docking is a computational procedure for fitting small molecule ligands into a macromolecular binding site by translation, rotation, and conformational sampling. Most often the binding site is kept rigid or semirigid while allowing the small molecules to flex. Pharmacophore searching is the matching of ligands to a set of spatially constrained key binding (pharmacophoric) features. These features

Mark A. Williams and Tina Daviter (eds.), *Protein-Ligand Interactions: Methods and Applications*, Methods in Molecular Biology, vol. 1008, DOI 10.1007/978-1-62703-398-5_19,

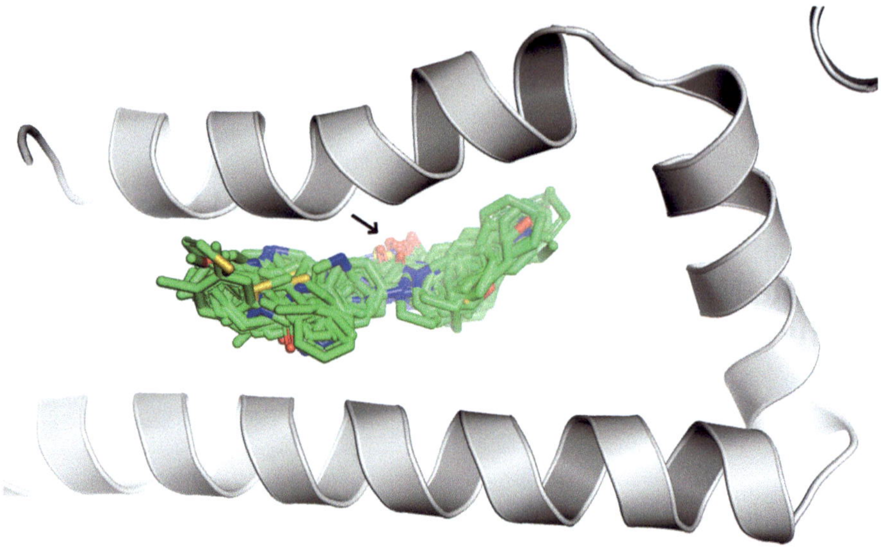

Fig. 1 An overlay of a subset of selected compounds from a HTD virtual screen. Note the presence of a common carbonyl oxygen hydrogen bond acceptor group near the middle of each molecule (*arrow*). Being able to form a key hydrogen bond in this region was one of the selection criteria used

can be, for example, hydrogen bonds, hydrophobic groups, or chemical substructures. In the methods described here, pharmacophoric features are either derived from the appropriate protein binding site structures or, preferably from the structures of protein–ligand complexes. We describe the HTD and pharmocophore approaches separately; however, they can straightforwardly and profitably be applied in combination. Both approaches can be computationally expensive when applied to large sets of compounds and simple calculated physicochemical property and text-based chemical substructure filters are a very useful means to reduce the search space prior to using these three-dimensional (3D) methods.

The success of a virtual screen is measured by hit rate, where a "hit" is a molecule for which binding or inhibition is subsequently experimentally confirmed by its ability to generate a dose-response curve. A 5 % hit rate is considered to be a good outcome for a virtual screen. However, the discovery of a single active molecule amongst the selected compounds might be all that is required. Below we describe in some detail the choices that need to be made and the procedures that need to be followed in structure-based screening. If these protocols seem too complex, it should be noted that many of the programs that are available to carry out these tasks have straight forward menu driven interfaces and sensible default options. Although many of the choices and procedures are software independent, the authors are most familiar with GOLD (1) (CCDC) and Glide (2) (Schrödinger) for docking, and Catalyst (3) (Accelrys), Unity (4) (Tripos) and Phase (5) (Schrödinger) for 3D pharmacophore searching, and additional comments are provided on special considerations when using these programs.

2 Materials

2.1 Software

2.1.1 High-Throughput Docking

HTD is typically performed when detailed 3D structural information is available for the protein target. Before attempting an HTD campaign there are many options and parameters that must be considered, many of which are based upon the scientific knowledge of the protein target.

The choice of a HTD software package should be made following consideration of its capabilities:

- *Ligand flexibility.* Many small molecules are able to adopt several distinct conformations and the bound conformation of a ligand may not be the same as the most common conformation found when it is free in solution. Most modern docking programs are able to generate conformations of the ligand in addition to sampling rigid body rotational and translational degrees of freedom. It is recommended that a program is used that is capable of full conformational sampling of the ligand (including flexible rings) during the docking process.
- *Protein flexibility.* It is also possible to sample protein conformational change during an HTD. The type of flexibility allowed by a program varies and may include rotatable polar side chain groups, side chain rotamer selection and even backbone flexibility. All these options are available in GOLD or GLIDE; however, these refinements are computationally impracticable in the high-throughput context unless very large computer resources are available.
- *Explicit waters.* Water molecules play an important role in the molecular recognition. Most modern programs will allow explicit waters to be included; moreover some programs allow water molecule rotation and toggling (rapid removal or addition of individual water molecules at particular locations) during the docking, for example GOLD.
- *Cost.* There are several docking software packages that are free to noncommercial users, such as AutoDock (6) (Scripps), AutoDock Vina (7) (Scripps), Dock (8) (UCSF), and FRED (9) (OpenEye). Commercial vendors also sometimes offer free or reduced-cost versions of their software to academic users and are often flexible on pricing. Some software vendors operate a "per node" or "per CPU" licensing strategy which must be considered in the design and budgeting of a virtual screen. However, the best choice of software will depend not only on cost considerations, but also upon the target. Some packages are more successful than others in identifying ligands for particular target proteins. We recommend using more than one package and comparing outcomes (preferably including those for some known active ligands).

- *Speed.* Docking software was often originally designed to predict accurately the binding modes of single small molecules and as such was too slow for HTD. However, many programs now also include a specific "virtual screening mode," in which a measure of accuracy is sacrificed for speed. Obviously, the more accurate the calculation, the longer it will take; a balance is needed between accuracy, the computer resources available, and the number of compounds to be screened.

For a list of docking software *see* **Note 1**.

2.1.2 Structure-Based Pharmacophore Matching

Structure-based pharmacophore matching is a complementary technique to HTD, which focuses only on the key interactions between protein and ligand. The choice of such programs is more limited and free versions are harder to find.

An important question is "when is it appropriate to run a 3D pharmacophore based search rather than using a HTD method?" If multiple known active compounds are available and the protein target structure is of poor quality or disordered, then a 3D pharmacophore approach might be more appropriate. However, a pharmacophore search can also be run in parallel to a docking-based virtual screen as it will likely provide a different selection of potential ligands and any ligands in common between the two methods might be thought to be particularly interesting for further study. Because pharmacophore searches do not take into account the full extent of a protein–ligand complex it is useful to dock resulting hits in order to check for full compatibility with the binding site.

2.1.3 Preparation and Post-processing

The preprocessing and post-processing steps of a virtual screening campaign are often overlooked; however, they are a key component of a successful protocol. Preprocessing steps include conversion of 2D chemical structures to 3D models of the compounds, molecule standardization, charging and generation of tautomers. Furthermore it is usual to apply functional group filters to reduce the total size of the library of compounds to be screened and select for chemical groups that are thought to be advantageous, or against those thought to be deleterious, in ligands for a particular target protein. The analysis of the outcome of a virtual screen can be very time consuming and can have a significant impact on the choice of docking program and how the screen is performed. The post-processing steps include filtering on the basis of desirability of particular interactions and chemical substructures and the visual filtering of high-scoring screened compounds by experienced medicinal chemists. The advent of molecular "pipelining" software such as Knime (10) (Knime) and Pipeline Pilot (11) (Accelrys) has greatly facilitated the process of filtering small molecules and it is highly recommended that one of these or a similar workflow tool is used for the purpose. Details of these and other useful software packages are given in **Note 2**.

2.2 Hardware

Computer hardware is constantly improving, but the current recommendation for a HTD screen would be a cluster of 100 CPU cores, capable of processing at least one compound per minute per core. This would allow hundreds of thousands of compounds to be screened with appropriate accuracy, and in reasonable time. If a computer cluster is not available, then stringent pre-filtering of compounds will be needed, either using knowledge of the binding site (2D pharmacophore feature type filtering) or by only considering a more limited number of very small compounds/fragments. Pharmacophore searching is typically less computationally demanding than HTD (even allowing for database construction) and may be the preferred option where computational resources are limited.

3 Methods

3.1 Sourcing and Selection of a Protein Structure

X-ray crystal structures of protein molecules are used for the majority of virtual screening studies. The coordinates of these structures usually come as a PDB (Protein Bata Bank) format text file. These files also contain other information such as the resolution of X-ray data used to determine the structure and B-factors (or temperature-factors) of individual atoms that are a measure of the disorder within a structure. Many PDB files can be downloaded from the official site of the Protein Data Bank (12) or obtained from your friendly crystallographer. If there are no X-ray structures of the target protein, but there are structures of related proteins available, it is possible to either use them directly for docking (as a so called surrogate structure), or to use them as the basis for building an homology-model.

Proteins are flexible molecules, often adopting distinct conformations as part of their function and in response to interactions with other molecules. A single crystal structure cannot reflect the dynamic properties of the binding site and the precise shape of the site may well be dependent, amongst other things, upon the co-crystallized ligand. Ligand-bound protein structures, consequently, usually give better results when used in a docking screen, but they may also bias the virtual screening hits toward ligands similar to the original.

Protein structures from crystallography are models based upon electron density data. In interpreting that data, crystallographers have many choices, and sometimes make errors or simply have insufficient data to obtain a unambiguous structure. Before working with a crystal structure, it should be inspected closely in a 3D viewer to check whether it is chemically sensible and complete. If there are multiple crystal-structures available for the target protein, it may be necessary to rank them, based on a number of criteria:

1. Completeness: Are all residues present in the structure? Do residues have missing side-chain atoms (e.g., does a residue look like alanine or glycine but is listed in the sequence another residue type)?
2. Resolution: Is the resolution good enough (<2.5 Å) to give confidence that the experimenter has likely been able to correctly determine the atomic positions.
3. B-Factor: The absolute values of B-factors are dependent upon how they were modelled in the refinement process. Color coding of atoms by their B-factors can be performed in most molecular graphics programs. This will provide an indication of the variability of each atom's position within the binding site.
4. Occupancy and alternative residue/atom positions: Sometimes a substructure (e.g., a side chain) will be modelled as occupying two or more alternative positions. If this occurs in the binding site it complicates matters, and although there is no general rule, it may mean screening against each protein conformation.
5. Crystal-packing: Protein crystals are held together by protein-–protein contacts. If these contacts are close to the targeted binding site, they may distort the protein in that region, and give the site an artifactual shape. It is good practice to check the crystal-contacts and if possible to choose a crystal-structure without any close to the binding site.

Another thing to note during visual inspection are the locations of water molecules (*see* Subheading 3.3) in and around the binding site that may be important for binding.

3.2 Protein Structure Preparation

The procedures needed to prepare a binding site model for docking can be carried out in most general-purpose protein modelling packages such as Sybyl or Maestro (*see* **Note 3** for a description of an example).

1. Regenerate any residues in the sequence that are missing from the structure, e.g., loops are frequently poorly defined in electron density and because reliable coordinates cannot be obtained from the experimental data they will be omitted from the PDB file. In general, loop modelling is less reliable than other modelling procedures and docking into pockets with modelled loops should be done with caution.
2. Model missing atoms and side chains. Choose the most common rotamer (*see* **Note 4**) that does not clash with the rest of the protein or block the binding site.
3. Choose one conformation for each residue if more than one is provided by the crystallographer.

4. Add all hydrogens.
5. Unless there is specific experimental or structural (e.g., hydrogen bonding) evidence to the contrary, ensure basic residues (LYS, ARG) and the amino-terminus are protonated and acidic (ASP, GLU) side chains and the carboxy-terminus deprotonated and that HIS residues are neutral.
6. Select HIS tautomers, ASN and GLN rotamers and water orientations that maximize the number of hydrogen bonds in the structure.
7. Take care that metals, cofactors, sugars and other nonprotein molecules in the binding site have correct bond orders and protonation.
8. Optionally energy minimize modelled portions of the binding site (e.g., hydrogens, side-chain positions, etc.) but not those parts taken directly from the crystal structure.

3.3 Treatment of Protein Flexibility and Structural Waters

Typically, information about protein flexibility is obtained by examination of B-factors and occupancies; and also by comparing different crystal structures for a protein, for example, the apo structure of the protein and complexes with different ligands. This allows any differences in side chain and backbone conformation within the binding site region to be identified. Again, care should be taken when working with low resolution structures as these may contain conformational inaccuracies. When direct experimental information on flexibility is scarce, an alternative approach is to perform a molecular dynamics simulation of the protein in water, using a program such as GROMACS (13) or Desmond (14) (Schrödinger), and to monitor conformational fluctuations. It is also possible to explore alternative theoretical models for the binding site regions of proteins using homology modelling software such as Modeller (15) (UCSF) or Schrödinger's PRIME (16) program.

Having identified flexibility issues that are likely to affect virtual screening results, what can be done about them? A "brute force" approach is simply to use multiple protein structures for screening (ensemble approaches), and then to combine the results of the screening runs. Alternatively, docking programs such as GOLD contain methods that can treat certain limited types of flexibility within a single protein structure. For example, GOLD automatically samples the orientation of protein OH groups during docking in order to optimize potential hydrogen bond interactions with ligands. It is also possible to allow selected side chains to change conformation during docking using GOLD's side-chain rotamer library approach. More sophisticated methods are available for performing flexible protein docking but are much more computationally demanding, and therefore currently unsuitable for large scale screening.

Another important consideration when devising virtual screening strategies is the role of structural water molecules in the binding site region. Water molecules are often determined by X-ray crystallography to be structurally conserved within binding sites, and are sometimes seen to be involved in mediating hydrogen bond interactions between protein and ligands. Also, structural waters, if not included in the binding site, may leave unlikely pockets that ligands can be docked into. Clearly, if waters are believed to be important, then they need to be included in the virtual screening process. In pharmacophore-based virtual screening approaches, waters can be approximately accounted for by adding additional hydrogen-bond donor and/or acceptor features to the pharmacophore query. The GOLD docking program allows water molecules to be included explicitly in binding sites, and the presence, displacement and orientation of these molecules can be sampled during the docking process.

3.4 Sourcing and Selection of Small Molecules

3.4.1 Sourcing Compounds and Compound Structures

Any search algorithm, no matter how good, can only find high-quality "hits" if they are there to be found; a high-quality source of screening compounds is therefore the most fundamental requirement of a successful search. Appropriate compound collections are readily available externally, although these may be augmented by in-house compounds. There are two requirements for a virtual screening source collection: the compounds must be physically available in suitable purity, quantity, and cost to perform the appropriate assays; and the chemical structures must be available in electronic form—fortunately many chemical suppliers are now able to fulfil these requirements. Many companies offer screening collections that may be selected from on a compound-by-compound basis; they have the advantage of being offered in quantities appropriate for assay purposes, and can often be shipped pre-weighed into 96-well plate format or proprietary enclosures on request. Such services might at first sight appear costly, but this cost must be considered against the considerable time and effort required to re-weigh and plate compounds in-house. Using a single commercial supplier has advantages in simplifying the virtual selection process and the logistics of purchasing the compounds. Many suppliers maintain large and diverse collections that make a single source a viable option. The ultimate solution is offered by some vendors, who will acquire compounds from other "third-party" suppliers as well as supply their own stocks and then consolidate them into a single order. This is the optimum solution, maximizing flexibility of sourcing without the attendant logistical complexities, but not surprisingly it comes at a cost; it is not unusual that the "per compound" surcharge for such a service will be comparable with the initial material cost of the compounds.

Vendors will almost always supply their catalogues free of charge in electronic format, usually as a 2D structural data file (sdf) (17). *See* **Note 5** for a list of such vendors. Often they are available from a download server via the corporate Web site; if not, many respond to a request for the file on CD. Unlike general chemical catalogues, screening collections tend to be quite dynamic, with compounds frequently becoming unavailable as well as new compounds added, therefore it is important to obtain the most up to date catalogues—most are updated at least 6-monthly. Even so, expect a certain amount of unavailability when it comes to the final selection.

The question of how many compounds to buy from those identified as potential binders in the virtual screen can be a difficult judgement; obviously the point of making a virtual selection in the first place is to reduce the number of compounds bought and tested. Unfortunately, we are seldom in a position to know the "hit rate" of the virtual screen in advance, it being dependent upon the suitability of the screened collection, the validity of the pharmacophore or binding site model and the sensitivity of the experimental assay, amongst other factors. For an enzyme/ligand structure-based search into a structurally diverse set of molecules, a 1–5 % "hit-rate" among the potential binders indicated by the virtual screen could be considered good, compared to the outcome of random screening of perhaps 0.05 %. In practice, the number of compounds purchased for assay will often be determined by availability of funds. However, unless there is reason to expect a higher than usual hit rate (selection from close analogues of a known active, for instance) an absolute minimum selection should be the "best" 100 compounds from the virtual screen.

3.4.2 Small Molecule Selection (Pre-filtering)

The purpose of pre-filtering is to eliminate unsuitable compounds and avoid wasting computational search time on compounds that would not be progressed. Filtering may be based on chemical substructure or calculated chemical or physical properties. The choice of pre- or post-search filtering will depend on the nature of the search and the condition of the initial list of compounds. Pre-filtering obviously improves the computational efficiency of VS. However, if searchable conformer databases have to be constructed (e.g., for pharmacophore searching), then this is likely be the most computationally demanding part of the process, and better done as few times as possible. In that case it is more efficient to build the conformer database once, pre-filtering only on universally undesirable properties and post-filtering those characteristics specific to each subsequent search.

Usually, measured properties of molecules will not be available, so filtering will depend on the structure and properties calculated from the structure. Substructure searching is the most practical way to eliminate undesirable functional groups; it may also be extended to eliminate a variety of generally undesirable chemical features. Many calculated properties of the molecule as a whole are readily generated "on-the-fly" by modern pipelining tools. It is worth noting that molecular weight is the most significant single property; not only is it easily calculated, but many other pharmacological properties correlate with it quite highly.

Depending on the source of the initial compound list, one or more of the following undesirable categories may be encountered.

1. Reactive functionalities (acid chlorides, alkyl iodides, anhydrides, etc.). Compounds containing these should in all cases be removed since they may interfere with assays, be toxic, and will decompose to other chemical species.
2. Chemical types associated with toxicity, insolubility or metabolic liability. Most medicinal chemists will have their own (perhaps somewhat subjective) list of disfavored chemical types (18–20).
3. Racemic chiral centers result in the presence of heterogeneity in the actual substance, often in uncertain proportion. Restricting such centers to a maximum of two is sensible.

The well-known Lipinski's "Rule of 5" (21) is often applied, as the properties to which the rules apply (molecular weight, LogP, and hydrogen bond acceptor/donor count) are readily calculable from the 2D structure. Strictly, Lipinski's rules are only appropriate to an oral drug candidate and it is debatable as to what extent a virtual screening hit is required to be "drug-like." It may be preferable in some cases to be more stringent than Lipinski, choosing for instance an upper molecular weight cut-off of 350 rather than 500 so as to give room for subsequent "growth" during ligand optimization. Depending on the initial dataset, it may be worth filtering on minimum values, since very small and/or featureless molecules are unlikely to make strong binders, and in any case will not be chosen by a search algorithm if they do not have a sufficient number of the appropriate features. It is also worth noting that the number of rotatable bonds in the molecule may exponentially increase computational time for some searching algorithms, and may have to be restricted for that reason.

The actual stringency of filtering may depend on the target. For instance, if finding "hits" for a particular target is expected to be difficult, then less stringent filtering may be appropriate, with an improved chance of finding "hits" offset against the acceptance of more challenging starting points. On the other hand, stringent filtering may not reduce the search input list to a manageable size, in which case molecular similarity algorithms can be employed to generate a structurally representative reduced set. This technique is useful if a particular chemical substructure is clearly over-represented (N.B. such filtering may be somewhat arbitrary in practice, since structural and functional similarity may not coincide, and in the worst case a "representative" structure may be the least active of a series).

For the sake of simplicity, we suggest the following general purpose pre-filters for a virtual screen:

- Molecular weight >150 and <400.
- Number of rotatable bonds <7.
- Number of undefined stereo-centers <3.
- No reactive substructures.

3.5 Preparation of Small Molecules

Once a set of small molecule compounds has been selected for structure-based virtual screening, the next stage is to convert this into a screening deck of 3D molecular structures of appropriate chemical form for input into a docking or pharmacophore search program. First to be considered are the treatments of charge states and tautomerism. Compounds containing acidic or basic functional groups may exist in either charged or neutral form in protein-bound environments. Similarly, for compounds containing hetero-aromatic ring systems, particular tautomeric forms can be stabilized upon binding to a protein. In order to account for such effects, we need to ensure that the different forms a compound can adopt are present in the virtual screening set. Several commercial programs such as Tripos' Protoplex (22) can be used to achieve this.

Often commercial compound collections contain racemic compounds. In order to use these compounds in virtual screening, the different chiral forms need to be enumerated. Expansion of these centers can be achieved with programs such as StereoPlex (23) (Tripos), LigPrep (24) (Schrödinger), and Corina (25) (Molecular Networks).

Isomer and tautomer enumeration are highly recommended because, if these steps are not carried out, the "false negative" rate will be inflated and potentially whole hit chemical series will be missed.

Subsequent conversion of 2D structures into 3D form is a straightforward task that can be accomplished using programs such as CONCORD (26) (Tripos) and Corina (*see* **Note 6** for a list of 2D to 3D conversion programs). These programs are very fast and usually produce reasonable low energy 3D structures that are suitable as starting points for large-scale virtual screening.

3.6 Docking Based Virtual Screening

3.6.1 Binding Site Setup

The first step in setting up the HTD job is to define the protein binding site into which the ligands will be docked. There are numerous ways to define the extent of the binding site, the two most common methods are:

- Selection of all atoms within a chosen radius of a point in space or residue at the center of the proposed binding site.
- Selection all atoms or residues in contact with a ligand which sits in the proposed binding site in a known structure.

Where a ligand-bound structure is available, it is recommended to select all atoms within 4–6 Å of the atoms of a bound ligand to define the binding site. Programs such as GOLD will then define a cavity, in which to dock molecules, calculated within the atom selection. It is highly advisable to visualize this cavity to check that it covers the area of interest (*see* **Note 7**). If water molecules or cofactors are located in the binding site and are to be included in the docking then these must be defined during binding site setup.

3.6.2 Treatment of Protein Flexibility

Additionally, in setting up the binding site, there are options in both GOLD and Glide to allow some degree of side-chain flexibility within the defined binding site during docking. Introduction of flexibility is computationally expensive, and we take the view that it's more effective to search more compounds with a rigid site than to include flexibility. Consequently, it is recommended to keep the protein rigid in a HTD run, at least in the first instance.

3.6.3 Uploading the Ligand Dataset

Next it is required to upload the prepared ligand dataset; this can be in any common structural file format such as structure-data file (sdf) or mol2. Fine control of ligand flexibility is allowed in GOLD, for example, pyramidal and planar conformations around nitrogens can be specially defined and the conformational flexibility of rings can also be included. We suggest you do sample both planar and pyramidal conformations around nitrogens as these atoms tend to form key hydrogen bonds within proteins and the geometry of the nitrogens can affect hydrogen bond formation. Where the docking software itself does not enable ligand ring flexibility, programs such as Confort (27) (Tripos) can be used for the fast generation of alternative aliphatic ring conformations for inclusion in the ligand dataset prior to docking. In addition, the CONCORD program contains an option to generate anilines with either planar or trigonal nitrogen atoms.

3.6.4 Docking Options

Having set up the binding site and ligands, it is now necessary to set some of the options for the docking procedure itself such as the choice of scoring function, application of any pharmacophoric constraints, and the number of poses (docked conformations) per ligand to keep for subsequent analysis. Where applicable, it is generally best to choose the virtual screening setting of the docking program (fast but less accurate). The docking program might choose the number of docking operations per ligand and the number of poses to generate. For more details on virtual screening settings, *see* **Note 8**.

The choice of scoring function (as well as the docking program) is protein system dependent. Ideally, each scoring function should be tested on the particular target protein in order to decide which produces the most reliable binding modes for known active ligands docked into that protein. At the time of writing, GOLD offers a choice of four scoring functions (Goldscore, Chemscore, PLP, and ASP) as well as a user-defined score whilst Glide offers two scoring functions (Glide SP and Glide XP). For further docking settings, *see* **Notes 8** and **9**.

It is useful to add pharmacophoric constraints to the docking procedure. For instance, if there are hydrogen bonds present within the binding site which are important for activity, or if a hydrophobic pocket needs to be filled in order to gain activity, then these can be set as constraints on the docking. Hydrogen bond or similar distance constraints limit the degrees of freedom that need to be sampled by the docking program and can really help in finding the right pose. Such constraints can also be applied as a post-docking filter (*see* Subheading 3.7), but then will obviously not influence the generated docking poses.

3.6.5 Pilot Run

It is highly recommended to carry out a pilot docking run on about 50 ligands in order to check whether the binding site is set up correctly and whether the scoring function and docking options are producing believable binding modes. It is particularly useful to include known ligands in this set of compounds, especially if their binding modes are known. This pilot run will also allow estimation of how long the virtual screen will take to run. For long runs, it is recommended to split the ligand dataset into multiple batches rather than running as a single large job. This will enable management of the process to be more effective and aid crash recovery.

3.7 Selection of Compounds for Testing

Selection of compounds from a virtual screen for purchase and biological screening is a filtration and enrichment process, starting with a large set of compounds from the docking or pharmacophore search, and reducing this to a much smaller set, while retaining as many potentially active compounds as possible.

In this section we outline our preferred compound final selection criteria. Docking programs such as GOLD and Glide also have tools which allow processing of virtual screening results and may be of help (Goldmine (28) (CCDC), Schrödinger Virtual Screening Workflow).

3.7.1 Use of Pharmacophoric Constraints

It is important to employ filters which will remove all compounds which do not fit key pharmacophoric points. Programs such as Goldmine/Hermes (29) (CCDC) are available for this purpose and their use can be pipelined within tools such as Pipeline Pilot or Knime.

When applying an H-bond or hydrophobic filter, it is worthwhile also including a functional group filter, as although the more general pharmacophoric constraint will indicate if the desired interaction is found, in the author's experience these interactions are not necessarily formed by "ideal" functional groups (*see* **Note 10** for further details).

3.7.2 Interpretation of Docking Scores

The scores that are produced by docking programs do not reliably rank compounds in terms of affinity. However the rankings produced are usually better than would be produced by a random scoring scheme. The simplest means of using a scoring function as a filter is to sort docking results according to their scores and simply take the number of highest scoring compounds equal to the target number of compounds you wish to purchase. In reality, this direct application of the scoring function is likely to lead to poor hit rates. One reason for this is that these scores are sometimes strongly correlated with molecular weight. If a plot of score versus molecular weight reveals a correlation, consider dividing the score by molecular weight of the compound and re-ranking on the basis of this normalized score.

There are numerous papers which describe the use of consensus scoring to enrich hit rates in compound selection in virtual screening (*see* ref. 30 for an example). Applying additional scoring functions to the previously generated docking poses can either be done rigidly or flexibly. Whereas rigid application simply applies each new scoring function to the existing pose, a flexible method re-optimizes the pose (and score) for each scoring function.

3.7.3 Docked Pose Descriptors

Pose-based filters can be defined using programs such as Goldmine. These descriptors fall into categories such as:

- Simple property measures including buried protein/ligand polar surface and exposed hydrophobic surface.
- Protein–ligand contacts including steric clashes and unfulfilled hydrogen bond potential.
- The extent to which available binding site cavities are filled (as measured by overlap with user specified spheres representing the cavities).

It is recommended to calculate numerous descriptors for the docked dataset and use data visualization tools such as Spotfire (31) (TIBCO) or Excel (Microsoft) to plot graphs of combinations of the various descriptors. The average and extreme results for the dataset and the impact of setting a specific cut off for a specific descriptor in terms of the number of poses rejected can then easily be seen. Employing the filters described above should reduce the docked dataset down to those binding in the desired pose and interacting with the protein via favorable functional groups.

3.7.4 Visual Inspection

The final and often most crucial filter in virtual screening is the visual inspection of poses. As protein–ligand binding affinity cannot be accurately or reliably calculated and factors known to be potentially important are often omitted in order to speed up the search, it is necessary to use judgment and experience to predict which compounds are likely to be hits. This will usually involve assessing the extent of complementarity of the binding partners. If any negative factor such as high energy ligand conformation, water molecules trapped in hydrophobic pockets, shape mismatch or electronic repulsion (to name but a few), is present it is unlikely that a compound will bind. This assessment is of course highly subjective, but can be aided by first studying protein–ligand crystal structures and structure–activity relationships (SAR) of related systems.

A 2D visualization, ideally in collaboration with a medicinal chemistry team, is the final stage of the filtering process. The aim of this is to remove compounds that would be difficult to progress due to synthetic or intellectual property issues.

It is a good idea to select small series of closely related compounds so that is possible to obtain confirmatory SAR during initial experimental testing. Alternatively, if funds are tight, a diverse set can be tested with later purchase of series derived from any hits obtained. Clustering and diversity selection of compounds can be done by eye or using ligand-centric chemoinformatics software (*see* Chapter 18).

3.8 3D Pharmacophore-Based Screen

The use of pharmacophore constraints in a docking protocol has been described in Subsection 3.6. An alternative approach is to use only pharmacophoric points to search through a database of compounds. We refer to this technique as a 3D pharmacophore-based screen and its stages are as follows.

3.8.1 Ligand Database Generation

Conformers of ligands (selected using the criteria in Subsectiom 3.4) are stored in a pre-calculated conformer database (some programs allow "on-the-fly" conformational search, but this is less efficient when several searches are to be made using the same ligand dataset). A pharmacophore-based screen is computationally simpler than docking and more conformers can be searched increasing the chance of finding matches.

3.8.2 Pharmacophore Definition

This can be based on a set of known active compounds or protein structure information. If active compounds are used, they should be aligned either manually or automatically. This allows the identification of common features such as hydrophobic or hydrogen-bonding groups in equivalent positions. It can be assumed that one or more of these common features are essential for binding and are therefore pharmacophoric points. The spacing and orientation are described in terms of distance, angle or spatial constraints. Pharmacophore features can also be derived from the structure of the protein binding site alone or, ideally, from that of protein–ligand complexes. Separate pharmacophores containing different combinations of features can be designed. Excluded volume features can be used to represent the steric limits of the binding site. These should be smaller in radius than Van der Waal's radii of the

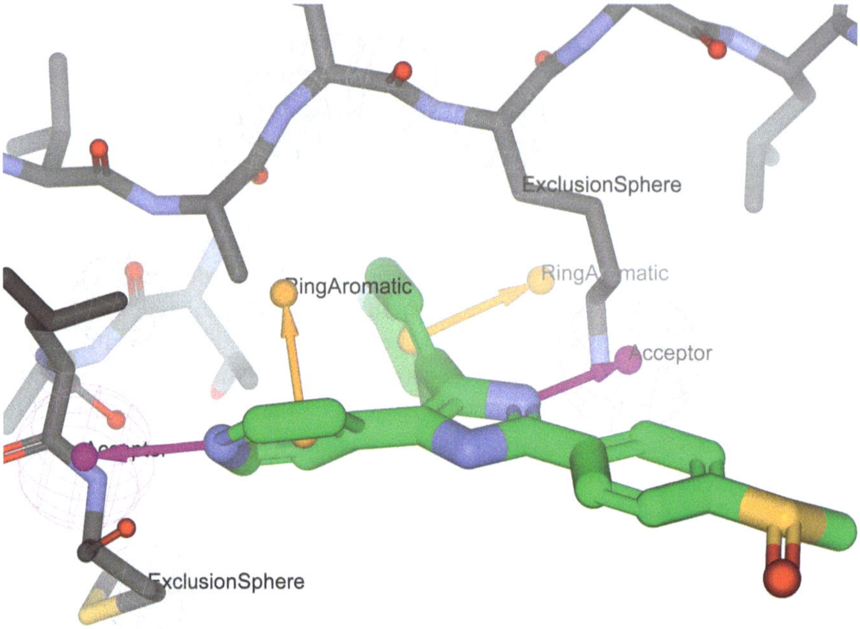

Fig. 2 An illustration of an example of a three-dimensional (3D) pharmacophore—based upon compound SB203580 bound to the ATP binding site of the protein kinase p38α (PDB code 1A9U). *Small meshed spheres* are excluded volume constraints; *larger meshed spheres* are spatial constraints for the pharmacophoric features they contain; *arrows* are hydrogen bonds and aromatic rings. It was created using DS Visualizer (Accelrys Inc.)

binding site atoms to allow for flexibility. A standard radius of 0.5 Å will suffice.

3.8.3 Running the Search

A pharmacophore search is usually quicker than a HTD run for a given size of ligand database. Furthermore, the analysis of results produces binary results, i.e., a "hit" or a "miss" against the pharmacophore, making the post-screen analysis simpler. If too few hits are identified, either reduce the number of pharmacophore features or loosen their geometric constraints. Experiment until the desired number and type of compounds are selected. Many of the post-screen filters described previously for final selection of virtual hits are themselves pharmacophoric in nature and consequently do not apply here. However, the 2D visualizations of suitability for medicinal chemistry and patentability and assessments of similarity diversity should be used.

An example structure-based pharmacophore is shown in Fig. 2.

4 Notes

1. Currently available software for structure-based HTD include Gold (1), Glide (2), Dock (8), FlexX (32), Fred (9), ICM (33), and Autodock (6).

2. Software available for pre- and post-processing include Corina (25) and Concord (26) for 3D structure generation. Pipeline Pilot (11) is a multipurpose tool. The Daylight Toolkit (34) allows chemical information to be processed in Smiles (35), Smarts (36), or Smirks (37) format. Maestro (38), Sybyl (39), and Goldmine (28) allow graphical visualization of docking results. Loop modelling software includes PRIME (16) and Modeller (15).
3. An example for a workflow to prepare a protein structure is the "Protein Preparation Wizard" that is integrated in Schrödinger's Maestro interface. In a stepwise process it allows the assignment of bond orders, addition of hydrogens, creation of disulfides, filling of missing side chains and loops, capping of termini, and deletion of waters far from the binding site. It can also generate charge- and protonation-states for the bound ligands and optimize the H-bond network, including the charging of GLU and ASP and the selection of appropriate states for HIS, ASN, and GLN. The wizard also allows the energy-minimization of the obtained structure.
4. Protein side chain dihedral angles are not evenly distributed. Rotamer libraries are collections of conformations for each residue type, containing a selection of the most common rotamers (side chains with specific sets of dihedral angles), found by statistical analysis of side-chain conformations in known protein structures.
5. Some suppliers of compounds for screening (the number in parentheses indicate the approximate size of the screening collections at the time of writing). Compound structure catalogues can be downloaded from the provided Web site unless otherwise stated. SPECS: (200,000) can source other suppliers by arrangement (40); ASINEX: (500,000) (41); ChemDiv: (one million) catalogues available as sdf files on request (42); ChemBridge: (450,000) (43); InterBioScreen: (400,000 synthetic plus 45,000 natural products) (44); Key Organics: (40,000) (45); Life Chemicals: (300,000) (46); Enamine: (one million) (47); Maybridge: (56,000) (48).
6. 2D to 3D conversion software includes: CONCORD (26) and Corina (25). Free software to do this includes Balloon (49) and Frog (50).
7. This can be done by reading the GOLD fit_pts.mol2 file into your molecular visualization package.
8. Within GOLD you can define the number of genetic algorithm (GA) optimization operations, the higher the number of operations the longer the docking per ligand will take, but the more likely the most optimal scoring pose will be found. We would recommend 10,000 operations per ligand for a HTD run. For

Glide we recommend selecting the HTVS (high throughput virtual screening) setting.

9. For HTD, we would recommend returning a maximum of five poses per ligand and include any early termination options. Retain only the top scoring pose for each ligand.
10. For example, functional groups such as exocylic carbonyls or non-morpholine ethers are often seen to form hydrogen bonds in the hinge region of kinases. As these have been shown to be undesirable in many kinase systems, it is sensible to filter them out in any screen against kinases. In another instance, you may wish that only aromatic groups fill a hydrophobic pocket on a target and, thus, alkyl chains filling this pocket should be filtered out.

Acknowledgments

Our thanks to Professor Tom Blundell, Dr. Sachin Surade, Sean Boyle, and Gengshi Chen for the use of Fig. 1.

References

1. Jones G, Willett P, Glen RC, Leach AR, Taylor R (1997) Development and validation of a genetic algorithm for flexible docking. J Mol Biol 267:727–748, http://www.ccdc.cam.ac.uk/products/life_sciences/gold
2. Friesner RA, Banks JL, Murphy RB, Halgren TA, Klicic JJ, Mainz DT, Repasky MP, Knoll EH, Shaw DE, Shelley M, Perry JK, Francis P, Shenkin PS (2004) Glide: a new approach for rapid, accurate docking and scoring: 1. Method and assessment of docking accuracy. J Med Chem 47:1739–1749, http://www.schrodinger.com/products/14/5/
3. Sprague PW, Hoffman R (2007) CATALYST pharmacophore models and their utility as queries for searching 3D databases. In: Computer-assisted lead finding and optimisation. Verlag Helvetica Chimica Acta. http://accelrys.com/products/pipeline-pilot/component-collections/catalyst.html
4. http://www.tripos.com/index.php?family=modules, SimplePage,,,&page=UNITY
5. Dixon SL, Smondyrev AM, Knoll EH, Rao SN, Shaw DE, Friesner RA (2006) PHASE: a new engine for pharmacophore perception, 3D QSAR model development and 3D database screening. 1. Methodology and preliminary results. J Comput Aided Mol Des 20:647–671, http://www.schrodinger.com/products/14/13/
6. Goodsell DS, Morris GM, Olson AJ (1990) Automated docking of substrates to proteins by simulated annealing. Proteins 8:195–202, http://autodock.scripps.edu/
7. Trott O, Olson AJ (2010) Autodock vina: improving the speed and accuracy of docking with a new scoring function, efficient optimization and multithreading. J Comput Chem 31:455–461, http://vina.scripps.edu/
8. Shoichet BK, Bodian DL, Kuntz ID (1992) Molecular docking using shape descriptors. J Comput Chem 13:380–397, http://dock.compbio.ucsf.edu/
9. http://www.eyesopen.com/fred
10. http://www.knime.org/
11. http://accelrys.com/products/pipeline-pilot/
12. Berman HM, Westbrook J, Feng Z, Gilliland G, Bhat TN, Weissig H, Shindyalov IN, Bourne PE (2000) The Protein Data Bank. Nucleic acids research 28:235–242, http://www.rcsb.org
13. Spoel DD, Lindahl E, Hess BA, Groenhof G, Mark AE, Berendsen HJC (2005) GROMACS: fast, flexible and free. J Comput Chem 26:1701–1718, http://www.gromacs.org/
14. Bowers KJ, Chow E, Xu H, Dror RO, Eastwood MP, Gregersen BA, Klepeis JL, Kolossvary I, Moraes MA, Sacerdoti FD, Salmon JK, Shan Y, Shaw DE (2006) Scalable algorithms for molecular dynamics simulations on commodity clusters. Proceedings of the ACM/IEEE conference on super computing (SC06), Tampa,

FL, 11–17 November 2006. http://www.schrodinger.com/products/14/3/

15. Marti-Renom A, Stuart A, Fiser A, Sanchez R, Melo F, Sali A (2000) Comparative protein structure modelling of genes and genomes. Annu Rev Biophys Biomol Struct 29:291–325, http://www.salilab.org/modeller/
16. http://www.schrodinger.com/products/14/14
17. http://accelrys.com/products/informatics/cheminformatics/ctfile-formats/no-fee.php
18. Rishton GM (1997) Reactive compounds and in vitro false positives in HTS. Drug Discov Today 2:382–384
19. Hann M et al (1999) Strategic pooling of compounds for high-throughput screening. J Chem Inf Comput Sci 39:897–902
20. Baell JB, Holloway GA (2010) New substructure filters for removal of pan assay interference compounds (PAINS) from screening libraries and for their exclusion in bioassays. J Med Chem 53:2719–2740
21. Lipinski CA et al (1997) Experimental and computational approaches to estimate solubility and permeability in drug discovery and development settings. Adv Drug Deliv Rev 23:3–25
22. http://tripos.com/data/SYBYL/protoplex_072505.pdf
23. http://www.tripos.com/index.php?family=modules,SimplePage,,,&page=StereoPlex
24. http://www.schrodinger.com/products/14/10/
25. http://www.molecular-networks.com/products/corina
26. http://www.tripos.com/index.php?family=modules,SimplePage,,,&page=Concord
27. http://www.tripos.com/index.php?family=modules,SimplePage,,,&page=Confort
28. http://www.ccdc.cam.ac.uk/products/life_sciences/goldmine/
29. http://www.ccdc.cam.ac.uk/Solutions/GoldSuite/Pages/Hermes.aspx
30. Feher M (2006) Consensus scoring for protein–ligand interactions. Drug Discov Today 11:421–428
31. http://spotfire.tibco.com/
32. http://www.biosolveit.de/flexx/index.html?ct=1
33. http://www.molsoft.com/docking.html
34. http://www.daylight.com/
35. http://www.daylight.com/smiles/index.html
36. http://www.daylight.com/dayhtml_tutorials/languages/smarts/index.html
37. http://www.daylight.com/dayhtml_tutorials/languages/smirks/index.html
38. http://www.schrodinger.com/products/14/12/
39. http://www.tripos.com/index.php?family=modules,SimplePage,,,comp_informatics&s=0
40. http://www.specs.net
41. http://www.asinex.com
42. http://us.chemdiv.com
43. http://www.chembridge.com
44. http://www.ibscreen.com
45. http://www.keyorganics.ltd.uk
46. http://www.lifechemicals.com
47. http://www.enamine.net
48. http://www.maybridge.com
49. http://users.abo.fi/mivainio/balloon/
50. http://bioserv.rpbs.univ-paris-diderot.fr/cgi-bin/Frog2

INDEX

Mark A. Williams and Tina Daviter (eds.), *Protein-Ligand Interactions: Methods and Applications*, Methods in Molecular Biology, vol. 1008, DOI 10.1007/978-1-62703-398-5, © Springer Science+Business Media New York 2013

D

T

U

V

W

X

MIX
Papier aus verantwortungsvollen Quellen
Paper from responsible sources
FSC® C105338

If you have any concerns about our products,
you can contact us on
ProductSafety@springernature.com

In case Publisher is established outside the EU,
the EU authorized representative is:
Springer Nature Customer Service Center GmbH
Europaplatz 3, 69115 Heidelberg, Germany

Printed by Libri Plureos GmbH
in Hamburg, Germany